MINING WITH BACKFILL

PROCEEDINGS OF THE INTERNATIONAL SYMPOSIUM ON MINING
WITH BACKFILL / LULEÅ / 7-9 JUNE 1983

Mining with Backfill

Edited by
S.GRANHOLM
Luleå University of Technology, Sweden

A.A.BALKEMA / ROTTERDAM / 1983

*The texts of the various papers in this volume were set individually
by typists under the supervision of each of the authors concerned.*

ISBN 90 6191 509 0

© 1983 A.A.Balkema, P.O.Box 1675, 3000 BR Rotterdam, Netherlands

Distributed in USA & Canada by: MBS, 99 Main Street, Salem, NH 03079, USA

Printed in the Netherlands

Table of contents

3 Geomechanics

4 *New development*

1. Fill properties

Engineering properties of cemented aggregate fill for Uludağ Tungsten mine of Turkey

ERGIN ARIOĞLU
Technical University of Istanbul, Turkey

SYNOPSIS : As a result of the need for reducing cement content in fill practice, a study was carried out to assess the effect of coarse aggregate addition on strength properties of cemented fill. This paper reveals laboratory test data related design parameters,such as amount of coarse aggregate leading to maximum density, total aggregate content,cement content and (water/cement) ratio to cemented fill strength. The results reported clearly pointed out that strength properties markedly increase with increased cement content and decreased (water/cement) ratio. The increase in strength as compared with cemented fill containing only tailing can be attributed to the fact that the coarse aggregate requires less mixing water. Dependences of strength properties upon cement content and (water/cement) ratio were also established statistically.

1 INTRODUCTION

The mining at Uludağ is a characteristic application of mechanized sub-level. The method generates vertical pillars which must be extracted from within stopes to be filled. Two conditions in recovering vertical pillars are fundamentally furnished. These are :
  1. to maintain dilution of ore by spalled wall rock and fill material to a minimum extent
  2. to ensure general stability of the panel
These requirements can be realized by filling stopes with a cemented fill. From cemented fill technology it is well known that cemented fills place a limit on the cost (Thomas, Nantel, Notley 1979). An attempt at cost reduction of cemented fill must be therefore made. On this basis addition of coarse aggregate to cemented fill is proposed for the Uludağ Tungsten mine operation of Etibank, Ankara.

Presently,there is a considerable increase in application of cemented aggregate fill in world mining practice since cemented aggregate fill posseses the advantages over common cemented fill from particularly cement requirement and stowing rate point of view. Most of the available studies in the literature were concerned with technological feature of fill only under consideration. Presented is a summary of investigation

(Arıoğlu, 1982 a,b) which was fullfiled to determine the mechanical properties of cemented fill made of coarse marble aggregate. Also, the effect of aggregate addition to cemented fill on the properties of strength, cohesion and elasticity was displayed clearly by comparing with common cemented fill.

2 BRIEF DESCRIPTION OF EXPERIMENTAL STUDY

2.1 Materials

Cemented aggregate fill produced in the test program was made of ordinary Portland cement, crushed coarse aggregate and Tungsten tailing.

Engineering properties of the tailing are summarized in Table 1 (Notation list was compiled at the appendix).
By examining Table 1 it can be seen that the tailing has ideal compaction characteristic. The used aggregates were obtained from marble outcrops around the mine. Table 2 exhibits main characteristics of the coarse aggregates. Although the particle shape of aggregate is classified as flaky and elongated particle the performance of aggregate appeared quite satisfactory during the producing process.

Attempt was made to obtain an ideal grading based upon the idea that the combined material (crushed marble aggregate-tailing)

TABLE 1. Some engineering properties of Uludağ Tailing

γ_s	γ_b	D_r	C_u	C_c	D_{10}	k	LL	PL	W_{op}	$\gamma_{max,d}$	$\gamma_{max,w}$	C	ϕ
3.01	1.129	0.496	4.7	0.81	0.052	0.013 – 0.025	15	0	6	1.96	2.13	0.3	38

TABLE 2. Characteristics of coarse marble aggregate

γ_s	γ_b	D_{max}	C_u	f_m	f_s	f_v	particle shape
2.65	1.97	45	4.89	3.8	3.35	0.18	flaky-elongated

is packed to achieve maximum density and thus maximum strength. Ideal overall grading, i.e. the grading of the aggregate and tailing was expressed by the equation expressed by the equation

$$P = 124 \left(\frac{d}{D_{max}}\right)^{0.47}$$

where P = percentage of material smaller than size d, D_{max} = maksimum particle size

The percentage of marble aggregate by weight of total material was found to be % 60 at which maximum density was achieved. This particle size analysis approaching ideal grading, is compiled in Table 3. Throughout these works material with the overall grading was tried to employ in order to maximise the effect of marble aggregate.

TABLE 3. Particle size distribution of combined material (% 40 tailing – % 60 marble aggregate)

Sieve size, mm.	Cumulative pertentage retained
45	0
30	1.85
15	37.15
7	47.52
3	52.79
1	65.79
0.63	81.02
0.149	95.34

2.2. Mix design and preperation

In the test program it was desired to determine the effect of cement content on the strength properties of cemented aggregate fill. Thus, the mixes were made with three differing total aggregate/cement ratios i.e. 5/1, 10/1, 20/1 (Total aggregate here is referred to as combination of tailing and marble aggregate). So the water/cement ratio was ranged 0.72 to 2.21 respectively. The basic mix designs are given in Table 4.

Cemented aggregate fills were mixed in a 50 lt revolving-pan mixer. Compressive strength and elasticity characteristics were determined by employing 15 cm by 30 cm cylinder specimens Indirect-tension tests was performed on 10 x 15 cm diameter discs. Four cylinders and six discs of mix corresponding to each total aggregate/cement ratio respectively for compression and indirect-tension tests were crushed at an age of 28 days. The specimens were kept in their moulds for the first three days and then demoulded and allowed to cure under normal laboratory condition. The average temperature was 20°C.

The standard equipment and well known techniques in testing of specimens were utilized. To economise space the description will not be included here.

3. RESULTS AND STATISTICAL EVALUATION

Table 5 summarizes the strength properties obtained. In figure 1, typical results have been also plotted against the cement content in the mix. It is evident from the results of the tests for compressive strength, indirect-tensile strength, cohesion and elasticity that cemented aggregate fill properties largely increase with increased cement

TABLE 4. Mix designs used in the tests

Components of mix	Total aggregate/cement ratio		
	5/1	10/1	20/1
Cement, kg/m³	352.2	195.1	103.1
Total aggregate, kg/m³	1761.0	1951.0	2062.0
Water, kg/m³	256.4	238.8	228.7
$\frac{Water}{cement}$ by weight	0.72	1.22	2.21

content. The most important feature of this figure is the indication of definite relationships between the strength properties and the cement content. Especially, the relationship between compressive strength and cement content obtained has a bearing from a mix design point of view.

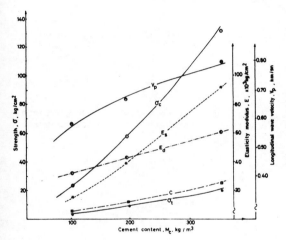

Fig.1. Variation of strength properties with cement content

The method of least square regression was performed to set up the relationships for the combination of variables used here in. Power and linear functions were interestigated and typical relationships acquired are reproduced in Table 6.

On passing it should be placed an empshasis on the derived relationship between the compressive strength of cemented aggregate fill and Schmidt rebound number (Figure 2). Since compressive strength is of primary concern when dealing with stability of the cemented block (Mitchell et al, 1982

TABLE 5. Strength properties of cemented aggregate fill

Strength properties	Total aggregate/cement		
	5/1	10/1	20/1
σ_c , kg/cm²	131.77	57.96	23.77
σ_t , kg/cm²	19.77	9.84	4.11
C , kg/cm²	25.52	11.97	4.95
τ , kg/cm²	44.3	20.45	8.42
ϕ , °	47.65	45.19	44.81
E_s , kg/cm²	92063	39062.5	14705.8
E_d , kg/cm²	60415	42842	31823
V_p , km/sn	0.79	0.66	0.58
R	24	20	14

TABLE 6. Regression equations, cement content, water/cement ratio versus strength properties

Relationships	Correlation coefficient
$\sigma_c = 0.037 \ (M_c)^{1.39}$	0.999
$\sigma_c = 80.27 \ (\alpha)^{-1.53}$	-0.999
$C = 0.01 \ (M_c)^{1.33}$	0.999
$C = 16.03 \ (\alpha)^{-1.47}$	-0.999
$E_s = 14.58 \ (M_c)^{1.49}$	0.999
$E_s = 54.55 \ (\alpha)^{-1.64}$	-0.999
$E_d = 49.56 \ (\alpha)^{-0.57}$	-0.996
$V_p = 0.17 \ (M_c)^{0.25}$	0.996
$\sigma_t = 0.143 \ \sigma_c + 1.068$	0.998
$\sigma_c = 480 \ (V_p)^{5.47}$	0.996
$C = 0.189 \ \sigma_c + 0.669$	0.999
$\sigma_c = 0.00255 \ (\gamma R^2)^{1.47}$	0.952

Arıoğlu 1983 c, Birön-Arıoğlu 1983). The Schmidt Rebound Hammer as nondestructive testing can be utilized to evaluate the strength of cemented fill. Specially, fill in place within a stope. Also pulse velocity mesurements may be employed in conjunction with the Schmidt hammer for a more comprehensive and meaningful assesment of in situ strength. This matter merits future investigations.

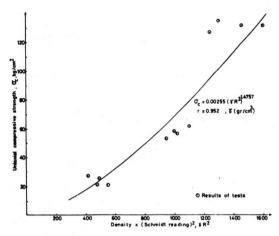

Fig.2. Variation of uniaxial compressive strength with (Schmidt reading x density)

4 COMPARISON OF CEMENTED FILL

In a attempt to evaluate the effect of coarse aggregate addition, these results have been compared with the results pertaining to cement mixtures made of tailing. The comparison made is summarized in Table 7. Reference to this table obviously indicates that cemented aggregate fill posseses higher strength properties (compressive, tensile strength, cohesion, elasticity moduluses) than common cemented fill. On the average they are 400 percent greater, i.e., four times as large as the strength in the compression tests, and 350 percent greater than in the indirect-tension tests. It is also of interest to note that the increase observed in static moduli is more tremendous. Having this feature cemented aggregate fill as a suitable support material can be successfully utilized to reduce or eliminate rock bursts which frequently take place in deep mines.

The above mentioned results may be due to the fact that coarse aggregates in general require less mixing water giving rise to increase strength. It can be verified examining cemented fills which demanded more water, i.e., greater water/cement ratio, for a given tailing/cement ratio (Table 7)when mix proportioning. On the basis of these results it can be stated that the amount of Portland cement can be decreased without lessening strength to below that of common cemented fill. Needless to say this result is of importance to bring cement cost to more economic level. This has been numerically illustrated below.

In case of cemented fill, the required uniaxial compressive strength to support 30 m high by 100 m across vertical fill faces at a safety factor of 1.5 was computed to be 9 kg/cm^2 at a age of 28 days (Arıoğlu 1982 c). From the relationship below, cement content meeting this requirement can be approximately determined.

$$\sigma_c = 0.0016 \ (M_c)^{1.765} \ , \quad kg/cm^2$$

as

$$M_c = 133 \quad kg/m^3$$

Using the equation below the same uniaxial compressive strength for cemented aggregate can be achieved at

$$\sigma_c = 0.037 \ (M_c)^{1.39} \ , \quad kg/cm^2$$

$$M_c = 52 \quad kg/m^3$$

As clearly shown, reducing cement content (% 60), in other words vital economy can be accomplished by employing cemented aggregate fill in place of common cemented fill.

5 CONCLUSIONS

The results from this investigation justify to derive the following conclusions
1. This work brings a modest contribution to the engineering knowledge of cemented aggregate. This is vital owing to the fact that technical information gained on strength properties of cemented fill made with coarse aggregate is limited in fill technology.
2. Fill containing coarse marble aggregate with 16.6-4.7 wt percent (dry basis)

TABLE 7. Comparison of cemented fills

Total aggregate/ cement ratio	5/1			10/1			20/1		
Fill type	A	B	increase %	A	B	increase %	A	B	increase %
Cement, kg/m^3	288	352	20	156.5	195	24.6	81.6	103	26
$\dfrac{Water}{cement}$ by weight	1.47	0.72	-50	2.72	1.22	-55	5.24	2.21	-57
Compressive strength, kg/cm^2	43.56	131.77	200	7.98	57.96	626	4.76	23.77	399
Tensile strength kg/cm^2	6.57	19.77	200	1.57	9.84	526	-	4.11	-
Cohesion, kg/cm^2	8.43	25.52	202	1.77	11.97	576	-	4.95	-
Static elasticity modulus, kg/cm^2	1812	92063	4980	541	39062	7120	-	14705	-

A : Cemented fill produced from only tailing

B : Cemented aggregate fill produced from coarse marble aggregate and tailing

Portland cement provide uniaxial compressive strength of the order (131.77-23.77 kg/cm^2) an age of 28 days. The lowerest value can provide supporting capacity for 300 m high by 100 m across vertical fill faces at a safety factor 10.

3. Strength properties of cemented aggregate fill are greatly affected by factors such as cement and water/cement ratio (Figure 1). There are definite statistical relationships between strength properties and cement content used in the mix (Table 6). The following expression can be employed for design purpose of fill mixes.

$$\sigma_c = A \ \alpha^{-n}$$

where σ_c = uniaxial compressive strength
α = water/cement ratio by weight
A,n = experimental constants

4. The relationship between cohesion and uniaxial compressive strength can be defined by the equation

$$c = A \ \sigma_c + B$$

where c = cohesion in kg/cm^2
σ_c = uniaxial compressive str strength in kg/cm^2

A,B = experimental constants.
Equation in view can succesfully used in Mitchell's approach to three-dimensional analysis of the cemented fill (Mitchell, et al 1982). So the required uniaxial compressive strength for stability of cemented fill block in panel can be more conveniently formulated.

5. Tensile strength of cemented aggregate fill are approximately 15 percent of the uniaxial compressive strengths

6. The derived relationship between uniaxial compressive strengths and production of (hammer readings^2 x density) suggests that in fill technology the rebound hammer can be utilized in determining relative strengths in a matter of minutes.

7. Addition of coarse aggregate to the mix gives rise to immense increase in strength properties of cemented fill (Table 7). This result provide essential decrease in cement cost.

ACKNOWLEDGMENTS

The study reported in this paper is part of a major project which was supported by Turkish Scientific and Technical Research Concil (TÜBİTAK, Ankara). The author wishes to thank the Tübitak for giving the funds for the project. His sincere thanks are expressed to Prof.Dr.Cemal BİRÖN, Dipl.Ing. Ali YÜKSEL and Dipl.Ing.Nihal ARIOĞLU and Rabia YARGIÇ who contributed extensively to the experimental part and typing. Etibank Uludağ Tungsten Mine supplied the tailing and marble aggregates is gratefully acknowledged.

APPENDIX

The following symbols were employed in the text, tables, and illustrations of this paper.

C = apparent cohesion of tailing, cohesion of cemented fill, kg/cm^2

C_c = coefficient of degree

C_u = coefficient of uniformity

D_{max} = maximum particle size, mm

D_r = relative density

D_{10} = diameter in milimeters of 10 percent finer size, mm

d = particle size, mm

E_d = dynamic elasticity moduli, kg/cm^2

E_s = static elasticity moduli, kg/cm^2

f_m = fines modulus

f_s = shape factor

f_v = volumetric shape factor

k = permeability coefficient of uncemented tailing, cm/sn

LL = liquid limit, %

M_c = amount of cement, kg/m^3

PL = plastic limit, %

P = percentage of material smaller than size d, %

R = Schmidt rebound number

V_p = ultrasonic longitudinal wave velacity, km/sn

α = water/cement ratio

γ = density of cemented aggregate fill, g/cm^3

γ_b = bulk density, g/cm^3

$\gamma_{max,d}$ = maximum dry density (at standart Proctor compaction tests), g/cm^3

$\gamma_{max,w}$ = maximum wet density (at standart Proctor compaction tests), g/cm^3

γ_s = specific gravity

σ_c = uniaxial compressive strength, kg/cm^2

σ_t = indirect-tensile strength, kg/cm^2

τ = shear strength, kg/cm^2

φ = angle of internal friction, °

W_{op} = optimum moisture content (at standart Proctor compaction tests), %

REFERENCES

Arıoğlu, E. 1982 a. As fill material investigation on useage of the Uludağ's tungsten tailing, thesis submitted for associate professorship, (in Turkish) İTÜ Maden Fakültesi, Turkey, 185 p.

Arıoğlu, E. 1982 b. Properties of the engineering materials to be used in the stopes for the Uludağ Tungsten Mine of Etibank, MAG:579, Turkish Scientific and Technical Research Concil (TÜBİTAK), Ankara, 258 p.

Arıoğlu, E. 1982 c. Engineering properties of stabilized tailing-cement mixtures with the Uludağ Tungsten tailing (in Turkish), Doğa Bilim Dergisi, Tübitak, 7(4) (to be publised).

Birön, C and E.Arıoğlu 1983. Design of Supports In Mines, New York : John Wiley and Sons, 256 p.

Mitchell, R.J. and R.S. Olsen, J.D. Smith 1982. Model studies on cemented tailings used in mine backfill. Can. Geotech. J. 19 : 14-28.

Thomas, E.G. and L.H. Nantel, K.R. Notley 1979. Fill Technology In Underground Metalliferous Mines. Ontario : International Academic Services Limited, 293 p.

Proceedings of the International Symposium on Mining with Backfill / Luleå / 7-9 June 1983

A study on the possible uses of waste calcium sulphate from chemical industry as a mine fill material

B.CONTINI
Mineraria Silius, Italy

G.IABICHINO
CNR, Centro di Studio per i Problemi minerari, Italy

R.MANCINI & S.PELIZZA
Politecnico di Torino, Italy

ABSTRACT: calcium sulphate, in different forms, is a commonplace waste product of many industrial chemical processes. In the case here discussed, anhydrous calcium sulphate from an hydro fluoric acid production plant (a material known as "fluor-gypsum" or "fluor-anhydrite") is investigated as a fill material for a fluorspar mine (the Silius mine in Sardinia). The mine plans to switch from the presently adopted sublevel stoping method to a cemented (or partly cemented) fill method, resorting on the fluor-gypsum from a nearby plant as a comparatively low cost filling and cementing material; it can be predicted that the new method will be safer and will allow to increase the orebody recovery. A laboratory study has been carried out on the long term setting properties, mechanical strength and chemical reactivity of fluor-gypsym and fluor-gypsum based mixtures, and filling tests have been started at the mine to evaluate the in situ properties of the mine fill made with fluor-gypsum. The results of the laboratory tests are exposed in detail and discussed in the paper.

1 SILIUS MINE

1.1 Geological features

Silius mine exploits a fluorspar orebody, in the group of lodes, outcropping along an E-W line in the Gerrei district (Southeastern Sardinia), between the small towns of Silius and S. Basilio. The exploited lode is regular in shape, with a sub-vertical asset (both at the outcrop and in the deeper parts) and is about 3 km long; mineralization is known to occur from surface to a depth of 450 m. Typical minerals are fluorspar, barite, galena, sphalerite, pyrite, marcasite, calcite and calcedonium. The orebody is composed of two units, that at places are separated, at places join together; the structure of the orebody is fairly uniform.

The country rock is the basal porphyroid of Caledonian age, a massive complex, known to the level of 204 m.a.s.l., which has a special significance for its geometrical correlations with the mining operations.

1.2 Mining scheme

The mine is divided into two units; we will take into consideration the "Genna Tres Montis" one. The present mining method is a modified "sub level stoping", where the ore chutes have been suppressed from the base floor, and the blasted ore haulage is done almost completely through the lower sub-level, by means of remotely controlled electro-hydraulic showels. A general schema of the present mining method is shown in fig. 1.

The remotely controlled showels and the well suited drilling equipment adopted give to the mining system a remarkable efficiency.

From the lower sub level the ore is conveyed by chute to the main road, and carried by means of self-unloading Hagglund cars to the bins, close to the shaft. hoisting is made by skips, and the same skips are also adapted to the personnel and material transport.

According to the working scheme, two sec-

tions are under exploitation, two in the vertical borehole drilling phase, and two more are under development: this scheme allows a section to be ready for exploita tion as the preceding one is nearly exausted. We term "section" a parte of the o-rebody comprised between two level drives (spaced by 50 m in heigth) and two raises (spaced by 100 m in horizontal).
The base and level floors of the level drives are left in place as supporting structures. The orebody recovery is about 80%, and we are currently studying a sche me that could improve it.

1.3 Filling

The mechanic characteristics of both the lode and the country rock are fairly good, however some caving occurs in the southern wall, at the contact between the lode and the porphyroid. Caving poses both stabili ty and ore dilution problems. Besides, the depth of the workings is considerable, and remarkable stresses are to be taken into account on the walls. An artificial support is to be provided to ensure a sa-tisfactory stability of the abandoned voids, and the mine resorts on filling to

achieve this objective. Presently only loose filling is in use, consolidated fil ling with "fluorgypsum" is a proposed new method, to which the present paper is chiefly devoted.

1.3.1 Loose filling

The scheme is shown in fig. 2. The fill material is conveyed to the working from outside by chutes. Hagglund cars are loa-ded with the fill material at the top le-vel and distribute it where needed, by di scharging the material in the voids through chutes. The angle of repose of the fill is about 45°, therefore the void can not be completely filled; another defect of the system is that the loose fill can not com pletely absorb the horizontal pressures acting on the walls and on the protective floors.

1.3.2 Filling with Fluorgypsum (see fig.3)

As the stresses get higer with the deepe-ning of the mining operations, a more ef-ficient stabilization of the fill is re-quired. The Fluorgypsum is the filling ma

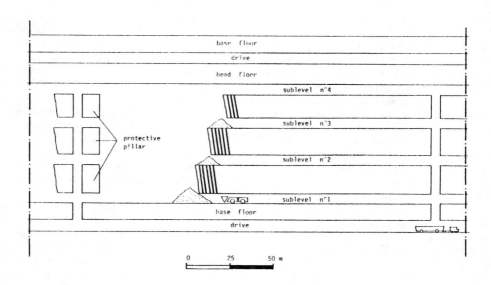

Fig. 1. The exploitation system.

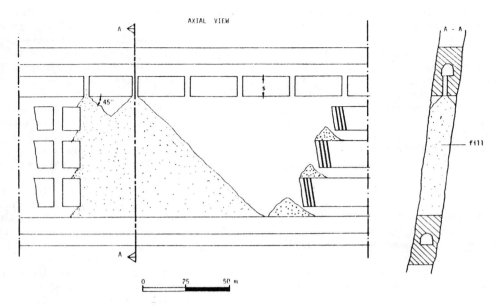

Fig. 2. Present filling method with loose fill.

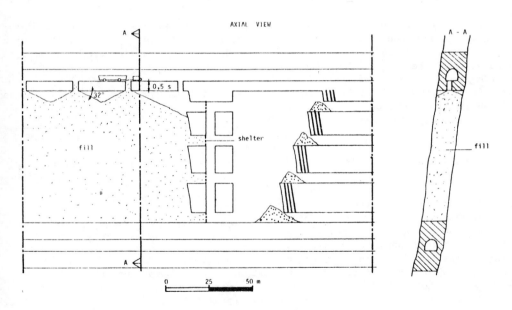

Fig. 3. Proposed filling method with fluorgypsum.

terial we are testing.

The Fluorgypsum is a waste product from chemical processing of the fluorspar, and its properties are exposed in detail in the next chapter.

Due to the resistance of the mortar resulting from the admixture of fluorgypsum and water, we expect a general improvement

11

of the stability of the workings; we also expect the partial or complete recovery of the top floor of the voids will be feasible, when fluorgypsum will be used as a filling material.

The emplacement of the fluorgypsum fill will be done according a method quite similar to the loose filling emplacement; the water mix will take place by spraying the water on the dry filling material, when it passes through the chutes.

1.4 Expected improvements

The achievement of favourable stability conditions, due to the new filling material, should give some additional advantages:
- lower ore dilution, as the rock lump caving from the walls will be less frequent
- lower stresses on the road supports and other structures of the mine
- increased recovery of the orebody, as the head floor can be recovered: with a complete recovery of the head floor the ore recovery should rise to 92%, with a 50% recovery to approx 87%.

On the basis of the expected benefits the research presently under development appears highly promising, notwithstanding the problems expected from the acid water percolation through the mortar.

2 LABORATORY TESTS ON FLUORGYPSUM

Fluorgypsum, a substance that should be termed more exactly Fluor-anhydrite, is a waste material hitherto devoided of practical use, whose properties as binding agent are practically unknown. A detailed laboratory research has been therefore carried out, to obtain quantitative data on Fluorgypsum mortars and Fluorgypsum-aggregate mixtures long term behaviour. The main points to be elucidated were:
- setting properties and mechanical strength attainable
- permeability
- chemical corrosion exerted by the Fluorgypsum mortars.

A great number (more than 400)of cylindrical testing bars have been prepared with Fluorgypsum mortars and Fluorgypsum/aggregate mixtures, with different water contents, aggregate type, modifying reagents; the cylinders have been subjected to aging under constant humidity conditions, at an average temperature of 18°C (with small seasonal variations); aging duration ranged between 1 and 8 months. The cylinders have been then subjected to mechanical tests and other measurements. 47,5 mm dia., cylinders have been used in tests on Fluorgypsum mortars and mixtures with fine aggregate, 97,5 mm dia. cylinders for tests on mixtures on coarse aggregate. It must be noticed that aging has been conducted under atmospheric pressure: in the actual use of the Fluorgypsum as binding agent higher values of the mechanical strength are to be expected than in the laboratory tèst, due to the compaction exerted by the filling material column.

2.1 Fluorgypsum properties and composition

Fluorgypsum, as received by the laboratory, is a fine, white-gray, freely flowing powder whose properties are summarized in the table 1.

Table 1. Fluorgypsum properties and composition.

Bulk s.g.	1,3 to 1,4
True s.g.	3 to 3,1
% of pores by volume	approx. 60
Angle of repose	31° to 32°

Grain size distribution:	
+ 0,6 mm	6,5%
0,6 to 0,07 mm	35,5%
- 0,07 mm	58,0%

Composition		
anhydrous $CaSO_4$ (anhydrite)	90 to 95%	
hydrated $CaSO_4$ (gypsum)	up to 5%	(1)
unreacted H_2SO_4	up to 2%	(1)
unreacted CaF_2	up to 1,5%	
barite and others	up to 1,5%	

Chemical reaction: strongly acidic

(1) depending upon the amount of lime added at the plant.

The chemical reaction that gives rise to the Fluorgypsum (CaF_2 attack with concentrated H_2SO_4, giving HF as useful product) takes place at high temperature, and therefore the calcium sulphate is obtained in the anhydrous form.

Fluorgypsum absorbs promptly the water, giving a plastic, easily poured mass that hardens upon drying; such a hardening is reversible, that is the mass is reverted to the plastic state when water is added again, and is unrelated to the setting effect (a slow, not reversible hardening caused by the hydration and re-crystallization of the sulphate) we investigated. Fresh, correctly mixed Fluorgypsum/water mortar (with 20-25% water by weight) has a s.g. of 1,7, a resistance to shear of 0,036 MPa (as measured with a Vane Test Equipment) and an angle of internal friction of approx. 32°.

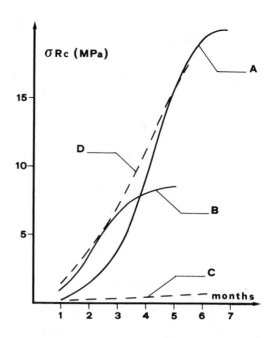

Fig. 4. The hardening of the Fluorgypsum mortar: A: with the stoichiometric amount of water; B: with 1/2 of the stoichiometric amount; C: with 1/4 of the stoichiometric amount; D: expected actual behaviour of the mortar in situ.

2.2 Fluorgypsum mortar setting and hardening

Fluorgypsum mortars with different amounts of water (the stoichiometric water amount for anhydrite hydration, that is 26 g of water per 100 g of dry Fluorgypsum, giving a mortar with 20,6% of water; 1/2 of the stoichiometric amount, that is 10,3% of water; 1/4 of the stoichiometric amount, that is 5,15% of water) have been formed in testing bars and subjected to aging, then tested for mechanical resistance, porosity, % of hydration and other properties. Testing bars were made with slightly compacted mortar, having an apparent specific gravity ranging between 1,75 and 2, depending upon the water contents.

The results of the tests (performed on un-dryed bars) are shown in the graph of fig. 4.

It can be seen that maximum hardening speed is obtained with an amount of water lower than stoichiometric, whilst maximum long term resistance is obtained with the stoichiometric water quantity. In the practical use of the Fluorgypsum as a fill binder it seems therefore advisable to add at the beginning a reduced amount of water, to hasten the hardening process; in due time the mine water will add the extra

amount of water required to obtain the maximum strength. When such a policy is followed, the expected time/strength relationship is shown by the line D of graph of fig. 4.

The setting process has been monitored also by measuring the amount of hydration water captured (that is, the amount of gypsum formed), that proved to be well related to the mechanical resistance. The results of the hydration measurements and of the mechanical tests are summarized in the graph. of fig. 5.

It is noteworthy that a different relationship between the amount of gypsum produced and the mechanical strength has been obtained when mortars containing hydration accelerating reagents have been tested, as shown in the next chapter.

It is known that in the anhydrite-gypsum transformation a volume increase of 61% occurs; the volume increase has been absor

13

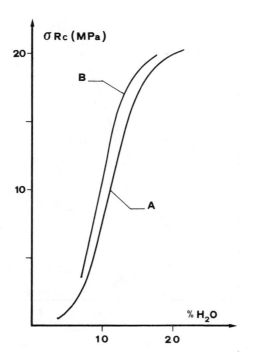

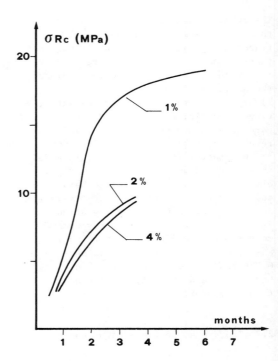

Fig. 5. Correlation between the amount of gypsum formed (expressed as hydration H$_2$O percentage) and the strength of the mortar: A: plain Fluorgypsum; B: Fluorgypsum with accelerating reagents.

Fig. 6. The accelerating effect of different percentages of K$_2$SO$_4$ + FeSO$_4$ mixture on the luorgypsum mortar.

bed by a pore volume reduction of the mass (from a 40% value at start to a minimum 20% value at the end of the hardening process).

2.3 Acceleration of setting by means of reagents

A mixture of K$_2$SO$_4$ and FeSO$_4$ is reported in the technical literature (Hyde N.: A study of the strength of anhydrite stabilized mill tailing – Colloque International sur l'utilisation des sous-produits et déchets, Paris, 28-30 nov. '78) as having an accelerating effect on anhydrite hydration; our results agree with the results of the above quoted report (results are shown in the graph of fig. 6).
We tested also other reagents, all of them belonging to the class of the soluble sul-

phates and comparatively inexpensive. The results are shown in the table 2.

Table 2. Effect of different soluble sulphates on Fluorgypsum mortar hardening

Reagent	Amount	Aging (months)	Resistance (MPa)
Na$_2$SO$_4$	1%	2	7
Na$_2$SO$_4$	1%	3	8
Na$_2$SO$_4$	2%	2	7
Na$_2$SO$_4$	2%	3	10
Na$_2$SO$_4$	3%	2	5
Na$_2$SO$_4$	3%	3	7
MgSO$_4$	1%	2	7
K$_2$SO$_4$	1%	2	8
By comparison:			
No reagents		2	3
No reagents		3	6

It can be seen that many reagents show an interesting accelerating effect; a choice could be made on the basis of an economic analysis (reagents that can be obtained as byproduct from other chemical industries, such as sodium sulphate, should be prefer red).

Under the action of the hydration accelerants, microscopic examination of the mor tar shows that a more finely crystallized gypsum forms; mechanical tests show that the mortar acquires an higer strength, the amount of hydration being the same, when hydration is accelerated by the reagents, as shown by the comparison of the curves A and B of the graph of Fig. 5, where the hydration and strength values of the plain mortars and of mortars containing accelerating reagents are represented.

2.4 Fluorgypsum mortar/aggregate mixtures

A number of aggregates has been tested:
a) finely crushed and sized microcrystalli ne limestone
b) finely crushed and sized porphyroid (the country rock of the Silius orebody)
c) Silius flotation tailing, partly desli med by cyclone (it contains calcite, quartz, silicatic rock debris, and tra ces of barite, sulphydes, fluorite)
d) Silius flotation tailing, undeslimed
e) Silius Sink-Float reject material (mos tly porphyroid, calcite and quartz)
f) Silicous, fine grained sandstone, from a quarry close to the mine site (alrea dy used as loose fill material), crushed

Grain size distributions of the aggregates are shown in table 3.

The mixtures hitherto tested have been pre pared with 50% by weigth of aggregate; com pression tests on aggregates a) to d) have been made with small diameter (47,5 mm) cy linders on the e) and f) with larger (97,5 mm dia.) cylinders. In cement bound fill lower amounts of binding agent are common ly used, but Fluorgypsum is obviously less effective than Portland cement, and an higher percentage is probably required if structural strength is expected from the fill. Lower amounts of Fluorgypsum are expected to be necessary in these parts of the fill where space filling rather

than structured strength is the desired result.

Table 3. Grain size distribution of the aggregates.

Fine crushed rock:

Size (mm)	Limestone	Porphyroid
+6	nil	nil
6-3,3	55%	53%
3,3-0,1	44%	38%
-0,1	1%	9%

Flotation tailings:

Size (mm)	Deslimed	Undeslimed
+1	0,7%	0,2%
1-0,8	0,5%	0,7%
0,8-0,6	1,0%	2,2%
0,6-0,5	4,0%	6,5%
0,5-0,3	12,9%	18,1%
-0,3	81,0%	72,3%

Coarse crushed rocks:

Size (mm)	S.F. reject	Sandstone
+19	0,4%	1,3%
19-12,7	6,6%	13,4%
12,7-9,5	20,8%	26,4%
9,5-5,6	27,0%	22,0%
5,6-2,8	17,6%	13,8%
2,8-1,4	10,1%	8,1%
1,4-0,7	6,7%	5,8%
0,7-0,35	4,1%	3,8%
0,35-0,17	2,4%	2,2%
-0,17	4,3%	3,2%

The main objective of the tests on the a) and b) materials was check a suspected dif ference of behaviour between silicatic and calcareous aggregates. The results of the tests, on mixtures with 50% of the stoi-chiometric water contents, are shown in the graph of fig. 7.

It can be seen that mixtures attain a far lower resistance (the aging duration being the same) than plain Fluorgypsum mortars, partly due to the higer porosity (40÷50% against 30-40%), partly to the following reasons:

- adhesion between Fluorgypsum mortar and silicatic aggregate is poor, perhaps be cause of traces of HF in the mortar
- calcareous aggregate undergoes some sul

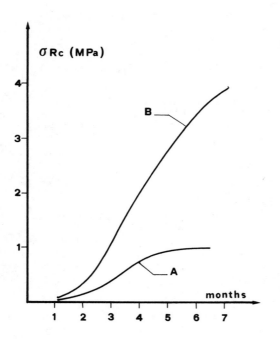

σRc (MPa)

Fig. 7. Behaviour of the mixtures of Fluor gypsum mortar with aggregates: A: silicatic aggregate; B: calcareous aggregate.

phatation (limestone grains recovered from the debris have shown a gypsum con tents of approx. 2%, after 2 months of aging)
- calcareous aggregate slows down the hydration because reduces the mortar acidity (the effect is more thoroughly discussed later).

However, the calcareous aggregate proved to be superior to the silicatic one.
The mechanical behaviour of the mixtures is noticeably improved when accelerant reagents are added: with 1% of $K_2SO_4+FeSO_4$ added to the Fluorgypsum, a mechanical resistance of $2\div2,5$ MPa is attained (with both aggregate types) with a 2 months aging duration, that is a 20 fold increase with respect to the plain mixtures.
It can be said, however, that a satisfactory strength of the Fluorgypsum mortar/aggregate mixtures requires a far longer aging duration to develop than with plain mortar.
Mixtures with flotation tailings have been studied in detatil because, due to their

anticipated low permeability, to the comparative cheapness, and to the environmental advantages offered as a tailing disposal means, are very promising.
The mixtures have been prepared with a stoichiometric amount of water and with different additives, and subjected to prolonged aging. Resistance values attained with a 7 months aging duration are shown in table 4.

Table 4. Resistances attained after 7 months of aging by 50/50 mixtures of Fluor gypsum mortar and flotation tailings

Fluorgypsum	Reagents type and amount	Resistance MPa
Mixtures with deslimed tailing:		
Plain	K+Fe suplh., 1%	3
Plain	Na+Fe sulph., 1%	2-2,5
Plain	Na sulph., 1%	3
Lime added	K+Fe sulph., 1%	3
Lime added	Na+Fe sulph., 1%	6
Lime added	Na sulph., 1%	3,5
Mixtures with undeslimed tailing:		
Plain	K+Fe sulph., 1%	9-10
Plain	Na sulph., 1%	3-4
Lime added	Na sulph., 1%	3,5-4
By comparison: mixtures without reagents, deslimed tailing:		
Plain	–	2
Lime added	–	0,8
By comparison: mixtures without reagents, undeslimed tailing:		
Plain	–	4-4,5
Lime added	–	1,5-2

Plain Fluorgypsum and lime added Fluorgypsum (lime was added at the plant, in a- mounts close to 1%), to obtain a partial neutralization of the unreacted sulphuric acid) have been used in the tests.
The following conclusion and tentative conclusions can be drawn from the results:
- breadky speaking, the undeslimed tailing proved to be superior to partly deslimed tailing (due, we suppose, to the lower porosity of the mass and to the lower tendency to segregation of the rea-

gents added)
- the accelerating reagents counteract the noxious effect of the lime (by lowering, we suppose, the Ph value of the mortar)
- the $K_2SO_4+FeSO_4$ mixture proved to be more effective than the plain Na_2SO_4 (again we suppose it to be due to the lower tendency to segregation of the former: in due time a partial hydrolisis of the $FeSO_4$ gives a colloidal hydroxide that efficiently hinders the reagent segregation)
- bentonite strongly improves the resistance (apart from porosity and reagent segregation reduction, we suspect other phoenomena unknown, that could deserve a separate investigation, are playing a role here).

Tests on mixtures with coarser aggregate (sandstone, Sink-Float reject) had to be made on larger cylinders; the results therefore can not be exactly compared with results from small diameter cilinders until now rewieved; it can be assumed however that the tests on large diameter more closely represent the actual fill.
The results are summarized in the table 5. It can be seen that accelerating reagents are necessary to attain a satisfactory resistance, even if a long aging duration is provided; results are very good when the accelerating mixture $K_2SO_4+FeSO_4$ is used, somewhat inferior but still acceptable when Na_2SO_4 is used. The noxious effect of the lime is satisfactorily counteracted. No important differences are found between the two aggregate types.

2.5 Elastic properties of the Fluorgypsum mortar

The elastic modulus and the Poisson ratio have been determined on mortars of plain and accelerating reagent added Fluorgypsum. The results are shown in the table 6.
The very low value of E, joined with a comparatively high resistance to compression, can be considered a favourable feature of the Fluorgypsum mortar in the intended use, showing that the mass can gield somewhat under load without breaking, absorbing a considerable deformation energy; it can be advantageous when dinamic

loads (for example, arising from roof failures) are feared.

Table 5. Resistances attained, after 7 months of aging, by Fluorgypsum/aggregate mixtures.

Fluorgypsum	Reagents type and amount	Resistance MPa
Mixtures with Sink Float reject:		
Plain	K+Fe sulph. 1%	12
Plain	Na sulph. 1%	10
Lime added	K+Fe sulph. 1%	12
Lime added	Na sulph. 1%	6
Mixtures with sandstone:		
Plain	K+Fe sulph. 1%	9
Plain	Na sulph. 1%	4
Lime added	K+Fe sulph. 1%	13
Lime added	K+Fe sulph. 2%	12
Lime added	Na sulph. 1%	4
Lime added	Na sulph. 2%	4

By comparison, mixtures without reagents:

Sink Float reject:		
Lime added	-	1,5
Sandstone:		
Lime added	-	1,0

2.6 Permeability

The permeability of the fill binder is obviously an important factor in any filling operation; in this case two reasons render the permeability knowledge and control more important than in conventional fill:
- gypsum is sligthly soluble in water (2g /dm^3 in distilled cold water)
- the water becames acidic and corrosive upon permeation of the mass.
Several permeation tests have been made on fluorgypsum mortars and on mortars containing different aggregates. The results are here summarized:
- plain Fluorgypsum mortar (aged by two months) shows a permeability to water ranging between 0,15 and 0,20 cm/day. Permeation tests have been protracted by about 4 months: some gypsum loss has

Table 6. Elastic properties of Fluorgypsum mortars, completely hardened

Composition	Aging (mon.)	E (MPa)	Poisson ratio
Plain Fluorgypsum with stoichiome-tric water	6	$1,9-25.10^3$	0,08-0,1
Same, with 1/2 stoichiometric water	6	$2-2,1.10^3$	nd
1% K+Fe sulph. added	4	$2,4-3,9.10^3$	0,1
2% K+Fe sulph. added	4	$2,4-2,5.10^3$	nd
4% K+Fe sulph. added	4	$2,9-3,5.10^3$	nd

been observed, but paradoxically the permeability shows a tendency to decrease with the elapsed time, apparently because the pore reduction due to the hydration of the still unreacted anhydrite and the gypsum re-crystallization compensates the lasses (see graph. of fig. 8).

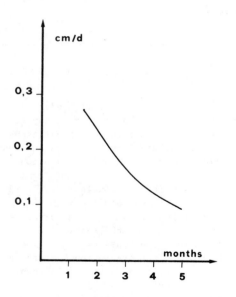

Fig. 8. Permeability decrease with permeation time, from tests on plain fluorgypsum mortar.

The resistance of the bars subjected to permeation does not show a noticeable reduction when compared to unpermeated bars (the aging being the same)
- lime added Fluorgypsum (again, upon 2 months of aging) shows a higer permeability (0,2÷0,25 cm/day)
- mixtures of plain Fluorgypsum mortars with aggregates of the silicatic (porphyroid) and calcareous type, are, as expected, more permeable than plain mortar: permeability values range between 1,17 and 3,50 cm/day
- bentonite (2%) added mortar shows a much lower permeability than plain mortar (0,025 against 0,15-0,20 cm/day). Bentonite does not reduce the strength of the mortar, on the contrary improves the long term resistance (a resistance of 12 MPa has been attained with 4 months of aging by the Fluorgypsum mortar containing 1% of bentonite; simultaneous addition of bentonite and accelerants gives an excellent long term resistance, as shown in the table 7, irrespective of the presence of lime)
- accelerating reagents show a very important permeability reducing effect, albeit not so strong as bentonite does. The permeability of the mixtures of Fluorgypsum mortar with aggregates is excessive, and apparently forbides the use of the fluorgypsum as a binding agent at places where an abundant groundwater seepage occurs. In any case permeability reducing additives are advisable.

2.7 Chemical corrosion exerted by the Fluorgypsum mortar

The Fluorgypsum imparts a strong acidity to the permeating water, hence a strong corrosion effect is to be expected on the metallic structures imbedded in the mass, or in contact with water that seeped through it. The most obvious way to counteract this effect is to neutralize the mass by adding alkaline substances. Our tests proved, however, that this simple way is not easily practiceable: alkaline substances slow down considerably the anhydrite hydration (and therefore the setting of the mortar) even if the

Table 7. Resistances attained (with 7 months aging duration) by Fluorgypsum mortars containing lime, bentonite and accelerants

Additives	Resistance MPa
Lime, 2% bent., 1% Fe+K sulphates (1)	19,5
Lime, 2% bent., 1% Na sulphate (1)	18,6
2% bent., 1% Fe+K sulph.	16,3
2% bent., 1% Na sulph.	16,3

(1) Lime was added in the amount required to obtain pH3 in the permeating water.

amount added is not sufficient to obtain a neutral or nearly neutral mortar.
Results of the tests on Fluorgypsum mortars neutralized (albeit incompletely) with quicklime are shown in the graph. of fig. 9.

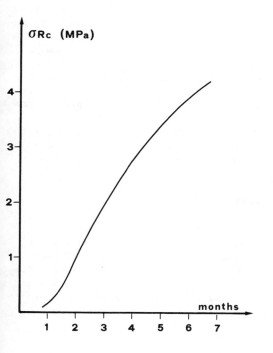

Fig. 9. The hardening of the partially neutralized Fluorgypsum (neutralizing reagent: lime; residual pH approx. 3)

It can be easily seen how unfavorably the results compare with plain Fluorgypsum mortar.
With caustic soda (NaOH) added, the picture is somewhat more complex. A very small amount (1%, leaving a pH of the mortar close to 3) of caustic soda acts as an accelerant, due to the immediate transformation of the reagent in Na_2SO_4, whose effect has been already discussed. When greater amounts, needed to fully neutralize the mortar, are added, the NaOH is deleterious. The results are shown in the table 8.

Table 8. Effect of NaOH on the Fluorgypsum mortar hardening

NaOH added	Aging (months)	Resistance MPa
1%	2	6
1%	3	7
2%	2	2
2%	3	2
4%	2	0
4%	3	0

We can therefore conclude that a complete neutralization of the mortar is not possible, and even a partial neutralization has a cost in terms of mechanical strength if other reagents (accelerators, bentonite) are not added to counter act the noxiuos effect of the alkaline substance.
The corrosion problem can be divided in two parts:
- corrosion caused by water that has been in contact, or seeped through, the Fluorgypsum bound fill, on pumps, pipes etc.: in our opinion this problem can be only sligthly reduced by adding tolerable amounts of neutralizer to the Fluorgypsum; rather, it can be solved by reducing the mortar permeabiliy (and, of course, by avoiding the use of Fluorgypsum as a fill binder in the sections of the mine where groundwater is abundant)
corrosion exerted on metallic structures imbedded in the mass: no way to counteract efficiently this effect can

be envisaged. To give an idea of the
foreseable service life of these structures, we measured, however, the corro
sion rate of iron bars imbedded in the
Fluorgypsum mortar.
In plain Fluorgypsum mortar we detected,
in tests run by several months, corrosion
rates ranging between 1.10^{-3} and 4.10^{-3}
grams of metal per square centimeter of
exposed surface and per day; in mortar
partly neutralized with lime the corrosion
rate was still high, but somewhat lower:
approx. $0,8.10^{-3}$ g/cm^2 d.
Reinforced Fluorgypsum structures, similar to reinforced concrete, are clearly
out of question.

2.8 Concluding remarks on the suitability of Fluorgypsum to the use as a fill binder

The laboratory study hitherto rewiewed
shows that a cemented fill that can com-
pete with some type of conventional cemen
ted fill (that is, a fill consolidated
with a low amount of portland cement, see
table 9) can be obtained using the Fluorgypsum as a binder, provided that the set
ting delay is not a critical factor. Plain
Fluorgypsum can be used in dry places,
whilst mixtures of Fluorgypsum with small
amounts of bentonite and accelerants are
advisable where some small water inflow
is known to occur or is suspected; on the
contrary no Fluorgypsum should be used in
the wet sections of the mine, the main
concern being the contamination of water
with acidic, corrosive substances.
It should be however noticed that the laboratory tests, being carried out under
atmospheric pressure, give forcibly a con
servative estimation of the mechanical
strength obtainable by the Fluorgypsum
mortars and Fluorgypsum bound fills in
the actual practice.

3. IN SITU TESTS

Until now only comparatively small voids
have been filled with Fluorgypsum mortar
(an abandoned ore chute). The site for
the small scale test has been chosen ha-

Table 9. Comparison between the resistance of Fluorgypsum bound and Portland bound
fills (1)

Mine	Cement amount %	Resistance MPa
Gavorrano (pyrite)	15	10
Raibl (Pb–Zn)	19,4	22,5
Fontane (talc)	20	20
By comparison: 50/50 Fluorgypsum S.F. reject, with additives		10–12
50/50 Fluorgypsum Flot.tailing with additives		9–10

(1) Data on conventional cemented fill are
from Italian mines.

ving in mind to core at intervals the fill
along all the filled depth, thus obtaining
the actual behaviour of the mortar in
the time and under different pressures
(given by the filling material column).
Other experimental drilling tests,with
reagent added fluorgypsum and with mixtures of Fluorgypsum with aggregates, will
be soon started, based on the preliminary
results.
At the date the setting process in the
experimental filling operation is in a
too early stage to provide reliable data.

AKNOWLEDGEMENT

The authors express their thanks to Miss
Marilena Cardu, who carried out a great
part of the laboratory tests as a part
of hers doctoral thesis in Mining Enginee
ring on the properties of Fluorgypsum as
a filling material, to be discussed at
the Politecnico di Torino.

Influence of tailings particles on physical and mechanical properties of fill

LIU KEREN & SUN KAINIAN
Northeast Institute of Technology, Shenyan, Liaoning, China

ABSTRACT: The recovery of tailings for underground backfill is low. Many mill plants are shortage of tailings for backfill. An approach to the utilization of fine tailings is presented in the article based on the study of physical and mechanical properties of fine tailings. The influence of fine tailings on coefficient of permeability, angle of inter-friction, modulus of elasticity, behavour of force transmission and strength of cemented fill was tested. An orthogonal experiment design and regression analysis were adopted.

Universal problems existed in using the way of tailings fill in mining are shortage of sand quantity, difficult to build the disposal dam by the fine tailings and easy to run off. So it is a very important subject to increase the recovery of tailings for backfill. An approach has been given concerning the influence of content of fine tailings on physcial and mechanical properties of fill in this article.

1 THE INFLUENCE OF FINE TAILINGS ON PERCOLATION RATE

Percolation rate has a very close relation with the content of fine particles in tailings. With the increase of the content of fine particles the percolation rate fast drops. This is a reason why a mine has to deslime the tailings as a kind of fill material.

The factors influencing the percolation rate would be concluded into three parts: Porosity (particle shape included). constituent of particles (specific surface and content of fine particles included) and composition of mineral (something important in thickness of watering film).

The relation between the percolation rate and the content of fines can be analysed by the data from lab using the tailings from four gold mines of China. The main composition of these tailings is quartz.

1.1 The criteria of percolation rate

General thought in the process of backfill, when $K_{10} > 10$ cm/h, water will be

leaked out on time. After the backfill work finished mining stope can be allowed in. But the percolation rate in practice (Fig. 1) K_{10} = 4-19 cm/h. The percolation rate in Fankou and tonglushan mines is K_{10} = 4-6 cm/h. So when $K_{10} > 5$ cm/h the production can

Table 1. Percolation rate of mines

Mine	K_{10} , cm/h
Fenghuangshan	15
Muding	6
Xikuangshan	9.7
Fankou	4-5.7
Huangshaping	14
Lingshan	10.6
Huongtoushan	19
Dongxiang	13.5
Tonglushan	4-6
Zhaoyuan	8-10

be allowed. In this case the value can be as a criteria in evaluating percolation rate for backfill.

1.2 The relation between percolation rate and porosity

There is a very closed relation between the percolation rate (k_{10}) and porosity (P). Fig. 1 is a K_{10} and P curve of Zhaoyuan tailings. With the increasing P, K increases directly. When the fine particles become less, a great change will take place in K_{10} in company with P. This because the greater P is, the smaller resistance will be when the gravity water passing the backfill and the bigger K_{10}.

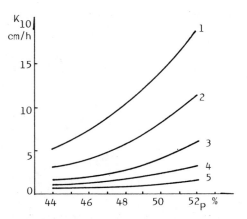

Fig.1 K_{10} -P curve of Zhaoyuan tailings

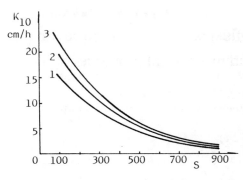

Fig. 2 Relationship curve of K_{10} and S
1-P=49%; 2-P=50%; 3-P=51%

1.3 The relation between percolation rate and specific surface

Because of the resemblance of the tailings composition in the four mining plants, so the data can be handled synthetically. And because of the difference of the specific gravity and the contents of +40 μm, it is adopted in using volume specific surface to indicate how much the fines have. After the fines increased in the same condition of porosity the specific surface is a main factor to influence the percolation rate. Tailings keep a layer of watering film on the grain surface. The film thickness mainly is decided by the mineral composition and content of electrolyte. In the case of the same conditions the bigger specific surface is, the larger watering film would be in proportion. In this way the resistance of percolating of gravity water would be increased. The more fines are, the larger specific surface and watering film will have, but K_{10} would be decreased.

In order to approach the influence of specific surface on percolation, a regression analysis between the percolation rate K_{10} and specific surface S is carried out. When the porsity P=49%, the regression equation is

$$K_{10}=21.563 \ e^{-0.0033 \ s} \qquad (1)$$

Correlation coefficient r=-0.7918. Degree of freedom N-2=18. When α =0.01, |r| =0.56. So eq. (1) on the level of α =0.01 is notable. The regression curve of eq.(1) is the curve 1 of Fig. 2.

When P=50%, the regression equation is:

$$K_{10}=26.978 \ e^{-0.0033 \ s} \qquad (2)$$

Correlation coefficient r=-0.8179 and notable on the level of α =0.01. The regression curve of eq. (2) is curve 2 of Fig. 2.

When P=51%, the regression equation is:

$$K_{10}=31,537 \ e^{-0.0033 s} \qquad (3)$$

In this case r=-0.8444, also notable on the level of α =0.01. The regression curve is curve 3 of Fig. 2.

It is obvious that eq. (1), (2), (3) are index functions of the type of $K_{10}=a \ e^{-bs}$. With the change of P coefficient a changes, but there is no change for b which remains at -0.0033. This indicates that the main influence factor on a is porosity, but main effect factor on b is mineral physical and chemical properties. For the similar composition of mineral b is a constant.

1.4 The relation between K_{10}, P and S

The relation between the coefficient a and P is basicly linear relation.

$$a=498.7P-222.6 \qquad (4)$$

Correlation coefficient r=0.9988. When N-2 =1, |r| =0.997 for α =0.05. It is notable on level =0.05.

Therefore when the main mineral is quartz the regression equation of K_{10}, S and P:

$$K_{10} =(498.7P-222.6) \ e^{-0.0033s} \qquad (5)$$

A percolation test of tailings from a copper mine which main mineral composition is skarn. The relations of K_{10} and S is the same as the type of $K_{10} =a \ e^{-bs}$. In addition, no matter now changes P has, value b is at average of 0.0036. That means index b is decided by the composition of mineral. When a is changed into a respect of a=b'p-a', the relationship equation is:

$$K_{10} =(1935P-818.65) \ e^{-0.0036 \ s} \qquad (6)$$

So there is also a relationship equation of

$$K_{10} =(b'p-a') \ e^{-bs}$$

1.5 The possibility of increase of fine particles

According to equation (5) when p=50% and $K_{10}=5$ cm/h, $S=500$ cm^2/cm^3. In this case the content of -20μm is about 8%. It overtakes the content norm of -20μm taken is practice 5%. That makes it possible to increase the content of fine paricles and improve the recovery of underflow of a cyclone.

It would further increase the utilization of fine particles if some chemical reagents, such as polyacrylamide or electrolyte would adopted. Eletrolyte can reduce thickness of watering film. Flocculant can make small particles be flocculent structure and en-large the percolation. Tested data of tail-ings from a copper mine are listed in Table 2. It is possible to increase usage of tail-ings to 60-70%.

Table 2 Influence of flocculant on the per-colation rate

Type of reagents	percolation rate,cm/h
no add	2.64
FeSo$_4$	2.78
FeCl$_3$	3.67
Al$_2$(OH)$_n$ Cl$_{6-n}$	3.69
Polyacrylamide	4.08

2 THE INFLUENCE OF FINE TAILINGS OF COEF-FICIENT OF INTERNAL FRICTION

2.1 The relation between content of fine tailings and coefficient of internal fric-tion

Factors influencing coefficient of internal friction tanφ are: particle size of tail-ings, shape, density, moisture and etc. Particle size and content of fine particles influence tanφ a lot. Whether there is a cohesive force will be decided by how much fine particles will have. When fines (-20 μm) under 10%, cohesive force is small or no exist at all. Taking Lingshan tailings for example, when the content of -20μm is 10%, cohesive strength $C=0.048$ Kg/cm^2, tanφ =0.672. When normal pressure is 1-2 Kg/cm^2, cohesive strength takes up only about 5% of shearing strength. So it can be neglected.

If there is not much fines, tanφ will reduce with increasing content of fine par-ticles. The curve in Fig. 3 is made of tailings from Zhaoyuan gold mine. The re-gression equation is:

$$y=0.762x^{-0.072} \qquad (7)$$

Where y-tanφ
       x-content of fines
The correlation coefficient of the equatior $r=-0.999$. $N-2=3$. When α =0.01, |r| =0.959. Eq. (7) is notable on the level of α =0.01

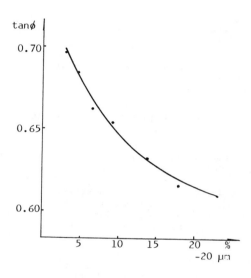

Fig. 3 Curve of fines content and tanφ
In eq. (7) the relation of coefficient of internal friction and content of fines is a power function. With the increase of the content of fines tanφ will reduce. The reducing tendency will become relax after a certain period arrived.

2.2 The relation between normal premal pressure and tanφ

From the test results obvious that tanφ becomes smaller with the increase of normal pressure. The relationship between tanφ, normal pressure and particle size is shown in Fig. 4.

In fact the void ratio is a variable. With the work going on, it will become smaller, in other word, tailings will be-come more dense. When void ratio reaches 0.65 about, normal pressure reaches 470-500KPa, the tanφ of the material above tends to the same, about φ=29.1°

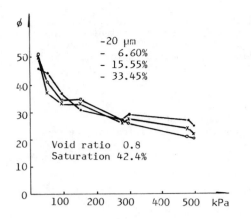

Fig. 4 The relation curve of tan∅ and normal pressure

3 THE INFLUENCE OF FINE PARTICLES ON THE MODULUS OF ELASTICITY

3.1 The compression curves of tailings with different particle size

In order to study the deformations of different tailings a compression test has been carried out with tailings from Laiyuan copper mine. The tailings were classified as following:

Table 3. Classification of the tailings

mesh	+60	+120	+160	+200	+300	+360
d,mm	0.455	0.203	0.113	0.086	0.061	0.045

More compression tests have been given to the 6 sorts of tailings. The loading and unloading curves are expressed in Fig. 5. The solid lines indicate the dry tailings and dotted lines saturated tailings.

3.2 The influence of moisture content on the modulus of elasticity E

From Fig. 5 obvious that for the same particle diameter of tailings there is a clear difference in the compression curves between dry sand and saturated sand. The E of saturated sand is between 85-115 Kg/cm^2, but of dry sand is 130-170Kg/cm^2. That means deformation of dry sand is smaller than saturated sand, i.e. tan∅ is lower.
Under the same compression strength though there is a big difference in the absolute value of deformation between dry sand and saturated sand, the ratio of plastic deformation to the elastic deformation remains the same (Table 4).

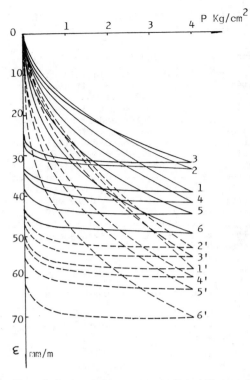

Fig. 5 Compression curves of tailings with different particle diameters
1. 2. 3. 4. 5. 6 respectively d=0.455, 0.203, 0.113, 0.086, 0.061, 0.045 mm

Table 4. Proportion of the plastic deformation

Mean diameter d, mm	Percentage of plastic deformation to the tatal deformation, %	
	dry	Saturated
0.455	84	86
0.203	82	83
0.113	80	82
0.086	83	84
0.061	86	88
0.045	87	85

3.3 The relation between wodulus E and mean particle diameter d

According to the compression curves of Fig. 5 the influence of particles diameter on the E was discussed. The compression curve can be devided into two periods, the primary period, pressure is 0-0.25Kg/cm^2, and the later period, pressure is 1-3Kg/cm^2.

24

1. In the primary period the relation between E and d is shown in Fig. 6.The regression equation of E_o is:

$$E_o = 30.32 + 6.91 \ln d \quad Kg/cm^2 \quad (8)$$

Eq. (8) is a logarithm curve. Correlation coefficient r=0.9813 N-2=4. When α=0.01, $|r|$ =0.917, Eq. (8) is notalbe on the level of α =0.01

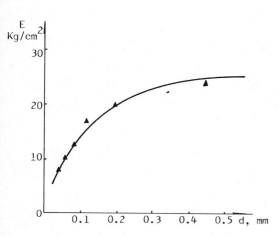

Fig. 6 Relationship curve of E and d measured data

The results indicated that with increase of particle diameter, the primary E_o increases, because E_o is related to porosity. With the decrease of particle diameter, the porosity increases. E value reduces.

2. E in the later period of compression. The pressure in this period were 1-3 Kg/cm^2. Moduli E were calculated from P_i=1 Kg/cm^2 and P_{i+1}=3Kg/cm^2. The E-d curve in the later period is shown in Fig. 7 The regression equation is:

$$E=79.19-11.25 \ln d, \quad Kg/cm^2 \quad (9)$$

r=0.9642. N-2=4. When α =0.01 $|r|$ =0.917. Eq. (9) is notable on the level of α =0.01

From Fig. 7 and equation (8) can be found that E of the later period opposites to the primary period. It is increased with the reducing of particle diameter. After the primary period of compression, with the reduce of particle size the bearing capacity of tailings is increased. Though the content of fine particles increased and tanϕ is reduced, the relationship changes according to the power function tanϕ = ax^{-b}, where x is content of fine particles. The quantity degree of b is 10^{-2}. Specific

surface will increase with particles diameter reducing in an inverse ratio of d. That is with the reduce of diameter d, the increase of specific surface is faster than the reduce of tanϕ. In other words diameter

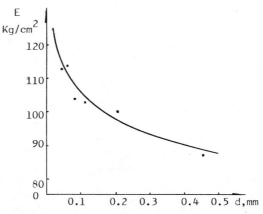

Fig. 7 Relationship curve of E and d measured data

reduces, friction force in an unit volume is increased, E increases and the bearing capacity is promoted. Besides this, when particle diameter reduced, diameter of the voids became smaller. Capillary force is inversely proportional to the square of hole diameter. Therefore the increased capillary force raises the modulus E.

It is instructive to expose the information of two periods of modulus. Due to this situation more fine tailings can be adopted in order to reduce the porosity of backfill and increase the modulus of elasticity. So a point of view of breaking away all the fine particles from tailings is open to question.

4 THE INFLUENCE OF PARTICLE SIZE ON THE FORCE TRANSMISSION OF FILL

4.1 Test model and results

Steel tubes with inside diameter 61 mm were used. Heights of test tubes were H=27, 50, 100 and 150 mm. Tested tailings were poured in the tube and pressed on a mechanical press. The compressive strength on the bottom was measured by a force transverter.

Table 6. Table of regressive orthogonal design

No	H	d	0	x_1	x_2	x_3	x_4	x_5	y
1	-1	-1	1	-1	-1	1	1	1	13.63
2	0	-1	1	0	-1	-2	1	0	5.30
3	1	-1	1	1	-1	1	1	-1	1.71
4	-1	0	1	-1	0	1	-2	0	13.52
5	0	0	1	0	0	-2	-2	0	5.19
6	1	0	1	1	0	1	-2	0	2.02
7	-1	1	1	-1	1	1	1	-1	12.81
8	0	1	1	0	1	-2	1	0	5.12
9	1	1	1	1	1	1	1	1	2.21
			d_j	9	6	6	18	18	4	
			B_j	61.51	-33.98	-0.46	14.68	-0.72	1.32	
			b_j	6.83	-5.67	-0.08	0.81	-0.04	0.33	y_i^2=625.79
			Q_j		192.44	0.04	11.97	0.03	0.43	Q=204.91

In order to investigate the relation between the strength on the bottom (q_z), mean diameter (d) and the height of fill (H) a regressive orthogonal experimental design method was adopted. The coding table of factor levels is shown in Table. 5.

Table 5. The coding table of factor levels

Factor	H, mm	d, mm
Code Sign	X_1	X_2
Datum level (0)	100	0.086
Variable interval	50	0.041
Up level (+1)	150	0.127
Down level (-1)	50	0.045

Test model and results are illustrated in Table 6. y stands for the strength on the bottom (q_z).

The results of variance analysis are summarized in Table 7 and they are notable on 0.01 level.

Table 7. Table of variance analysis

Origin	sum	regression	deviation
Sum of	205.40	204.91	0.49
Degree of freedom	9-1=8	5	3
Mean value of variance		40.98	0.16
F ratio		256	
Notability $F_{0.01}$=		$Fa > F_{0.01}$	
131.6		(5.3)	
(5.3)			

The regression equation is

$$y=5.29-5.67X_1-0.08X_2-0.33X_1X_2+2.43X_1^2-0.12X_2^2 \quad (10)$$

It's an equation between $H(X_1)$, d (X_2) and q_z (y) in case of a constant pressure (700 Kg). Putting the H into eq. (10) the relation between q_z and d can be found, such as H=150mm, X_1^z=1. The eq. (10) becomes:

$$Y=2.05+0.25X_2-0.2X_2^2 \quad (11)$$

The curve of eq. (11) is shown in Fig 8. In case of H=150 mm q_z decreases with the decrease of particle size (d).

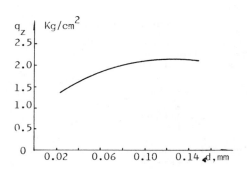

Fig. 8 d-q_z curve in case of H=150mm

In case of H=50mm, X_1= -1 Eq(10) becomes:

$$Y=13.39-0.41X_2-0.12X_2^2 \quad (12)$$

The d-q_z curve of eq. (12) is in Fig. 9.

In case of H=50mm q_z is increased with decrease of d, i.e. the efficiency of force transmission is raised.

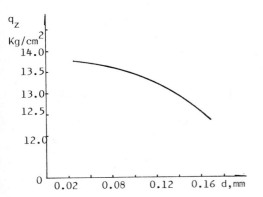

Fig. 9 d-q_z curve in case of H=50mm

In the cases of H=50mm and H=150mm the results turned out coutrary. This can be explained by the formation of pressure curve under certain H/D ratio.

4.2 The influence of H/D on the force transmission

1. H/D ratio under which the pressure curve formed. Experimental work was carried out under confined compression apparatus. After the loading the horizontal passive force produced on the tube wall. Because of the big friction coefficient between the particles and the tube wall, the friction resistance is concentrated on the wall. The farther from the wall the faster decreasing of friction resistance.

When the height is large enough, due to the wall effect a pressure surface in form of paraboloid of revolution will be produced in the loose material. In order to simplify the analysis the problem is discussed as a pressure curve in a vertical plane through the tube center.

The arch rise (h) of pressure curve can be calculated by eq. (13)

$$h = \frac{1}{4\tan^2(45 - \frac{\phi}{2})} \, d \qquad (13)$$

Angles of internal friction of five tested tailings are shown in Table 8.

Table 8. Angle of internal friction

d,mm	0.455	0.127	0.086	0.061	0.045
ϕ	35.8	35.0	34.2	33.9	33.6
$\tan\phi$	0.721	0.700	0.680	0.672	0.664
$\tan^2(45-\frac{\phi}{2})$	0.262	0.271	0.280	0.284	0.288

Put the data into eq. (13) and get h=55mm

Pressure curve will be formed if h>55mm. Settlements due to consolidation of tailings are about 15mm High, and above the pressure curve must keep some tailings, so the real altitude of forming the pressure curve is H>80-90 mm. Therefore H/D>1.3-1.5 can be taken as the criterion of forming pressure curve.

2. Force transmission, when H/D<1.3-1.5. There no pressure curve appears. The uniform loading makes tailings to consolidate. The resistance is mainly caused by $\tan\phi$. From Table 8 it is obvious that $\tan\phi$ reduces with decrease of particles, that means resistance decreases. So compressive strength on the tube bottom increases. That is why the q_z increases with reducing the particle size (Fig. 9).

3. Force transmission, when H/D > 1.3-1.5. There appears a pressure curve in the loose fill material. The normal force acts mainly on the contact between pressure curve and tube wall, forming a ring load similarly in the loose material. The active force of loose material to the tube wall is $q_x = q_z \cdot \tan^2(45-\frac{\phi}{2})$. The resistance from this horizontal pressure $T = q_x \cdot f$. Where f-coefficient of friction between tailings and wall of steel tube (Table 9).

Table 9. Coefficient of firction

d,mm	0.455	0.127	0.086	0.061	0.045
f	0.488	0.512	0.601	0.625	0.675

f increases with reducing the particle diameter. The value of inerease is larger than the decrease of $\tan^2(45-\phi/2)$. For example, comparing d=0.061 mm to d= 0.045 mm, the decrease of $\tan^2(45-\frac{\phi}{2})$ is 0.004, but increase of f-0.05. That is why the bottom pressure q_z (Fig. 8) decreases with decrease of particle diameter.

4. Coefficient of vertical pressure (K_z). According to the soil mechanics in case of uniform pressure the strength on certain

depth is calculated by eq. (14)

$$q_z = K_z q_{zo} \qquad (14)$$

Where q_{zo} -strength under uniform pressure.
In this case loading is confined, not semi-confined loading. K_z can be obtained by experimental method. X_2 (d), X_1 (H) and H/D may be found from eq. (10). K_z-H/D curve is shown in Fig. 10

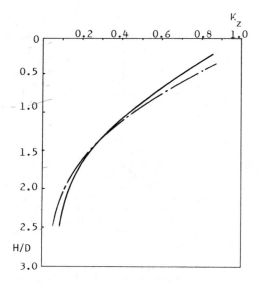

Fig. 10 K_2-H/D curve

Solid line-Kz curve for d=0.127mm.dot dash line-Kz curve for d=0.045mm

q_z can be calculated from Fig. 10. This negative exponential curve is similar to the law in a silo of loose material. The two curves got an intersection point on H/D= 1.5. This is identical with the theory of pressure curve.

5. THE INFLUENCE OF FINE PARTICLES ON THE STRENGTH OF CEMENTED FILL

5.1 The influence of fine particles content on the compressive strength

Three kinds of tailings were tested. The contents of -20μm tailing are 6.60%, 15.55% and 33.48%. The tested results indicated that fill with more fines got larger strength, specially in the primary period and low cement content. See Fig. 11.

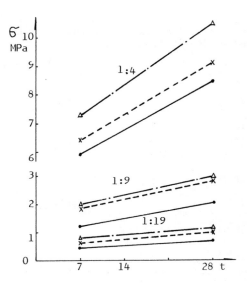

Fig. 11 Strength curve of cemented fill
——— 6.60% (of-20μm fines)
- - - - -15.55%
— - —33.48%

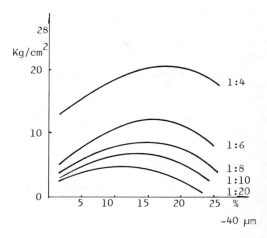

Fig. 12 Strength curve of cemented fill

The influence of fines may be different for different mineral tailings. For example, cemented fill of tailings from Jiaojia gold mine got higher strength when the fines content increases within a certain extent (Fig. 12).

5.2 Elastic modulus E and content of fine tailings

Elastic modulus mainly indicate the relation between the loading and deformation. For backfill the elastic modulus depends not only on the strength but also on the porostiy. Modulus of cemented fill increases with increasing cement content and fine tailings content. The results of E of Lingshan mine tailings are illustrated in Fig. 13.

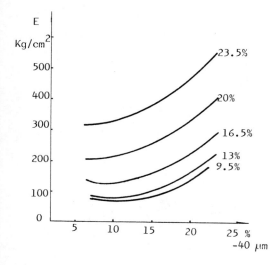

Fig. 13 Curve of E and fines content

6. DISCUSSION AND CONCLUSION

1. In order to increase the recovery of tailings for backfill the fine tailings must be utilized as much as possible. The results of test and practice indicated that K_{10} 5cm/h can be accepted as criteria of the fine tailings content for backfill. The relation between K_{10}, P and S is illustrated by following exponential function:

$$K_{10}=(b'P-a')\ e^{-bs}$$

Where exponent b changes with the mineral composition.

2. Coefficient of internal friction (tanϕ) decreases with increase of normal compressive strength. Tanϕ (y) and fine tailings content (X) have a functional equation as:

$$Y=ax^{-b}$$

3. The modulus (E) of fill with different particle size (d) behaved itself different-

ly. In the primary period of loading the relationship between them is

$$E_0=a+b\ \ln d$$

and in the later period is

$$E=a'-b'\ \ln d$$

That means E_0 decreases with the decrease of d, but E increases in progress of loading. Keeping the $K_{10} > $ 5cm/h the increase of fine particles reduces the porosity and improves the E value of fill.

4. The results of simulation test of force transmission indicated that there was an intersection point on H/D=1.5 of the curve of vertical pressure coefficient (Kz). When H/D < 1.3-1.5, qz increased with the decrease of diameter of particles. When H/D > 1.3-1.5 qz decreased.

5. For cemented fill the compressive strength increases with increasing the content of fines in a certain extent. The compressive strength possible gets higher with an optimum content of fine particles. And it depends on the mineral composition, cement content and slurry concentration.

ACKNOWLEDGEMENTS

The authors with to acknowledge the assistance of colleagues of Mining laboratory of Northeast Institute of Technology. They would also like to thank Mr. Li Jun for helping is transtating the article into English.

Strength of cemented rockfill from washery refuse
Results from laboratory investigations

WALTER KNISSEL & WOLFGANG HELMS
Technische Universität Clausthal, Germany

ABSTRACT: Laboratory tests of uniaxial compression strength of cemented rockfill from washery refuse of coal mines and portland cement have shown, that the main factors of influence are cement content, age and water-cement ratio. For cement contents up to 15 % strength will increase overproportionally with increasing content of binding agent. Moisture content respectively water-cement ratio are of greatest influence to strength. For a definite aggregate type and cement content an optimum water-cement ratio can be calculated to achieve maximum possible strength. Density respectively porosity of cement aggregate mixtures depend linear on moisture content, if the rockfill is not compacted. Measuring the consistence of fresh mixtures is difficult. The vane shear test seems to be appropriate for in situ tests.

1 INTRODUCTION

Underground coal mines in the Ruhr District of Western Germany have hoisted almost 126.8 mio t of raw coal in 1980. About 57.7 mio t or 47.1 % of this amount have been separated as washery refuse in the washing plants. 68.1 % of this tailings have been thrown on dumps. 24.7 % have been used on the surface for roadway and dam construction. Only 7.2 % have been transported back into the mines for backfilling. It becomes more and more difficult to obtain ground and permission for tailing dumps.

In opposition to these facts 350,000 t of building material such as sand and gravel have been consumed by the collieries. Therefore investigations for the economic use of washery refuse mixed with binding agents are necessary. Possible ranges of application could be for
- backfilling, especially for mining steeply dipping seams,
- roadway dams,
- packing behind roof supporting steel arches, and
- road surfaces for trackless transport.

2 PROGRAM OF INVESTIGATIONS

Informations about the properties of cemented rockfill and the factors of influence are needed for economic use of this material. Although a lot of investigation results on cemented hydraulic backfill has been published, there is a lack of informations about properties of cemented rockfill (Dight & Cowling 1979, Gonano 1979). One possible reason may be, that according to the grain size of the aggregate material used for cemented rockfill, large specimens for testing are required. Furthermore the petrographic conditions of the aggregates, which are in use, are very different. So a comparision of test results from different mines is not easy.

Important properties of cemented rockfill are:
- strength, for example uniaxial compression strength,
- deformation behaviour,
- cohesion and angle of internal friction,
- density and porosity, and
- consistence of fresh mixtures.

A program for laboratory investigations of cemented refuse from coal mines has been set up to obtain datas about these properties.

The following parameters are of interest and have been varied during the test program:
- Binding agent. Mainly portland cement PZ 35 F and blast furnace cement HOZ 35 L (corresponding to German standard DIN 1164) have been used. Standard mixtures of these cements and sand must reach an uniaxial

cube compression strength of 35 MPa after a curing time of 28 days.

- Cement content. The cement content z was in the order of magnitude of 1.5 to 15.0 % by mass referred to the dry mass of aggregate.
- Water-cement ratio (w/c-ratio). The water-cement ratio was varied from 0.6 to 6.0. The moisture content w (mass of water/dry mass of solids) can be calculated from the w/c-ratio ω and the cement content z:

$$w = \frac{\omega}{1 + 1/z}$$

- Compaction. After filling into the moulds only a part of the mixtures has been compacted by hand. Most of them have not been treated anyway.
- Curing time. Compression strength of the specimens was normally tested in the age of 1, 3, 7, 14 and 28 days.
- Specimen shape and size. Cube shaped specimens with 100 mm edge length and cylindrical specimens with diameters of 100, 150 and 315 mm and high/diameter-(h/d)-ratios of 1 and 2 have been tested.

3 PROPERTIES OF WASHERY REFUSE

Refuse from washing plants is always a mixture of grains from different rock types. The composition of the washery dirt, which has been used for the tests, was as follows:
- 4 % (by mass) sandstone,
- 46 % sandy shale,
- 44 % sandfree shale, and
- 6 % carbonaceous shale and coal.

Fig. 1. Typical grain shapes of washery refuse.

Uniaxial compression strength of these rock types varies from 10 MPa to 100 MPa. The density of the grains was in average 2.35 g/cm^3. The bulk density of the air dried washery refuse has been measured to 1490 kg/m^3 in average. The shape of the grains was longish to flat with different degrees of roundness (Figure 1). The length :thickness ratios were in the order of magnitude of 2:1 to 6:1, the length:width ratios were measured between 1:1 and 6:1.

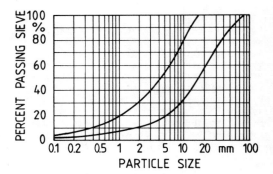

Fig. 2. Grain size distribution curves of washery refuse used for tests.

The maximum grain size of the aggregate collections has been limited in dependance on the size of the specimens. Figure 2 shows the grain size distribution curves of the aggregate. Stability of rocks against water is important. Especially shale tends to loose its strength after long contact with water.

4 TEST PROCEDURE

Air dried washery refuse, cement and water for the mixtures have been measured by weigth. Cement and water have been mixed to cement lime, before adding the aggregate and mixing for five minutes in a customary concrete mixer. The mixture has been filled into moulds made of steel and plastics. Density and water content have been measured.

After one day of curing the moulds could be removed. The specimens have been stored under plastic foil at room temperature. The end surfaces of the cylindrical specimens have been made parallel with plaster of Paris before testing the compression strength. Load rates have been held constant at 0.5 MPa/s. Load and axial deformation have been measured and registrated.

5 TEST RESULTS
5.1 Consistence

Obviously there is no consistence test
method suitable for all investigated mixtures. The moisture content in most cases
was too high ot too low for standard tests.
For the controll of mixtures with high
contents of fine grained solids a modified
vane shear test seems to be suitable. The
apparatus (Figure 3) is sticked into the
mixture and turned slowly by hand.

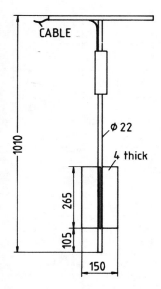

Fig. 3. Vane shear test apparatus
(All dimensions in mm).

The torque is measured electronically by
strain gages. An apparent shear strength
can be calculated from the maximum measured
torque and the geometric datas. Depending
on the grain size distribution of the
solids in the mixture a linear correlation
between water content and shear strength
has been found (Figure 4). The vane shear
test seems to be appropriate for in situ
testing of the consistence of rockfill mixtures with high moisture contents and
grain size under 30 mm.

5.2 Density and porosity

The density of the fresh mixtures is depending on moisture content and compaction
apart from grain size distribution, grain
density and grain shapes.

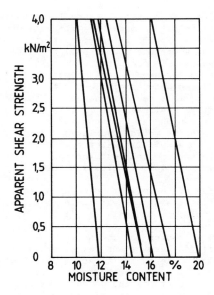

Fig. 4. Apparent shear strength of
fresh cement-aggregate mixtures with
different grain size distribution of
the solids versus moisture content.

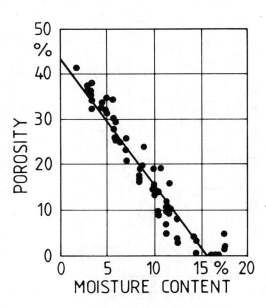

Fig. 5. Porosity of fresh cement-washery
refuse mixtures versus moisture content
(without compaction).

Density and also porosity depend linear on
the water content (Figure 5). With moisture
contents from 2 to 25 % (referred to the
dry mass of aggregate and cement) the density was measured in the order of magnitude
from 1500 to 2300 kg/m^3 and the porosity
was calculated between 42 % and 0 %.
Greatest possible density has been obtained
with moisture contents over 15 %.

Mixtures, which had been compacted after
filling into the moulds, achieved maximum
density with a definite water content,
similar to the Proctor test for soil mechanical investigations (Ingles & Metcalf
1972).

Condition of the mixture depends on content of fine grains and water. In mixtures
with low moisture contents the finer grains
are adhered to the surface of the coarser
grains. In mixtures containing large
amounts of fine grains and low water contents pellets are formed after long mixing
procedures. If the water content is too
high, the fine grains and cement lime will
settle down to the bottom of the mould.

5.3 Uniaxial compression strength

The hardened aggregate-cement mixtures
consist of skeletons from coarser grains.
The voids are completely or partially filled
with a matrix of cement and finer grains.

The strength increases overproportional
with the cement content, but strength can
not be calculated from the cement content
alone (Figure 6).

Strength increases also with curing time
as for normal concrete, but ultimate
strength is achieved only after long periods
up to 90 days (Figure 7).

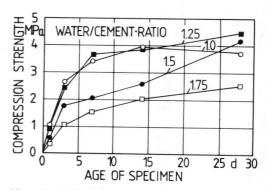

Fig. 7. Uniaxial compression strength of
mixtures with different w/c-ratios versus
age of specimens (Cylindrical specimens,
d = 100 mm, h/d = 2).

The moisture content respectively the
water-cement ratio are of greatest importance to the strength. Specimens with the
same size, age and cement content have very
different compression strength, depending
on the moisture content (Figure 8). To obtain the maximum strength for a definite
cement content, a definite moisture content
is needed. This optimum content respectively
water-cement ratio is not the same as for
normal concrete, for which w/c-ratios of
0.5 to 0.7 should be preferred (Walz 1971).

The shape and the size of the specimen
are also of influence to the measured
strength. Figure 9 shows measured compression strength values of cubic specimens in
comparision to the strength of cylindrical
specimens with the same composition and age.

5.4 Deformation behaviour

Figure 10 shows typical stress-strain
curves for cemented rockfill from washery
refuse. The deformation behaviour can be
described by a deformation modulus, which
should be defined here as the tangent modulus to the stress-strain curve at begin of
the loading process. These modulus is depending on the same factors as the compression strength. Moduli up to 10,000 MPa have
been measured. A correlation between deformation modul and strength cannot be given.

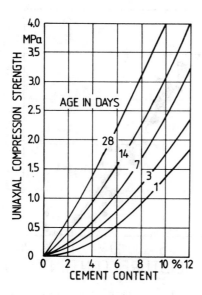

Fig. 6. Uniaxial compression strength of
mixtures with optimum water-cement ratio
versus cement content (Cylindrical
specimens, d = 315 mm, h/d = 1).

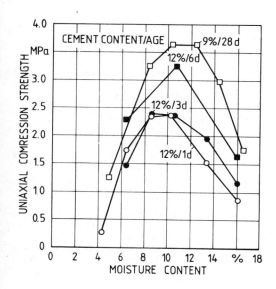

Fig. 8. Uniaxial compression strength of mixtures with different cement content versus moisture content (Cylindrical specimens, d = 315 mm, h/d = 1).

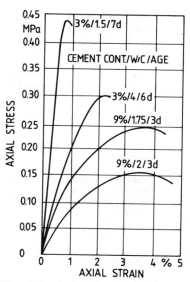

Fig. 10. Unconfined uniaxial stress-strain relationship of specimens from cement-waste mixtures.

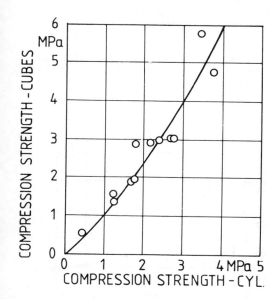

Fig. 9. Uniaxial compression strength of cylindrical specimens (d = 100 mm, h/d = 2) versus uniaxial compression strength of cubic specimens (edge length 100 mm) with different water-cement ratios and cement content of 9 % portland cement.

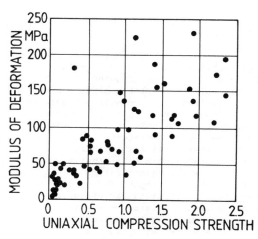

Fig. 11. Modulus of deformation versus uniaxial compression strength (Cylindrical specimens, d = 315 mm, h/d = 1).

5.5 Failure mechanism

Only specimens with relative high strength broke by sudden failure. Most were deformed slowly and steadily with vertical cracks on the surface. After the test procedure almost all specimens showed a typical double cone shape.

This can be seen as a confirmation, that despite of the inhomogeneous structure of the specimen the stress distribution had been uniform.

Cracks passed in most cases through the fine grained matrix and through the grains of shale. Sandstone and sandy shale grains were destroyed only in specimens with relative high strength.

6 DISCUSSION
6.1 Density and porosity

The voids between coarser grains are partly filled by a fine grained matrix. In specimens with high moisture content these voids are filled up completely. Furthermore friction between the grains is lower, therefore a denser package is possible.

A definite water content of the mixture is essential to moisten the surface of the grains completely. A surplus of water easily leads to a separation of cement lime and rocks.

Washery refuse normally contains a lot of water when leaving the washing plant. This fact has to be taken into consideration when calculating the composition of a mixture.

6.2 Strength

The strength of normal concrete decreases with increasing water-cement ratio. Water filled voids are lowering the strength. Theoretically a w/c-ratio of not more than 0.4 is sufficient for complete hydration of portland cement. Practically higher w/c-ratios are used to abtain a suitable consistence of the mixtures. On the contrary laboratory tests with washery refuse as aggregate have shown here, that mixtures with small water-cement ratios do not lead to the highest strength. The curves constructed from the measured datas in the strength-versus-moisture content diagram (Figure 8) have significant maxima, especially for mixtures with high cement content.

Before a possible interpretation will be given, the difference between normal concrete and cemented washery refuse should be explained:
- Rock types, which are common in the roof and floor of coal seams have in most cases lower strength than rock types, which are used for aggregate in normal concrete. Rocks of shale contain large amounts of phyllosilicates, especially minerals of the illite group.
- The binding agent content of cemented rockfill, here 20 to 200 kg/cm^3, is lower than in normal concrete with 150 to over 400 kg/m^3.
- Values of porosity can be high, especially if compaction methods are not used.

The hardened mixture of refuse and cement consists of
- coarser grains of waste,
- a fine grained matrix of cement lime and finer tailings, and
- air or water filled voids.

The strength of the entire mixture is depending on the portions of coarser grains, matrix and voids, furthermore on the strength of these coarse grains and of the matrix and at last on the adhesion between them.

Strength of the rock particles depends on petrography, moisture content and shape. Strength of the matrix is influenced by type and content of binding agent, water-cement ratio and curing conditions and age. Petrographic properties, grain size distribution and surface conditions can also be of importance.

Objects of the addition of water are as follows:
- A definite amount of water is necessary for wetting the surfaces of cement and aggregate grains. The water demand especially of fine grains is high, because the surface area is large.
- There is a relative stable adhesion of water especially to grains, which contain clay minerals. Another portion of the water is soaking into the grains and cannot be used for hydration.
- Last not least a definite amount of water is necessary for complete hydration of the cement mass.

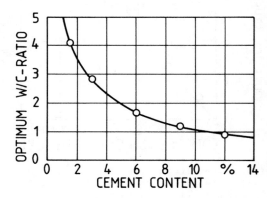

Fig. 12. Optimum water-cement ratio versus cement content for cemented rockfill from washery refuse - 80 mm size (See Figure 2).

With increasing water content of the mixture a greater amount of water is necessary for the hydration. Figure 12 shows, . that the optimum water-cement ratio to achieve the maximum strength for a mixture of washery dirt and portland cement is decreasing with increasing cement content. The necessary mass of water m_w consists of the amount of water m_{wh}, which is essential for the hydration process, and the mass of water m_{wg}, which is bound by the aggregate:

$$m_w = m_{wh} + m_{wg}$$

The quantity of water m_{wh} can be calculated from the cement mass m_z and the water-cement ratio ω_h, which is theoretically sufficient for complete hydration of the cement mass m_z:

$$m_{wh} = \omega_h \cdot m_z$$

With m_k standing for the dry mass of aggregate it follows:

$$\frac{m_w}{m_z} = \frac{m_{wh}}{m_z} + \frac{m_k}{m_z} \cdot \frac{m_{wg}}{m_k}$$

Therefore it can be written:

$$\omega_{opt} = \omega_h + a'/z$$

with

$$a' = m_{wg}/m_k$$

and

$$z = m_z/m_k$$

The factor a' should be called the water-demand factor of the aggregate. When taking a theoretical water-cement ratio ω_h of 0.33, which will be sufficient, and a water demand factor of 0.08 or 8 %, a curve can be constructed, which fits the measured values for the optimum w/c-ratio ω_{opt} well (Figure 12).

The specific water demand factor will be different for each type of aggregate and has to be determined by tests.

Knowing this factor, the optimum water content for mixtures with different binding agent content can be calculated.

7 REACTIONS BETWEEN AGGREGATE AND CEMENT

During cementation of clay mineral yielding rocks or soils reactions between phyllosilicates and hydration products of the cement can take place (Herzog & Mitchell 1963, Moh 1965). The washery refuse used for the tests contained mainly illite minerals.

Investigations of cemented specimens of different age by X-ray diffration have given the idea, that calcium hydroxide, which is formed during the hydration process, reacts with the clay minerals and is consumed gradually. This fact is also known from the stabilization of soils with portland cement and should be taken in consideration for cementation of clay mineral rich rockfill.

8 CONCLUSIONS

When preparing cemented rockfill from washery refuse of coal mines, the specific water demand to achieve optimum strength of the mixture has to be taken into account. The optimum water-cement ratio can be calculated from the cement content and a water demand factor, which is depending on the aggregate type. Strength will decrease, if the moisture content of the mixture is too low or too high. With excessive amounts of water there is the danger of separation between cement lime and rockfill.

An other aspect is, that often a definite consistence of the mixture will be necessary for transportation and handling. Therefore a minimum of mixing water will be required.

Both aspects have to be considered, when calculating the composition of a cemented rockfill mixture.

REFERENCES

Dight, P.M. & R. Cowling 1979. Determination of material parameters in cemented fill. Proc. 4 Int. Congr. Rock Mech., Vol 1: p. 353-359.

Gonano, L.P. 1977. Cemented rockfill in mining - a review. CSIRO Austral., Div. Appl. Geomech., Techn. Report 38: p. 1-7.

Herzog, A. & J.K. Mitchell 1963. Reactions accompanying stabilization of clay with cement. Highway Research Board Record 36: p. 146-171.

Ingles, O.G. & J.B. Metcalf 1972. Soil stabilization - principles and practice. London: Butterworth.

Moh, Z.-C. 1965. Reactions of soil minerals and chemicals. Highway Research Board Record 86: p. 39-61.

Walz, K. 1971. Beziehungen zwischen Wasserzementwert, Normfestigkeit des Zementes (DIN 1164 Juni 1970) und Betondruckfestigkeit. In: Betontechnische Berichte 1970, p. 165-177. Wiesbaden: Betonverlag.

Proceedings of the International Symposium on Mining with Backfill / Luleå / 7-9 June 1983

Mill tailings and various binder mixtures for cemented backfill:
Analysis of properties related to mining problems

P.P.MANCA, G.MASSACCI, L.MASSIDDA & G.ROSSI
Universita' di Cagliari, Italy

ABSTRACT: The need to dispose of large amounts of mill tailings often poses serious environmental problems: on the other hand, stoping methods employing cemented backfill require considerable amounts of gravel and Portland cement which represent an additional financial burden on mining costs. These concomitant costs prompted research efforts aimed at finding a single solution to the problem of their reduction. The use of mill tailings as aggregates for cemented backfill along with some less expensive substitutes for Portland cement is considered the most attractive solution. The paper reports on the outcome of a laboratory investigation carried out on various mixtures of different types of binders with the fine-size tailings of an Italian mine with a view to their utilization in the stopes. Portland cement, mixtures of Portland cement and fly-ash, activated blast-furnace slags as well as mixtures of lime with two types of natural pozzolanas or with fly-ash were used as binders. In order to throw some light on the effect of mill tailings characteristics, parallel tests were conducted using standard sand conforming to Italian specifications for cement testing. The effect of various compositions, water contents and curing times were simultaneously investigated. The following observations can be made:

1) the substitution of Portland cement with amounts of fly-ash ranging from 20 to 40% by weight produces a slight decrease in strength only for short curing times. For curing times of 12 weeks or more strengths as high as 14 MPa were observed, higher than those achieved with straight Portland cement;

2) the results obtained with binders other than cementitious ones are similar for lime/fly-ash and lime/pozzolana mixtures which develop strengths lower than 0.20 MPa even for long curing times;

3) when the binder is a blast-furnace slag activated by 10% by weight of Portland cement, strengths are of the same order of magnitude as those achieved with straight Portland cement.

In conclusion, the importance of a suitable laboratory testing standardization is discussed.

1 INTRODUCTION

The benefits deriving from the development of cohesion within a backfill have long been recognized and were exploited for instance in some sulphur mines in Sicily back in the early 1900 s (Gerbella 1964: 305). Therefore there is no need to emphasize the growing importance that cemented backfill has been gaining in mining technology and economy over the past two decades. The numerous reports which have appeared in mining journals and in symposia proceedings (Massacci 1980, Carta & Rossi 1981, Carta et al. 1980, Almgren 1976, Willoughby 1981, Singh & Hedley 1981) as well as a recently published book (Thomas et al. 1979) substantiate the firm position now held by this solution to the problem of support in large underground cavities produced by stoping.

However, one limitation to the utilization of cemented backfill can be imposed by economical constraints in that it is practically a concrete prepared according to the conventional specifications of concrete mixes.

Even though the proportion of binder is usually far lower than that of concrete mixes employed for structural purposes, its cost can still be prohibitive in small mining operations and/or where the assay of the run-of-mine ore is close to the cut-off grade. Furthermore, even the supply of a suitable aggregate may represent, in some areas, an insurmountable economic problem.

This state of things prompted the search for alternative solutions, aimed at providing low-cost materials which could suitably replace the sometimes highly expensive conventional components. As far as the aggregate is concerned, the utilization of sands forming the tailings of flotation plants may represent a very promising solution. The increasing amounts of low-grade run-of-mine ores processed in modern concentrators entail, in fact, the disposal of amounts of sand often of the order of several thousands of tons per day. The increasingly stringent regulations enforced by environmental protection laws have, on the other hand, imposed so many constraints on the surface disposal of concentrator tailings that mining enterprise economies are sometimes severely affected.

The utilization of mill tailings sands as uncemented hydraulic backfill represented, for a time, a satisfactory technical solution (Marzocchi 1968) and has undoubtedly contributed to the economic soundness of many mining enterprises. However, one serious drawback inherent in this type of hydraulic backfill is that it is practically lacking in stiffness and compressive strength. In certain instances this has created serious difficulties (Carta & Rossi 1981). As far as Portland cement is concerned, the considerable progress achieved over the last few decades in the understanding of the physico-chemistry of its hydration, has revealed the possibilities offered by other materials such as natural pozzolanas, blast-furnace slags and other metallurgical slags, and pozzolana-like materials (e.g. fly-ash) as partial - or, sometimes, total-substitutes. However, whereas the utilization of natural pozzolanas depends, apart

from on their composition and quality, on the availability in the proximity of mining operations, slags and fly-ash may constitute in themselves an environmental problem for their producers as far as disposal is concerned. A convergence of interests between slag or fly-ash producers on the one hand and mining companies on the other, could be fruitful, where these materials can be profitably employed as low-cost substitutes for Portland cement in cemented backfill preparation.

It should be pointed out that the above considerations have already evoked the interest of researchers in both industrial and academic spheres and numerous investigations have already been carried out on the possibilities of utilizing mixes with mill-tailings sand as an aggregate and fly-ash or various other types of slag partially substituting Portland cement (Thomas et al. 1979, Kheok 1980, Thomas & Cowley 1978, McGuire 1978, Barsotti 1978, Corson 1970, Weaver & Luka 1970, Hull 1978, Askew et al. 1978). However, as anticipated, the results of these investigations cannot be easily generalized since the physico-chemical behaviour of a certain material is related to its origin. In addition, the mechanical properties of a concrete with a given composition of solid components can vary considerably, depending on several parameters. For instance, the procedure of backfill emplacement may require different workabilities thus affecting the water-to-binder ratio on which strength depends considerably. Moreover, the environmental conditions prevailing in the stopes have a strong influence on curing times.

Last but not least, the lack of any standardization in experimental procedures has contributed to a great extent to the difficulties involved in comparing data furnished by different investigations.

The above observations, in conjunction with the fact that in Italy several mines are already facing problems of mill tailings sands disposal in conformity with environmental regulations and the need to improve existing stoping methods with backfill, are the main motivations underlying the investigation, whose results are reported in the present paper.

2 INVESTIGATION PROGRAMME AND ITS MOTIVATION

Mining with backfill has been practised for over a century in Italian mines in its different variants. In particular stoping methods with cemented backfill have been employed for the last twenty years (Salle et al. 1970, Berry 1981).However these have been restricted up to the present time to those methods employing conventional mixes of sand and gravel as aggregates and Portland cement as binder. Many materials, such as natural pozzolanas, slags and fly-ash are available in Italy that could be profitably utilized as a partial (or total if mixed with lime in suitable proportions) substitute for Portland cement. The production of fly ash, in particular, is destined to increase considerably in the near future, due to the growing number of electric power stations being converted from liquid fuel to coal.

A joint research programme has thus been planned by the Dipartimento di Ingegneria Mineraria e Mineralurgica and by the Istituto di Chimica Applicata e Metallurgia of Cagliari University with several objectives. One of these is to devise laboratory testing techniques which, by taking into account the emplacement and curing conditions in the stopes, may provide reproducible results and in the meantime be used as a basis for the development of a suitable testing standardization.

3 MATERIALS AND METHODS

3.1 Materials

- The aggregates. The tailings sands produced by a flotation plant processing the run-of-mine ore of a Sardinian mine were used. The valuable components of the ore are lead and zinc sulphides; the gangue is mainly quartz with minor amounts of calcite, siderite, ankerite, etc.(Cavinato & Zuffardi 1948). Up to beginning the present research, the tailings were used as uncemented hydraulic backfill and were passed through a hydrocyclone with the aim of separating their finest classes (-20 μm) (Marzocchi 1968). Table 1 and Figure 1 show, respectively, the grain size distri-

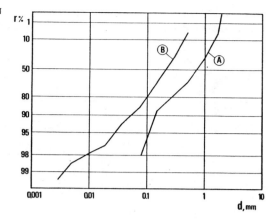

Fig.1. Size distributions of (A) standard and (B) mill tailings sand. Rosin Rammler Bennet graph.
r = cumulative weight percent retained;
d = grain size

bution and cumulative weights curve of the sand. For comparative purposes a standard siliceous sand conforming to Italian specifications for cement testing was employed (1968). The size distribution curve of this sand is plotted in Fig.1

- Portland cement. A commercial Portland cement type 425, kindly supplied by Cementerie di Sardegna S.p.A. was used.

- Fly-ash. Fly-ash coming from the electric power plant of Porto Vesme (Sardinia) was employed. This fly-ash is the by-pro-

Table 1. Particle-size analysis of mill tailings

Size μm	Cumulative percent Weight retained
500	4.60
300	29.83
212	49.18
150	65.93
100	80.00
75	86.54
37.5	92.98
18.7	96.97
9.6	97.81
4.7	98.58
2.8	99.28

duct of the combustion of a South African coal, and its composition is shown in column 4 of Table 2.

Table 2. Chemical composition of pozzolanas, fly-ash and slag (percent by weight)

	Pozzolana from Bacoli	Pozzolana from Sardinia	Fly-ash	Slag
SiO_2	57.8	64.0	41.7	34.6
Al_2O_3	18.0	17.0	28.3	12.0
Fe_2O_3	2.2	3.9	5.2	1.1
CaO	3.1	1.4	9.1	43.9
MgO	1.1	0.3	1.5	5.5
Na_2O	4.1	2.9	1.0	–
K_2O	7.9	4.7	0.6	–
MnO	–	–	–	0.8
SO_3	–	–	3.0	2.5
C✱	–	–	7.0	–
l.o.i.	6.0	4.8	6.8	–

✱ unburnt; l.o.i. = loss of ignition

– Blast-furnace slag. A granulated blast-furnace slag, supplied by Italsider S.p.A., was dried in an oven at $110^{o}C$ and subsequently dry ground to -32 μm in a ceramic ball mill. The chemical composition of this slag is shown in Table 2, column 5.
– Pozzolanas. Two types of pozzolanas, one coming from Bacoli (Naples) and the other from Sardinia, were used. Their compositions are shown in columns 2 and 3 respectively of Tab.2. These pozzolanas were also dried at $110^{o}C$ and dry ground in a ceramic ball mill to -32 μm.

3.2 Methods

– Specimen preparation for strength tests. The outcome of an investigation carried out in the framework of the programme outlined in the foregoing produced exhaustive evidence that the specimen preparation procedure may remarkably affect the strength characteristics of the mixes as well as the reproducibility of the results. Therefore, in a preliminary stage of the research, several specimen preparation procedures were tested and eventually the one described below - which appears to be the most suitable - was adopted.

The mixes were prepared according to the standard schedule (1968) in a Hobart mixer. Once ready they were poured into PVC cylinders, 42 mm in diameter and about 110 mm long, sealed at the bottom by a sheet of Mylat foil. Once this operation was completed the top of the cylinder was also sealed in an identical manner. Both paper seals were completely waterproof. The samples thus prepared were left in the curing oven for five days at the desired temperature. The upper seals were then removed and the specimens were completely immersed into plastic trays full of water and returned to the oven for the fixed curing time. After curing the specimens were carefully removed from their PVC cylinders and trimmed at both ends - in order to obtain smooth ends perpendicular to the cylinder axis - measured, weighed and loaded to failure in compression under uniaxial testing conditions according to the above mentioned specifications (1968). The ratio between length and diameter of the cylinder was always kept close to 2.5.

In order to achieve statistically reliable results with acceptable confidence intervals, the number of samples for each series of testing conditions generally ranged from 5 to 15.

– Strength tests. Strength tests were performed in an M58 200-KN compression machine manufactured by Giazzi R.M.U. Italy. The rate of advance of the platens was fixed at $1.04 \, mm \cdot min^{-1}$.

– Experimental conditions. Throughout the whole series of tests the temperature of the curing oven was maintained at $25 \pm 1^{o}C$, with curing times of 2, 4, 12 and 24 weeks.

The following mixes were tested:
a1 mill tailings sand, Portland cement and fly-ash;
a2 standard sand, Portland cement and fly ash;
b1 mill tailings, lime and pozzolanas or fly-ash;
b2 standard sand, lime and pozzolanas or fly-ash;
c1 mill tailings and blast furnace slag activated by Portland cement;
c2 standard sand and blast furnace slag activated by Portland cement.

Mixes b and c were cured according to two different procedures which are described in the following.

4 RESULTS AND DISCUSSION

For the sake of brevity, the data concerning the different mixes and the experimental procedures are summarized in Tables 3,4, 5 and 6, where the results of strength tests in terms of means and 95% confidence limits, calculated according to Student's t distribution, are also shown.

Table 3 and Figures 2,3 and 4 illustrate the behaviour of mix a1, whereas Table 4 that of mix a2.

Water contents has a pronounced effect on the strength of all the mixes; as it increases, the strength decreases considerably for all curing times.

The presence of fly-ash appears to reduce the strength for short curing times. However, only from the fourth week onwards do the differences in strength between mixes containing fly-ash and those prepared with Portland cement tend to diminish. For longer curing times these differences become insignificant and the mixes where fly-ash substitutes 20% of Portland cement even exhibit higher strengths. This behaviour can be explained by the reaction of the $Ca(OH)_2$ - produced by the hydrolisis of Portland

cement - with fly ash; the higher the $Ca(OH)_2$ content of the former, the faster the reaction rate. Thus, the mixes containing fly-ash and Portland cement in the ratio 0.2 : 0.8 approach the most favourable $Ca(OH)_2$-to-fly-ash ratio and will therefore have faster recoveries and will probably react more completely.

The substitution of Portland cement with fly-ash therefore merits the due attention, always in relation to the particular use of the backfill. It is clear that the utilization of fly-ash may become particularly attractive in those instances where the time lag between two stoping sequences - during which the cemented backfill must remain undisturbed and develop its strength - is longer than the curing time required to reach the predicted strength.

The data of Table 4 show the importance of the effect of the type of aggregate. The strengths attained by the mixes prepared with standard sand are remarkably higher than those exhibited by the mixes prepared with tailings sand.

This can be attributed, on the one hand to the considerably lower specific surface of the aggregate and on the other to the possibility of obtaining mixes having the same workability with lower water-to-binder ratio.

The paramount importance of aggregate grain size is also corroborated by the data

Table 3. Uniaxial compression strengths, in MPa, of samples formed with mixes prepared with binder-to-aggregate ratio 1 : 5 at different curing times. Aggregate: mill tailings sand; Binder: type 425 Portland cement (PC); Substitute for binder: Fly-ash (FA); Water-to-binder ratio: W : B.

PC : FA	W : B	Curing time			
		2 weeks	4 weeks	12 weeks	24 weeks
1 : 0	1.25 : 1	6.41 ± 1.09	8.66 ± 1.08	10.31 ± 0.61	11.59 ± 0.94
	1.50 : 1	3.73 ± 0.68	5.36 ± 0.54	6.77 ± 0.53	8.29 ± 0.30
	1.75 : 1	2.84 ± 0.35	3.83 ± 0.30	6.13 ± 0.04	6.45 ± 0.33
0.8 : 0.2	1.25 : 1	3.22 ± 0.41	5.90 ± 0.31	13.49 ± 0.63	13.90 ± 0.53
	1.50 : 1	2.21 ± 0.39	3.96 ± 0.36	9.07 ± 1.16	8.97 ± 0.55
	1.75 : 1	1.47 ± 0.17	2.66 ± 0.15	5.63 ± 0.26	6.62 ± 0.50
0.6 : 0.4	1.25 : 1	3.30 ± 0.15	6.04 ± 0.49	10.40 ± 1.31	11.32 ± 0.27
	1.50 : 1	2.00 ± 0.25	3.69 ± 0.28	8.11 ± 0.70	8.67 ± 0.35
	1.75 : 1	0.97 ± 0.15	1.64 ± 0.16	5.41 ± 0.23	5.78 ± 0.14

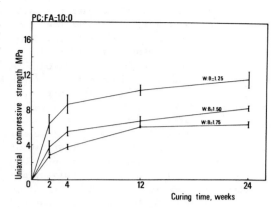

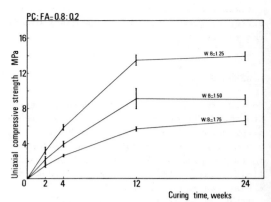

Fig.2. Uniaxial compression strength vs. curing time at various water-to-binder (W:B) ratios. Aggregate: mill tailings sand; binder: Portland cement.

Fig.3. Uniaxial compression strength vs. curing time at various water-to-binder (W:B) ratios. Aggregate: mill tailings sand; binder: 0.8 Portland cement (PC), 0.2 Fly-ash(FA

of Table 5, in which the performances of pairs of mixes formed using binders other than Portland cement are compared (the main role of the 10% Portland cement contained in the mix with slag being that of activating slag hydraulicity).

The difference between the components of each pair consisted in the aggregate, which was either standard sand (SS) or mill tailings sand (MV).

After 4 weeks curing all the mixes containing lime in the binding mixture were so weak as to not exhibit any strength, with the exception of the mix containing fly-ash or a pozzolana and standard sand.

After longer curing times, strengths rang-

ing from 5 to 10 MPa were achieved by all the mixes containing lime and standard sand as aggregate, the highest values - of the order of magnitude of 10 MPa - being exhibited by those containing natural pozzolanas. Practically no strength was developed by the same mixes when mill tailings sand was used instead of standard sand. On the other hand, the mixes containing activated slag already show appreciable strengths after 12 weeks curing. These strengths continue to increase with time and at 24 weeks they even exceed those of the mixes prepared with Portland cement (see Table 3). Of course, in this particular instance this phenomenon can be attributed to the higher percentage of binder which, although reacting more slowly than Portland cement may, after long curing times, attain even higher strengths.

The results of experiments summarized in Table 6 demonstrate to what extent the curing procedure may affect the strength characteristics of the mix.

At 4 weeks curing time the strength of the samples prepared with activated slag and cured in water baths is less than one half that of the corresponding samples sealed in bags, but catches up after 12 weeks and is slightly higher at 24 weeks. Examination of the other columns of Table 6 reveals similar results, although curing in

Table 4. Uniaxial compression strength in MPa of samples formed with mixes prepared with binder-to-aggregate ratio 1 : 5 at 4 weeks curing time. Aggregate: standard sand; Binder: Type 425 Portland cement (PC); Substitute for binder: Fly-ash (FA); Water-to-binder ratio: W : B.

| W : B | PC : FA | | |
	1 : 0	0.8 : 0.2	0.6 : 0.4
0.55:1	25.34±2.37	14.37±5.91	11.32±2.55
0.66:1	25.24±2.83	20.62±4.49	19.96±1.29
0.77:1	19.65±0.92	23.64±1.37	15.25±0.92

Table 5. Uniaxial compression strength, in MPa, of samples formed with mixes b1, b2, c1 and c2 in the ratios shown at different curing times. Aggregate: mill tailings sand (MV) or standard sand (SS); Binder-to-aggregate ratio = 1: 3; Water-to-binder ratio = 1 : 1. Fly-ash = FA; lime = $Ca(OH)_2$; Pozzolana from Bacoli = P1; Pozzolana from Sardinia = P2; Slag = S; Portland cement = PC.

| Mix | Aggregate | Curing time | | |
		4 weeks	12 weeks	24 weeks
0.75 FA + 0.25 $Ca(OH)_2$	MV	n.s.	0.23 ± 0.013	0.28 ± 0.38
	SS	0.17 ± 1.23	0.57 ± 0.049	4.26 ± 0.48
0.75 P1 + 0.25 $Ca(OH)_2$	MV	n.s.	0.22 ± 0.013	0.52 ± 0.15
	SS	n.d.	6.39 ± 0.63	10.24 ± 0.35
0.75 P2 + 0.25 $Ca(OH)_2$	MV	n.s.	0.19 ± 0.11	0.14 ± 0.05
	SS	2.16 ± 0.12	n.d.	11.69 ± 0.99
0.90 S + 0.10 PC*	MV	1.78 ± 0.43	11.88 ± 1.03	17.10 ± 0.54
	SS	3.18 ± 0.29	n.d.	n.d.

* Water-to-binder ratio = 0.9 : 1; n.d. = not determined; n.s. = no strength developed.

Table 6. Uniaxial compression strength, in MPa, of samples formed with mixes b1 and c1 in the ratios shown, at different curing times. Aggregate: mill tailings sand (MV); Binder-to-aggregate ratio = 1 : 3; Water-to-binder ratio = 1 : 1; Fly-ash = FA, lime = $Ca(OH)_2$, Pozzolana from Bacoli = P1; Pozzolana from Sardinia = P2; Slag = S, Portland cement = PC Curing in water bath = W; curing in sealed bags = B.

| Mix | | Curing time | | |
		4 weeks	12 weeks	24 weeks
0.75 FA + 0.25 $Ca(OH)_2$	W	n.s.	0.23 ± 0.013	0.28 ± 0.38
	B	0.21 ± 0.14	0.27 ± 0.029	0.40 ± 0.054
0.75 P1 + 0.25 $Ca(OH)_2$	W	n.s.	0.22 ± 0.013	0.52 ± 0.15
	B	0.18 ± 0.014	1.83 ± 0.12	4.73 ± 0.34
0.75 P2 + 0.25 $Ca(OH)_2$	W	n.s.	0.19 ± 0.11	0.14 ± 0.050
	B	0.15 ± 0.016	0.18 ± 0.025	0.28 ± 0.019
0.90 S + 0.10 PC*	W	1.78 ± 0.43	11.88 ± 1.03	17.10 ± 0.54
	B	4.18 ± 0.35	11.12 ± 0.36	14.64 ± 0.28

*Water-to-binder ratio = 0.9 : 1; n.s. = no strength developed.

water baths seems more beneficial than in sealed bags.

5 CONCLUSIONS

The results of the present investigation confirm the possibility of using mill tailings sand for cemented backfill. The most serious drawback inherent in this type of sand is probably the often very fine grain size distribution, and can only be partly overcome by resorting to mixes containing higher proportions of binder.

The partial substitution of Portland cement in the mixes with less expensive pozzolana-like materials such as fly-ash appears technically feasible. This confirms

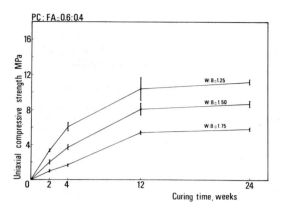

PC : FA=0.6:0.4

Fig.4. Uniaxial compression strengths vs. curing time at various water-to-binder (W:B) ratios. Aggregate: mill tailings sand; Binder: 0.6 Portland cement (PC), 0.4 fly-ash (FA).

and broadens the results obtained by other researchers (Weaver & Luka 1970; Thomas et al. 1979; Thomas 1971; Hull 1978; Askew et al. 1978).

Amongst the industrial by-products which, on account of their environment-degradation potential and the consequent need of their safe disposal, might be available at acceptably low costs, fly-ash and blast-furnace slags seem particularly interesting.

In addition to confirming the suitability of fly-ash as an efficient substitute for Portland cement, evidence is produced that this material can in fact replace Portland cement in proportions of up to 30% (or more, depending on its reactivity) without detrimentally affecting the strength properties of mixes.

It can thus be concluded that the use of fly-ash in the mixes should be considered permissible and even beneficial, provided that the stoping cycles are consistent with the strength development times required by the mixes. Mining method design and mix design must therefore become two closely interrelated stages of the production cycle.

The use of pozzolana-lime or fly-ash-lime mixtures as binders appears to be unfeasible on account of the low strengths reached. Adequately activated blast-furnace slag appears, on the other hand, to offer interest-

ing possibilities as a binding agent.

The changes in experimental conditions, such as water-to-binder ratio or the curing procedure appear to remarkably affect strength development and may explain the sometimes very different results obtained by various researchers.

It is unquestionable, however, that the right choice of mix components and their proportions should be made with a view to the economy of the mining method as a whole

ACKNOWLEDGEMENTS

The present investigation was financed by the Consiglio Nazionale delle Ricerche, Centro Studi Geominerari e Mineralurgici, Engineering Faculty, University of Cagliari.

REFERENCES

Almgren, G. 1976. Cut-and-fill mining in Sweden: Development trends and applications. Trans.Am.Inst.Min.Engrs. 260:84-88

Anonymous 1968. Norme sui requisiti di accettazione e modalità di prova dei cementi. In Gazzetta Ufficiale 180 17/07/1968 e successive modificazioni.

Askew, J., P.L. McCarthy & D.J. Fitzgerald 1978. Backfill research for pillar extraction at ZC/NBHC. In Mining with backfill, p.100-110. Sudbury: Can.Inst.Min. Metall.

Barsotti, C. 1978. The evolution of fill mining at the Ontario Division of Inco Metals. In Mining with backfill, p.37-41. Sudbury: Can.Inst.Min.Metall.

Berry, P. 1981. Geomechanical investigation for the design of cemented fill. In O. Stephansson & M.J. Jones (eds.), Applications of rock mechanics to cut and fill mining, p.79-92. London: I.M.M.

Carta, M., R. Cotza, S. Giuliani, P.P. Manca G. Massacci & G. Rossi 1980. Rock mechanics topics in cut and fill mining of veins of varying dip. In O. Stephansson & M.J. Jones (eds.), Applications of rock mechanics to cut and fill mining, p.49-54. London: I.M.M.

Carta, M. & G. Rossi 1981. Le coltivazioni con ripiena cementata della miniera di Bleiberg: osservazioni e considerazioni.

Res.Ass.Min.Sarda. 86(1): 29-54.

Cavinato, A. & P. Zuffardi 1948. Geologia della miniera di Montevecchio. In Notizie sull'industria del piombo e dello zinco in Italia, p.427-464. Montevecchio: S.A.P.E.Z.

Corson, D.R. 1970 Stabilization of hydraulic backfill with Portland cement. U.S.B.M. R.I. 7327.

Gerbella, L. 1964. Arte mineraria. Vol.2. Milano: Hoepli.

Hull, B. 1978. Magma's sandfill system as employed at the Magma Mine, Superior, Arizona, U.S.A. In Mining with backfill, p.75-83. Sudbury: Can.Inst.Min.Metall.

Kheok, S.C. 1980. Pozzolanic smelter slag and related additives in cemented hydraulic mine fill. Ph.D. Thesis, Un. N.S.W., Kensington, Australia.

Marzocchi, G. 1968. Le ripiene idrauliche nella miniera di Montevecchio. Res.Ass. Min.Sarda. 73(9):46-64.

Massacci, G. 1980. Note sulla coltivazione con impiego di ripiena cementata discreta della miniera di Saint Salvy, Francia. Atti della Fac. D'Ingegneria, Univ. Cagliari. 13:303-320.

McGuire, A.J. 1978. Falconbridge slag as a cementing agent in backfill. In Mining with backfill, p.133-138. Sudbury: Can. Inst.Min.Metall.

Salle, P.I., A. Bonetti & G. Vaiani 1970. Cimentation totale de remblai: facteur déterminant pour l'exploitation économique d'un amas de pyrite. In 6th World Mining Congress, Paper III-D.3. Madrid.

Singh, K.H. & D.G.F. Hedley 1981. Review of fill mining technology in Canada. In O. Stephansson & M.J. Jones (eds.) Application of rock mechanics to cut and fill mining, p.11-24. London: I.M.M.

Thomas, E.G. 1971. Cemented fill practice and research at Mount Isa. Proc. Australas.Inst.Min.Metall. (240):33-51.

Thomas, E.G. & R. Cowling 1978. Pozzolanic behaviour of ground Isa Mine slag in cemented hydraulic mine fill at high slag/cement ratios. In Mining with backfill, p.129-132. Sudbury: Can.Inst.Min.Metall.

Thomas, E.G., J.H. Nantel & K.R. Notley 1979. Fill technology in underground metalliferous mines. Kingston: International Academic Services Ltd.

Weaver, W.S. & R. Luka 1970. Laboratory studies of cement-stabilized mine tailings. CIM Bull. 63:988-1001.

Willoughby, D.R. 1981. Rock mechanics applied to cut and fill mining in Australia. In O. Stephansson & M.J. Jones (eds.) Applications of rock mechanics to cut and fill mining, p. 1-10. London: I.M.M.

The use of blast-furnace slag and other by-products as binding agents in consolidated backfilling at Outokumpu Oy's mines

P.NIEMINEN
Tampere University of Technology, Finland

P.SEPPÄNEN
Outokumpu Oy, Mining Technology Group, Finland

SYNOPSIS : For the last three decades Outokumpu Oy has successfully used consolidated backfilling which has increased the recovery of ore and made working conditions safer in the stopes of its mines.

Traditionally stabilisation of filling materials has been done by using portland cement. However, the high price of cement forced Outokumpu Oy to look for cheaper binding materials. First tests with iron blast-furnace slag were done during the years 1966-70 in the Keretti mine. The tests were continued in the 70´s and regular use of iron slag as a binding agent was started in 1978 at Pyhäsalmi mine and in 1979 at Vihanti mine and will start at Keretti and Vammala mines in 1983.

The other binding agents tested were coal and peat fly ash and some metallurgical slags. Some of them are suitable as a partial replacement for cement, but they have not proved to be as economical as blast furnace slag.

In cement production energy is mostly used in the kiln preparation though by using slag that energy can largely be saved. Outokumpu Oy´s mines saved about 4.6 million FM (about 1 MUSD) in 1981 by using iron furnace slag as a binding agent in consolidated backfilling.

1 INTRODUCTION

The very first laboratory tests to consolidate classified tailings using portland cement were done at Outokumpu mine in 1949. A mining method based on consolidated backfill (Concrete Pillar Stoping) started at Outokumpu in 1952. The Concrete Pillar Stoping became the main stoping method in Keretti mine in 1957. Soon after consolidated backfilling applied in various stoping methods was used also in some of Outokumpu Oy other mines.

Up to the present about 2 million m^3 of stopes have been filled by consolidated (cemented) backfill.

At first the hydraulic tailings backfilling method was changed to consolidated backfilling simply by adding a portion of cement to it. Then there was not any possibility of preparing the cemented backfill, following the rules of concrete technology. In order to obtain the required strength of backfill, an extra amount of cement had to be used. Not even then could the quality of the backfill be certain, due to the classification and segregation of the filling components, before stoping the adjacent stope.

During the last decade efforts have been concentrated on developing the backfilling method by paying attention to :

1 Proper grain size distribution
2 Feasible filling process, in other words: what is the most economical way to mix and transport the filling materials to the stopes.
3 Optimize the ratio of water/cement and
4 consider the possibility of replacing cement by (other) cheaper binding agents.

The high cost of consolidated backfilling due to the high price of Portland cement, especially after the oil crisis of 1974-75, the cost of using cement in backfill became uneconomical. This was one of the main reasons why Outokumpu Oy again started to study the possibilities of replacing portland cement partly or totally by blast furnace slag or some other by-product substitutes.

The aim of this presentation is to give some information about the results of these studies and tell about some of the applications and uses of by-products as binding agents in practice. Blast-furnace slag has

totally replaced portland cement in Vihanti
mine since 1979, partly in Pyhäsalmi mine
since 1978 and 90 % in Keretti and Vammala
mines starting from this year.

2 INITIAL TEST WORK

In initial test work, done in 1975, looked
at the possibilities of finding out what
kind of waste products from other mines and
industries nearby could be used to replace
portland cement in consolidated backfill.
This work and also the main part of the
laboratory tests were later done by Technical University of Tampere. Of the large
number of by-products some were more
accurately investigated (see table 1).

Initial test results and economical
calculations indicated that blast-furnace
slag, coal fly ash and peat fly ash could
be considered as feasible binding agents.
A accurate account of the characteristic
features of Finnish blast furnace slags
and fly ashes follows in the next part of
this report.

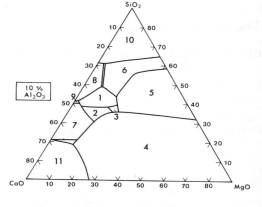

Figure 1. $CaO-MgO-Al_2O_3-SiO_2$ -phase
diagram (Karjalahti 1981)

1. Melitite	7. Calcium orthosilicat
2. Mervinite	8. Pseudo wollastonite
3. Montisellite	9. Rankinite
4. Periclase	10. Christobalite
5. Forsterite	11. CaO
6. Pyroxene	

Table 1.

Subject	Chemical Composition						
	CaO	SiO_2	Al_2O_3	MgO	Fe_2O_3	S	Fe
Blast furnace slag	37	32	10	8	0,5	0,7	-
Coal fly ash	3	51	26	2	11	-	-
Peat fly ash	10	38	12	-	10	-	-
Pulp industry waste ash	39	+ 54 % $CaSO_4$					
Steel industry slags	58	28	3	4	0,7	-	-
Chrom furnace slag	5	28	25	25	-	-	-
Ni Rever Furnace slag	2	33	-	5	+ Fayalite Fe 42 %		

3 BLAST FURNACE SLAG

Iron smelters in Finland are yearly producing about 600 000 tons of blast-furnace
slag as a by-product. Blast-furnace slag is
produced from the fusion of host rock
minerals in iron concentrate with coke ash
and so-called slag makers. Blast-furnace
slag consists of calcium-, magnesium- and
aluminium silicates. The theoretical optimum
of minerology for slag is melitite-mervinite-
montiselite-area (see fig. 1), the CaO-MgO-
$Al_2O_3-SiO_2$ -phase diagram. Table 2 shows the
average composition of blast-furnace slag
in Finland. Rautaruukki slag is placed in
the melitite area and Ovako slag in the
mervinite area.

Table 2. The average composition of Blast
Furnace Slag in Finland.

	Rautaruukki Oy	Ovako Oy Ab
CaO	37 %	41 %
MgO	11 %	8 %
SiO_2	35 %	37 %
Al_2O_3	9 %	9,5 %
TiO_2	2,5 %	-
MnO	0,8 %	0,6 %
K_2O	1,0 %	0,8 %
Na_2O	1,0 %	0,8 %
S	1,5 %	1,1 %

4 BLAST-FURNACE SLAG AS A CEMENTING AGENT

Granulated slag is produced by cooling blast-
furnace slag rapidly with large quantities
of water, which causes the vitrivication of
the slag. Glass content of granulated slag
is 90-97 % and it is extraordinarily hard
and consequently difficult to grind.
Another way of treating blast-furnace slag
is by applying a pelletizing process.
The glass content of pelletized slag varies
from 80 % to 95 % and it is easier to grind
because of its brittle property. Blast-
furnace slag quickly cooled and ground to
the fineness of cement has hidden hydration-
al characteristics, which can be reactivated
e.g. with portland cement, calsium hydroxide
or gypsum. Blast furnace slag it has been
developed and various indices like the
hydration index in concrete technology have
been ascribed to it. For example in Germany
blast furnace slag mixed with cement has
to qualify the formula

$$H\ I\ =\ \frac{CaO\ +\ MgO\ +\ 1/3\ Al_2O_3}{SiO_2\ +\ 2/3\ Al_2O_3}\ >\ 1$$

It is also assumed that the high glass
content of blast furnace slag would increase its hydration.
  The mineralogical composition of portland
cement and blast furnace slag are slightly
different concerning mainly tricalsium-
silicate and tricalsium aluminate. In portland cement these two minerals produce a
rapid curing effect, while in blast furnace
slags these minerals are almost totally
absent and thus have very slow curing
properties. Dicalsiumsilicate, which belongs
to boath binders, produces long-term curing.
  By using activated blast furnace slag for
stabilazing backfilling materials, which can
be kept in a mobilised liquid state for a
long time after mixing, the great advantage
is that curing starts only when the filling
material isn´t in motion any more.
One disadvantage of blast furnace slag is
its hardness and quite high grinding costs.
In some cases it is possible to grind slag
granules by wet grinding (as can later be
seen in the Vihanti case) which cuts down
on energy consumption. In figure 2 the
consumption of energy in dry and wet grinding of granulated blast furnace slag is
presented.
  A great advantage of slag is its good
resistance to corrosion due to the lack of
aluminate minerals and reactive calsium-
oxide.

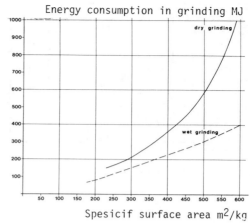

Figure 2. Energy consumption in the
grinding of granulated blast furnace slag.

5 FLY ASHES

Finnish industry is currently producing
coal fly ash, peat fly ash, pulp industry
gypsum chalk and communal waste ash.

Coal Fly Ash

The power plants, which burn coal are
situated along the coasts of Finland. They
produce yearly about 400 kton of fly ash,
which goes almost totally to the cement
industry, though 3-4 years ago it was waste.
Chemical composition varies according to the
quality of the coal. The main constituents
of coal ash are SiO_2, Fe_2O_3 and Al_2O_3, with
smaller amounts of CaO, MgO, Na_2O, K_2O and
SO_3. The burning temperature usually exceeds
the melting point of the minerals, so in the
ash they are not in their original form.
Up to 90 % of the ash can be glassy. The
grain size distribution in coal fly ash is
presented in figure 3. The specific surface
area of coal fly ash, measured by nitrogen
adsorbtion method, varies from 1000-2500 m^2/
kg. The coal content of the ash adds to the
spesific surface area. Coal ash grains are
more or less round balls of varying sizes.
  The microstructure of coal fly ash is
presented in figure 4.

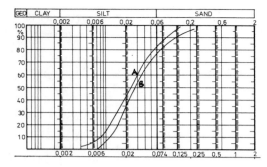

GEO	CLAY	SILT		SAND	

Figure 3. The grain size distribution in coal fly ash (A) and peat fly ash (B).

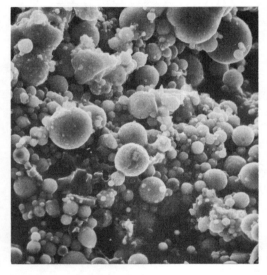

Figure 4. Microstructure of coal fly ash, thousandfold.

Pozzolan

Fly ash is a pozzollanic material. ASTMC219 norm defines pozzolan in following way :

Pozzolan, a siliceous or alumino-siliceous material that in itself possesses little or no cementitious value but that in finaly divided form and in the presence of moisture will chemically react with alkali and alkaline earth hydroxides at ordinary tempe-ratures to form or assist in forming compounds possessing cementitious pro-perties.

Because coal fly ash often contains so called free lime it in moist conditions under pressure can take on a form similar to that of concrete. If coal fly ash does not contain free lime the same effect can be achieved by mixing some lime or cement with it before packing. Reaction kinetics demands that at least 70-75 % of pozzolan should contain SiO_2 + Al_2O_3 + Fe_2O_3 and th Si-acidic constituent should be in the amo phous form because chrystallic Si reacts very slowly at normal temperatures. Reacti vity also depends on the fineness of the fly ash. Fine ash is more reactive than coarse ash.

Utilization of Coal Fly Ash

It is possible to use coal fly ashes as a partial portland cement replacement in consolidated backfill. Laboratory tests indicate that in some cases even 70 % of portland cement can be replaced by coal fl ash.

Peat Fly Ash

We have in Finland many power plants, whic burn peat. Peat fly ash can also be pozzol ic depending on the quality of the peat an the burning temperature. Grain size distri bution is variable but belongs to the geo-classification according to silt, figure 3

Peat fly ash is a very light material. Solid specific weight is 2400-2800 kg/m^3 a loose specific weight is 400-1100 kg/m^3. While the corresponding specific weights o coal ash are 1900-2500 kg/m^3 and 800-1600 kg/m^3. The specific surface area of peat f ash as measured by nitrogen adsorbtion met varies according to the coal content in as between 5000 - 30 000 m^2/kg. The optimum water content of peat fly ash can then be very high 43-54 % by weight. In mine backf peat fly ash can partially replace portlan cement.

6 BINDING AGENT TESTS AND PRACTICAL USE AT OUTOKUMPU OY´S VIHANTI, PYHÄSALMI AND VAMMALA MINES

Figure 5 shows the location of Outokumpu Oy mines in Finland.

VIHANTI MINE

The main stoping method used in Vihanti mi is sublevel stoping with ore pillars. For post pillar stoping adjacent stopes are

52

filled with cemented hydraulic backfill. The solid content of hydraulic backfill during transportation is 60 % by weight. The filling material is of classified tailings. Grain-size distribution is presented in table 3 and the chemical composition in table 4.

Table 4. Chemical Composition of Vihanti mine classified tailings.

Item	CaO	SiO$_2$	Al$_2$O$_3$	MgO	S	Fe
Wt %	14.4	61.3	6.2	9.2	6.7	6.4

Portland cement was the traditional binding agent used in consolidated backfill in Vihanti mine until 1979. The composition of Finnish portland cement is presented in table 5.

Table 5. Chemical Composition of Finnish Portland Cement.

Item	CaO	SiO$_2$	Al$_2$O$_3$	MgO	FeO	S
Wt %	64.1	22.0	5.5	1.4	3.0	2.1

The binding agent tests for the replacement of portland cement started in 1977. The result from these tests is presented in figure 6.
    The binding agents tested were :
- blast furnace slag (MK), ground to 300 m^2/kg
- portland cement (sem), comparing and activation agent
- gypsum lime (TK) activation agent
- steel slag (RIKU)
- peat fly ash (TT)
    The chemical composition of these "binding agents" is presented in table 1.

Figure 5. Location of Outokumpu Oy mines in Finland.

Table 3. Grain-size distribution of Vihanti mine classified tailings.

Size (mm)	Percent passing
0.420	99.4
0.297	97.2
0.210	84.2
0.149	72.8
0.105	46.4
0.074	28.5

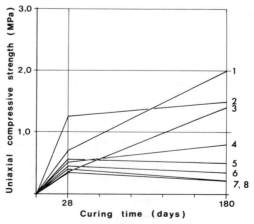

Figure 6. Curing curves of consolidated backfilling tests at Vihanti mine using different binding agents according to the next list :

53

1. MK   55 kg/filled m³ + sem   5 kg/filled m³   was also possible to use wet grinding and
2. Sem 120 "   -   connect the grinding process directly to the
3. MK   55 "   + TK   18 "   hydraulic backfilling system. The flowsheet
4. Sem 60 "   -   of the grinding plant is presented in figure
5. Sem 30 "   + TT   60 "   8.
6. Sem 30 "   + TK   30 "
7. Sem 15 "   + RIKU 15 "
                                 + TT   60 kg
8. TT   60 "   + sem 15 kg/filled m³

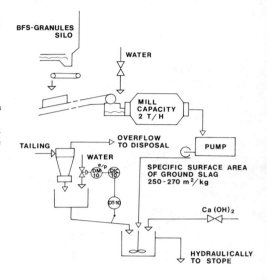

   The results from using blast-furnace slag
activated with cement and gypsum lime were
very interesting.
   The results from the later tests are
presented in figure 7. These results confirm
that blast furnace slag activated with the
correct amount of Ca(OH)$_2$ or portland cement
is a very good binding agent and is feasible
in Vihanti mine.

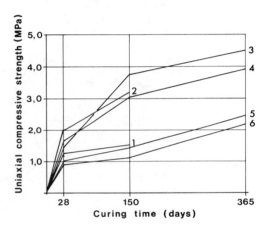

Figure 7. Curing curves of consolidated
backfilling tests of Vihanti mine with
blast-furnace slag as binding agent.
1. Cement 117 kg/filled m³
2. Slag   106 "   + cement 5.3 kg/m³
3. Slag   106 "   + Ca(OH)$_2$ 2.1 "
4. Slag   106 "   + Ca(OH)$_2$ 3.3 "
5. Slag   106 "   + Ca(OH)$_2$ 5.3 "
6. Slag   106 "   + Ca(OH)$_2$ 11.0 "

Figure 8. Flowsheet of Wet Grinding Blast-
furnace Slag Granules Plant at Vihanti Mine

With a grinding capacity of 3.2 t/h a
correct cement/sand -ratio of 140 kg of slag
per m³ filled was obtained. Slag is activated
by using lime milk, which is taken from the
lime milk process of the concentrating plant.
Activating Ca(OH)$_2$ is needed ; about 1.5 w
of slag. So, from 1979 it was possible to
apply consolidated (cemented) backfill in
Vihanti mine without using any portland
cement.

The investment cost of the slag grinding
plant was payed back after 3 months, because
of the price difference between cement and
slag.

The quality of consolidated backfill bound
with slag has been tested by drilling holes
though the fill and also one tunnel has been
driven into the fill. In figure 9 x-ray
diffraction graph of one test piece can be
seen. Uniaxial compressive strength for in
situ specimens has been between 1.5 and
2.0 MPa. Mining engineers in Vihanti mine
calculated that during the last 3 years
about 7.5 million FIM have been saved through
using consolidated backfilling with ground
granulated blast-furnace slag instead of
portland cement.

Slag Binder in Practice at Vihanti Mine

The plans for replacing portland cement with
ground blast-furnace slag went into practice
at Vihanti mine in 1978.
   Constructing a grinding plant at the mine
for granulated slag was found to be the most
economical way of operation at Vihanti.
Because the nearest granule producer is
situated in Raahe, only 70 km from Vihanti
mine, and because hydraulic backfilling at
Vihanti mine is running continuously, it

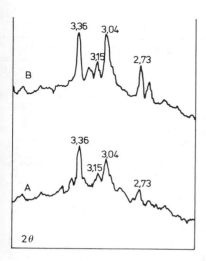

Figure 9. X-ray diffraction graph of Vihanti mine backfilling material at the age of 8 months binded by blast furnace slag (graph B). Graph A is a test sample stored in laboratory condition.

PYHÄSALMI MINE

For the moderate recovery of the sulphur-copper-zinc -orebody of Pyhäsalmi mine the most reasonable stoping method is sub-level stoping and consolidated backfilling. The only economical filling material that can be used are the classified tailings which have a high sulphur and iron content. The chemical composition and grain size distribution of filltailings are presented in tables 6 and 7.

Table 6. Chemical Composition of Pyhäsalmi Mine Classified Tailings.

Item	CaO	SiO_2	Al_2O_3	MgO	S	Fe
wt %	2.1	43.1	4.7	2.1	32.9	22.9

Hydraulic backfilling is used in Pyhäsalmi mine and it is possible to apply it with or without a binding agent. All of the operations are automated.

As well as the high price of cement, the question of sulphurcorrosion was one additional reason for making the research to discover a new binding agent to replace portland cement.

The effect of sulphur-ion-corrosion can be seen from pictures 10, 11 and 12.

Table 7. Grain-size Distribution of Pyhä-salmi Mine Classified Tailings.

Size (mm)	Percent Passing
0.59	
0.42	99.97
0.297	99.14
0.210	91.27
0.149	81.17
0.105	59.47
0.074	43.07

Sulphur-ion-corrosion has not proved to be a effect in largescale backfill operation but sulphur corrosion is a problem in shot-creting. Using blast furnace slag binding agent in backfill and in shotcreting no signs of sulphur corrosion has been found.

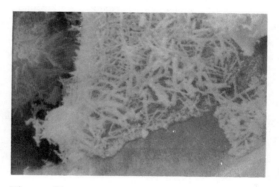

Figure 10. Microstructure of Pyhäsalmi backfilling material. Ettringite on the surface of pyrite grain.

Figure 11. The test samples of Pyhäsalmi backfilling materials. PS18 binder mixture formed by lime-blast furnace slag. PS20 and PS21 binder is cement.

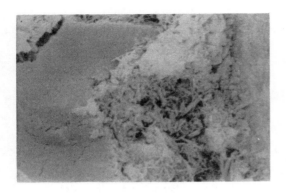

Figure 12. Ettringite in Sample PS21, thousandfold.

In figure 13 there is a summary of the test results made at Pyhäsalmi mine using portland cement and blast furnace slag ground to a specific surface of 300 m^2/kg. Slag as a binding agent gave much better compressive strengths than cement. Deviation in the slag tests is mainly due to various amounts of activation lime present in the slurry.

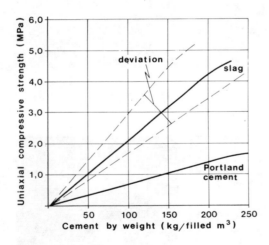

Figure 13. Compressive strengths of Pyhäsalmi mine filltailing bound with cement and blast furnace slag activated with Ca(OH)$_2$ after 6 months.

VAMMALA MINE

The most economical stoping method for

Vammala Ni-orebody is sublevel stoping with consolidated backfill pillars. Vammala mine is comparatively new mine and the backfilling system for the cycloning and hydraulic placement of classified tailings and the consolidated backfilling system was ready in 1981. Consolidated backfilling there differs from that in Vihanti and Pyhäsalmi. The filling material is transported to the stopes by a garvimetric method (no extra water used) with an optimum water content. The optimum water content for Vammala filltailings (fig. 14) is 13 % weight, whereas the water content of filltailings for hydraulic placement is about 49 %.

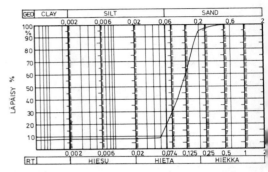

Figure 14. The Grain-size of Vammala Mine Filltailings.

By adding 110 kg of binding agent for every m^3 filled the optimum water content increases by 22-24 % by weight. The grain size distribution of Vammala filltailings is presented in figure 14 and the chemical composition in table 8.

Table 8. Chemical Composition of Vammala filltailings.

Item	CaO	SiO$_2$	Al$_2$O$_3$	MgO	Fe	S
wt %	3.0	41.1	4.7	20.7	14.2	4.

Binding Agent Tests

The strength needed for the backfill at Vammala is 1.5 MPa. Four different binding agents were tested as follows :
1. Portland cement (PS) 100 %
2. PS 50 % + coal fly ash (KHLT) 50 %
3. PS 50 % + peat fly ash (TLT) 50 %
4. Blast furnace slag (MK) 90 % + Ca(OH)$_2$ (KH) 10 %

After 90 days' curing at a water content
of 49 % by weight the compressive strength
of each of the binding agents was tested.
The results of these tests are presented
in figure 15. From figure 15 it can be seen
that the strength of 1.5 MPa was obtained
by either mixing 150 kg/m^3 of cement or
over 150 kg/m^3 of blast furnace slag activated with Ca(OH)$_2$ (10 % by weight of the
slag) with backfill. In order to find out
the effect that the water content had on
the filling material a test was made using
an optimum water content of 23 % by weight.
The results of this test are presented in
figure 16. In the test three types of

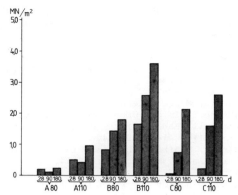

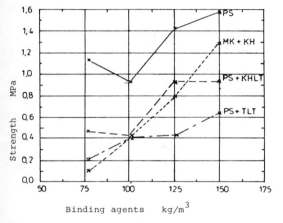

Figure 15. Compressive strength test of
Vammala mine produced using portland cement
(PS), blast furnace slag (MK) activated with
Ca(OH)$_2$ = KH, and 50/50-mixtures of coal fly
ash (KHLT) and cement and peat fly ash and
cement.

binding agent were used :

A   Portland cement
B   Blast-furnace slag activated with
     10 % wt cement
C   Blast-furnace slag activated with
     10 % wt Ca(OH)$_2$

Each of the binding agents was mixed with
backfill in the proportions of 80 kg/m^3 and
110 kg/m^3.

Uniaxial compressive strengths are
measured after 28, 90 and 180 days' curing
respectively. Because of the correct water
content compressive strengths were much
higher in this test than in earlier ones.
Slag as a binding agent gave much better
compressive strengths than cement. According
to the tests made in Vammala case it is

Figure 16. The compressive strength deve-
lopment of Vammala filltailings with a water
content of 23 wt % produced by
   A - 80 and 110 kg of portland cement
   B - 80 and 110 kg of slag activated with
       10 % portland cement
   C - 80 and 110 kg of slag activated with
       Ca(OH)$_2$.

possible to save about 25-40 % of the
binding-agent through using the correct
amount of water in backfilling.

Blast furnace slag has proved to be both
technically and economically the best binding agent for use in consolidated backfill
at Vammala mine.

SUMMARY

Numerous tests have been carried out in
order to find cheeper binding-agents to
replace portland cement in Outokumpu Oy's
Vihanti, Pyhäsalmi and Vammala mines. Many
by-products have been studied, but only
fly-ash and blast furnace slag were found
to be suitable. Ground blast-furnace slag
proved most economical in these cases, and
it was possible to replace portland cement
totally or 90 % of it in the Vihanti,
Pyhäsalmi and Vammala mines.

REFERENCES

E. Hakapää, H. Tanner and V. Vähätalo,
   Finland's Outokumpu Mine, The Mine -
   The Shaft - The Mill. Mining Engineering,
   July 1955: 623-633.
E. Pihko, Evolution of Underground Mining
   Methods Outokumpu Part I, World Mining,
   March 1966: 34-39.

E. Hakapää, De finska Bergverken (föredrag,
26.11.1965), Meddelanden från Svenska
Gruvföreningen 117, Volym 8, 1966: 54-68.

P. Kupias, Abbauverfahren, Bohr- und
Schiessarbeit in der Kupfererzgrube Outo-
kumpu in Finland, Entwicklung und heutiger
Stand, Erzmetall, Band XXI (1968),
Heft 6: 252-261.

J. Porkka, Mechanisierung der Bohr- und
Ladearbeit in den Gruben der Outokumpu Oy
Erzmetall, Band 24/1971, Heft 9 (Septem-
ber 1971) : 416, 429-434.

P. Särkkä, Mining Methods in Finland,
World Mining, March 1978: 49-54.

I. Autere, Experiences of Pillar Stoping
in the Mines of Outokumpu Oy, 1979,
Proceedings 10th World Mining Congress,
Istanbul, Turkey, III/18.

P. Nieminen, On the Chemical and Physical
Properties of Peat Fly Ash and its
Utilization, Proceedings of the Symposium
of Commissioning II', Kouvola Finland,
7-11 Aug, 1978.

P. Nieminen, Use of Slags and Fly Ash
Cements as binders in Soil Stabilization
and Mining Construction.

K. Karjalahti, Masuunikuona ja sen hyödyn-
täminen, 1981, Kemia-Kemi nr, 4/81.

ASTM C219, Book of ASTM Standards. Part 9,
p. 232, Philadelphia. American Society
for Testing and Materials, Philadelphia,
1967.

R. Matikainen, P. Särkkä, Cut-and-Fill
Stoping as Practice at Outokumpu Oy,
Underground Mining Methods Handbook,
AIME 1982, Sec. 3, Sub. 2, Chap. 9.

Characteristics of cemented deslimed mill tailing fill prepared from finely ground tailing

E.G. THOMAS
University of New South Wales, Sydney, Australia

ABSTRACT: For the purposes of maximising metallurgical recovery, it is sometimes necessary to grind ore very finely, hence producing a very fine tailing product. Such tailing has traditionally been regarded as unsuitable for preparation of mine fill material, for two reasons.

1. insufficient percentage recovery of fill from tailing, and
2. absence of coarse fill particles.

This paper describes a laboratory test study conducted on behalf of Roxby Management Services Pty. Ltd. on tailing produced from metallurgical tests on drill core from Olympic Dam, South Australia.

The test tailing was characterised by being very finely ground, with approximately 90 wt. per cent finer than 53 μm and 30 wt., per cent finer than 10 μm.

The paper describes preparation of fill from this tailing, with recoveries ranging from 45 to 65 wt. per cent. Also presented are the results of tests on the fill product, both cemented and uncemented.

Results show conclusively that the fill product was quite suitable and recoveries quite adequate. Results are considered to be highly significant to fill practice worldwide, in that they indicate techniques to allow satisfactory fill preparation from tailing materials hitherto considered too fine for such utilisation. A broad, new area of deslimed mill tailing fill technology is opened up, with significant implications for both fill preparation and underground disposal of mining wastes.

1 INTRODUCTION

During the past 20 years, the author has visited many mines in many countries to study mine fill technology and has frequently been told that the tailing from the particular operation visited is too fine to allow preparation of a suitable mill tailing fill. A "suitable" deslimed mill tailing fill infers, in the main,

1. adequate percentage recovery of fill from tailing,
2. adequate fill permeability, and possibly
3. satisfactory performance when additions of Portland cement are made.

The result of a tailing being considered as too fine for production of a deslimed mill tailing fill is frequently that

1. some other form of fill is used, such as surface sand or rock, or
2. a mining method not requiring fill is adopted.

These alternatives in turn have the effects of

1. increasing filling cost by requiring a separate mining operation to provide fill, and/or
2. increasing the environmental impact of the operation by deposition of all tailing on surface and requiring a separate mining operation to provide fill.

It is considered that development of
fill preparation techniques to produce
suitable deslimed mill tailing fill from
very finely ground tailing would represent
a significant step forward, on a worldwide
basis, in improving mining economics and
minimising environmental impact of mining.
It could be that mining could actually
be made possible in particularly environmentally sensitive areas where at present
this would not be so, though this comment
should not be taken as applying directly
to the study described within this paper.

2 PREVIOUS TESTWORK

It has already been shown by Thomas (1978)
that the main control over fill permeability is that of the very fine particles,
less than say 5 µm. If these particles
can be very effectively removed, all
particles above say 5 µm can be included
in the fill, giving adequate percentage
recovery from tailing without excessive
reduction of permeability.

(Selection of 5 µm as the critical cut-
off point cannot at present be justified
on the basis of actual test results. It
must be regarded as an estimate only.
Experimentation to date is inadequate in
this respect in that the Warman Cyclosizer
used for sub-sieve sizing does not allow
a split below about 8 or 9 µm and for less
dense materials below say 10 µm, under
normal operating conditions).

It is the object of the current paper
to introduce experimental results aimed
at
1. supporting results of previous investigations, and in particular to
2. assess the performance of cemented
   fill prepared from tailing with very
   minor amounts of coarse (plus 53 µm)
   material.

This second step is regarded as a significant step forward in fill technology
investigations.

3 TEST MATERIAL

The material employed in the test programme
currently being reported was a tailing
produced from metallurgical testing of
drill core from Olympic Dam, South Australia. The programme was conducted on behalf
of Roxby Management Services Pty. Ltd. as
a Unisearch Ltd. project within the School
of Mining Engineering, University of New
South Wales.

The test tailing was supplied dry. Its
relative density was measured at 3.45.

A particle size analysis was conducted
by
1. wet screening at 38 µm,
2. dry screening oversize, and
3. cyclosizing undersize,
as detailed by Thomas, Nantel and Notley
(1979).

Results are shown in Table 1 and Fig. 1.

Table 1. Particle size analysis of tailing
sample.

Fraction µm	wt. %
+150	0.04
106–150	0.26
75–106	2.26
53–75	8.58
38–53	14.68
35.7–38	4.49
26.4–35.7	6.95
18.7–26.4	12.39
12.8–18.7	11.30
10.5–12.8	5.16
–10.5	33.89

4 FILL PREPARATION FROM TAILING

Work conducted by the author in the past
has shown that it is possible to produce
a fill of adequate permeability and with
sufficient recovery from a material as
finely ground as the Olympic Dam tailing
sample, only by a very sharp split at a
nominated micrometre value. Ideally,
all material above this value is retained
and all material below is rejected. It is
possible thus to split at quite low micrometre values and still maintain adequate
fill permeability. In practice of course
such a split is not strictly possible.

Fill was prepared from tailing by a
repetitive process of
1. slurrying, using mains water,
2. settling for pre-determined periods,
   (three fill grades, three settling
   periods), and
3. syphoning of unsettled solids.

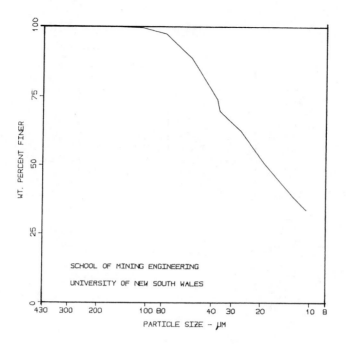

Fig. 1. Particle size analysis of test tailing.

At the completion of the required number of settling cycles, each fill was dried and percentage recovery calculated, as reported in Table 2.

Table 2. Percentage recovery of fill from tailing.

Fill grade	% recovery
Coarse	44.58
Medium	52.55
Fine	64.53

It is not claimed that the desliming process used in the test programme exactly duplicates the use of hydrocyclones, which would most likely be used in practice. However, the separating mechanisms are the same and the settling approach is certainly adequate for preparation of test fills, where the objective is more to study the fills produced than their methods of preparation.

5 FILL PROPERTIES - UNCEMENTED

5.1 Particle size analysis

Particle size analyses for the three pre-pared fill grades are reported in Table 3 and Figs. 2 to 4. All cyclosizer fraction limits are based upon a (mean) fill solids relative density of 3.55.

The main features of these size analyses are

1. the very minor amounts of material plus 75 μm,
2. the concentration of material in the 12-75 μm range, and
3. for the coarse and medium fills, the very minor amounts of minus 12 μm material.

The increased proportion of minus 10 μm material indicated for the fine fill is probably due to inclusion of 5 to 10 μm material rather than 0 to 10 μm material.

61

Table 3. Particle size analyses of prepared fills.

Fraction	wt. %		
μm	Coarse	Medium	Fine
+150	0.11	0.12	0.20
106-150	0.54	0.52	0.51
75-106	4.99	4.74	3.58
53-75	17.80	16.96	12.83
45-53	16.27	12.45	6.25
38-45	14.15	12.02	14.34
33.7-38	9.46	8.02	7.65
25.0-33.7	14.76	12.77	11.78
17.7-25.0	18.90	21.40	18.66
12.1-17.7	2.48	9.46	13.51
10.0-12.1	0.09	0.72	4.30
-10.0	0.45	0.82	6.38

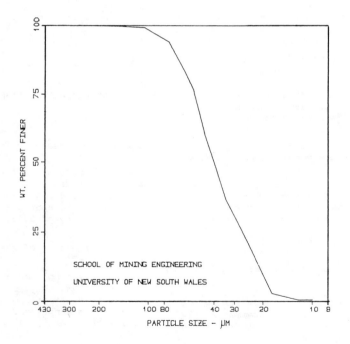

Fig. 2. Particle size analysis of coarse test fill.

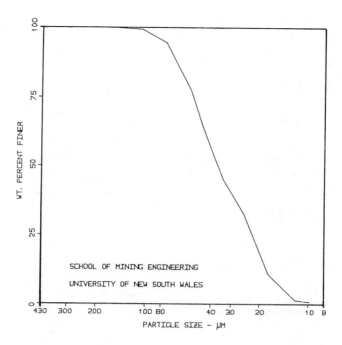

Fig. 3. Particle size analysis of medium test fill.

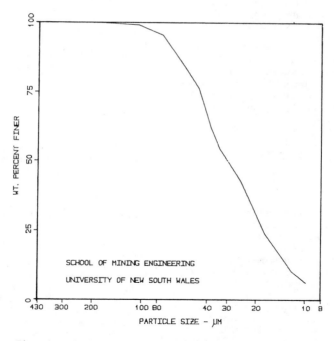

Fig. 4. Particle size analysis of fine test fill.

5.2 Permeability

Percolation tests for each of the three prepared fill grades were run, in parallel with a test on a reference fill. Results are reported in Table 4.

Table 4. Percolation rates of prepared fills.

Fill	Percolation rate, as compared with reference fill
Coarse	600
Medium	530
Fine	370
Reference	100

The reference fill subsequently proved to be somewhat finer and therefore less permeable than would be typical of it, bringing a typical value more in line with the Olympic Dam values.

Lack of time and test material for replicate testing precluded a totally thorough investigation of the question of relative permeabilities. The overall and unquestionable conclusion was however that each of the three fill grades exhibited more than adequate permeability. This is in agreement with earlier work by the author that recoveries can be high without excessive permeability reduction provided the very small particles (say minus 5 μm) are effectively removed.

6 FILL PROPERTIES - CEMENTED

6.1 Batch preparation

Nine batches of cemented fills were prepared, being three fill grades at each of three Portland cement levels, as detailed in Table 5. A replication of three was aimed at, though in approximately 20 per cent of cases this had to be reduced to two because of lack of test material.

For each batch, pouring pulp density was 71.6 wt. per cent solids, this giving a pulp viscosity comparable with that used for placement of the reference fill referred to above. For each batch, pouring temperature was 25°C. Curing temperature

Table 5. Cemented fill batch details.

Batch	Fill	Portland cement wt. %
1	Coarse	3
2		6
3		9
4	Medium	3
5		6
6		9
7	Fine	3
8		6
9		9

was 40°C, selected on the basis of anticipated mine stope temperatures. (It is not really known at this stage how important it is to pour at the same temperature as the curing temperature. Curing at a reasonably elevated temperature is reasonably straightforward, though pouring at say 40°C introduces significant practical experimental problems).

Standardised pouring, curing and testing procedures used by the author for many years were used throughout the programme, as detailed by Thomas, Nantel and Notley (1979).

6.2 Results - strength

Compressive strength results at a cell pressure of 0.1 MPa and a curing temperature of 40°C are reported in Table 6 and Fig. 5.

Each value reported is the mean of three (usually) tests but on occasion only two samples were available. The value reported is the total stress on the specimen at failure, that is, the applied stress plus the cell pressure.

The main conclusions from Table 6 and Fig. 5 are as follows.

1. Strength results compare favourably with those from other deslimed mill tailing fill studies. This means that the lack of coarse material (say 75 to 425 μm) in the fill is not affecting strength development, at least in a gross fashion. This is a most significant finding.

Table 6. Compressive strengths of prepared fills

Fill	Portland cement wt. %	Compressive strength - MPa		
		14 days	28 days	56 days
Coarse	3	0.490	0.543	0.578
	6	0.744	0.747	0.894
	9	1.077	1.178	1.402
Medium	3	0.520	0.574	0.600
	6	0.676	0.737	0.801
	9	1.104	1.177	1.361
Fine	3	0.634	0.688	0.720
	6	0.785	0.898	0.956
	9	1.201	1.304	1.482

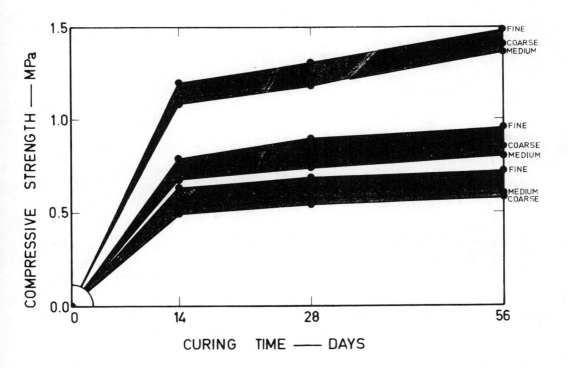

Fig. 5. Compressive strength results (curing curves) at a curing temperature of 40°C and cell pressure of 0.1 MPa.

65

with possible far reaching consequences.

2. Strength increases quickly and rapidly flattens out. This behaviour is typical of elevated temperature (40°C) curing.

3. The fine fill is in all cases superior. Coarse and medium fills are comparable, as would probably be expected from the particle size analyses, especially in that each has almost negligible minus 10 μm material.

4. Fine fill improvement is proportionally greatest at the lowest Portland cement content, showing the grading modifying effect of only minor additions of fine material - the Portland cement itself.

5. The significance of the superiority of the fine fill is probably not fully shown in Fig. 5, because of the necessary selection of the vertical scale. It could be equivalent to between one and two per cent of cement, in the 3 to 6 per cent addition interval.

ACKNOWLEDGEMENTS

The testing programme described herein was fully sponsored by Roxby Management Services Pty. Ltd., as a Unisearch Ltd. project within the University of New South Wales. Permission to publish these results is acknowledged, with appreciation.

REFERENCES

Thomas, E.G. 1978. Fill permeability and its significance in mine fill practice, Proc. Symp. Mining with Backfill, Sudbury, Canada.

Thomas, E.G., J.H. Nantel, and K.R. Notley 1979. Fill technology in underground metalliferous mines, International Academic Services Ltd., Kingston, Canada.

2. Technology

Cemented gravel fill for a small underground mine

J.R.BARRETT
Barrett, Askew, Fuller & Partners, Melbourne, Australia

J.E.STEWART
Aberfoyle Ltd., Melbourne, Australia

J.G.BROCK
Que River Mining Pty. Ltd., Tasmania, Australia

ABSTRACT: Many small underground metalliferous mines produce ore rather than concentrates and hence do not have a ready source of tailings for fill. Consideration must then be given to other fill sources especially if cemented fill is required. This situation was faced in the development of the silver lead zinc deposit at Que River in north western Tasmania. The fill material chosen for the mine is a well graded gravel. A simple fill plant was developed based on concrete batching technology. A truck distribution system is used in which trucks dump mixed fill directly into stopes via holes from the surface. The mixture flows from the drop zone almost horizontally across the stope without segregation. Large scale testing of the final product has been used to check strength expectations. The gravel supply, fill plant and fill distribution are run intermittently and are manned on a contract basis. The paper describes the development of the fill system, aspects of which are novel solutions to the problem of cemented fill in small underground mines.

1 INTRODUCTION

The Que River mine in north western Tasmania is a new underground base and precious metals mine producing 200,000 tpa of high grade ore, for sale to the Electrolytic Zinc Company of Australasia Ltd for treatment in their Rosebery concentrator. The mine is located in dense rain forest, about 700m above sea level. Early development proposals included construction of a concentrator. They were economically and environmentally less attractive than the current project based on underground mining of the largest and richest orebody, PQ lens, and haulage of ore 40km by road to an expanded Rosebery concentrator.

The combination of a high grade competent ore and less competent country rock indicated that mining methods using fill would be preferable as they would allow high ore recovery with minimum dilution. In the wider parts of PQ lens, mining is by primary long hole stoping between transverse pillars which will extend almost the full height of the near vertical lens. These stopes require cemented backfill to facilitate safe and efficient mining of the valuable pillars.

In the absence of concentrator tailings it was not obvious where this fill would come from or how it would be prepared and placed. Many different sources and techniques were considered before the present system was conceived and developed.

2 GEOLOGY

PQ lens is a fairly typical sulphide deposit, enclosed in altered and pyritised volcanics rocks (Wallace and Green 1982). Its shape is lenticular with a maximum ore thickness of more than 30m near the centre. Its strike length is 600m compared with 200m down dip. As the ore extends to the surface at its southern end, this is also the maximum depth (see Figure 1).

Ground conditions in PQ lens itself are generally good except where joints are well developed or where some deterioration has occurred due to weathering. Ground conditions in the enclosing rocks are variable, but generally poorer than in the massive sulphides and deteriorate towards the surface. Several faults and shear zones are sub-parallel to PQ lens and there is well developed cleavage within the volcanics.

The mineralisation is pyrite, spalerite and galena, with associated silver, gold

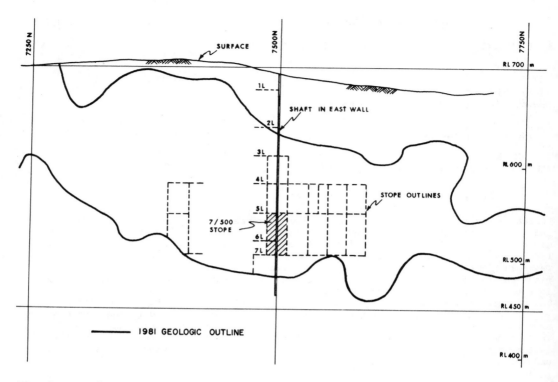

Fig. 1. Longitudinal Section, Que River Mine.

and minor chalcopyrite. The transition
from economic grades to virtually barren
pyroclastics is generally very sharp,
although disseminated ore boundaries occur
locally. This abrupt transition from rich
to barren material provides strong economic incentives to match mining limits to
geological boundaries. At the same time
the relative incompetence of wall rocks
represents a serious threat to stope
stability and safety. Fortunately, these
boundaries are generally continuous and
sufficiently regular to permit longhole
open stoping.

3 MINING METHODS

The average grade is in excess of 20%
combined lead and zinc with additional
silver and gold. This justifies use of
mining methods which ensure high ore
recovery. Any pillars or remnants which
become unsafe or uneconomic would represent expensive losses. Transport and
treatment charges are also relatively high
and metal recovery depends on diluted
grades. Hence there are also strong
incentives to minimise dilution.

The modern techniques of "high lift"
open stoping with "secondary" stoping of
pillars between cemented fill offers many
advantages where wall geometry is regular
and ore widths and dip suit longhole open
stopes, (eg. Hornsby and Sullivan 1977).
Planning for Que River Mine has aimed at
exploiting these techniques at the smaller
scale imposed by the dimensions of PQ lens.

3.1 Ground Conditions

The poor structural properties of the rocks
enclosing PQ lens are the most obvious
constraint on safe mining of large open
stopes. The advantage of such stopes could
be quickly eroded if overbreak into the
barren walls led to ore losses, dilution
and mining delays.
To minimise this risk, stopes (including
the secondary stopes planned for pillar
recovery) will have a maximum strike length
of 20m. The level interval was set during
exploration at 30m. Thus the minimum
stoping block within PQ lens is 20m long
by 30m high. Production commences in the
bottom block of each stope and it is
possible to extend these stopes upward

through successive levels if ground conditions prove encouraging.

In the expectation that high wall exposures could be untenable, it was planned to introduce fill to the lower section(s) of each stope before resuming blasting of the ore above. When the fill reaches the floor of the next extraction level, stoping would resume.

3.2 Orebody Width and Depth

In the wider parts of PQ lens the ore broken in all but the bottom block has to be safely and efficiently loaded off the fill. Remote control diesel loaders will be used to clean up stopes before resumption of filling.

The shallow depth of PQ lens has three advantages which have been important in the development of the new fill system. Firstly, initial production could be economically established at the bottom of the lens. This simplified filling by eliminating the need to transfer fill around and under fill stopes. Secondly, the problem (recognised at Mount Isa and Geco mines) of severe size degradation of rockfill falling to great depths, is largely eliminated.

Finally, fill passes direct from the surface to each stope are economical and capable of handling a wide range of fill materials.

4 FILL MATERIALS

The decision to sell ore to a distant concentrator meant that tailings are not readily available for use as fill. In addition, even if the PQ lens tailings were available, the volume of fill recoverable after concentrates and slimes had been extracted would be below mining requirements. A detailed account of such volume balance relationships is given by Stewart (1980).

The volume of cemented fill required during the first years of production also far exceeded that which might be produced from mine development. Also waste material would require crushing to make it suitable for the production of cemented fill. Thus, even if development waste was supplemented by surface mining, the resultant product was expected to be expensive in both processing and cement requirements.

A number of potential sources were recognised. These included products from existing quarries, waste products of other mines and established gravel pits and sand deposits.

Subsequent investigations were concerned with three aspects of these sources:
- long term supply and availability,
- plant and processing aspects, and
- the mechanical properties of the final product.

4.1 Supply and Availability

The products from existing quarries and other mines were not pursued as the purchasing and haulage costs were higher than the cost estimates of the sands and gravels. The available natural sand deposits were found to contain up to 50% of silt and clay. They were poorly located for year round access, and handling problems were anticipated.

Fluvioglacial gravel deposits are relatively common in the area and a number of sites with adequate reserves were available. Some of these had already been worked and abandoned so that legal and environmental requirements for development could be easily satisfied. Those adjacent to the ore haulage route were examined and two selected for detailed evaluation.

4.2 Plant and Processing Requirements

The fluvioglacial gravels all contain significant proportions of clay and gravels, both of which could create handling problems. Gravel pits should have easy (loose) digging conditions, good drainage for moisture control during year-round operations, minimal overburden and environmental disturbances, and minimal oversize.

The gravels and the product which was required bore similarities to low grade concretes used in civil engineering. A similar approach to fill preparation would require the determination of the maximum particle size permissible in a plant and the allowable clay and water content for efficient handling. These aspects were considered through discussions with plant designers, inspections of similar operations and field trials using an existing concrete batch plant. Trial extraction was undertaken at one of the pits which had a uniform particle size distribution and suitable handling characteristics were noted.

71

4.3 Fill Segregation and Strength

In parallel with the above investigation, a laboratory test program was undertaken to determine if the gravels alone could be stabilised economically with cement or whether additions of sand were required. The test program had to develop samples which reproduced the conditions which were anticipated in the future stopes. As the majority of the potential fill materials had coarse particles in them (up to 200mm size) reports on the behaviour of rockfill at Mount Isa, Kidd Creek, Hoits, and Juno were examined (Barrett and Cowling, 1980).

At each of these sites segregation of the rockfill occurred with a cone of rockfill forming below the inlet point and the finer, more fluid fractions reporting to the lower portions of the rill. The size distributions for three of these fills are shown in Figure 2. At Mount Isa the ratio of rockfill of hydraulic fill can altered between stopes. Figure 2 includes size distributions for extremes of the Mount Isa fill ratio. In all cases segregation occurs.

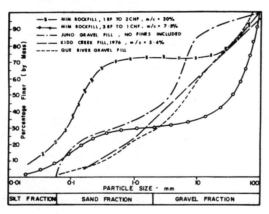

Fig. 2. Mine Fill Particle Size Distributions.

A typical size distribution of the gravels available near Que River is shown in Figure 2, superimposed on the size distribution of the other fills. It can be seen that the gravel size distribution contains the same broad distribution of sizes as those of the other materials. Its overall moisture content of 8-10% is within the range used at other mines. It was therefore expected that a cemented gravel mine fill would segregate during placement. Thus, the laboratory test program was undertaken based on preparing

fill specimens that mirrored the possible material variations that could be formed in segregated and non segregated fills.

Two groups of gravel samples were prepared and tested to represent a uniform gravel. The first group was of scaled-down gravel prepared from a blend of five bulk samples taken from the preferred gravel pit. The size distribution was scaled-down (Gonano, 1978) with the 200mm to 13.2mm gravel particles being replaced by particles in the 4.7 to 13.2mm size range. The second group of samples consisted of the scaled-down gravel with a nominal 6% addition of the available sand. This material had a higher moisture content and lower density than the specimens of simulated gravel alone.

The variation in unconfined strength with cement content after 56 days curing is shown in Figure 3. A 50% difference in unconfined strength was noted for the same cement content between the two groups, that is, the gravel strength was halved when the sand was added. Negligible gain in strength was noted from 28 to 56 day curing.

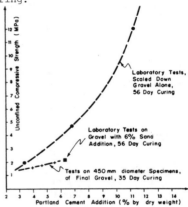

Fig. 3. Fill Strengths from Laboratory and Large Scale Uniaxial Tests.

Three additional groups of materials were also prepared to represent the zone of finer, hydraulically transported material which could be generated at the edge of a stope if segregation occurred. The first group of fine materials were those derived from the gravel deposits and one from an available stockpile of quarry reject. The gravel fines were nominally defined as the minus 4.7mm material. The specimens of quarry material did not cement at all as they remained open textured and the cement flowed out from the specimens as a slurry. These specimens of fines showed strengths approximately one tenth of the specimens of gravels alone.

The second group of fines was a 60/40 mixture by mass of natural sand and of minus 4.7mm gravel material gained from all the bulk gravel samples. The test results indicated that the strength of the gravel fines decreased further when combined with sand.

The third group of fine material to be tested was the sand alone. The sample strengths were slightly higher than for the sand/gravel fines blend but were still significantly lower than for the gravels alone. In both cases, that is for the sand/gravel blend and the sand samples, there was no increase in strength from 28 to 56 days curing. In fact a small decrease was often noted.

Thus, the laboratory tests showed that a fill based on gravel alone would be a more satisfactory material providing segregation could be prevented. Model work undertaken by Ingles et al (1973) suggests that the moisture content can be varied to control fill segregation and this concept was pursued at Que River through field trials.

5 FIELD TRIALS

Many of the important questions concerning mechanical behaviour of a cemented gravel fill cannot be answered by laboratory scale tests because of scale effects. Also, aspects of detailed plant design would depend on the bulk handling characteristics of the fill materials. Field trials were therefore undertaken:
- to determine if oversized rocks could be easily removed from the gravel,
- to achieve experience in handling and mixing the gravel fill, with or without sand,
- to investigate whether the gravel fill required additional fines to produce a stronger mix,
- to find out how the fill moisture content and placement conditions interacted to control segregation of the material,
- to estimate approximate cement requirements for fill plant design, and
- to determine mixing and quality control parameters for final plant design.

A test site and a concrete batch plant for fill preparation were arranged and a bulk sample of gravel obtained. Thirty tonnes of gravel were mined, screened and transported to the concrete batching plant. Operations at the gravel pit were observed to determine the amount of reject and if handling problems were encountered. A simple inclined grizzly screen was used successfully. A maximum particle size of 100mm did not develop significant reject and, as any larger particles were expected to create problems both in the fill plant and in fill passes, scalping on a 100mm grizzly was adopted for the final fill. The gravel was initially at 7% moisture content.

Three large trenches were excavated at the base of the trestle, 7m long with the drop point at one end. This compared favourably with the anticipated 10m run from the stope drop point to the stope edge in future stopes at Que River. Each batch of fill used in the tests was approximately of 7,000kgm. They were prepared in the concrete batching plant and were mixed and transported to the test site in a concrete agitator truck.

The test site consisted of a large timber trestle built out over an old reclaimed tailings pond. Drop heights of 10 to 12m were available. A 200mm diameter steel chute through which the fill was to be dropped was suspended over the side of the trestle. The test arrangement is shown in Plate 1.

Plate 1. Field Test Arrangement

The first mix of 7 tonnes of gravel plus water was placed as dry as could be ejected from the concrete agitator to see if segregation would occur or if a uniform fill was formed. The fill was dropped into a trench and a gradual expansion of the fill occurred from the inlet point down the trench without segregation as shown in Plate 2. The trench had a base inclination varying from 7° to 15°, and the overall rill angle of the fill was from 21° to 25°. This suggests that almost the full height of a stope wall would be covered by fill placed from a central inlet point.

Plate 2. Gravel Fill Rill in Field Test.

The low (12%) moisture content fill was almost unworkable, however, in terms of discharging it from the agitator truck but the fill that was produced did not appear to have a deficiency in fines. Therefore no sand additions to the mixes were attempted in the remainder of the trials. Rather, revised objectives were immediately formulated after the first mix. These objectives define the bounds in moisture content which limit workability if too dry and which would produce segregation if the moisture content was too high.

A batch of gravel was then prepared and discharged in three lots with varying water additions to check how the fill fluidity changed with moisture content. It appeared that a change in moisture content from 12% to 16% was sufficient to alter the rill angle from approximately

25° to 10° and that segregation still did not occur. Excess water however would reduce strength.

Additional 7 tonne mixes of gravel with 3% to 6% cement plus water were then prepared. The water was batched into the agitator truck both before and after the gravel at the concrete plant. A very uniform fill was produced in 7 to 10 minutes mixing. Moisture contents of 15% to 16% respectively were required to achieve the same workability as the non cemented fill at 12% to 13% due to the presence of cement. Samples of the 6% cement mix hardened in 24 hours and samples of the 3% mix hardened in 4 days.

6 LARGE SCALE TESTING

With the fill material selected and its condition in the stopes assessed, testing was required to determine the minimum cement levels in future stopes. Better results are obtained with the actual fill material rather than with smaller, scaled-down specimens but the 100mm sized gravel fractions necessitated large diameter samples. In addition, as there was potential for the use of cement substitutes to reduce the cost of stabilisation, alternative cementing agents were to be evaluated. Silica fume, but not fly ash, was available. Silica fume is used as a partial cement replacement by some local concrete contractors. It is a by-product of silicon and feuro-silicon manufacture and is much finer than fly ash or cement.

When the fill plant was first commissioned, trial mixes of the gravel fill were prepared in 0.9m high, 450mm diameter cylinders. Six tonne batches of each mix were prepared in the fill plant and were poured into moulds with a concrete agitator truck. This batch size was adopted to ensure that representative material and standard preparation techniques were used. The trial mixes included portland cement alone and portland cement with silica fume. Final moisture contents range from 13 to 15%. Testing of the samples was undertaken on site using a 30 tonne press from the mine workshop adapted as shown in Plate 3.

The strength of the large scale specimens with portland cement alone was approximately 60% of the laboratory strength of specimens of scaled-down gravel (Figure 3). There again appeared to be no significant difference between 14 day and 35 day curing. Silica fume proved ineffective as an additive and currently is not being used in the fill.

Plate 3.   Large Diameter Uniaxial Test.

The lack of segregation in this fill contrasts with the experience at other mine sites as discussed earlier. Based on these field trials, the lack of segregation could be due to the low moisture, more cohesive fines, rounded particles or pre-mixing of the fill. Other trial mixes with crushed, angular, development waste showed a similar lack of segregation suggesting that moisture content and/or thorough mixing may be the critical factors.

7 FILL PLANT DESIGN

The parallels between the proposed cement gravel fill and bulk concrete has led to the consideration of concrete mixing and handling technology. Numerous options were considered, even to contracting the whole project to a pre-mix concrete organisation prepared to construct and operate its own plant. The use of such a plant in the field trials had demonstrated the feasibility of this, and the anticipated demand pattern was compatible with portable batch plants.

The cemented fill plant which was eventually built is essentially a concrete batch plant, but with numerous features intended to match it to specific aspects of the materials and the mine. The plant layout is shown in Figure 4. It consists of four gravel bins with individual scales, twin concrete silos, a cement weigh hopper, a mixer and truck dispatch. The design allows some flexibility in operations as follows:

- A two way chute allows batching direct into transit mixers when required,

- Four ground bins allow batching with several materials (for example sand and aggregate to make concrete),

- A valve regulates discharge from the cement weigh hopper to match the cement discharge time to the gravel batching time and so yield improved mixing,

- The water metering and recording system is designed to handle dirty water from the mine and also resist damage from freezing, and

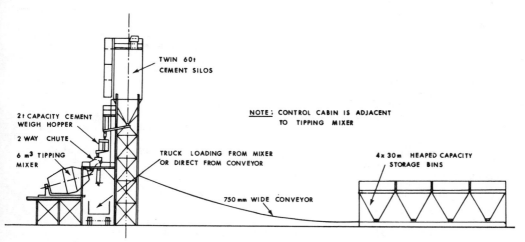

TWIN 60 t CEMENT SILOS

2 t CAPACITY CEMENT WEIGH HOPPER

2 WAY CHUTE

6 m³ TIPPING MIXER

NOTE: CONTROL CABIN IS ADJACENT TO TIPPING MIXER

TRUCK LOADING FROM MIXER OR DIRECT FROM CONVEYOR

4 x 30 m HEAPED CAPACITY STORAGE BINS

750 mm WIDE CONVEYOR

Fig. 4.   Schematic Layout of Fill Plant.

- The semi-automatic batching control system can be manually overridden for various purposes (for example to combine gravel from near empty bins to make a full batch).

7.1 Plant Capacity

For PQ lens, the long term demand for cemented fill is about half the mining rate. Initial dependence on primary stopes and the multi-lift stoping sequence have meant that the fill system would have periods of urgent demand separated by delays while subsequent stopes are emptied. Because of poor wall stability it is not appropriate to accumulate empty stopes as a buffer between production and filling ·schedules.

The nominal plant capacity selected for the Que River fill system is 40m^3 per hour. Even on a one shift basis, this is far higher than the long term usage and the system will be idle for extended periods. One consequence of this "campaign" filling strategy is the use of contractors to operate the fill system and gravel operation. Another is the use of substantial gravel stockpiles which buffer filling operations from gravel haulage based on backloading of ore trucks.

7.2 Batch Mixing and Transport

It was evident from early investigations that a batch rather than continuous process would be appropriate. Firstly, a batch system has the ability to handle large gravel particles. This in turn meant maximum recovery of fill from the natural gravel and would also contribute to lower cement usage.

Secondly, batching is also very compatible with the use of conventional dump trucks to distribute fill to numerous fill passes. Finally, process and quality control were considered superior with a batch system with a mixer in the plant rather than allowing mixing in agitator trucks.

The short length of PQ lens and the road above it meant that the use of agitator trucks would also be more expensive than the current system.

7.3 Gravel and Cement Storage

Most small concrete batching plants do not have live storage bins and require a loader to feed a weigh hopper from sand and gravel stockpiles. The Que River batch plant includes covered ground bins which can store up to three hour's gravel supply. The extra capital cost of this facility was justified by the following:

o A surge capacity was available to allow for differences in fill delivery and use,

o The fill trucks could service the fill plant directly, removing the necessity for a loader and operator, and

o Moisture content of gravel dumped direct from trucks into the covered ground bins was expected to be less variable than that handled via surface stockpiles exposed to the weather.

The Que River plant has two cement storage silos. The second silo was provided in anticipation that it would be used for a pozzalanic cement substitute.

7.4 Process and Quality Control

Success in controlling segregation during the first field trials meant that quality control aspects, particularly cement and water additions, became far more significant. It was recognized that concrete industry experience was very relevant to this and is reflected in the system which has been installed at Que River. The system also incorporates features which facilitate operation of the plant by an independant contractor who is paid according to plant throughput.

Batch plant controls include the following:

- load cells for each ground bin and the single cement weigh hopper,
- semi-automatic opening and closing of ground bins and silos to discharge preset batch weight,
- manual control of water addition,
- a revolution counter to ensure minimum mixing, and
- a printer to record batch weights.

8 OPERATING PRACTICE

The first major stope, No.7/500 stope as shown in Figure 1, was filled in July 1982. Currently, early 1983, a third stope is being filled.

8.1 Gravel Pit

Quarrying of gravel is carried out at Tullah some 20km from the mine site. Top soil is cleared off from the in-situ gravel and stored for future land rehabilitation. The gravel is mined using a bulldozer and rubbed tyred loaders, and sized over an inclined parallel bar grizzly to 95% passing 100mm. The material is pushed up by the dozer so that proper intermixing of the material occurs.

Oversize material is stockpiled for possible outside sale and undersize material is stockpiled for haulage to the mine site. Gravel is hauled by semi-trailers with 12 metre cubic tipping bodies. Because the gravel pit is on the same highway route on which ore is hauled to the concentrator, gravel is used for back loading, thus reducing the haulage cost by approximately 20%.

8.2 Cement

Cement is the most costly component of the cemented fill and currently comprises 75% of the total fill cost. The cement is purchased from the manufacturer in Tasmania and is road freighted to the mine site in bulk tankers and stored on the site in two 60 tonne elevated silos (Ref. Fig. 4).

8.3 Mixing Plant

Operation of the fill mixing plant is carried out by off-mine contractors on a 12 monthly contract. The contractor supplies and maintains all mobile equipment used and a complement of labour. Gravel is loaded from the stockpile into the feed bin where it is automatically batching into the mixing bowl with the required amount of cement and water. Mixed fill is then transported in $8m^3$ tip trucks from the batch plant to the appropriate fill hole where it is tipped into a launder located on top of the hole leading to the stope to be filled. The contractor who operates the fill plant is paid for each tonne of fill placed.

8.4 Fill Holes

To convey the fill from the surface to the required stope, raisebore holes, 1.2m in diameter, were initially used. From the experience of filling the first stope and testwork involving different diameter pipes, the raisebore holes were subsequently reduced to 600mm diameter and have now been replaced by 381mm diameter percussion drilled holes. To date these holes have been satisfactory for conveying fill over a drop of 150m.

8.5 Fill Placement

Nominally stopes have a length of 20m and a width of 10 to 30m. Two fill holes are used for each stope to ensure that the fill spreads evenly throughout the stope and as a backup in case one hole should become blocked. The fill is reasonably fluid and its workability can be appreciated by comparing its probable slump with that of normal construction concrete. The cone formed in a slump test is anticipated to be 60%-70% as high as that of normal construction concrete, that is, the fill is more fluid. This allows the fill to flow throughout the stope and not leave a large cone under the fill hole. Within a stope, the fill flows out from the cone in layers. A load haul dump unit is used to flatten out the slight rill after the stope is filled to enable mining of the next lift to take placed on a level floor.

8.6 Stope Preparation

Prior to filling, drawpoints are sealed using timber barricades. These are built on a conventional basis with three vertical uprights rockbolted to the floor and backs and horizontal boards bolted to the uprights. Because of the high consistency of the fill it is not necessary to plug small holes in the barricades, as is the case with hydraulic fill. Observations to date have shown that there is very little loading on the barricades. Drainage pipes have been installed to allow ground water which enters the stope to drain. No water drains from the fill itself.

8.7 Operating Problems

There have been some operating problems but these have not been significant enough to stop the operation of the plant.

Accuracy of the weigh system used to meter the gravel from the bins has not been within the required limit. The design of the bins with respect to how they are filled has caused the inaccurate weighing. To compensate additional cement has been added to ensure the desired fill strength has been obtained adding to operating costs.

Wear on the mixing bowl was greater than anticipated because of the size of material being mixed. This has necessitated the replacing of the original bowl after the mixing of some 50,000 cubic metres of fill. The replacement bowl was redesigned for heavier duty.

The weather pattern at the mine site of high rainfall then dry periods has prevented a uniformly low moisture content being maintained. This then produces variations in the flow characteristics of the gravel so that material flow from the storage bins to the belt feeder has been a problem. To overcome this, large bin vibrators have been installed, but this has increased feed bin maintenance.

Considerable concern existed about the segregation of fill while being transported, because of the high slump factor required. This concern has proved to be unfounded and fill is well mixed in-situ.

8.8 Quality Control

The quality of the fill is checked in a number of ways. Inspections are regularly made of the gravel pit and bulk samples obtained weekly for particle size analysis. The fill plant is also regularly inspected, the workability of the fill visually assessed and cement usage checked. Once or twice a week samples of fill are poured into 450mm diameter moulds and unconfined compression tests undertaken after 14 day curing to ensure that target strengths are achieved. Finally, the surface of the fill is inspected underground during placement to ensure that a uniform material is being placed.

9. FINAL SELECTION OF CEMENT CONTENT

The development of the fill system has been undertaken in conjunction with initial experience in the mining method itself. From the test work during fill development it was projected that a cement content of 3%-6% would be sufficient for the range of anticipated exposures. To date a higher cement content has been used until the details of fill plant operation, material variability and the size of the final fill exposures are resolved. Currently a test program is in progress to determine the change in fill strength with the likely variations in composition. A reduction in the cement content should follow after this work.

10. CONCLUSIONS

The development of a system to prepare and deliver cemented fill in small underground mines requires the solution to a large number of problems. At Que River:

- A lack of mine waste material was resolved by using natural gravel,
- Haulage costs for fill were minimised by using back haulage by ore trucks,
- A fill plant was based on concrete batching technology at comparatively low capital cost,
- Fill holes lead directly into stopes allowing truck haulage using standard trucks and a coarse fill,
- The intermittent filling of stopes is resolved using contractors, and
- A non-segregation coarse fill was developed to minimise cement addition levels.

Details of the stages of development, as described herein, should assist those adapting the system to other sites.

11. ACKNOWLEDGEMENTS

The authors wish to thank Aberfoyle Limited and Que River Mining Pty Ltd for permission to publish this paper.

12. REFERENCES

Barrett, J.R. & R. Cowling, 1980. Investigations of cemented fill stability in 1100 orebody, Mount Isa Mines Ltd, Queensland, Australia. Transactions, Section A, I.M.M., Vol.89, A118-A128.

Gonano, L.P., 1977. Sample preparation techniques for the triaxial testing of cemented rockfill. CSIRO, Division of Applied Geomechanics, Tech. Report No:37.

Hornsby, B. & B.J.K. Sullivan, 1977. Excavation Design and Mining Methods in the 1100 orebody, Mount Isa Mines, Australia, AIMMGM, Mem. Tech., Xll Conv.

Ingles, O.G., R.C. Neil & T.J. Richards, 1973. Assessment of Mine Fill Properties Warrego Mine, N.T., CSIRO, Division of Applied Geomechanics, Tech. Report No:19.

Stewart, J.E., 1980. Mining a 100 million tonne orebody without subsidence, Aus I.M.M., Conf., New Zealand, May 1980.

Wallace, D.B. & G.R. Green, 1982. Que River Mine and Aspects of the Mt. Read Volcanics in the Pieman River Area. Geology, Mineralization and Exploration: Western Tasmania, Geological Society of Australia, Tasmanian Division.

Proceedings of the International Symposium on Mining with Backfill / Luleå / 7-9 June 1983

Sub-level vs cut and fill stoping at the Funtana Raminosa Mine: An economic comparison

C.BERLINGIERI, M.CONGIU, A.MEDDA, L.MUSSO, C.SANCILIO & I.TRUDU
SAMIM S.p.A., Roma, Italy

ABSTRACT: Unexpected detachments from the footwall of the deposit, causing dilution in the Funtana Raminosa mine, have raised the issue of replacing the present sub-level stoping method with cut and fill stoping. Laboratory investigations have focussed on cemented hydraulic fills and not on the utilization of the tailings from the treatment plant, as the disposal of the tailings is already per se a large problem for this mine. On the basis of these studies a technical and economic analysis has been made of the suggested alternative as compared with the stoping method currently used.

1. THE FUNTANA RAMINOSA DEPOSIT

1.1 Introduction

The Funtana Raminosa deposit is located in the Gadoni municipality (Nuoro), bordering the Barbagia di Belvì and Sarcidano. Known ever-since the Nuraghi era, this mine involves a lentiform deposit of mixed sulphides (copper, lead, zinc, cadmium and silver)(Fig.1).

Industrial activities were started by a French group in 1910; in 1916 the first Italian plant for beneficiating a copper ore by means of flotation was set up.

At present, the available ore reserves to be mined are approximately 3 million tons.

1.2 Some information about the geology and the deposit characteristics

The mixed sulphide ore bodies are housed inside a thick skarn seam (green rock" in local jargon) between Silurian carnubianites as hanging wall and Hercinian porphyry as footwall (Fig.1).

The cornubianites originally consisted of an alternation of thick layers of shales, grey limestones and thin interbedded layers or clayey material.

The finding of graptolite fauna in carbonatic facies is evidence that the entire formation may be attributed to the Silurian.

The effect of metamorphism related to Hercinian intrusions is quite evident in these sediments.

The resulting cornubianites always consist of silicate rocks, whose mineral composition varies according to the original composition and to the variability of their components. The black graphite shales have been transformed preferentially into fine-grained, hard, black or violet rock. Quite often there are cornubianites of various shades of red called 'flesh-coloured cornubianites" in the local mining jargon.

The intense magmatic activity of the Hercinian which is to account for the metamorphism of the sediments, is responsible for the presence of a dense network of porphyry and porphyrite seams in this area.

These porphyry seams and the innumerable faults and fractures caused by their intrusion have allowed the circulation of late magma solutions whose metasomatic action on the hosting rocks was very active.

The green cornubianites deriving from the metamorphism of calcareous rocks (called "green rock") are of great importance from a mining point of view, as the orebodies are exclu-

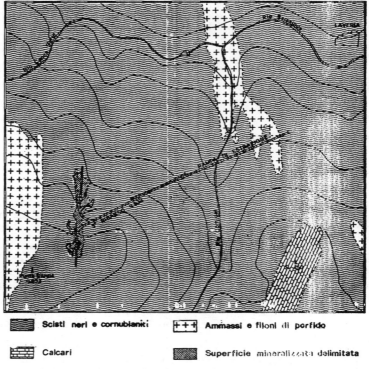

Fig. 1 – Carta geologica della zona di Funtana Raminosa

sively housed inside them. It is a
fine-grained rock consisting of epidote, chlorite and a little quartz,
or the grain-size may be a little
larger, basically garnetiferous,
with amphibole, pyroxene and epidote,
scarce quartz and carbonatic resi-
dues.

The mineralizations at Funtana Raminosa are constantly linked to this
rock abd the useful ore, mixed sulphides of Pb, Zn, and Cu, are often
accompanied by variable quantities
of magnetite, hematite, pyrite and
pyrrhotine.

It must however be pointed out
that the distribution of the metals
inside the orebodies points to a
likely relationship between the orebodies and the faults of fractures
that accompanied the intrusion of
the igneous rocks.

The silicate-metal ratio is extremely variable thus showing a highly erratic distribution both in the
horizontal and vertical directions;
this means that there are sterile
zones where dissemination ranges
from nil to very poor, and zones
where the concentration of sulphide
is very high.

Lenses of industrial interest
range in thickness from a few metre
to a few tens of metres; also the
length (30-300 m) and the vertical
development vary within a wide rang
Longitudinally it develops mainly i
the N-S direction with a 45-50° Eas
dip. (Figs. 2a and 2b).

2. THE FUNTANA RAMINOSA MINE
2.1 Presentation

The Funtana Raminosa deposit "stric
tu sensu" is undoubtedly the most
important amongst the various orebodies in the area.

Since 1950 it has produced over 1
million tons of sulphide tout venar
useful for treatment; the availabl
reserves are on the order of 3
million tons, with grades of 0.8 of
copper, 0.6 lead and 2.1 zinc.

2.2 Outline of the large developmer
    works

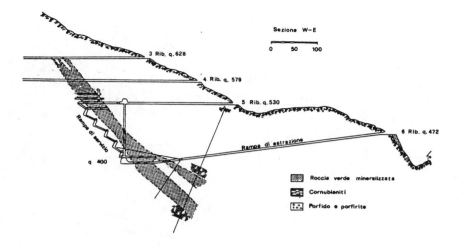

Sezione W-E

0   50   100

3 Rib. q. 628

4 Rib. q. 579

5 Rib. q. 530

6 Rib. q. 472

Rampa di estrazione

Rampa di servizio

q 400

▓ Roccia verde mineralizzata

▭ Cornubianiti

⊡ Porfido e porfirite

Fig 2a   Schema di sezione verticale del giacimento di
Funtana Raminosa

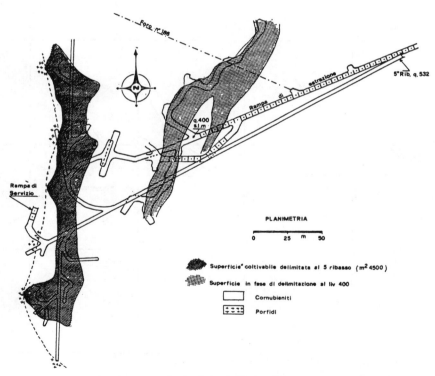

Foro n° 188

N

5°Rib. q. 532

Rampa di estrazione

q.400
s.l.m

Rampa di
Servizio

PLANIMETRIA

0   25   m   50

▓ Superficie' coltivabile delimitata al 5 ribasso (m² 4500)

▒ Superficie in fase di delimitazione al liv 400

▭ Cornubianiti

⊡ Porfidi

Fig. 2b – Planimetria della lente 273

81

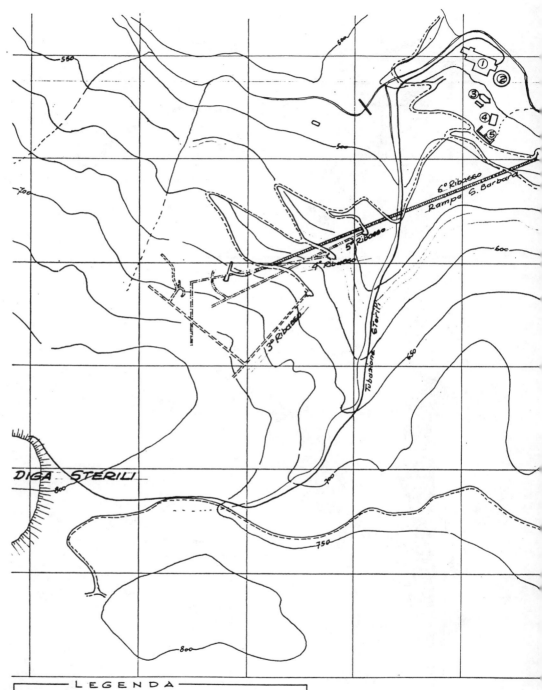

LEGENDA

① Impianto di Tratt.   ④ Frantumazione
② Decantatore   ⑤ Silos Tout_Venant.
③ Stock_Pile   ◯

fig. 3

The orebody is reached from the surface (el. 472) by means of a straight hauling ramp, 550 m long, a 13.50% gradient, with a 12 m^2 cross-section, partially reinforced with TH steel ribs, with a 15 cm thick concrete lining at the foot of the ramp.

From the 5th level (el. 532), located at the footwall, a service ramp with a 20% slope and 12 m^2 cross-section provides easy access to the sub-levels and ensures rapid displacements of personnel and of vehicles. This ramp, extending for 640 m, provides communication between the 5th level and the tunnel at el. 400 (bottom of the mine). This new site is completed by two central chutes, each having a capacity of 1000 t, which is supposed to act as an appropriate stock between the mining stages and the ore-processing (Figs.2a and 2b).

2.3 The treatment plant

Differential flotation produces three products having the following final characteristics:

- chalcopyrite 27.5% Cu and 800 g/t Ag
- galena 48.0% Pb and 1200 g/t of Ag
- blende 47.0% Zn and 8.2 kg/t of Cd

Crushing consists of three stages and the product is stocked by means of a stock-pile system. The material obtained is riddled at - 12 mm. Grinding consists of one stage and is carried out with a ball cylinder mill. For classification, a cyclone is used which feeds $d_{80} = 0.14$ mm grain-sizes to the flotation section. There are two scavenging lines (Pb+Cu and Zn), two lines for rewashing Pb+Cu and then separation of Pb and Cu; there is also a line for rewashing the Zn.

For process control, an on-stream analyser is used, and the grain-size is controlled with an Autometric PSM.

2.4 Water recovery and pumping of tailings

Strict ecological legislation and an insufficient availability of fresh water to be used for treating the minerals have led to the development of scrupulous recovery circuites of the waters discharged with the concentrates and the tailings.

70% of the water is recovered in the vicinity of the plant; this allows a considerable reduction of the solid-fluid tonnage to be pumped to the tailings deposit (dam) located at 1700 m away, with a geodesic line of about 400 m.

Landscape and ecological constraints together with the orography of the region have been critical factors in deciding the siting of the tailings dam at such a high elevation. Fig. 3 shows the siting of the mine, the treatment plant and the tailings dam.

The tailings are taken to the dam by means of two pumping systems which are both automated to a sufficient degree. One consists of a piston-pump which lifts the material in a single stage; the other carries the tailings to the dam through two pumping stations consisting of 4 aligned centrifugal pumps placed at 800 m from each other.

2.5 Mining method

The mechanical characteristics of the hosting rocks and of the skarn which houses the orebody have been so far conducive to sub-level stoping. Data obtained through drillings showed that it was possible to go on using this method. And so, a sub-level stoping programme was worked out.

The programme envisages ten sub-levels with 9 slabs; the irregular nature of the orebody and the value of the tout-venant do not make larger sizes advisable, especially if one wishes to avoid heavily diluting the ores.

To the north and to the south two slots will serve to start exploitation, progressing towards the central chutes.

The drilling system for the vertical drill-holes having a diameter of 64 mm pneumatically charged with AN-FO. The blasted ore is loaded at the foot of the sub-level by a diesel scoop-tram (TORO 350), having a 3.5 m^3 bucket, and thrown into the central silos.

At el. 400- bottom of the hauling ramp - the tout venant is tapped from the silos and loaded on to dumpers. These vehicles drive up the 550 m long ramp to the surface and then 200 m down a road and unload the ore into the topmost silos for

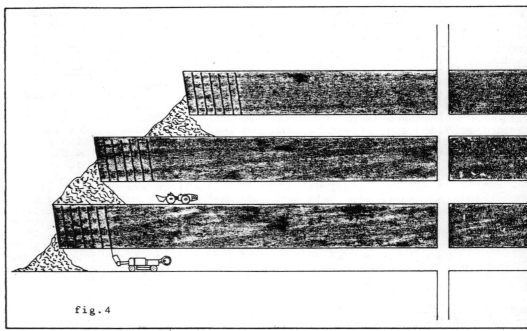

fig. 4

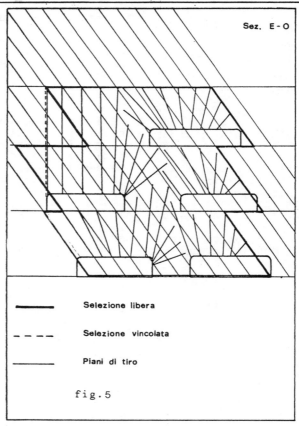

Sez. E - O

———— Selezione libera

– – – – Selezione vincolata

———— Piani di tiro

fig. 5

primary crushing.

2.6 Technical data about the sub-level stoping method

2.6.1 Excavations for the haulage and service ramps

Length of ramp	m	584
Cross-section (3.50x3.50)	m^2	12 - 14
Gradient	%	13.50
Type of rock ($= 2.5 - 3$ t/m^3)		Schists Porphyry Carnubianites skarns
Standard length of volley	m	2.7
Average advance rate of volley	m	2.5
Drill-holes	n.	35-40 $\emptyset$ 38mm
Drilling rate	cm/min	80
Type of explosive		Nitrogelatine 1 cartridge $\emptyset$ 25x400 mm
Consumption of explosives	kg/m	23
Detonators		Electric - series 50 ms
Reinforcement where necessary		Hardened and tempered st. steel ribs T.Heintzmann 16.5 kg/m
Concrete bed		15 cm thick

2.6.2 Excavations for the chutes Carried out by a Subterranean Raise-Borer 009/L

Type of rock	Porphyry,carnubia nite, skarn
Size	$\emptyset$ 2.40 m
Length of chutes	130 x 2 m
Inclination	90°
Incidence	0.013 cm/t

2.6.3 Development works (10 N-S sublevels)

Type of rock	Skarn
Vertical distance	12 m
Cross-section	Varies according to thickness of deposit
Length	250-300 m
Incidence	0.125 cm/t

2.6.4 Blasting

2.6.4.1 Drilling

The drilling of the vertical mine drill-holes will be carried out with an electric-hydraulic fandrill.

Technical data:

Average thickness of deposit	10 m
N° of drill-holes (along two rows)	9
Total drilled length	90 m
Spacing at bottom of hole	3 m
Distance between firing plane planes	1.5 m
Length of holes	10.0 m
Blasted volume (along 2 rows)	300 m^3
Tons produced	900 t
Mine yield	10 t/m
Diameter of holes	64 mm
Rods (I" 1/2)	1830 mm
Button bit	100cm/min
Drilling rate	100 cm/min
Drilling yield	75 m/shift

2.6.4.2 Mine charge

Unit consumption	250 g/t

2.6.4.3 Mucking at the base of the slot

A TAMROCK Toro 350 D scoop tram will be employed having a 3.5 m^3 bucket.

Maximum transport distance	150 m
Average distance	80 m
Slope gradient	0%
Average yield	120 t/h
Yield per shift	600 t

2.6.4.4 Haulage

Contractors will take care of hauling the material from the bottom of the haulage ramp by using 12 ton dumpers up to the stocks at the top of the plant.

2.7 Main underground machines and equipment

TAMROCK TORO scoop-tram with a 3.5 front bucket	1
TAMROCK TORO scoop-tram with a 2.00 m^3 bucket	3
Two-arm BBUD 141 Jumbo Alimak with pneumatic E 400 TAMROCK hammers. 3200 mm long 1" drilling bits	1

Two-arm TAMROCK MINIMATIC Jumbo
with pneumatic E 400 TAMROCK
hammers. 3200 mm long 1" drilling bits   2

ATLAS COPCO SIMBA H221 fan drill 1

GHH MK A-12 Dumper: 8 m^3 - 15 t 1

250 Atlas Copco Diamec probe
powered by a hydraulic engine   2

Tyred vehicle for transporting
personnel   2

2.8 Mining costs for sub-level stoping

BLASTING: Drilling	1284	Lit/t
Mine charging	650	Lit/t
Mucking	1500	Lit/t
	3434	Lit/t
HAULAGE: (Contractor)	1150	Lit/t
MINE SERVICES	2090	Lit/t
DEVELOPMENT WORKS	1950	Lit/t
TREATMENT & TAILINGS DISPOSAL	10809	Lit/t
TRANSPORTATION OF CONCENTRATES	1031	Lit/t
TECHNICAL AND ADMINISTRATIVE SERVICES	4000	Lit/t
TOTAL	24464	Lit/t

3. PROBLEMS REGARDING THE DILUTING OF THE TOUT VENANT

Excavation of the sub-levels below
el. +580 (4th level), provided evidence as to the irregular geometry
of the deposit, and furthermore,
evidence of weakness of the footwall
was noticed due to the presence of
mylonized faults and to alterations
in the porphyry. Such geomechanical
difficulties are not important to
the extent that they jeopardize the
feasibility of adopting sub-level
stoping, but quite certainly they
do suggest that the dilution of the
tout venant is greater than had been
expected.

Indeed, the evaluation of the
tonnage and of the average grade was
made by selecting krigged blocks
whose geometry was made to be compact and stable as required by sublevel stoping, which, as compared
with the grades obtained by unconstrained selection of the krigged

blocks, provides a dilution effect
of the order of magnitude of about
12% (unconstrained selection: 4% Pb+
Zn+Cu; constrained selection: 3.5%
Pb+Zn+Cu). Any detachments from the
footwall would therefore further
worsen the grades down to estimated
values of about 10%.

This new situation, in a multi-
metal sulphide deposit of considerable value, even if the grades are
not high, raised the issue of examining, all over again, the optimization
of the production cycle in the light
of the cut and fill mining method,
which is undoubtedly more selective

Besides regulating the grade during mining, this method is better
also for the tailings disposal, which
as pointed out previously, must be
carried to el. 400 m above the treatment plant with considerable technical problems.

4. DESIGN HYPOTHESIS FOR CUT AND FILL MINING

As the mineralized skarn does not
present stability problems, and considering that the service ramp has
already been built for the whole panel between el. +400 and +532, cut
and fill mining was chosen where
each slice is 3.5 m high and the
width varies according to the thickness of the deposit.

A protection slab will be left between el. +400 and +414, to act as
support for the fill and to allow
the development works for the underlying part to be carried out. Access
to the deposit will be provided by
means of adits or cross-cuts starting from the service ramp and having
variable gradients. The deposit
will be divided into two equivalent
parts so as to provide a greater
number of blasting faces. The acce
adits mentioned above will serve the
two halves baricentrically.

In order to attain the expected
output (300,000 t/year) stoping will
have to be carried out on two diffe-
ent levels, blasting in one, and
filling in the other. The first cu
in the upper panel will be filled
with a high strength fill so as to
avoid instability and dilution when
the diaphragm between the two panel
will be cut.

For blasting, horizontal mines
will be used, having a diameter of
48 mm, 4.5 m long and charged pneu-

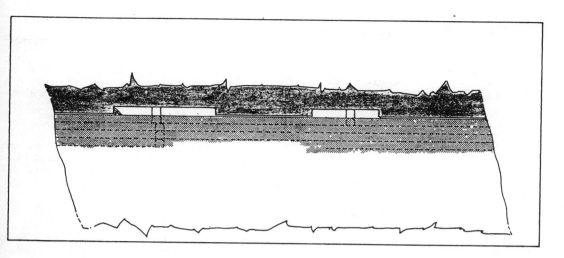

fig.6a

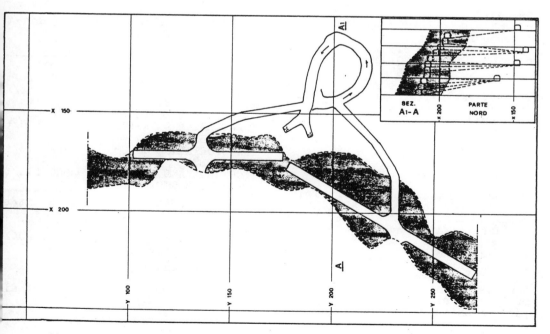

fig.6b

matically with AN-FO.

A 3.5 m front bucket will be used for mucking. Flotation tailings will be pumped as they come from the plant and used as fill.

Haulage will be carried out by the existing structure which had been designed for sub-level stoping(Fig.6a and 6b).

4.1 Technical data on cut and fill mining

4.1.1 Development work

4.1.1.1 Layout of the base tunnel

Type of rock:	skarn
Cross-section:	varies according to thickness of deposit
Length:	250-300 m

4.1.1.2 Layout of adit tunnels

	1st adit	Blasting at back
Length	50 m	50 m
Cross-section	3.5x3.5 m	3.5x3.5
Gradient	0% - 10%	0%-10%
Type of rock	schists, porphyry cornubianites	
Standard length of volley	2.7 m	2.7 m
Advance rate per volley	2.5 m	2.7 m
N° of holes	30-40 ∅ 38 mm	min 10 max 20 ∅ 38
Type of explosive	nitro-gelatine 1	
Amount of explosive used	23 kg/m	10 kg/m

4.1.2 Blasting

4.1.2.1 Drilling

Average thickness of deposit	9 m
N° holes per volley	32
Total drilled length	144 m
Drilling grid (metres)	0.8x1.2
Length of holes	4.5 m
Blasted volume	142 m^3
Tons of output	425 t
Mine yield(t/m of mine)	2.95 t
Diameter of holes	48 mm
Drilling yield (m of mine per man/shift)	220 m

4.1.2.2 Mine charge

AN-FO (loose) will be used; pneumatic loading.
Specific consumption of expl.400g/t
Charged mine yield (m/man-day) 200.

4.1.2.3 Mucking

A 3.5 m^3 front bucket will be normally used.

Transportation distance	150 m
Average yield	90 t/h
Yield/shift (t/man-shift)	450 t

4.1.2.4 Haulage

Contractors will take care of haulage. The material will be carried from the foot of the haulage ramp to the top of the treatment plant by means of 12 t dumpers.

4.1.2.5 Filling

Tailings as they come from the treatment plant are to be used with the addition of some cement (2.5% in weight).

Pumping density is to be 1.75 t/m
The envisaged flow is 36 m^3/h.
Curing is expected to take 4 days where the adopted compression strength is 5 kg/cm^2.

4.2 Cut and fill mining costs

BLASTING:	Drilling	1960	Lit/t
	Mine charging	1140	"
	Mucking	1600	"
		4700	Lit/t
FILL		2270	Lit/t
HAULAGE		1150	Lit/t
MINE SERVICES		2090	Lit/t
DEVELOPMENT WORKS		1740	Lit/t
TREATMENT AND TAILINGS DISPOSAL		9809	Lit/t
TRANSPORTATION OF CONCENTRATES		1184	Lit/t
TECHNICAL AND ADMINISTRATIVE COSTS		4000	Lit/t
TOTAL		26943	Lit/t

5. LABORATORY TESTS ON THE HYDRAULIC FILL

5.1 Introduction

The aims of the tests were to obtain data as to:

- the optimal grain-sizes;
- the density;
- water seepage time;
- density in relation to amount of cement used;
- time needed for acquiring a determined compression strength in re-

CLASSE mm	TAL QUALE		SLIMATO A 0,45 mm		TAGLIO A 0,04 mm		TAGLIO A 0,08 mm	
	% parz.	% cum.	% parz.	% cum.	% parz.	% cum.	% parz.	% cum.
+0,315	–	–	–	–	–	–	0,37	0,37
+0,160	14,90	14,90	19,60	19,60	21,41	21,41	34,13	34,50
+0,080	23,43	38,33	31,94	51,54	40,50	61,91	57,37	91,87
+0,040	17,03	55,36	22,93	74,47	27,99	89,90	7,70	99,57
+0,040	44,64	–	25,53	–	10,10	–	0,43	–
TOTALE	100,00	–	100,00	–	100,00	–	100,00	–

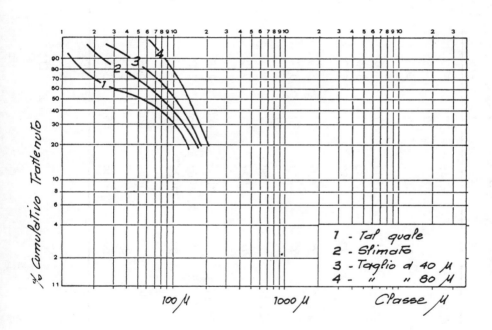

1 - Tal quale
2 - Slimato
3 - Taglio d 40 μ
4 - " " 80 μ

TABLE II

DENSITY kg/l	CEMENT (%)	SOLID (%)	COMPOSITION OF 1 m³ PULP Cement kg	Tailings kg	Water l	SEEPAGE WATER l/m³	TOTAL SEEPAGE TIME (min)	N° DAYS FOR ATTAINING COMPRESSION STRENGTH OF 2.5 kg/cm²	5.0 kg/cm²	10.0 kg/cm²	GRAIN SIZE (microns)
	0	48.61%	–	729	771	510	180	negative			as from plant
	5		36	693	771	465	123	7	9	11	
	10		72	657	771	460	113	3	5	7	
	2.5	48.61	18	711	771	466	23	2	4	6	+15 microns
	5.0		36	693	771	525	32	2	3	5	
1.50	10.0		72	657	771	517	34	1.5	2	4	
	5	48.61	36	693	771	525	11	1	1.5	2	+40 microns
	10		72	657	771	514	18	0.5	1.0	1.5	
	5	48.61	36	693	771	506	12	4	7	9	+80 microns
	10		72	657	771	515	20	false slab			
	0	62.30	–	1090	660	278	77	negative			as from plant
	2.5		27	1063	660	242	55	2	4	7	
	5.0		54	1036	660	226	46	1.5	3	6	
1.75	0	62.30	–	1090	660	334	23	3	4	negat.	+15 microns
	2.5		27	1063	660	319	20	1	2	6	
	5.0		54	1036	660	322	23	1	2	4	

lation to the machines used;
- volume of seepage water to be removed from the mine.

5.2 Preliminary operational conditions

The following preparation work was deemed necessary:

a. Preparation of the tailings required for the tests: the tout venant was sampled directly in the mine and it was prepared in the labora tory for reproducing – as close as possible – the grain-sizes as required by the new plant.

b. The accepted unit limit load was inferred from the maximum load of a 5 m^3 front loader (determined on the anterior axis, having load weight: 20,060 kg/cm , surface area of front wheels: 0.2 m^2) and that is 10 kg/cm ; this is the condition under load; unloaded it is about 5 kg/cm^2.
The strength tests were carried out in any case for three different values (2.5 – 5-10 kg/cm^2).

c. The cement amounts have always been referred to the total solid (cement + tailing); 10 and 5% values were tested during the first stage, and due to the findings of the latter, during the second stage 2.5% was adopted.

d. The chosen densities are those of the outcoming material from the plant, that is 1.5 kg/dm^3 and 1.75 kg/dm^3 .

e. The values of seepage time were determined through classical tests, but they can be used as comparative data; the real values are to be tested at a later stage with in situ jets. The same holds true for the sedimentation tests, both for the observations on the amount of seepage water, and for the residual humidity of the in situ jet, and for the difficulties in reproducing the in situ operational conditions (drainage medium, aeration, humidity, sedimentation surface in relation to the fill volume).

5.3 Performance of the tests

Four grain-sizes were chosen for testing; starting from the tailings as they come from the plant (base

testing, the other three grain-sizes are 15, 40 and 80 microns.
The grain-size analyses and their relevant curves are shown in Table 1.
It can be pointed out that while d_{20} (= d_{80} passing) of the various curves is included between a narrow grain-size range (between 130 and 180), d_{80} (= d_{20} passing) is included between 15 (as from plant) and 95, with intermediate situations. Each of these four products have been, in turn, investigated carefully.

5.3.1 Tailings as from plant (curve 1 in Tables I and II)

As can be seen, the amounts of cement used per m^3 in situ vary between 27 and 72 kg.
Seepage water having a 1.5 density is estimated to be between 44% and 51% of overall inflowing water per cubic metre; at a density of 1.75, instead, it varies between 34% and 42% of the overall water running in.
It must be pointed out that for a density of 1.50, the volume of water account for 77% of the overall pulp which drops down to 66% at a density of 1.75.
It is also important to point out that at density 1.5 the behaviour of the pulp undergoes quite visibly the laws of sedimentation, and so there is a grain-size selection and a certain separation of the cement which, together with the fines stratifies at the surface or, if there is no barrier or filtering diaphragm, it just flows away together with the surface water. In any case, there is a tendency to create a buffer of fines which tends to greatly slow down seepage. The consequence is eloquently evidenced by the time required to acquire the expected compression strength values.
Table III shows a diagram which describes this behaviour.

5.3.2 Materials selected according to grain-size

These preparations were obtained by cutting grain-sizes aroung the 15-40-80 micron classes (curves 2-3-4 in Table 1).
The amounts of cement ranged between 0, 2.5 and 10% at a density of 1.5, and given the results obtained, between 0, 2.5 and 5% at a den-

TAV. III

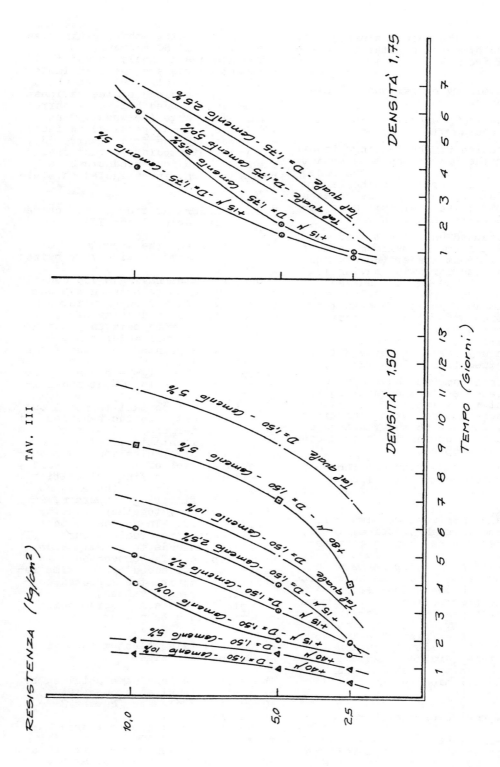

RESISTENZA (Kg/cm²)

TEMPO (Giorni)

DENSITÀ 1,75

DENSITÀ 1,50

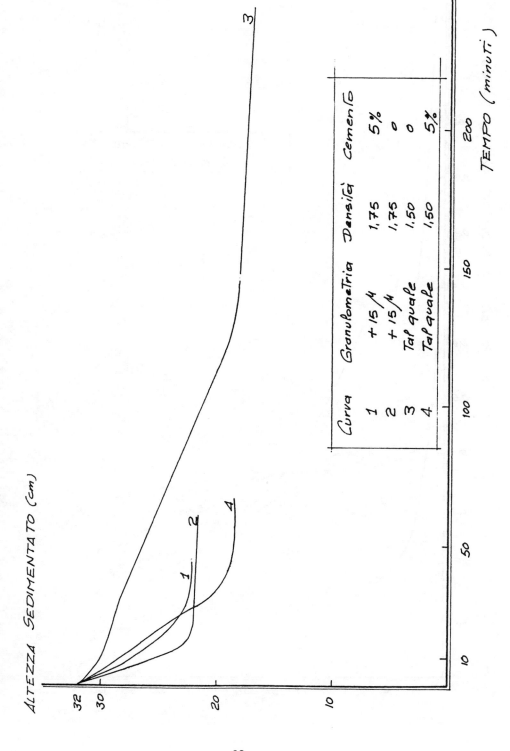

Comportamento della sedimentazione in diverse condizioni

TAV. IV

ALTEZZA SEDIMENTATO (cm)

Curva	Granulometria	Densità	Cemento
1	+ 15/4	1,75	5%
2	+ 15/4	1,75	0
3	Tal quale	1,50	0
4	Tal quale	1,50	5%

TEMPO (minuti)

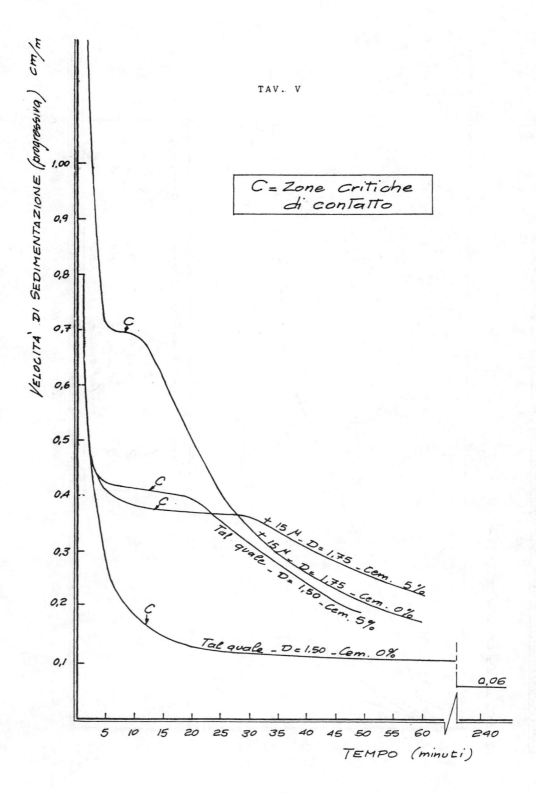

TAV. V

C = Zone critiche di contatto

+ 15 M. - D = 1,75 - Cem. 5‰
+ 15 M. - D = 1,75 - Cem. 0‰
Tal quale - D = 1,50 - Cem. 5%
Tal quale - D = 1,50 - Cem. 0%

0,06

94

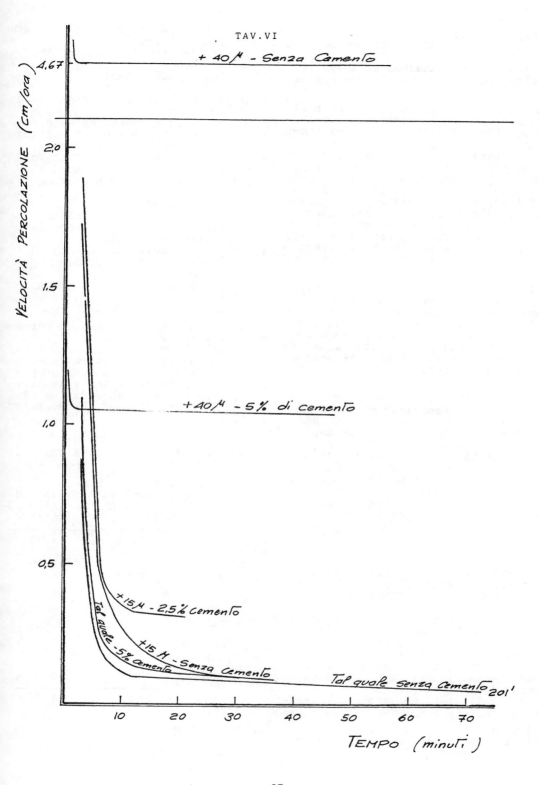

TAV. VI

95

sity of 1.75.

The overall results are summarized in Tables II and III.

From the examination of these results, the following remarks may be made:

a) In terms of grain size, the best results are to be attributed to the 40 μ size. Indeed, besides attaining highest strength over a very short interval (max. 2 days), they also present the most rapid seepage time. Also the +80 μ fraction shows a rapid seepage time, but the lack of fines does not allow good aggregation, unless the cement percentage is 10. The same occurs with the +40 size if only 2.5% of cement is used. Thus, there appears to be a direct relationship between grain size and cement.

b) From the standpoint of density, there is an inverse relationship between density and cement. When density is 1.5, grain-size classification occurs, and thus a segregation of the cement; after drainage the underlying material is all the more incoherent the smaller the amount of cement.

c) If the finest particles are eliminated, the behaviour of the sedimentation changes sizeably, as can be seen from the diagrams in Tables IV and V. The same thing is suggested by looking at the seepage time values (Table VI). The cement accelerates sedimentation (thus behaving as a flocculant), but it does not greatly affect the seepage (but at low densities is does favour a rapid removal of the water which gathers above the precipitated solid).

6. SELECTION OF THE TYPE OF FILL AMONGST THE VARIOUS ALTERNATIVES TESTED IN THE LABORATORY

By assuming that by using 3.5 m^3 buckets the strength of the fill does not need to exceed 5 kg/cm^2 in order to allow it to be moved, all the assumptions for which this value was reached in over ten days were discarded.

The following parameters were analysed for the remaining alternatives:

1) depletion time of a cut
2) time required for developing the fill site
3) time required to reach a strength of 5 kg/cm^2 (curing time)
4) available time for jetting the fill
5) drainage coefficient
6) amount of pulp required
7) amount of sand to be transferred underground
8) correlation between the sand required and the output of tailing from the treatment plant
9) operation time of the plant for producing the required sand
10) hours required for executing the fill
11) stocking needs
12) rate of supply to the fill
13) degree of self-sufficiency in providing sand fill
14) degree of self-sufficiency in providing sand for setting up the dam
15) presence or absence of preventive cycloning
16) amount of cement used.

Table VII summarizes and compares these parameters. A close look at this Table shows that the solution which better fits the operational needs of this mine is a pulp as produced by the plant having a density of 1.75 kg/dm with an addition of cement equal to 2.5% of the overall solid weight.

7. ECONOMIC ANALYSIS OF THE TWO METHODS

7.1 Production costs

	Sub-level stoping Lit/t	Cut and fill Lit/t
BLASTING	3434	4700
FILL	---	2270
HAULAGE	1150	1150
MINE SERVICES	2090	2090
DEVELOPMENT WORKS	1950	1740
TREATMENT	9029	9029
DISPOSAL OF TAILINGS	1780	780
TRANSPORTATION OF CONCENTRATES	1031	1184
TECHNICAL AND ADM. SERVICES	4000	4000
TOTAL	24464	26943

7.2 Investments

The initial estimated investment

	Cement as % weight of solid	Site preparation (days)	Fill curing (days)	5 kg/cm² Jetting total fill	Jetting fill (work days underground)	Total	m³ left after drainage / jetted m³	m³ jetted / cut	Solid content in pulp	Solid jetted/solid produced per cut	Solid jetted/solid produced (%)	Solid to be produced by plant to fill one cut	Hours of plant operation to produce necessary solid	Hours to fill 1 cut working at 3 t/d and 8 h/shift	Hours of plant operation to produce stock materials	Solid outcome of operation hours for producing stock t	Solid used as stock (t)	Rate of supply to fill m³/h	Sand required (t/y x 1000)	Sand that can be produced (t/y x 1000)	Self-production of sand is sufficient	Sand for dam sufficient	No stock	No cycloning	Cement consumption in ranking order
	5	6	9	13	28	10	0.535	15264	10578	100	10578	274	240	34	1312	1312	64	113	280			X	X	
	10	6	5	17	28	13	0.540	15122	9935	100	9935	257	312	-	-	-	59	106	280	X	X	X	X	5
+15μ	2.5	6	4	18	28	14	0.534	15292	10873	80	13591	352	336	16	617	494	46	116	224	X	X	-	-	1
	5	6	3	19	28	15	0.475	17192	11914	80	14893	386	360	26	1003	802	48	128	224	X	X	-	-	3
	10	6	2	20	28	15	0.483	16907	11108	80	13885	360	360	-	-	-	47	119	224	X	X	X	-	5
+40μ	5	6	2	20	28	15	0.475	17192	11914	55	21662	561	360	201	7755	4265	48	128	154	X	-	-	-	3
	10	6	1	21	28	16	0.486	16802	11039	55	20071	520	384	136	5247	2886	44	118	154	X	-	-	-	5
+80μ	5	6	7	15	28	11	0.494	16530	11455	40	28638	742	264	478	18443	7377	63	123	112	-	-	-	-	3
	2.5	6	4	18	28	14	0.758	10773	11452	100	11452	297	336	-	-	-	36	123	280	X	X	X	X	2
	5	6	3	19	28	15	0.774	10550	10930	100	10930	283	360	-	-	-	37	117	280	X	X	X	X	4
+15μ	0	6	4	18	28	14	0.666	12261	13364	80	16705	433	336	97	3743	2998	36	143	224	X	X	-	-	0
	2.5	6	2	20	28	15	0.681	11991	12746	80	15933	413	360	53	2045	1636	33	137	224	X	X	-	-	2
	5	6	2	20	28	15	0.678	12044	12478	80	15598	404	360	44	1698	1358	33	134	224	X	X	-	-	4

δ = 1.5 (top group); δ = 1.75 (bottom group). Calendar days/cut 24,500 t.

T A B L E   VII

(during the design stage) for the fill equipment is roughly 1000 million Lit. to which 300 million Lit. are to be added every two years for further displacements in the subsoil.

By adopting the cut and fill method, about 100 million Lit. per year can be saved in relation to the construction of the tailings dam.

7.3 Proceeds

7.3.1 Value of the concentrates

Independently from the feeding grades to the treatment plant, constant values, taken from the Prices and Exchange rates of the 1982 SAMIM Budget, have been assumed, and that is:

Chalcopyrite: 27.5% Cu with 800 g/t of Ag   737250 Lit/t

Galene: 48% Pb with 3% Cu and 1200 g/t of Ag   743100 Lit/t

Blende: 47% Zn with 8.2 kg/t of Cd   345100 Lit/t.

7.3.2 Production of concentrates

	Grades of the tout venant (%)			Output of tout venant t/year	Treatment yield (%)			Production of concentrates t/year		
	Cu	Pb	Zn		Cu	Pb	Zn	Cu	Pb	Zn
Sub-level stoping	0.72	0.54	1.89	300000	70.0	66.0	74.5	5498	2228	8988
Cut and fill	0.80	0.60	2.10	300000	72.5	68.0	77.0	6327	2550	10321

7.3.3 Proceeds

Sub-level stoping   8811 million Lit. per year

Cut and fill   10122 million Lit. per year.

8. DIFFERENTIAL CASH FLOW BETWEEN CUT AND FILL AND SUB-LEVEL STOPING

(Years)	1	2	3	4	5	6	7	8	9	10
INVESTMENTS:										
Fill equipment	(1000)		300		300		300		300	
Savings on dam construction		100	100	100	100	100	100	100	100	100
Running costs		(744)	(744)	(744)	(744)	(744)	(744)	(744)	(744)	(744)
Proceeds		1311	1311	1311	1311	1311	1311	1311	1311	1311
Cash-flow	(1000)	667	367	667	367	667	367	667	367	667

DCF - IRR = 53.88%

NPV i=5% = 2798 million Lit

Pay-back = 1.9 years

9. CONCLUSIONS

As shown in the foregoing, the cost of cut and fill mining is clearly higher as compared with sub-level stoping, but it does avoid diluting the output by detachments at the footwall, and thus difference in yield generously offsets the greater costs involved in the investments.

It is obvious that early estimates made for sub-level stoping did not take into account any dilution due to detachments, but only the dilution caused by the blasting geometry, and thus they pointed to values identical to those provided by cut and fill mining, and consequently at that stage it did prove to be the more convenient method of the two.

Proceedings of the International Symposium on Mining with Backfill / Luleå / 7-9 June 1983

Back filling operation at I.C.C. group of mines

D.BISWAS
Sand Stowing, H.C.L., I.C.C., India

ABSTRACT : This paper briefly deals with planning and operational aspects of backfill at I.C.C. group of mines. Starting with the selection of fill material, its processing, transportation & placement in voids, with particular reference to its strength, development, after it is poured.

1 INTRODUCTION

1.1 Production of copper ore in I.C.C. group of mines averages about 1.2 million tonnes per annum from five mines. Thus every year 0.4 million cubic metres of void must be filled for ground stability and for providing a suitable working platform in cut and fill stopes. Most of the areas are generally filled by classified mill tailings, although limited quantity of waste rock & river sand have been also used in some parts of the mines.

1.2 Waste rock from the development ends is transferred by mine cars and directly dumped into the worked-out stopes. This process is more of a waste disposal than ground stability. River sand and deslimed mill tailings are reticulated underground hydraulically for filling the voids. River sand is mainly used in some shallow mines since adequate quantity of mill tailings is not available to stabilise old worked-out areas.

2 HISTORY

2.1 Indian Copper Complex group of mines, a Unit of Hindustan Copper Limited are operating in the Singhbhum copper belt which is India's most important source of copper metal both in ancient times and during the present century. Mosaboni mine, located along the south-eastern flank of this belt, has been in operation for about 50 years & has reached a vertical depth of about 1200 metres from

surface; the other mines are comparatively shallow.

2.2 Narrow rich veins of copper ore, delineated at 1.8% Cu pay limit were mined out at a comparatively high run of mine grade. Mining was mostly carried out by open breast stoping with timber support. The need for a change in mining method and a fresh look on all the aspects of ground control was felt when the mining operations moved in depth and wider ore bodies were established at 1.0% Cu pay limit.

2.3 In 1966 hydraulic back filling was introduced by means of river sand. Modified breast stopes were filled by river sand against timber barricades lined with bamboo mats. This could not resist the strata movement efficiently, because of high rate of compression and nonconfinement of the fill material. A turning point in the history of mining at I.C.C. was the establishment of milling facilities at the pit head and availability of mill tailings as the fill material in 1973.

3 MINING METHODS

3.1 The copper lodes occur as massive sulphides, stringers and disseminations in quartz-chlorite schists, quartz-biotite-sericite schists and chloritic quartzites. The shear zone dipping at 25° to 40° and the width varies from 1.5m to 30m.

3.2 After considerable thinking and series of experiments, three stoping methods are adopted :   (Fig. I, II, III)

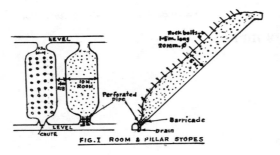

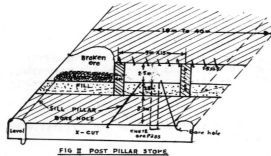

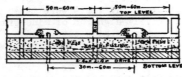

FIG.I ROOM & PILLAR STOPES

FIG II POST PILLAR STOPE

FIG. III HORIZONTAL CUT & FILL STOPE

i) Post Pillar Method
ii) Horizontal Cut & Fill Method
iii) Room & Pillar Method

3.3 For immediate support, timbers are replaced by means of grouted type rock-bolts and for mass support by complete filling of the worked-out stopes with river sand and deslimed mill tailings. In case of Post Pillar and Horizontal cut and fill stopes, the filling progresses vertically up-wards whereas post-filling is done in Room and Pillar stopes. The above mentioned stoping methods gave better safety, increased productivity and higher recovery of ore than the conventional breast stopes.

4 MILL TAILINGS AS FILL MATERIAL

4.1 A small fraction (5% by weight) of economic mineral in the form of copper sulphide is being recovered after crushing, grinding and concentration of copper ore. The remaining rock mass having a wide range of size distribution, is used for back-filling. Invariably the total tailing obtained from the concentrating plant,

undergoes certain processing before it is used as fill material. An important aspect of the fill processing is maintenance of a suitable relation between the fill quantity and quality, the one in general varying inversely with the other.

4.2 Relevant data was collected to determine the suitability of our mill tailing for use as backfill material. The underflow and overflow samples of mill tailings from hydrocyclones were collected and subjected to size analysis.

Size analysis fractions

Mesh size	Mill tailing	Underflow	Overflow
+ 45	0.6	5.7	0.0
+ 100	22.6	42.2	1.6
+ 200	34.6	30.5	16.8
+ 325	18.7	11.4	29.2
- 325	23.5	10.2	52.4

4.3 The underflow has fairly a large preparation of coarser fraction and the overflow contains mostly slime and bulk of the water. Further tests were carried out with the underflow fraction.

5 MINEROLOGICAL EXAMINATION

5.1 Minerologically these samples show that quarts, chlorite and biotite are the most dominant constituents, followed by tourmaline, muscovites, epidote & magnetite.

Mesh size	+45	-45 to +100	-100 to +120	-120 to +200
Cuboidal	11	12	21	24
Tubular	17	36	29	18
Platy	69	45	34	39
Triangular	1	6	11	16
Cylindrical	2	0	5	3

5.2 Shape analysis by and large has revealed that the "platy" particles mostly composed of micas and chlorites have the major share. "Tubular" form is contributed by feldspars and epidotes, while 'cuboidal' form is contributed by quartz, tourmaline, magnetite, apatite, sulphides and ilmenite. The 'cylindrical' forms are exclusively exhibited by acicular tourmaline & epidotes. The 'triangular' particles could not furnish any specific mineral composition.

5.3 The platy minerals which constitute bulk of the under flow tailings have a tendency to get settled with the longer axis parallel to the slope of the floor.

Hence water in the slurry moves relatively more freely in lateral direction rather than vertical direction after it is poured.

6 PERCOLATION TEST

With 10 percent (in an average) of -325 mesh fraction in the under flow, the percolation rate of 15cm per hour is obtained, it has good settling property.

7 RECOVERY OF FILL MATERIAL

7.1 With a single stage hydrocyclone, about 50 percent by weight of ore milled is recovered as fill material. Rubber lined hydrocyclones with the following specifications are used :

1) Cyclone diameter   – 38.1cm
2) Vortex finder   – 12.7cm
3) Spigot apex opening   –   5.08cm
4) Feed inlet diameter   – 10.16cm
5) Feed pressure   –   0.727 k/cm^2
6) Feed density   – 35 percent
                                 of solids by
                                 wt.

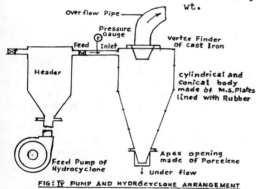

FIG: IV PUMP AND HYDROCYCLONE ARRANGEMENT

7.2 The underflow of the hydrocyclones has a pulp density of 70% by weight, after proper mixing with water, it is taken down

the mine through bore holes by gravity action with solid concentration of 60% by weight. The overflow (slime) is taken to the dam area, by means of 20cm dia seamless steel pipes.

(Fig. V)

7.3 Deslimed tailing is also collected in paddocks near the fill plants. Here the water is drained out and the dry tailings loaded on to the ore carrying dumpers by means of three drum slushers and transported to the mine head bunkers during their return trips, for back filling.

(Fig. VI)

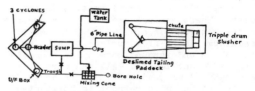

FIG: VI SURFACE GATHERING AND TRANSPORTATION ARRANGEMENT FROM PADDOCK

8 FILL TRANSFER

8.1 Efficient fill operation, with its major cost aspect, depends much in its reticulation system. The size of pipe and full bore stowing with minimum possible wear-effect is most desirable. In our system, the hydraulic fill material is transported down the mine through a combination of bore holes and pipes.

(Fig. VII)

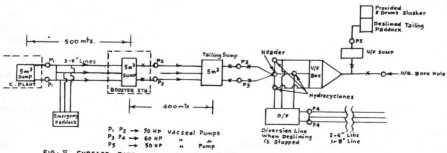

FIG: V SURFACE PIPE LINE FROM CONCENTRATING PLANT TO DESLIMING STATION

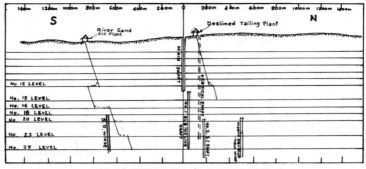

FIG: VIII RETICULATION OF MOSABONI MINES

8.2 The surface bore holes are drilled near the fill plants/bunkers and are provided with casing up to the unconsolidated ground, till hard rock formation is met with. These casings are regularly changed after running for about 38,000 tonnes of tailing. Apart from surface bore holes, there are underground bore holes, connecting different levels. All the bore holes are "NX" (7.5cm) size drilled at an angle of 40° to 70° with respect to horizontal. Replaceable attachments are provided to connect the bore holes with pipes. Three types of pipes are mainly used in our fill reticulation.

8.3 Galvanised iron pipe

Two sizes, 80mm and 100mm I.D. pipes are in use. The 80mm pipes with welded flanges have service life of 50,000 tonnes of tailing flow. 100mm pipes have slightly increased life than that of 80mm pipes.

8.4 High density polyethelene pipes

110mm O.D. pipes, with a permissible working pressure of 10 kg/cm^2, are extensively used in shallow mines. These have service life of about four times that of galvanised iron pipes.

8.5 Seamless steel tubes

In deep mines we have started using 114mm O.D. with 6.35mm wall thickness flange welded pipes. Service life of these pipes are expected to be about 0.5 million tonnes of mill tailings.

9 COMMUNICATION

The reticulation system can be compared with the arteries of the human body whereas the communication system is like the nervous system. Communication is essential for timely signalling between the feed and discharge points. Portable telephones are used extensively by the stowing crew for efficient communication with the surface fill plant operators. Five core telephone cables are laid all along the reticulation, having number of joint boxes to connect the branch lines. The main telephone cable is also connected to surface and underground exchanges.

10 PLACEMENT OF FILL MATERIAL

The bottom openings of worked-out stopes are first sealed by means of bulk-heads or barricades made up of masonry walls, to act as retainers of fill. The walls are designed to allow the stowing water to drain out quickly.

In case of "Room & Pillar" stopes, the only opening (2.4 x 1.8m) at the bottom, is closed by means of 25cm concrete walls, provided with three perforated pipes. These perforated pipes are wrapped with wire-net and jute barlap to act as mouse-trap.

In cut and fill stopes, 2.4m high cuts are taken and the backfilling is done, thus gradually extending the stope vertically upwards from the sill pillar. The panels are divided by means of timber bulk-head covered with bamboo mats and hessian cloth. A number of perforated pipes wrapped with wire-net and jute barlap are grouted on to the small bore holes drilled from the sill pillar to the level below. After each cut, before the filling operation, perforated pipes are extended vertically and after every 2-3 cuts in horizontal direction as well as for the quick removal of decanted and also percolated water from the hydraulic fill.

11 STRENGTH DEVELOPMENT

11.1 Primarily, strength development of the fill material for wet fill, depends upon its confinement and removal of water, after it is poured. This phenomenon of strength development takes place in two distinct stages, primary or short term and long term strength.

11.2 Short term strength

It is the period, for men and machinery to move over the fill after it is poured in the stope. Generally, the mining cycle is dependent on the rate of primary strength development. In addition to the facility of vehicular movement, it strengthens the crown/rib pillar and resists penetration of blasted ore in the fill. At this stage the strength development is mainly due to mechanical friction and the mutual interlocking of the adjacent particles, which are purely physical mechanism. In order to assess the primary strength of our fill material, a number of plate bearing tests at various stages of drainage, age and confinement were conducted over the actual fill inside the mine. (**Fig. VIII A,B**)

Age of fill	5 days old	3 Months old
Bearing capacity	4.09 kg/Cm^2	5.6 kg/Cm^2

All the machineries have been so selected that the mining cycle can be repeated after 5 days of the fill, in cut & fill stopes.

Another experiment was conducted to know the pillar deformation in cut and fill stope, by means of straingauge bore-hole extensometer with remote readout facility. The extensometer was installed in a horizontal hole of 33mm diameter and 2.5 metre long in one of the post pillars, when 3rd cut was in progress. The readings were taken periodically up to 4th fill and was plotted in a graph. Close observation of the graph revealed that the pillar deformed considerably while the cut was in progress, but became fairly stabilised after the fill, perhaps due to the confining action of the fill material.

(Fig. IX)

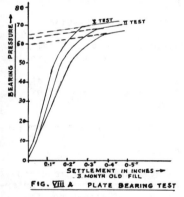

FIG. VIII A   PLATE BEARING TEST

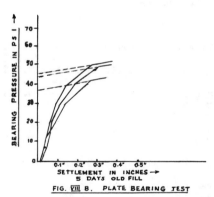

FIG. VIII B.   PLATE BEARING TEST

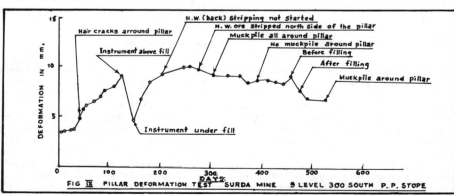

FIG IX   PILLAR DEFORMATION TEST   SURDA MINE   9 LEVEL 300 SOUTH P.P. STOPE

11.3 Long term strength

Long term strength develops progressively after the completion of the mining cycles and provides ground support for the whole life of the mine. Apart from physical mechanism, cementation within the fill by the process of oxidation of certain active minerals add to long term strength. Physical mechanism in the form of static load of the strata pressure and superincumbent fill material, increase the strength very slowly with time, provided the individual grains are not crumbled due to such pressure. In cut and fill stopes the constant vibration caused by the movement of the machineries over the fill, improves the compaction and hence increases support capacity of the fill. Oxidation within the fill material improves its cohesion which ultimately helps in the development of long term strength. At the same time, some of the minerals present in the fill material may be leached, creating unnecessary gaps in the fill. Such minerals should be identified and removed during the fill preparation.

Samples collected from our old fills were chemically analysed.

Chemical analysis

1)	Loss of ignition	–	1.17%
2)	$Si O_2$	–	65.59%
3)	R_2O_3 (Fe, AL, Ti, Mn & P)	–	22.24%
4)	$Cu O$	–	3.80%
5)	$Mg O$	–	4.00%
6)	Na_2O	–	1.38%
7)	$K_2 O$	–	1.41%
8)	S	–	0.34%
9)	Cu	–	0.07%

Detailed analysis of No. 3 fraction :

$Fe_2 O_3$	–	13.40
$P_2 O_5$	–	1.26
$Mn O$	–	0.08
$Ti O_2$	–	0.75
$Al_2 O_3$	–	6.75
		22.24%

After careful study of the chemical analysis it has been observed that $Fe_2 O_3$ proportion have increased to 13.40% in the old fills instead of 12.00% in fresh fill material. This gives an indication that oxidation does take place within the fill material and hence we can expect cohesion action of the fill particles. We have identified that minerals like apatite and magnetite are vulnerable to leaching in presence of acidic water.

Investigations are being carried out to determine the leaching action of these minerals.

12 ECONOMIC EVALUATION

It is extremely difficult to evaluate the economic aspect of the fill support system, as some hidden costs are involved in it. Basically the cost of fill support system can match, when the cost to place the fill material equals the installed costs of the support replaced.

Tailing filling in our mines has drastically reduced the timbering cost (both men and material) and improved rock handling capacity of the shafts. In addition, fill system can claim the following advantages, having some economic benefit.

1) Ability to work greater stoping widths with high rate of advance.

2) Higher percentage of ore recovery.

3) Concentrated stoping activities.

4) Ability to work at greater depths.

5) Possibility of mechanisation.

6) Application in case of geologically difficult or unusual ore bodies.

7) Improved and effective ventilation control (with proper control of humidity).

8) Increase in capacity of the surface tailing dam and reduced cost of tailing disposal.

With wet fill system, some extra costs are involved in pumping water from underground and cleaning of slime from the sumps.

13 CONCLUSION

Over a period of fifteen years hydraulic filling has been used as mass support for the ground control. It has been observed that deformation of the crown pillars has reduced to a great extent. To determine the exact pattern of ground movement and behaviour of the fill, investigations by means of bore-hole extensometers, convergence indicators and load cells are being carried out by our Rock Mechanics Engineers. It is

expected that its use will continue
throughout the life of our mines with
probable modification to allow the use of
cement additives and possible change in
mining methods at depths.

The recommended cemented fill material
will be 7% portland cement and 93% deslimed tailings; with 12.5% of boiler
ash and 6% portland cement by weight we
can achieve equivalent strength of 7%
portland cement i.e. 11 k/Cm^2 after 28
days. The introduction of cemented fill
can be considered only after a careful
analysis of the overall economics so that
the extra cost involved in cemented fill
is compensated by higher recovery of ore
and stable ground condition.

14 ACKNOWLEDGEMENT

The author thanks the Management of Indian
Copper Complex, a unit of Hindustan Copper
Limited for the permission to present this
paper. The valuable guidance given by
Sri M.A. Khan, Director (Operations),
Sri T.M. Chickabasaviah, Dy. General Man-
ager (Mines) and Sri P.A.K. Shettigar,
Mine Superintendent (Planning) while pre-
paring the paper is gratefully acknowledged.

Thanks are also due to all my colleagues
for the kind cooperation without which it
would have been difficult to present this
paper.

Reference

1. Back fill process - By A.L. Guise -
   could lead to im-   Brown
   proved strata   P.A.L. Hinds
   control

2. Fill support   - By S.J. Patchet
   system for deep
   level gold mines

3. Effect of back   - By G. Zahari
   fill on a stope
   wall

4. The important pro- By E.G. Thomas
   perties of Hydrau-
   lic fill with par-
   ticular reference
   to Mechanised Cut
   and fill mining
   operation

5. A Review of back   - By R.A. Ford
   fill mining
   practices and
   technology at
   the Fox Mine

6. "Appropriate   - By M.A. Khan
   Technology -
   Trend of Mecha-
   nisation in
   I.C.C."

7. Ground control   - By A.K. Singh
   at Mosaboni Mine

8. Petrological &   - By C.D.P. Singh
   Mineralogical
   observations of
   Mill tailings

9. Bearing capacity   - By D.K. Nag &
   of Mill tailings   D. Biswas

10. Experimental   - By S.V.S. Rao &
    studies on the   T.N. Gowd.
    deformation of a
    Post Pillar in a
    hydraulically
    filled Post
    Pillar Stope

Cut-and-Fill mining at Boliden, methods and economy

BENGT-OLOV CENTERVÄRN, BIRGER KOLSRUD & MATI SALLERT
Boliden Mineral AB, Sweden

ABSTRACT: The paper describes the mining activities at the Boliden Company and the application of various Cut-and-Fill mining (CFM) methods with regard to productivity and costs. Boliden's mining strategy is stressed, with special emphasis on ore recovery and waste dilution. Viewpoints are presented which indicate how Boliden intends to improve the CFM technique.

BOLIDEN

Boliden is a publicly quoted Swedish Company. It is the major Scandinavian industrial group as regards the mining of sulphide ores and the production of non-ferrous metals and heavy inorganic chemicals.

The activities of the Boliden Group also comprise contract management and consultancy in mining and other heavy industries, international trading in industrial goods, chemicals and oil products. The Group has a special organization for the international marketing of techniques applied at Boliden. During the last 50 years Boliden has developed and exploited more than 40 mines.

At present, Boliden operates 15 mines in Sweden. The size of the mines varies from those with a capacity of less than 100,000 tonnes of ore per annum to a large open pit mine, where 20 million tonnes of low grade copper ore and waste are removed annually. Figure 1.

The turnover of the Boliden Group in 1982 was about SEK 5500 million and the number of employees was 8000. The parent company of the Group is Boliden AB, Stockholm. Boliden Mineral is the mining company in the Group, and is a subsidiary of Boliden AB.

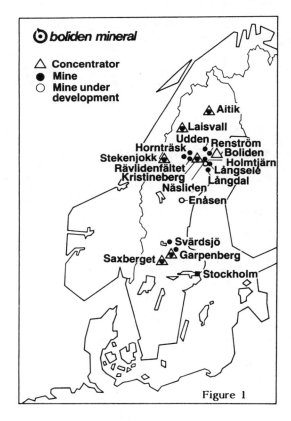

Figure 1

1. THE DEVELOPMENT OF CUT-AND-FILL MINING

Boliden's mining activities started in the Skellefteå area more than 50 years ago with the discovery of the Boliden ore-body.

The discovery resulted in further exploration in the area. The gradually expanding activities identified small ore-bodies containing very high grades of copper, zinc, lead, gold and silver forming the base for subsequent operations such as concentrating, smelting and refining.

The establishment of a system of central concentrators fed by ore from several small and closely located mines, improved the economy of those small mines previously unable to carry the burden of separate concentrating facilities.

High grade and small sizes characterised these mines and this forced an early development of mining methods which gave almost 100 percent ore recovery. This, in combination with poor rock stability, led to the application of CFM.

Furthermore the development of central research and development capabilities made it possible to concentrate research activities on areas which would improve the productivity and efficiency of CFM.

The technical development of CFM has passed through several stages:

- jack-leg and slushers

- mechanized drilling and pneumatic loaders

- pneumatic drill-jumbos and diesel LHD

- hydraulic drill-jumbos and electric LHD

At present Boliden has 15 operating mines and the mining methods used are open pit in one mine, room-and-pillar in one mine and CFM in the other mines. Although if the open pit and room-and-pillar methods represent the major part of ore production, the CFM method contributes most profit. Figure 2.

Production is compared with sales revenue for the different methods in the table below.

Method	Production Mtonnes per year	%	Sales revenue %
CFM	3.4	23	66
Room-and-pillar	1.5	10	12
Open pit	9.7	67	22

Three different CFM methods are used in Boliden. They are:

1. Horizontal cut-and-fill

2. Drift-and-fill

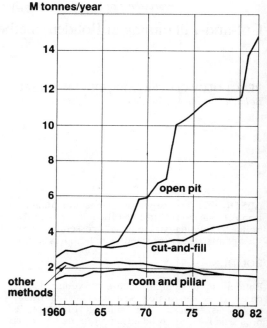

M tonnes/year

Figure 2 – Development of mining methods

3. Undercut-and-fill

Three million tonnes per year are produced using these three variations of the method and a break-down of these figures is shown below:

	Tonnes per year	%
Horizontal cut-and-fill	2,600,000	86
Drift-and-fill	260,000	9
Undercut-and-fill	150,000	5

The method chosen depends on the rock conditions. Figure 3.

Fill material consists of cyclonized tailings, natural sands and waste rock from development work. Cement is used as an additive depending on the method used.

	Kilo tonnes	%
Cyclonized tailings	1,100	64
Natural sands	360	21
Waste rock	260	15

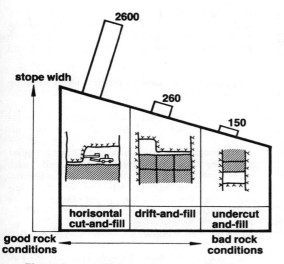

Figure 3 – production volumes k.tonnes/year

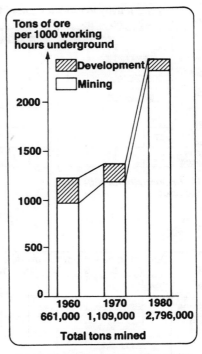

Figure 4 – Cut-and-fill productivity

The following productivity figures have been obtained for the different methods:

Productivity	Average hours per manshift	
Horizontal cut-and-fill	18	
Drift-and-fill	11	
Undercut-and-fill	9	Figure 4

Productivity and operating costs have, through extensive research work, been developed to a level comparable with other underground bulk mining methods. Figures 5, 6 and 7.

2. WHY CUT-AND-FILL MINING?

When studying international mining operations, we at Boliden often conclude our reports by saying that "if Boliden had been operating that mine, Boliden would have used CFM instead of the method used by the actual operator".

The same type of comments are made by foreigners when they visit Boliden's mines. "Why are Boliden using that CFM technique here, when another mining method would have in our opinion been better?"

Boliden's mining principles are, among other things, based on a few simple company policies:

- The mining operation shall generate

a profit at least of such magnitude that the minimum rate of return on invested capital stipulated by the company, will be obtained.

- Every tonne of ore mined shall normally have an ore value not lower than its specific gross operating costs.

- Ore is a non-renewable commodity. Hence, if possible every tonne of ore should be excavated, taken into consideration the two economic limitations presented above.

- If possible, the life of the mine for a Boliden operation should be not less than 15 years.

- The mining and milling cost shall be minimized per unit metal recovered and not per tonne ore.

With this policy, it's obvious that for underground mining of polymetallic ores, where both the beneficiation costs and the ore value are high, CFM is a method which accords well with the Boliden policy for the mining of mineral deposits. It should however be stressed that if other criteria

109

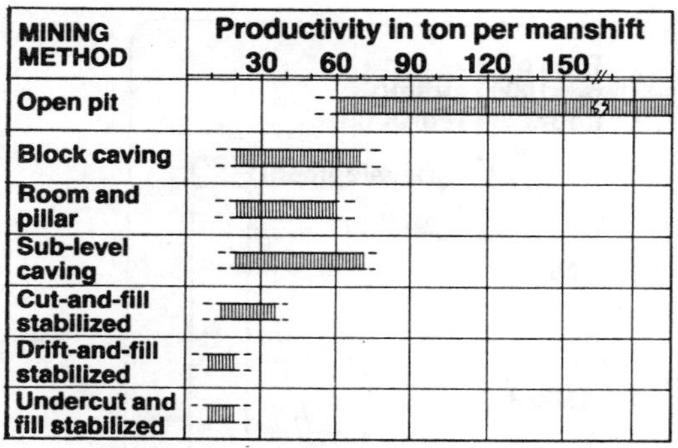

MINING METHOD	Productivity in ton per manshift				
	30	60	90	120	150
Open pit					
Block caving					
Room and pillar					
Sub-level caving					
Cut-and-fill stabilized					
Drift-and-fill stabilized					
Undercut and fill stabilized					

Figure 5 – Productivity for different mining methods

MINING METHOD	Operating cost $ per ton crude ore					
	3	6	9	12	15	18
Open pit						
Block caving						
Room and pillar						
Sub-level caving						
Cut-and-fill stabilized						
Drift-and-fill stabilized						
Undercut and fill stabilized						

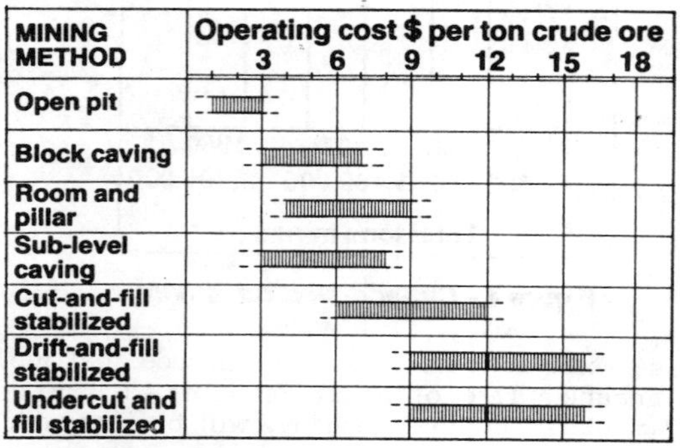

Figure 6 – Operating costs for different mining methods

MINING METHOD	Waste rock dilution into crude ore & ore losses					
	10%	20%	30%	40%	50%	60%
Open pit						
Block caving						
Room and pillar						
Sub-level caving						
Cut-and-fill stabilized						
Drift-and-fill stabilized						
Undercut and fill stabilized						

▭▭▭ Waste rock dilution into crude ore
▥▥▥ Ore losses

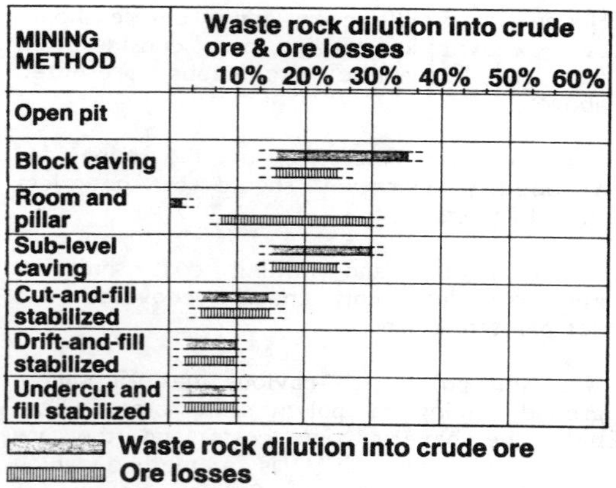

Figure 7 – Waste rock dilution and ore losses for different mining methods.

than Boliden's are used when planning a mine, this can then result in other mining methods being referred to for instance the CFM technique. Such criteria could be:

- optimum economic profit when mining a mineral deposit

- winning of all mineralized material in the deposit, whether it's ore or not

CFM methods do normally have higher cost per ton ore mined compared with other underground bulk-mining methods. Anyway, these higher costs are well compensated by the two big advantages of the CFM methods.

2.1 High ore recovery - low waste dilution

The economic consequences of these two factors have been well stressed in a paper by Professor G. Almgren, compiled for the Näsliden symposium here in Luleå in 1980.

In that paper, the technique for calculating the "net-present-value" (NPV) of an ore deposit, is presented. By varying the operating costs, the waste dilution factor and the ore recovery etc, of different mining methods, the effect on the NPV of the deposits can be studied.

The impact of waste dilution and of ore loss in an actual case, are calculated for a given scenario. Figure 8.

Assuming in this case that by changing the existing mining method to a new method where both the dilution and the ore loss percentages will be reduced from 20 % to 10 %, the operating costs for the new mining method can then be as much as 25 % higher than for the old method and still result in a higher NPV for the deposit.

It should be noted that:

- a 25 % increase in operating costs is a substantial increase for a new method

- a 10 % decrease in waste dilution is a small decrease when changing from a bulk-mining method to CFM. Possibly 20 % seems more realistic.

Beside the above mentioned two main advantages, some other advantages characterize the CFM method.

- The use of waste rock from development

110

work and of tailings for backfilling underground, results in reduced requirements for disposal areas on surface.

- Backfilling will consolidate the underground excavations, thus preventing subsidence and uncontrolled "cave ins" from surface.

- The mining equipment used is conventional tunnelling equipment.

- The amount of waste-rock to be hoisted is very limited compared with other mining methods, due to the fact that waste rock from development work to a large extent can be used directly for backfilling.

- In cases where the ore and host rock has low competence, CFM is the only mining method which can be applied.

- The method offers good selectivity between ore and waste when mining narrow veins or in ore-bodies of irregular shape.

- The method can be well adapted to different situations as regards rock mechanics.

The main disadvantages of the CFM methods can be summarized as follows:

- Other methods are less expensive, calculated per unit volume excavated underground.

- The methods consist of a series of linked operations, limiting the productivity calculated as volume excavated per man hour.

- The production capacity from each mining area is limited.

- Mining takes place all the time in stopes where only temporary roofs are established. Rockbolting later results in the problem of rockbolts in the broken ore when mining advances upwards.

Figure 8.   Scenario

Dilution and ore losses

Ore reserve underground, undiluted	12 Mt
Investments over three years	420 MSEK
Production per year	1 Mt
Concentrate per ton undiluted ore	0.2 t
Product price fob smelter	2,100 SEK/t of concentrate
Production costs fob smelter	245 SEK/t of ore
Annual inflation	15 %
Interest	20 %

Effect of waste rock dilution

Rock dilution	%	0	10	20	30
Material mined:					
ore	Mtpy	1	0.9	0.8	0.7
waste	Mtpy	0	0.1	0.2	0.3
Production period	Years	12	13	15	17
NPV* at start of investment	MSEK	896	678	420	109
NPV* difference from 0 % dilution	MSEK	0	- 218	- 476	- 787

Effect of ore losses

Loss of ore	%	0	10	20	30
Loss of ore	Mt	0	1.2	2.4	3.6
Ore mined	Mtpy	1	1	1	1
Production period	Years	12	11	10	8
NPV* at start of investment	MSEK	896	799	697	590
NPV* difference from 0 % of ore losses	MSEK	0	- 97	- 199	- 306

Combined effect of waste rock dilution and ore losses

Rock dilution	%	0	10	20	30
Ore loss	%	0	10	20	30
Material mined:					
ore	Mtpy	1	0.9	0.8	0.7
waste	Mtpy	0	0.1	0.2	0.3
Ore loss	Mtpy	0	0.1	0.2	0.3
Production stage	Years	12	12	12	12
NPV* total difference from 0 % ore losses and 0 % dilution	MSEK	0	- 295	- 592	-887

* before taxation

3. HOW DOES BOLIDEN INTEND TO IMPROVE THE CFM METHODS?

As previously mentioned, the productivity of CFM has increased substantially during recent decades. It is Boliden's opinion that this progress will also continue in the future.

These are new ideas and some of them are being tested today in order to improve this specific mining technique. These ideas can be summarized as follows:

- The studies carried out in the Näsliden project, have provided Boliden with valuable knowledge of the rock mechanics with regard to CFM. Boliden has today a new tool for optimal planning of the layout of mining, the sequence of mining and of reinforcement work, in relation to actual rock mechanics conditions.

- In order to reduce drainage costs in hydraulic backfilling and at the same time to shorten the time for the backfill operation, new ways of distributing the backfill material have to be developed.

- Methods of mining with backfill, but in unlinked sequences are being studied and tested.

- The ongoing R & D work on conventional tunnelling equipment, will benefit the CFM method.

- Applications of remote controlled equipment can result in a reduced number of unit operations in the CFM sequence.

- Defining the strength required in cemented backfill and then evaluating the optimal ratio between sand, cement, ballast-material and water.

Hydraulic filling – An effective way of ground control

B.B.DHAR, K.V.SHANKER & V.R.SASTRY
Banaras Hindu University, Varanasi, India

ABSTRACT: Development of safe mining practices have been one of the most important aspects that a practicing mining engineer often desires for. Therefore, for any successful mining operation the key is effective ground control so that the closure of walls is minimum. To keep the walls at its place and minimise its tendency of closure, the best way is to keep them apart as far as possible. Unit type of support is not successful for such purposes, however, mass support plays a vital role. The mass support therefore, can either be pillars or hydraulic fillings. To achieve an effective ground control, the hydraulic back-fill as mass support has been discussed here with special reference to a few Indian hard rock mining practices where an effective ground control has been achieved successfully even in one of the deepest metal mines of the world, where the depth of workings are over 10,000 feet.

1 INTRODUCTION

It is not always desirable to follow the caving methods of mining. In case of thick seams/veins, lying at great depths or at very shallow depths, closure of wall rocks and rock fracture may take place resulting displacements in the above sub-surface strata causing damage to surrounding rock structures and surface infrastructure, if the voids are not properly back filled. The stowing of refuse materials in mined out areas of underground mining operations provides a way to stablize the surface (Dhar, et al 1971 and Dhar and Khoda, 1972) and also eliminates environmental problems like air and land pollution caused by waste dumps. Stowing of waste material underground permits to recover ore pillars, thus ensuring higher recovery Hydraulic filling is one of the efficient methods of stowing and proved to be economical in many cases.

Classified mill tailings when placed hydraulically, fills all the irregularities in the walls of the stope, thus offering higher resi-stance compared to other types of fills. The use of bore holes, standard weight steel pipes and gravity transportation make it cheaper and effective. But it lacks cohesion and can not maintain verticality in the stopes without expensive timber support. If some cement is added, it becomes more consolidated offering good operating floor. Hydraulic filling has played a glamorous role in the introduction of cut-and-fill mining system with complete mechanisation,such as the application of rubber wheeled LHD's, a variety of drilling jumbos and reduced the overall cost of extraction of ore and made deeper mining economical, safer and efficient particularly while tackling deposits at greater depths.

2 PARAMETERS OF HYDRAULIC FILL

There are several factors which effects the efficiency of hydraulic filling. Some of the important parameters are briefly discussed here.

2.1 Cement tailings ratio

Knowledge of the variation of strength properties with cement content is an important factor for economical development of mining operations using hydraulic fill. The unconfined compressive strength increases with increase in cement content. A typical data of cemented fill is given in the Table. 1 (Espley, et al. 1970).

Table 1. Typical compressive strength of cemented backfill after 28 days of curing (Espley, et al. 1970).

Cement tailing ratio	Compressive strength (MPa)	Principal use
1:2	17.24	Structural
1:5	3.45-4.14	
1:6	2.76-3.45	
1:7	2.07-2.76	Flooring
1:8	1.38-2.07	
1:10	0.69-1.38	
1:20	0.48-0.69	
1:30	0.28-0.48	Back-filling
1:40	0.00-0.28	

From the table it can be noticed that wherever hard wearing qualities are required (such as for making floors) cement-tailing ratios should range from 1:5 to 1:10 and for back-filling, lower ratios may be generally used. The addition of portland cement improves the cohesive component of shear strength, provides tensile strength and increases stiffness of the fill.

2.2 Pulp density

It is a key factor in hydraulic filling as it affects the operations right from pulp preparation to the satisfactory strength development. Experience has shown that strength of the fill increases with increase in pulp density. The permeability to air and dewatering rate decreases with an increase in pulp density. Observations (McCready and Hall, 1966) have shown that pulp with 55%

solids by weight had an unconfine compressive strength of only 0.14 where as with 75% solids by weight it was 0.34 MPa, after 90 days of curing. Blockage of transportatio lines, wear of pipes and permeabil of the fill must also be taken int consideration for the selection of pulp density. A pulp density of 68% - 72% solids by weight is gene rally suggested to minimise the se regation and loss of fine cement during placement and dewatering.

2.3 Mixing time

For lower solid contents (60% - 75 mixing time is relatively unimport whereas for higher solid contents mixing becomes difficult and suffi- cient mixing time should be allowe

2.4 Effect of grain size

Grain size of the fill influences the density of the pulp, thereby affecting permeability of the cons lidated fill, jamming of pipes, we ing of pipe ranges.

2.5 Curing time

For cut-and-fill system where earl strength of fill is important, add tion of cement enables it to offer resistance early. In sections whe long-term strength is essential, lesser amount of cement may be use or some other cheaper ingradients may be used in place of cement. T deformation characteristics vary significantly with curing time. I has been observed that the strength gain over first twenty days is rela tively rapid and the rate decreases with time (Thomas, 1973).

2.6 Other factors

Factors like pH of water, curing temperature, flocculants etc. also affects the consolidation of fill. For example in mines where sulphide ore is being mined, oxidation will produce acidic water. If the pH value of the environment is consta ntly maintained below 6.0, the ceme bond will be bleached. The main damaging agents therefore are ammonium, aluminium sulphates, ammonium nitrate, hydrochloric, sulphuric and sulphurous acids, and other chloride salts. Similarly sulphate salts will attach the cement bond. Also the presence of limestone or dolomite particles in the fill will protect the cement bond. The water normally used should not be too acidic or too alkaline.

The curing temperatures do not seem to have much influence on the ultimate strength, although better strength has been reported for lower curing temperatures (Weaver and Lucha, 1970).

Use of flocculants improves the qualities of the fill and reduces the problem of accumulation of the fines in the drainage system. The addition of flocculants greatly reduces contamination of ore with cement and tailings, by keeping away slime and cement fractions in the fill, thereby reducing the problems due to cement in the milling process. The reductiin of slime on the surface of the fill reduces risk of injury due to slipping.

2.7 Economical factor

Portland cement is costlier (50% of the overall operating mining cost and depends on the availability and transporting distances). Now a days search is on to find cheap cementing materials locally available to make working of low grade ores economically feasible. Some such studies have suggested that materials obtained from fuel ash, copper reverberatory furnace slag, lead blast furnace slag, quenched copper converter slag may be considered as pozzolans (Thomas, 1971; 73; Weaver and Lucha, 1970). Pozzolans are silicious materials which by themselves possess no or little cementing value but which by reacting with calcium hydroxide at ordinary temperature forms compounds having cementing properties. Results have shown that these materials can be used in place of portland cement and considerable savings could be achieved. Table 2 gives the comparative compressive strength of pozzolans after 224 days of curing for difference combinations of portland cement and furnace slags.

Table 2. Properties of Pozzolans (after Thomas, 1973).

Material	Slag content (%Wt)	Port-land cement (% Wt)	Compressive strength after 224 days of curing, MPa
Copper reverberatory furnace slag	15	8	5.210
Lead blast furnace slag	12 16	4 8	2.361 9.571
Quenched Copper converter slag	12 16	4 8	1.360 3.559
Quenched dezincea gelby slag	12 16	4 8	2.721 8.892

It has been found that portland cement can be replaced by 20% fly ash or 50% ground slag for equal strengths. But for operational and safety reasons a minimum of 3% addition of portland cement is recommended. To give a cementing fill of 1-1.4 MPa strength, three combinations are advised; viz. (a) 5.5% portland cement, (b) 4.5% portland cement with 20% ground slag and (c) 3% portland cement with 60% ground slag. By using the last combination, the cementing cost may be reduced to half.

3 FILL BEHAVIOUR UNDER LOAD

The ultimate success of any filling mechanism is the reaction of fill under strata pressure or load. Significant work has been done in this area. The obervations for example made by Zahary et al.(1972), On fill pressure measurement in a cut-and-fill stope are that: (i) average strain in the fill is of the order of 2% at the end of two year period, (ii) wall convergence shows an increase with width of the stope, and (iii) the maximum pressure in the hydraulic fill in a narrow part of the stope is the order of 4.83 MPa, whereas in loose fill in a wide part of the stope, it is only 1.04 MPa.

Similar observations by Rawling et al. (1966), have reported that: (i) with the increase in moisture content the compressive strength of the cemented backfill decreases, (ii) with the increase in cement content permeability decreases, and (iii) permeability decreases as curing progresses with time.

An interesting observation by McCrudy and Hall (1966), revealed that the ultimate compressive strength of the cemented fill does not get much affected by the percolation of acid mine water through the fill.

The compressive strength of the fill increases with increase in cement content in general but largely depends upon the properties of sand used. Triaxial tests have shown that the strength increases with the increase in cement content. A sand to cement ratio of 40:1 (finer material) showed a higher cohesion than well graded material, but at the same time well graded material showed larger angle of internal friction, that is capable of supporting higher normal stress (Corson, 1970). Whenever additional strength is needed on the exposed floor of the fill to resist any machinery movement, it can be achieved by incorporating reinforcement in the form of protective cladding, or in the form of flexible plastic sheeting.

The general observations have shown that the capacity and stiffening of the fill will reduce the stress levels imposed on the other support media and thereby decreasing the stress build-up resulting in lesser incidence of rockbursts in pillar remnants. High quality fills obtained by compaction, cementation or by some other means, have the potential for reducing hazards and cost of mining through reduced stope wall closure and lower pillar stresses. The benefits of high quality fills would be especially significant for burst prone stopes when the stope nears completion, the time when the burst probability is the highest. High quality fills are more costlier. It is, therefore, significant to know, whether the

potential benefits could be realis and thus be able to predict the su port performance of hydraulic back fill in advance of the actual mini

The state of stress in a pillar a complex three dimensional proble According to Mohr-Coulomb criterio the failure occurs near the pillar face whenever the stress component acting normal to the vein equals t unconfined compressive strength of the rock. A basis for evaluating influence of the fill quality on pillar safety, therefore, is the reduction of the stress near the pillar face(or surface). The maxi mum value of stress acting normal to the vein always occurs here and the minimum value may occur near t core of the pillar which usually serves as a convenient reference. The danger of spalling and burstin of the pillar would appear to be the highest when the difference between stresses near the pillar face and core is maximum. Results have shown that improved fill qua lity not only reduces peak stress at the face, but it also promotes a more uniform distribution of str which favours yielding and gradual crushing in contrast to the spalli and bursting.

These results could be used as guidelines for evaluating the pote tial benefits of improved fill qua lity.

4 STOPING PRACTICES USING HYDRAUL FILL IN INDIA

In India hydraulic stowing was fir attempted in 1913 at the Ballapur colliery in the state of Madhya Pradesh and it soon picked up mome tum in Indian mines (Devaraj, 1980 Stability analysis of large underground excavations have shown the need of either reducing the size o the mine stopes or fill them up. This observation has been recently made by Dhar et al. (1982) while analysing one of the metal mines in India. At present due to the deeper and wider mining problems, hydraulic filling is being practic in most of the metal mines. Some the important filling practices followed in India are briefly discussed below for ready reference.

4.1 Hydraulic filling at Mysore mine (Kolar Gold Fields)

One of the deepest gold mines in the world is situated at Kolar Gold Fields of India where extraction goes to a depth of over 3 km. This depth has natural hazards and one such major hazard has been frequent rock brusts in the past. But with local knowhow, a modified cut-and-fill method known as Chatty stoping was developed to prevent wall closure and thereby reducing the occurrance of rock bursts. Figure 1 gives the general layout of this method (Shanker, 1980). This stoping method is basically dependent on hydraulic fill and solid granite rock blocks. This method is briefly discussed below.

The cut-and-fill method practiced in these mines consists of two stages, viz. (i) preparation of stopes and (ii) actual stoping. Stopes are developed for a suitable length of 21.0 m. Before stoping a total level of distance of 9.0 m. on either side is Chatty stopped. First preparatory stoping is done, in which the block of the stope is opened up and narrow slice is cut behind the arches above the lower level and another set of arches (baby arches) are provided over these arches with ore passes built at regular intervals. Chatty stope is built in the floor of the upper level to about 1.5 m. Then the operation of actual stoping starts. Generally filling is done after achieving an advance of 4.6 m. to 6.1 m. of adequate height of 3.7 m. in which 1.5 m. is left for mining operations and remaining 2.1 m. is filled. Mill tailings are used as hydraulic fill. The tailings of the mill feed consists of quartz and hornblende with a variable content of sulphide minerals which gets oxidized and cements the

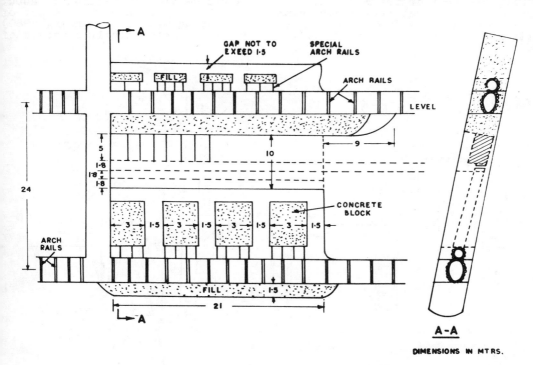

Fig. 1. Layout of Chatty stoping at Mysore mine, Kolar Gold Fields.

the sand, thus providing of firm fill without reducing the permeability qualities of the fill. The cyanide content in the fill is maintained below 0.0004%. Pulp density is about 65%. The rate of filling is about 30 tons per hour. Water from the fill is allowed to pass through the ore passes to the lower levels.

4.2 Jaduguda uranium mine

Hydraulic filling practice was started here in 1970. Previously shrinkage and open stoping methods were practiced. As the workings became deeper, strata control problems arose and finally they have adopted cut-and-fill stoping. Mill tailings from the surface plant are being used as hydraulic fill. Due to the excess requirement of the material, as the mill wastes amounts for 35% only, the rest of the material requirement is met from nearby mines. Pulp density is 50% – 60%. Pipe lines are of 75 mm diameter, G.I.

pipes, at the surface, in the boreholes and main levels. In stopes, PVC pipes of 70 mm diameter and heliflex pipes of 60 mm diameter are being used. Figure 2. shows the method of working with hydraulic filling. Figure 3. shows water drainage arrangement.

Excess water is drained out by syphons to the lower level via ore chutes, which in turn will be sent to the sump via channels(Figure 3 A minimum distance of 2.5 m. is maintained between the floor formed by fill and back of the stope.

4.3 Post-Pillar method using hydraulic fill at Mosabani Copper Project (Hindustan Copper Limited)

This is very much similar to the normal cut-and-fill method and used for wider ore bodies of more than 6 m.width. This layout is not cost and yields high extraction of about 85%. Figure 4. shows the layout of the stoping method (Shanker, 1980 This method essentially consists of

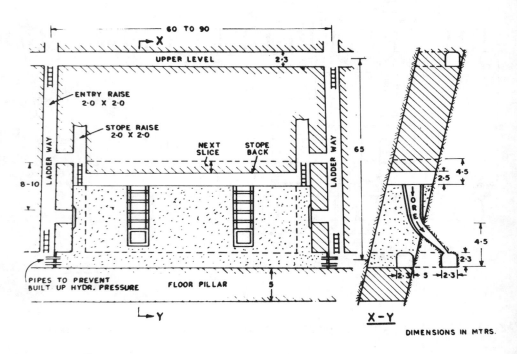

Fig. 2. Layout of cut-and-fill stoping practice at Jaduguda uranium mine

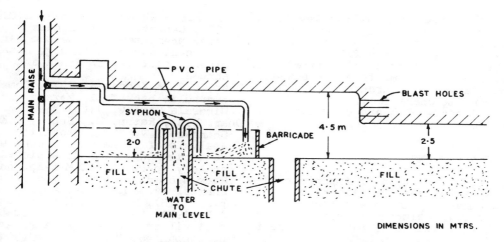

Fig. 3. Layout of filling arrangement at Jaduguda Uranium mine.

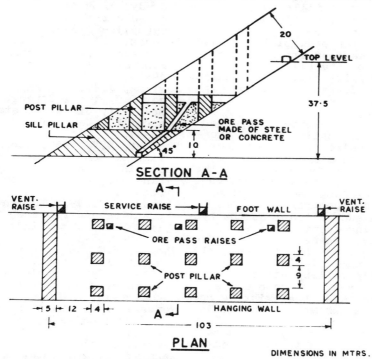

Fig. 4. Layout of Post-Pillar stoping method at Mosabani Copper project.

the following operations:

(i) ore passes are driven at 130 m. interval inclined at 50°, within the orebody or in the footwall.

(ii) one service raise is driven in the centre of the stope along the footwall contact. Two ventilation raises are made at two extremities of the stope along hanging wall contact.

(iii) development at sill floor starts after leaving a pillar of

8 - 10 m. over the haulage level.

(iv) back stripping is carried out in two stages upto a maximum of 4.7 m. height before filling.

(v) first layer of classified mill tailings of 2.2 m. thick is placed. This leaves a clearance of 2.5 m. for mining operations.

(vi) rock-bolts are used for support.

5 CONCLUSIONS

For the protection of important surface structures the filling of voids hydraulically is more suitable than other filling methods because of its high coefficient of filling (about 95%).

Even a moderate increase in quality of low modulus fills substantially lessens the stope clousure and peak pillar pressure. The difference between the peak pillar pressure and the maximum stress occuring at the core of the pillar is also reduced significantly. High quality fills are preferred in case of highly stressed areas, prone to spalling and bursting.

Since the ore reserves near the surface and at shallow depths are getting exhausted, mining of deeper deposits is becoming increasingly significant. As the economy of the country needs more production, extraction of such deposits often creates ground control problems. To overcome such situations the hydraulic filling seems to be an efficient remedy. Therefore in any deep-level mining practices in the world in general and India in particular, the hydrualic filling has a wider scope in the years to come and at the same time keep the existing rock burst prone mines under control.

To further improve the technology and economics of the system, more meaningful research in this area is called for.

REFERENCES

Carson, D.R. 1970, Stabilization of hydraulic fill with portland cement , U.S.B.M., R.I.7327.

Devaraj, V.G. 1980, Sand transportation for sand filling at Nandydroog mine, Kolar Gold Mines Centenary souvenir, 1980, 104-106.

Dhar, B.B., Khoda, R.,& Ray, S.C. 1971, Cemented fill as mass suppo paper presented in the seminar on Recent advances in u.g. metal mining practices in India and abroad.

Dhar, B.B. & Khoda, R. 1972, Cement hydraulic back fill, Metals and Minerals Review, 11(8):3-9.

Dhar, B.B., Ratan, S., Behra, P.K. Shanker, K.V. 1982, Stability analysis of large underground excavations using three dimensional analogy approach, ISRM symp. on Rock Mechanics-Caverns and Pressu shafts, Aachen, W. Germany.

Espley, G.H., Beattli, H.F. & Pasie A.R. 1970, Cemented hydraulic-bac fill within the Falconbridge grou of companies, CIM Bull., 63(701): 1002-1010.

McCrudy, J. & Hall, R.J. 1966, Ceme ed sand fill at Inco,, CIM Bull., 59(651):888-892.

Patchet, S.J. 1977, Fill support systems for deep level gold mine J.S. Afr. Inst. Min. Met., 78(2) 34-46.

Rawling, J.R., Toguri, T.M. & Ceri go, D.G. 1966, Strength and permeability of cement stabilized backfill, Can. Min. J., 87(12): 43-47.

Shanker, K.V. 1980, A critical review of some of the important stoping practices in metal mines (India), undergraduate project report.

Thomas, E.G. 1971, Cemented fill practice and research at Mount Isa, Proc. Aust. Inst. Min. Met. No. 240:33-51.

Thomas, E.G. 1973, A review of cem agents for hydraulic fill, Proc. Jubilee Symp. mining filling, North West Queensland Branch, Australia, 65-75.

Weaver, W.S. & Luka, R. 1970, Labo ratory studies of cement stabili ed mine tailings, CIM Bull., 63(701):988-1001.

Zahary, G., Zorychta, H. & Zaidi, S. 1972, Fill measurements in a cut and fill stope, Mines Branch report, 72/119.

State-of-the-art of pneumatic backfilling and its application
to a nuclear waste repository in salt

F.DJAHANGUIRI
Battelle Memorial Institute, Columbus, Ohio, USA

M.A.MAHTAB
Columbia University, New York, USA

ABSTRACT: Backfilling of a nuclear waste repository is required to provide a long-term barrier to radionuclide migration. Backfilling technology, though in use in Europe for about 50 years, is relatively new to the U.S.A. Recent applications of pneumatic back-filling of underground mines in North America suggest that the technique is potentially applicable to backfilling of a repository in salt. In this paper, we examine the various aspects of the pneumatic backfilling technology with particular reference to a conceptual repository in bedded salt. Some problem areas in the application are defined and some areas requiring further investigation are outlined.

1 INTRODUCTION

In the United States, as in most other countries concerned with the problem of isolation of nuclear waste from the environment, disposal of the nuclear waste in deep repositories, mined in geologic media, is the preferred solution. One of the host media being considered is salt (bedded or domal), which is attractive for its containment potential and other favorable characteristics. The discussion in this paper will generally relate to the disposal of high-level, nuclear waste in salt in the United States where backfilling of the underground openings is a requirement, with the objective of providing an engineered barrier to radionuclide migration and to reduce the long-term creep effects.

No specific sites have been designated for repositories in salt. However, both bedded and domal salt formations have been investigated since 1957 as potential sites for disposal of nuclear waste. The Permian Basin (Fig. 1), which contains extensive, bedded salt, appears to be one of the several likely candidates for housing a nuclear waste repository. For our discussion of backfilling of a salt repository, we will refer to a conceptual design for a repository in the Permian Basin (to be called the reference repository) that was prepared by Kaiser Engineers (1978). It is recognized that alternative designs for a salt repository are under consideration and that the final design of the repository and its location may be quite differ-

ent from the reference repository that is considered here for the purpose of illustrating the application of pneumatic back-filling.

2 CONSIDERATIONS FOR BACKFILLING A SALT REPOSITORY

The principal consideration for backfilling a salt repository is the regulatory requirement of providing an engineered barrier to radionuclide migration. Other important considerations are the layout of the mined repository, the thermomechanical response of the crushed-salt backfill, and the rationale for using a particular back-filling system. Further description of the above considerations is given below.

2.1 Regulatory requirements

The function of a high-level, nuclear waste repository is to contain or isolate the waste by a system of natural and engineered barriers so that certain environmental standards are satisfied. Backfill placed in the underground repository in salt (or other geologic media) is required by the proposed rules of the U.S. Nuclear Regulatory Commission (1981) to be an engineered barrier with the following functions:
(1) provide a barrier to ground water movement; (2) reduce creep deformation of the host rock; and (3) retard radionuclide

Figure 1. Index map showing locations of the Permian basin
(after Office of Nuclear Isolation, 1976,
Fig. 1.7)

migration. In addition, the proposed rules require that the backfill be selected to allow for adequate placement and compaction in underground openings.

2.2 Mine Layout

The reference repository (Kaiser Engineers, 1978) is located in bedded salt at a depth of 610m and covers an area of 3070 by 2620 m^2 (see Fig. 2). The repository is excavated as a room-and-pillar mine. The 1220 m-long rooms are laid out on either side of the five main entries. The rooms are 5.3m-wide by 5.8m high, are separated by 43.4m-wide pillars, and are interconnected by cross cuts at 305m centers. It is assumed that, with the exception of the initial portions of the entries in the shaft pillar, excavation of rooms and entries will be by a continuous miner. (Note that if drill-and-blast type of excavation is used, the production rate and the size of the excavated salt will be different.)

The waste canisters will be placed in vertical drill holes in the floor of the rooms. The drill holes (0.5m in diameter and 6.7m long) will be located in two rows spaced at 1.7m, with an in-row spacing of 1.2m.

2.3 Thermomechanical considerations

Another important consideration in sealing or backfilling a repository in salt is the reduction of creep deformation against the background of the thermomechanical (and chemical) perturbations created by the excavation and the heat generated by the waste. In what follows, we are concerned with backfilling of the larger openings (rooms, entries, and shafts) and, from among these openings, we will specifically refer to backfilling of rooms, assuming that the discussion will apply as well to the other openings. Although retrieval of the waste canisters is an option which can be exercised during a period of upto 50 years from backfilling of the rooms, our discussion will not refer to the condition or methods for retrieval of the waste.

The reference repository considers backfilling the rooms, shortly after waste emplacement, with crushed salt that is mined from the repository. We note that, in various studies of backfill for a salt repository, the effectiveness of mixtures of crushed salt and other sealants (bentonite) and desiccants (CaO, MgO) has been examined, for instance, Claiborne (1982). However, for the purpose of this paper, the backfill is assumed to be composed of crushed salt only.

Of significance to isolation of the nuclear waste is the potential of the salt-backfill to recrystallize under the influence of pressure and elevated temperature and form a relatively impermeable mass. The consolidation of crushed salt has been examined in the laboratory by Wagner (1980 and Holcomb (1982), among others, and the

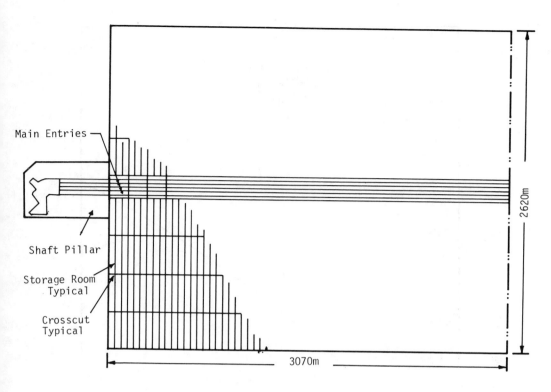

Main Entries

Shaft Pillar

Storage Room
Typical

Crosscut
Typical

2620m

3070m

Figure 2. Layout of the reference repository
(after Kelsall et al., 1982, Fig. 2-1)

subject has been discussed in some detail
by Kelsall et al. (1982). These studies
show that consolidation of the backfill
occurs in response to creep closure of the
rooms. The degree of consolidation achieved in a given period is a function of the
initial porosity of the backfill, intact
salt properties (creep rate), the thermal
load, and the stress concentrations due to
the applied in-situ stress and geometry of
the rooms. For an example configuration of
a repository room, the roof-floor convergence as a function of time is given in
Fig. 3 (Wagner, 1980) for two cases:
(1) no backfill and (2) 100 percent,
crushed-salt backfill. It can be seen
that the closure is a non-linear, decreasing function of time and, in the case of
the 100 percent backfill, the closure rate
decreases more rapidly as the backfill consolidates and supports an increasing amount
of the applied load (due to increased
modulus and strength).

2.4 Rationale for using pneumatic back-
      filling

The conditions for application of backfil-

ling techniques that are present in a salt
repository are: (1) the backfill material
(salt) is excavated from the repository
and the material is largely unprocessed
and can be used underground without the
necessity of taking a bulk of the material
to the surface; (2) the openings to be
backfilled are horizontal; and (3) addition
of water to the backfill is basically un-
desirable. The available backfilling techniques that can be used under these conditions are the mechanical techniques such
as conveyor belts, load-haul-dump equipment,
and pneumatic conveying. Of these techniques, pneumatic conveying offers the
following distinct advantages for application to backfilling a salt repository.

1. The same equipment is used both
for conveying and placing the backfill material. With any other method
of backfilling, for example, conveyor
belts, additional equipment is needed at
the discharge to pack the material to the
roof.

2. The conveying pipeline takes up very
little room in the entries and vehicles can
conveniently pass over or under the pipeline. Changes in direction can be readily

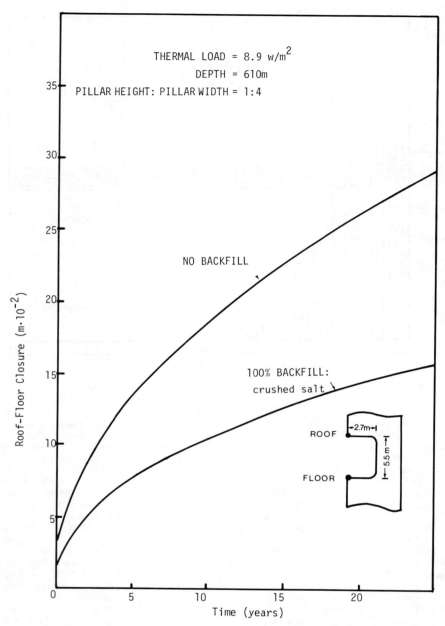

Figure 3. Roof-floor closure as a function of time in a salt
repository (after Wagner, 1980, Fig. 5.2)

negotiated by inserting elbows into the
pipeline.

3. The material when placed with a pneumatic backfill system has a high relative density (about 70%) in comparison with
other methods such as gravity, hydraulic,

or placing with a blet conveyor. In addition, the pneumatic system can fill irregular cross sections or cut-outs in the
entries or rooms, such as manholes in the
pillar sides.

4. The technique is cost effective in

124

terms of the capital and maintenance costs and, especially, in providing a ready means for conveying as well as stowing the mined salt.

5. Additives to the crushed salt are also readily handled by the pneumatic system and, if introduced at a constant rate at the infeed, will produce a thorough mix with the backfill material as it traverses the pipeline.

3 PNEUMATIC BACKFILLING TECHNOLOGY

Pneumatic backfilling (or stowing) was developed in European coal mines in the early 1930s for backfilling longwall panels. It has also been used to increase extraction from room-and-pillar mines and to reduce or minimize surface subsidence. Pneumatic conveying is relatively new to North American mines where one early example of its use was in hoisting coal mine waste vertically to the surface (Ball and Tweedy, 1975).

Pneumatic backfilling is also a novel method for backfilling abanadoned mine openings. During backfilling, the material (a mix of various sizes of particles) is ejected from a nozzle at varying velocities; upon impact, the material creates a high-density fill that will support the mine roof and ribs (Maksimovic and Lipscomb, 1982). A case history of pneumatic backfilling with crushed rock at the Sullivan mine, Canada, has been described by Reynolds (1972).

3.1 Description of backfilling systems

A pneumatic backfilling system consists of four essentials (see Fig. 4): a source of pressurized air, a regulated infeed to introduce the material into the pressurized air, a pipeline to convey the air and material, and a controlled discharge to direct the material as it leaves the open pipe (Maksimovic and Lipscomb, 1982; Powell, 1982).

For mine backfill applications the rotary positive displacement blowers are considered most suitable as these supply the quantity of air required at pressures of 8000 to 16000 Pa (12 to 22 psi). The blower and motor are mounted on a skid base for easy manouverability through the mine, and are equipped with intake filters to reduce the dust drawn into the blower and an enclosure to reduce the noise.

There are several techniques for introducing material into a conveying pipeline. The most suitable arrangement for mining applications is a variable-speed rotary air lock feeder. This consists of a drum, with pockets rotating between adjustable stator

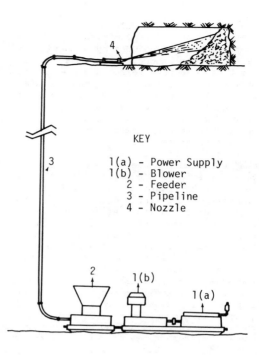

KEY

1(a) - Power Supply
1(b) - Blower
2 - Feeder
3 - Pipeline
4 - Nozzle

Figure 4. Schematic diagram of pneumatic backfilling equipment (after Maksimovic and Lipscomb, 1982, Fig. 1)

halves, which transfers the material from the infeed at the upper side and discharges this into the air stream at the lower side with the minimum loss of air. A hydraulic motor provides the variable speed (Powell, 1983).

The selection of the pipes that make up the pipeline depends not only on the volume of the material mix to be handled but also on other factors, such as the abrasiveness of the material and the method by which the pipes are to be handled in the mine. At the discharge of the pipeline, a nozzle is required to direct the material into the fill area and to build this up in layers from side to side to achieve the maximum compaction.

3.2 Potential application to a salt repository

In looking ahead for application of the pneumatic backfilling technology to a salt repository, there are a number of steps and procedures which can be suggested. At the same time, since site-specific information is not available, certain assumptions have to be made regarding the mining method and the backfilling schedule.

We are assuming that mining of the rooms and entries will be done by a continuous miner with a production rate of 100 tons per hour. We have selected a pneumatic system that can handle 200 tons of material per hour. However, the actual output is assumed to be approximately 100 tons per hour of material being placed, allowing for delays due to pipe removal as the backfill progresses in the room.

Initially, the salt which is excavated in the development phase will be transported to the service shaft by means of shuttle cars, feeder-breakers and conveyor belts. When the backfilling program starts, the salt is diverted (for delivery to the infeed of the pneumatic backfill system) from the main conveyor belt onto a conveyor belt running along the main entry of the panel where the repository rooms are located.

Assuming that the length of the room in the reference repository is 305m, one room can be backfilled from one set up. Following the completion of a room, the infeed equipment, blower-compressor, and belt discharge are retreated along the panel entry and the process repeated for the next room.

Stowing will start by directing the material to the corner, at the floor level, on one side of the room and forming a shelf approximately 1m deep and 0.5m high by traversing the nozzle accross the room. On reaching the opposite side of the room the nozzle is lifted and a second 0.5m high shelf started with the nozzle now moving in the opposite direction across the room to its original starting place. The nozzle will be lifted again and a third layer put down. The fill is continually placed in this way until the last layer to be placed is in contact with the roof.

When the fill has retreated along the room for a distance corresponding to a predetermined optimum length of the pipe section (say, 6m), the system is shut down, the pipelength removed, and the nozzle reattached for the next backfill-section.

An important consideration in the design of the conveying system would be the diameter of the pipe to be used, which will depend on the maximum size of the material to be conveyed, the quantity of pressurized air needed to accelerate the material, and the length of the pipe. Powell (1983) suggests that, assuming a particle size of 7.5cm and a length of pipe of the order of 300m, a 20cm-diameter pipe could be selected for this example to operate at about 0.15 MPa.

The costs associated with pneumatic backfilling have been shown to be less than the costs when using conventional stowing methods (Reynolds, 1972). For a salt repository, the cost advantages may be even more significant because of the low abrasion potential of the broken salt being conveyed in the pipes.

3.3 Problem areas

The two significant environmental problems in backfilling underground repository openings with crushed salt are noise and dust. The noise generated by the positive displacement blower can be controlled in several ways. The intake air is passed through a filter-silencer and, at the discharge, a silencer is also fitted or the air is discharged into a tank where the pulsations are dampened. Noise is also emitted from the open pipe, when the air is flowing without the material being conveyed, but is much reduced when conveying is taking place.

The generation of dust is a major drawback of the pneumatic backfilling system. Dust control procedures fall into three categories (Jerabek and Hartman, 1966): prevention, isolation, and suppression.

Dust prevention is possible by reducing air consumption and maintaining the equipment in good condition. Isolation of the dust can be accomplished by placing the stowing machine in an enclosure, and by collecting and removing the dust through an exhaust system. A retaining wall (a bulkhead in a repository) may also be used to confine the backfill material and the dust. Suppression of dust normally requires addition of water. However, this may not be an acceptable procedure in a salt repository.

Some minor problems relating to the pneumatic backfilling system are: the variable density of the stowed material which tends to decrease away from the point of impact of discharge; the necessity to ensure a continuous supply of backfill material for maximum efficiency—this problem may be solved by providing a surge bin; and the need for examining for each site, the most suitable combination of particle size, pipe diameter, pipe length, air velocity, and power requirements.

4. CONCLUSIONS AND RECOMMENDATIONS

Pneumatic backfilling of a nuclear waste repository in salt is a potentially viable concept. The most significant advantage of the technique, which is well-developed, is the complete backfilling of the openings, a condition which enhances the role of backfill, as an engineered barrier against radionuclide migration and, as a structural support for the openings.

Most ot the concepts used in pneumatic backfilling technology are based on experience and empirical relationships. Therefore, in applying this technology to backfilling a specific repository in salt, further investigations are necessary. The following are recommended as two broad areas of investigation for further development of pneumatic backfilling concepts for application to a salt repository.

1. Material characteristics

(a) Size and mix of the broken salt with reference to the mining method and requirements for further crushing or sizing.

(b) Strength and consolidation characteristics of the backfill, assuming a placement density.

(c) The influence of temperature and room closure rates on the long-term behavior of the backfill (salt plus any additives).

2. Backfilling technique

(a) Development of pneumatic backfilling system components for site-specific application.

(b) Use of pneumatic conveying in combination with backfilling.

(c) Control of dust, including choice of additives (water or other materials).

(d) Control of noise.

REFERENCES

Ball, D. G. & D. H. Tweedy 1975. Pneumatic hoisting from undergound. CIM Bulletin. 5(68):59-63.

Claiborne, H. C. 1982. The efficacy of backfilling and other engineered barriers in a radioactive waste repository in salt. ORNL/TM-8372: Oak Ridge National Laboratory, Oakridge, TN.

Holcomb, D. J. 1982. Consolidation of crushed salt backfill under conditions appropriate to the WIPP facility. SAND 80-0558: Sandia National Laboratory, Albuquerque, NM.

Jerabek, F. A. & H. L. Hartman 1966. Mine backfilling with pneumatic stowing. Mining Congress Journal, June: 69-79.

Kaiser Engineers 1978. A national waste terminal storage repository in a bedded salt formation for spent unprocessed fuel, vols. 1 and 2, KE 78-57-RE: Kaiser Engineers, Oakland, CA.

Kelsall, P. C. et al. 1982. Schematic designs for penetration seals for a reference repository in salt. Project No. NM 79-137: Office of Nuclear Waste Isolation, Battelle, Columbus, OH.

Maksimovic, S. D. & J. R. Lipscomb 1982. Sealing openings in abandoned mines by pneumatic stowing. U.S. Bureau of Mines RI 8730.

Office of Nuclear Waste Isolation 1976. National waste terminal storage program progress report, April 1-September 30, 1976, Report Y/OWI-8:11. Union Carbide, Oakridge, TN.

Powell, J. E. 1982. Pneumatic conveying equipment for the mining industry. Bulk Solids Handling. 2(4):781-787.

Powell, J. E. 1983. Reposiotry backfilling. Consultant's report to Office of Nuclear Waste Isolation, Battelle, Columbus, OH.

Reynolds, J. W. 1972. Pneumatic backfilling with crushed rock at the Sullivan mine. CIM Trans. 75:115-120.

U.S. Nuclear Regulatory Commission 1981. Disposal of high-level radioactive wastes in geologic repositories, proposed rule 10CFR Part 60. Federal Register 46(133): 35280-35296.

Wagner, D. J. 1982. Preliminary investigation of the thermal and structural influence of crushed-salt backfill on repository disposal rooms. ONWI-138: Office of Nuclear Waste Isolation, Battelle, Columbus, OH.

Proceedings of the International Symposium on Mining with Backfill / Luleå / 7-9 June 1983

Mining method with hydraulic stowing in the Austrian lead-zinc mine Bleiberg

ERWIN ECKHART
Bleiberger Bergwerks Union, Bad Bleiberg, Austria

ABSTRACT: In the eastern part of the lead-zinc mining district of Bleiberg-Kreuth there are veinlike small ore deposits. The difficult stratigrafical mining conditions give many technical and economical troubles. The application of hydraulic stowing, the mechanization of the stopes, the improvements in hoisting and hauling led to a very increased output and much lower costs.

INTRODUCTION

The lead-zinc mining district of Bleiberg-Kreuth in Carinthia in southern Austria has been worked a couple of centuries. Today still exists the Kreuth-mine in the west, which is the more important one, and the Bleiberg-mine in the east. The todays mining district covers a length of obout 7 kilometres. According to the different conditions of the mineral deposits of the two mines they are working with completely different mining methods. In the Kreuth-mine with its steeply inclined, big, massy ore body, with large horizontal areas in the cross section, sublevel stoping with concrete waste fill has been developed. This method is already known in the mining literature. The Kreuth-mine ist therefore a highly mechanized mine.

The eastern mine, Bleiberg, shows small lodes, unconformable and bedded veins, mainly combined with each other. The horizontal area for mining exceeds rarely a few hundred square metres. The difficult conditions of the ore deposit of this underground mine with existing long and complicated hauling distances, give many technical and above all economical troubles. Therefore during the last years steps were taken to increase the

efficiency of the Bleiberg-mine. The improvments in haulage and backfill will be explained in the following chapters, but manly the backfill technique.

The Bleiberg-mine

The Bleiberg-mine with its two sections Rudolf and Stefanie is situated 2 kilometres eastward of the todays centre of production. The mainly dyke-shaped lodes locally changing to layers, are varying in a wide range concerning shape as well as grade. The common mining method is overhand-stoping with waste fill. When larger horizontal width appears, the cross working method in benches, 3 metres high, going from the bottom to the top is being applied. The mined out stopes are being filled with waste.
Additionally there are secondary stopes. These are old workings without waste fill after mining in former years, but still containing considerable quantities of ore which could not be mined out by reasons of rock stability.

Before starting up the rationalizing measures in the nineteen-seventies the waste fill problem had to be solved. The waste fill method, as it was done until 1974, has clearly limited the output of the stopes. The waste fill material had to be transported in mine trucks from surface to the underground stopes covering long hauling distances with many difficulties caused by discontinous transport methods. Besides the enormous waste of time and effort of labour for transport, the necessary tunnelsystem neither in lay-out nor in size, did not exist in many cases.

Hydraulic stowing

Since 1960 the hydraulic waste fill had been in use in the Kreuth-mine and practical experience could be gained with doing so until 1972. The pipelines for the hydraulic transport of the filling material were short, only a few hundred metres. The ratio between vertical length to horizontal length of the pipelines was not more than 1 : 4. With the application of sublevel stoping with concrete fill the hydraulic stowing in the Kreuht-mine has lost its importance. The experiences from Kreuth were encouraging to try the hydraulic fill in the mining sections Rudolf and Stefanie.

The filling material plant in Kreuth consists of two systems with two-grade cyclons followed by dewatering screens. After dewatering remains a residual moisture of about 17 percent in the sands.

From 350.000 tons tailings of the flotation with particle size from 0 to 250 mycrons it is possible to recover 50 percent for backfill, when the separating-caesura is set at 50 mycrons.

The sandfill material is brought from the plant in Kreuth to surface by 3 cubic metre mine trucks. Here it is reloaded on road transport trucks and transported two kilometres eastward to the Rudolf-shaft. At the Rudolf-shaft a bunker plant containing 180 cubic metres had been built. At the bottom of this plant, which is 25 metres underground, the sandfill distributor is installed. Here the water-sand-mixture is made in funnel-like chutes, which are covered with screens. The sand from the flotation is washed out from the bunker by a high pressure water jet. When the water supply

is correct the level of the pulp stabilizes in the mixing funnel at a certain position. This water-sand-mixture is fed by soft PE-pipelines into the stopes often covering a distance of a couple of kilometres.

The nearer stopes of the section Rudolf are connected with the distributor by pipes with 42 millimetres inside diameter and 10 atmospheres nominal pressure. The stopes farther away are operated by pipes with 47 and 80 milimetres inside diameter and also 10 atmospheres nominal pressure. The longest single pipeline reaches under these conditions a length of 3.500 metres with a difference of hight of 620 metres. The ratio vertical pipeline to horizontal pipeline is in the most unfavourable situation 1 : 10.

This hydraulic transport is done by gravity only, without any pumps. The content of solids in the pulp figures out at 46 to 58 percent of mass or 600 t 750 grammes of solids per litre. The flowing velocity of the pulp in the pipes ranges from 1,8 to 2 metres per second. Initial difficulties were caused by too little strength of extension of the flange-joints at the position of the highest dynamic pressure of 27 atmospheres. These troubles could be eliminated by fixing clip-baskets. In case of pipe burst supervising equipments are installed. These are flow-metres which make it possible to measure the pulp without contact by measuring the change of frequences of ultrasonic signals reflexion in accordance with the Doppler-effect.

Concerning the wear of the pipes we could state, that straightly laid pipelines show only little wear after 7 years. In parts with turbulences in the flow (curves, bad couplings) the wear can be bigger.

Today are 14.000 metres PE-pipelines laid for transporting the sandfill material. On an average 560 tons of solids per day are filled into the stopes. Since the starting of the hydraulic fill operations in the middle of 1975 a total of 650.000 tons flotation tailings were refilled into the mine, while the plant has been extended during the last years. The capacity of the hydraulic transport method of the filling material can be shown by two figures: The productivity of the old waste fill method was 8 cubic metres per man and shift, now the productivity with hydraulic stowing has increased to 69 cubic metres per man and shift. Caused by the hydraulic sandfill in the Bleiberg-mine it was possible to cut the labour effort for backfill to a less important dimension. On the other hand the chance was given to apply the cut and fill method in the lower parts of the deposit, where conventional waste fill has been impossible. Long distances, depth, shape of ore bodies, as well as situation of the stopes became unimportant criterions. With this efficient backfilling method the chance was given to mechanize the stopes.

Mechanization of the stopes

Before application of the hydraulic backfill scrapers were used in the stopes. An important increase of output could be obtained by Atlas Copco T 2 GH loaders driven by compressed air. The distance between the ore passes could be extended from 3o to 1oo metres. Taking into account the increased output the diameter of the ore passes had to be enlarged from 4o centimetres to 11o centimetres. The ore passes consist of semicircle segments made of steel. Two segments are screwed together to a ring and put on top of the existing ore pass going through the sandfill. A travelling way is put up by the same way. Its diameter is 8o centimetres.

In 1976 the first experiment was done with a diesel-driven wheel-loader of the 1 cubic yard types. Subsequently further diesel-loader came into use wherever the difficult circumstances of the ore deposit gave a chance. Today are working in the Bleiberg-mine 1 Eimco 911, 1 Eimco 911 B, 2 GHH-LF 2 and 2 Eimco 912. Problems caused by working with these diesel-loaders on the fill do not exist. The percolation rate has been measured and stated that it is 14,4 centimetres per hour. The filling can be stepped on after a few hours and on the next day already the diesel-loaders can work on the new floor. The drilling operations in the stopes have been rationalized by taking away the 16 kilogrammes Hammer-drills and replacing them by 23 kilogrammes airleg mounted rock-drills. Half a year ago a small diesel-driven Alimac drill jumbo with one arm has been installed improving the drilling efficiency.

All these measures caused, that the output in stoping increased in the time from 1975 to 198o from 8,4 to 21,3 tons per man and shift. In the same time the ore production could be doubled with an increase from 69.ooo tons to 14o.ooo tons.

Improvements of hauling

With the increase of production the existing bottle-necks in hauling had to be removed. The ore coming from the section Stefanie must be hoisted 4oo metres through the Stefanie-blind-shaft. After that it has to be transported through a 4,2 kilometres long, very narrow tunnel to the main hoisting shaft at Kreuth.

In 1975 the cage hoisting at the Stefanie-blind-shaft, which was very expensive by personal costs, has been replaced by skip-hoisting in one shaft compartment. By this change the shaft capacity could be trebled.

For the ore transport in the main haulage mine cars as big as possible came into use. The old o,8 cubic metres Granby cars were replaced by 2,5 cubic metres bottom discharging wagons.

Efficiencies

The better filling-technique, the mechanization of the stopes, the application of LHD-technique, the improvements in hoisting and hauling definitly led to an increased output and lower costs. The following summary shows how the efficiencies in the stopes and underground developed since 1974 till 198o. It also shows the costs per ton ore hoisted at the ore dressing plant, basing on 198o values.

Table 1.

	tons output per man and shift	
	stope	underground
1974	8,4	3,4
1975	11,4	4,2
1976	14,5	5,9
1977	17,3	7,2
1978	19,5	7,7
1979	19,9	8,4
1980	21,3	7,6

Table 2.

	costs per ton hoisted AS/t
1974	880,--
1975	798,--
1976	614,--
1977	489,--
1978	451,--
1979	4o3,--
1980	438,--

We can see, that it is possible to
cut costs and to increase output
by rationalization even in rather
small ore deposits, characterized by
difficult conditions as irregularity
of the ore bodies and a mine lay out,
which is a couple of decades old.

Some aspects of using cemented filling in the ore mining of the GDR

H.GERHARDT
Bergakademie Freiberg, GDR

ABSTRACT: The German Democratic Republic has got in view of underground mining much experience in using the method of cemented filling since about 20 years.
Mixtures of filling material favourable in cost expenditure, control methods which were aspecially developed for the filled mass in the basis of ultrasonics, and for these reasons, reductions of the timber work in the working area had led to considerable increases in performances and to an increase in mining safety, too.
In the same time, the conditions for working in health of the miners had been improved. Reserves in performance may be further given by a greater use of Diesel driven loading and haulage devices, of ramps serving as connections between relatively small ore bodies, and effective drilling machines.

INTRODUCTION

The productive zones – it deals of palaeozoic sedimentes – in which mining methods with cementes filling are used in the ore mining of the GDR, possess complicated forms and tectonic structures. Their thickness varies between a few centimeters and several meters. The danger of a spontaneous combustion is given by organic components as well as by the presence of pyrite and marcasite. The compactness of the ores amounts to about six to eight after an evaluation according to the Protodjakonov scale.

DEVELOPMENT OF MINING METHODS WITH CEMENTED FILLING

Proceeding from these above-mentioned geological and mining conditions and under consideration of international experience, the first investigations for using cementes filling were made in 1959. Systematic investigations in a larger scale for clarifying problems of filling began in 1962 (1), so that we can look back now to an experience of more than twenty years. The first plant producing filling material came into operation in 1963.
Room-and-pillar work used to this date and sublevel caving led to difficulties in mining the pillars due to the collapsing of the rooms and due to occasional endo-

gene fires. The breasted-out working rooms were therefore slushed by reasons of safety, and the mining losses were relatively high. Other mining methods which were used, such as block mining with square-setting, variations of longwell mining and ascending sublevel caving with filling also showed a large number of technical and economic disadvantages which meant that they did not offer any possibilities of improving the technical and economical parameters. For this reason, after some successful tests, the development of the mining method: "descending sublevel stoping with cemented filling" was put in the sixties more and more into the foreground in such a way that is became dominant in the considered deposit zone (fig. 1).

The results obtained and the advantages to be seen can be summarized as follows:

– protection of the surface by means of avoiding or reducing surface influences caused by underground mining,
– reduction of the ore losses and the ore dilution,
– avoiding endogene fires,
– good adaption to different geological conditions,
– increase of safety in carrying out ope-

133

rations in the working room,
- improvement of the conditions of
  the ventilation technique and the mine
  climate by reducing air leakages,
- reduction of the timbering work in the
  working area, better mechanization of
  timbering work, saving of timbering material,
- greater utilization of effective trackless devices due to larger working rooms
  possible and due to working on a firm
  level,
- increase of the concentration of workings,
- increase in the labour productivity in
  the working area and reduction of mining
  cost.

author will report in the following chapter
on the complexes : filling mixture and filling technology, variations of timbering
and the undermining of artificial roofs
as well as the influence of the drilling
and blasting operations on the filled mass.
This report is given under full consideration of the obtained results and the
gained experience.

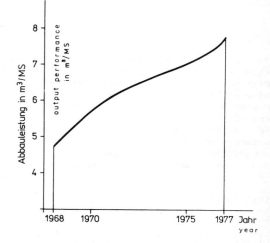

Fig. 2. Development of output performance
         by using the mining method "Des-
         cending sublevel stoping with
         cemented filling" (2)

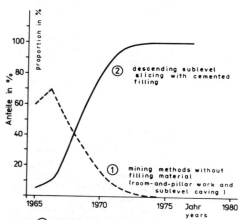

① Versatzlose Abbauverfahren (Kammerbruchbau
   und Teilsohlenbruchbau)
② abwärtsgeführter Teilsohlenbau mit erhärtendem
   Versatz

Fig. 1. Range of utilisation of the mining
         method "Descending sublevel
         stoping with cemented filling"
         in one deposit region (2)

The period between 1968 and 1977 shows a
continuous increase of the performance in
the working area by using the mining
method: "descending sublevel stoping with
cemented filling" and by realizing a series of detailed measures (fig. 2).
Therefore, the performance in the working
area has increased in average, under consideration of all used combinations of mechanization devices, from 4,8 m³/MS to
about 8 m³/MS, which corresponds to international scales.
After a short presentation of the fundamental variations of the m ining method, the

FUNDAMENTAL VARIATIONS AND MAIN PARAMETERS
OF THE MINING METHOD

In descending sublevel stoping with cemented filling the ore is won slice by slice
from above, downwards. A characteristic
feature is therefore the work under an artificial and under a natural roof. Mining
itself is done within a single slice section by section. To obtain a filling of
the working area without residual excavations in a satisfying way, the slices to be
worked and the transverse galleries made
from the sublevels are driven with an inclination of 3 gon. The height of the
slices amounts up to 3,5 munder a natural
roof andup to 4,5 m under an artifical one.
The different technological variations of
sublevel stoping with cemented filling used
today are characterized by the position of
the sublevels and the roads for transpo-
ting filling material in the working block
as well as by the performace of exact
transverse galleries and rises. They also

134

differ in the choice of the dimensions
of the working sections, the selection of
the timbering and the used technical
equipment. The utilisation of the different variations depends in the first line
on the geometrical form and the nature of
the deposit. In all cases exploratory
drillings have to be made from underground
due to the complicated deposit structure.
They are done in the base roads, and after
obtaining the results, the division into
the single working blocks is performed.
In relatively thick and large-faced mine-
ral-forming parts a variation is used
which is shown in figure 3. The working
within a single slice is hereby done with
room-and-pillar mining.

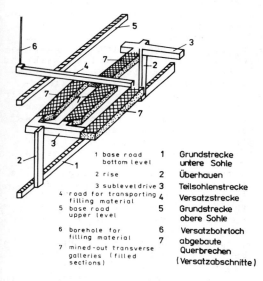

1 base road bottom level	1	Grundstrecke untere Sohle
2 rise	2	Überhauen
3 sublevel drive	3	Teilsohlenstrecke
4 road for transporting filling material	4	Versatzstrecke
5 base road upper level	5	Grundstrecke obere Sohle
6 borehole for filling material	6	Versatzbohrloch
7 mined-out transverse galleries (filled sections)	7	abgebaute Querbrechen (Versatzabschnitte)

Fig. 3. Variant of the descending sublevel stoping method with cemented
filling in relatively thick and
large-faced deposit zones

FILLING MIXTURE AND FILLING TECHNOLOGY

In the first years of the development of
this method greatefforts were made to
plan in a financially favorably manner
and to produce filling mixtures with low
cost. Table 1 shows a filling mixture used
in an industrial pilot test, and which
possessed the following three fundamental
components : cementing materal, additive
and water.

As cementing materials are used today brown
coal filter ashes and cement, as additives
sand-gravel-mixtures (granulation $\leq$ 40 mm)
as well as the fundamental component, that
is water with a p_h-value of 5 - 8. The relationship between cementing material and
additive amounts to 0,26 t to 1 t with the
use of brown coal filter ash and 0,14 t to
1 t with the use of cement (2). The factor
water/cementing material amounts to about
0,8. The washing components of the sand-
gravel-mixtures won in open-cut mines have
a proportion of 5 to 15 %. All components
of the filling mixture are controlled continuously. The consistency of the filling
mixture is measured after mixing with the
help of standard cones. A standard cone
with a height of 150 and a diameter of
80 mm as well as a mass of 300 g shall dip
into the mixture up to a depth of 9 to
10 cm.
The plants producing filling material have
an average capacity of 500 to 1000 m³ filling material per shift. The transport
underground is done either directly by
pipelines coming from the named plants or
by decentrally arranged boreholes for the
filling material in using vehicles which
transport and mix this material simultaneously. The pipeline diameter for the
transport of filling materials amounts to
150 and 220 mm, respectively. The horizontal lengths for the transport of filling
materials amount, in general, to $\leq$ 1000 m,
and the vertical haulage ways lie in the
range of $\leq$ 500 m.

Table 1. Filling mixture used in 1963 (1)

Filling mixture	Cementing material		Additives	Water
1 m³	liberated granulated blast-furnace slag ($>$ 9 % Al_2O_3, λ 35 % CaO)	Portland cement (1-5 % of the slag mass)	sand, preparation residues, waste rock	
	400 kg	1-4 kg	1400 kg	280-380 l

135

VARIATIONS OF TIMBERING IN THE WORKING ROOM, THE UNDERMINING OF ARTIFICIAL ROOFS

An important problem in increasing mining output is the use of an effective technical equipment. The dimensions of the devices are adapted in particular to the possible parameters of roof clearing. If there are better successes in essentially increasing clearing width, both timbering material and timber work may be reduced and larger devices (drill carriages and loading machines) can be used. Regarding the interpretation of the research work in the deformation behaviour of hanging cemented filled and rock masses, in the development of control methods for the filling material (quality and structural composition!) and in dimensioning clearing parameters under an artificial roof, the following results can now be given here (2):

- The filled mass does not accumulate any tensions, but tends to a deformation and to settlements. The danger of destresses like rock bursts is thus excluded.
- There are three forms of deformation to be distinguished:
  1) elastic zone that does not require a timbering and
  2) transition to the plastic zone without a fissuring in the filled mass, and
  3) appearance of fissures in the lower part of the filled mass without the danger of a caving-in.

If we introduce the term "limiting zone" after Slesarew (3) who understands by this term the possible clearing width under conservation of a limitless section length, the three deformation zones and the corresponding limited widths of the working area to be observed may be described as follows:

$$B_{limit\ 1} = 4,05 \cdot \sqrt{\sigma_D \cdot h}$$

$$B_{limit\ 2} = 3,32 \cdot \sqrt{\sigma_D \cdot h}$$

$$B_{limit\ 3} = 5,37 \cdot \sqrt{\sigma_D \cdot h} \ .$$

σ_D (MPa) designates the monoaxial compressive strength of the mass, and h (m) designates the height of the lowest layer being free from interfaces.

The basis of these explanations is a compressive strength of the filled mass between 1,5 and 2,5 MPa and a freedom from interfaces of $>$ 1,0 m. With the help of a variety of detailed investigations was proved that the relations between the tensile and the compressive strength of the filled mass are so stable that the use of the compressive strength in the above-cited formulae is possible. Within this connection a certain safety coefficient had been considered additionally. The nomogram shown in figure 4 serves as a working basis and presents the connections between the compressive strength, the structure, the kind of working, and the width of clearing. If one reduces the drivage lengths, an exceeding on the "limiting width" by a value of 1,4 is possible. Thereby the real values for reducing the drivage lengths depend on the magnitude of the load of the filled mass on the faces. For this phenomenon working diagrams (nomograms) can also be given.

After having laborated adequate methods for determining the compressive strength of the filled mass and for controlling its structural composition as well as for constructing the necessary devices for their practical use in the mining enterprises, the following rationalisation effects in mining under an artificial roof could be achieved up to the end of the seventies (4):

- with compressive strengths of the filled mass of $>$ 1,5 MPa and a freedom from interfaces of $>$ 1,0 m, the expenditure for timbering is limited to the placing of the warning support (simple setting of props with headboard without using caps, figure 5). Up to a width of 7,0 m a support in the working room ist in no way necessary; with drivage widths of 7-10 m, a row of props with a prop spacing of 6,0 m is set in the middle, and with drivage widths of $>$ 10 m, a row of props with a prop spacing of 4,0 m in the row is brought in (figure 6).
- if the filled mass does not correspond to the cited quality requirements, frame timbering must be used.
- due to the vast introduction of the saving of timbering material is considerable. The timber consumption could be reduced from 22,6 m³/1000 m³ excavation to 5,0 m³/1000 m³ excavation. Through this measure alone, an increase of the performance in the working

room by 15 % could be proved.
- the use of trackless devices made possible
by the larger rooms now available, led to
a further increase in performance of 20 %.
Figure 7 after (2) shows the performances
achieved by using the different device
combinations for drilling, loading, and
haulage. One can expect further increases in performance lying between 3 and
4 m³/MS owing to the now used Diesel mobile loaders.

The methods used earlier for determining
the compressive strength of the filled
mass had been the evaluation of the cube
compressive strength (edge lengths of the
cubes 7 cm) and the drilling out of cores
with subsequent determination of the compressive strength, respectively. Both
methods show certain disadvantages (5).
In determining the cube compressive
strength, they are, especially, as follows:

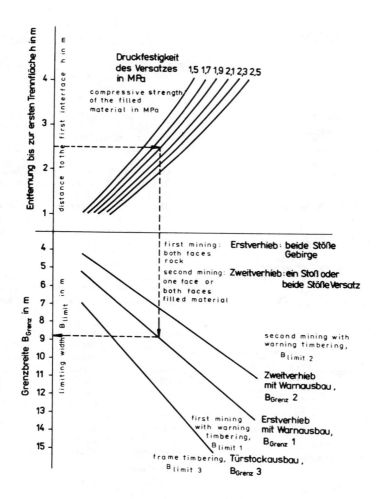

Fig. 4. Determination of the admissible limiting width of the
working rooms at different compressive strengths,
structural conditions in the filled mass, and different load conditions on the filled mass

137

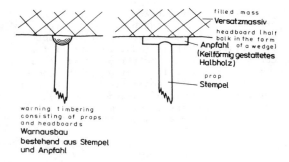

Fig. 5. Warning timbering consisting of props and headboard

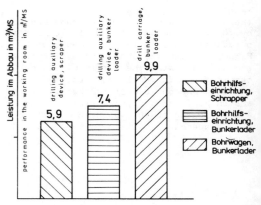

Fig. 7. Achieved output performances when using different device combinations

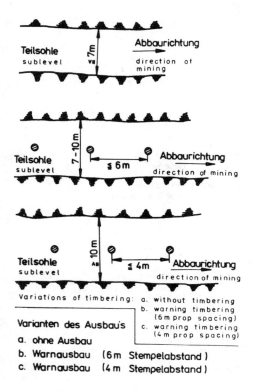

Variations of timbering: a. without timbering
b. warning timbering (6m prop spacing)
c. warning timbering (4m prop spacing)

Varianten des Ausbau's
a. ohne Ausbau
b. Warnausbau (6m Stempelabstand)
c. Warnausbau (4m Stempelabstand)

Fig. 6. Placing-in of the warning timbering under consideration of the different working widths

- a too low edge length compared to the largest possible grain diameter in the filling mixture (the really required edge lengths of 20 to 25 cm would lead to very considerable increases in cost),
- subjektive influence factors in taking out the filled mass and in the storage of the cubes,
- different conditions in the hardening process of the test cubes and during the binding action of them in the filled mass, and
- the fact that the undermining of the filled masses is already necessary in certain cases before the ultimate storing time of the test cubes of 14 and 28 days, respectively is reached.

With a borehole length of > 1 m, the determination of the compressive strength of the filled masses from the drilled cores is rather expensive, the economically reasonable core diameters of < 7 cm are too small, the slenderness ration being $\lambda < 1$ is partly too low, and interfaces in the filled mass are not present. On the other hand, a quality determiantion at the filled mass is absolutely necessary, because it is influenced by the following objective and subjective factors:

- the granulation band of 0 to 40 mm is relatively wide,
- quality determinations for provoding filling material in the surface area (e. g. standard cone control for determining the consistency) may be influenced subjectively,

vertical and horizontal haulage ways as well as interruptions in placing the filling material lead to changes in the mixtures of the filling material, which influences the proporties of the filled mass.

The control of the filled masses is nowadays done by geophysical methods with the help of ultrasonics. Thus sounds are sent in parallel through boreholes - maximum distance between emitting and receiving borehole 0,5 m - making use of the correlation factors between the propagation time of the longitudinal waves and the compressive strength, in order to determine the compressive strength of the filled mass (figure 8 after (2)). Considerable detailed investigations have proved that between the compressive strength and the tensile strength of the filled mass there exists a relation of about 4 to 1 with absolute amounts of the compressive strength of 1,5 to 2,5 MPa and that, for this reason, the decisive magnitude of the tensile strength with respect to the deformation of the filled mass can be determined in a satisfactorily exact way.

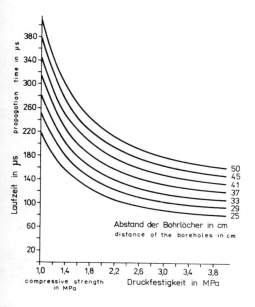

Fig. 8. Relations between the compressive strength of the filled mass and the propagation time of longitudinal waves with regard to different distances of the boreholes

INFLUENCE OF THE STABILITY AND THE CONSISTENCY OF THE FILLED MASS BY BLASTING OPERATIONS

Besides the composition of the filling material, the technology of placing in this material and the repartition of it in the mine excavation, also the drilling and blasting technologies have an influence on the quality of the filled mass. So long as the dynamic tensions which appear during the blasting operations are lower than those admissible for the filled mass, this mass will conserve its consistency properties also under the actions caused by the blasting operations. The dynamic tension σ can be determined by means of the propagation velocity of the longitudinal waves in the filled mass v_L and the velocity of the mass displacement v_M according to (5) after the following formula:

$$\sigma = \frac{10 \cdot \rho}{g} \cdot v_L \cdot v_M \text{ (MPa)}.$$

Within this formula the density ρ of the filled massivs given in kg/dm^3, the earth acceleration g in cm/s^2, v_L in cm/s, and v_M in m/s.

The velocity of the mass displacement and the longitudinal waves was determined by test blastings. The density of the filled mass was determined from the taken samples.

The specific impulse given by the columnar charge to the neighbourhood can be described as follows according to (6):

$$i = 0,1 \cdot d_L \cdot \rho_l \cdot v_D \text{ /g/cm.s/,}$$

whereby d_L designates the charge diameter /cm/, ρ_l is the charge density /g/cm³/ and v_D is the detonation velocity of the columnar charge /cm/s/. The product $\rho_l \cdot v_D$ is also designated as sound hardness or acoustic impedance of the explosive, which determines the detonation pressure of the explosive. In approximation the formula

$$p_D = \frac{\rho_{exp} \cdot v_D^2}{4} \text{ /Pa/}$$

is used for determining the detonation pressure, whereby the density of the explosive ist given in kg/dm^3, and v_D is indicated in m/s (7). It ist evident that gelatin explosives like gelatin-donarite 1, for example, or gelamon 22 with detonation pressures about sevenfold greater than those of ANFO-explosives, will have a more apparent influence on the filled mass.

But the sound hardness of the filled mass or of the solid rock, respectively, is also of importance. According to the general blasting theory a pressure wave that approaches a free face in the rock, is modified by it into a tensile wave and thus extensively reflected. There are similar phenomena between media with very sharply differing consistency proporties and showing a seperation by an interface. With a threefold greater sound hardness Mosinez (8) indicates up to 65 % of the stress waves energy being reflected in the rock, but is not transferred into filled mass.

If the drilling and blasting operations are executed with an adequate quality (that is consideration of a minimum distance of the last blasthole from the filled mass of 55 to 76 cm – when using gelatin explosives this is preseribed –, in different rocks and with a blasthole diameter of 36 mm (5), furthermore, exclusion of "cutting effects" in direction to the filled mass, and so on , one can surely prove that no negative effect on the filled mass occurs. The interface between rock and mass shows a screening effect. For safety reasons a distance of 70 cm should be kept. With the use of ANFO explosives, where poropolystyrol is added (smooth blasting), a further reduction of the minimum distance of the last blasthole from the filled mass can be expeeted. Practically, the conservation of a distance being 10 to 12 fold that of the blasthole diameter might be sufficient. In these cases, the stresses occuring in the filled mass during the blasting operation lie under the admissible values in order that the filling mass retains its consistency values.

REFERENCES

(1) Woloschtschuk, S. N.; Sereda, B. D.; Zygalow, M. N.; Kummer, G. 1965, New knowledges in filling large mine excavations with cemented filling (concrete filling). Bergakademie 16, part 4/5, pages 222 to 228

(2) Bär, H.; Konietzky, B. 1980, Sublevel mining with cemented filling - a productive mining method for winning without losses minerals of considerable economic use for society. Freiberger Forschungshefte A 623 VEB Deutscher Verlag für Grundstoffindustrie

(3) Slesarew, W. D. 1952, The effect of the rock pressure in mining coal seams in the Donezk basin. Ugletechizdat, Moscow–Leningrad

(4) Bär, H. 1981, Rationalisation of timbering in the basic mixing processesgetting and horizontal driving. Neue Bergbautechnik 11, part 11, pages 615 to 619

(5) Frenzel, M.; Tormyschow, L. M.; Saweljew, J.; Vàupel, K. H.; Tscherepanow, G. S. 1977, Examination of the explosive action of blastholes on the artificial roof in sublevel mining with filling. Neue Bergbautechnik 7, part 4, pages 271 to 273

(6) Mindeli, E. O. 1974, Rock breaking. Verlag Nedra, Moscow, page 270

(7) Jendersie, H.; Harzt, D.; Dietze, R. 1977, Small manuals: Getting and blasting technique 1 and 2, page 22

(8) Mosinez, W. N. 1965, Fundamentals of the process of rock breaking trough blasting and methods of reducing blasting energy. Akademie der Wissenschaften der Kirgisischen SSR, Verlag Ilim, Frunse

Usage of downward tailings-fill in the ore No.5
of Xiang-II Mine

HUANG WENDIAN & MEI RENZHONG
Jiangxi Metallurgical Institute, China

ABSTRACT: The ore No 5 of Xiang-II Mine was one of its lead-zinc orebodies with the highest grade. It had slope angle of 30-70 degrees, thickness of 4-12m and length of 50-60m. The ore was very crumbly, not resistant and partly oxidized. The hanging and foot walls adjoining rocks were not extremely steady. The surface allowed of no caving. The geological conditions were quite complicated. According to these, the mine had utilized the downward tailings-fill method.

The sizes of block for using this method: The block height was 36m, block length was 50-60m and width was equal to horizontal thickness of the orebody. Development: Developing with footwall fringe-drift and cross drift was at the main horizon. At the two ends and the middle of a block reef raises were arranged for the use of pedestrianizing, ventilating and drawing. Stoping: The whole orebody was not divided into interchamber pillar and cap-floor pillar, slicing from top to bottom with slice height of 2.6-3.0m. Slice cutout drift was driven along the orebody strike from the two sides or one side of the middle raise, and then the extraction drift and fan-shaped face were advanced on concrete circumstances. When slicing, shallow hole drilling equipment and ammon-dynamite were used for breaking ore , the ores were carried away with scraper, timber set posts and timber posts were utilized for support (concrete brick posts were successfully tested,too) and mat was built with steel concrete. Classifying-desliming tailings were used to fill every worked-out slice.

This mining method was constructed in 1971, first meeting with a successful experiment, and leading to good technical & economical results in our country.

1 GEOLOGICAL FEATURE

The ore No 5 of Xiang-II Mine was one of its lead-zinc orebodies with the highest grade, being the distance of 37m from the surface. There were great changes in the trend, slope and thickness of ore No 5. In configuration, its higher and lower parts were small and middle part was large. The orebody trended approximately from south to north, slanting towards the east. It had the length of 20-60m and deepened 40-50m, with dip angle of 30-70 degrees and thickness of 4-12m. The ore was very broken, partly oxidized and not extremely steady. Its Protodyakonov coefficient f was equal to 4-5. And there were old workings on its higher part. The hanging wall was yellow mud through which the water seeped. The foot wall was decayed igneous rock. The contact band between the country rock and the orebody was carbonaceous & argillaceous shale and kaoline. The hanging wall and foot wall were not extremely steady. The surface was a depression in which the drain area was 7.58 sq. m. It allowed of no caving.

2 THE BASIS OF SELECTING MINING METHOD

According to the geological feature stated above and the requirements for that ore loss & dilution in the process of stoping were little, the mining intensity was great , the operation was safe and the ventilating condition was good , many kinds of schemes had been worked out to compare with each other about its mining methods. For example, using square-set and back-fill method, the wood consumption was great , the support work was more difficult, and the skilled workers for mining were lacking, while, using top-slice caving method, the surface would

make an appearance of caving. After making
repeated on-the-spot investigations, it
had been decided that downward tailings-
fill method should be used. The usage of
this method was the first attempt in our
metal mines. Through the experiment, a
certain experience had been gained. This
can be used for reference in mining the
similar orebodies in future and offered
to the persons of same occupation in
internationality.

3 MINING METHOD

3.1 Development

The two raises at the off and port sides
of ore No 5 had been originally driven.
For the drawing at the intermediate section of the stope, one raise was added in
the middle of the orebody. All the raises
were separated into two divisions. One
was used for drawing and another was used
for ventilating & pedestranizing. With
the downward stoping of the orebody, the
support erection was being carried out
slice by slice from above to below at the
extracted-out parts of the raise wall.

3.2 Cut and stope

Interchamber, cap and floor pillars were
not left in the whole orebody, slicing
from top to bottom. The length and width
of every slice were respectively equal to
the length and horizontal thickness of
the orebody. The height of every slice

Fig.1. downward flat-back
tailings-fill

1— slice drift
2— filled long extraction drift
3— filled short extraction drift
4— fan-shaped long slot
    stoping face
5— steel concrete floor
6— concrete pre-made brick post
7— pine post with crossbeam
8— fir set post
9— separated water-filtering
    bamboo mat
10— sand gate

was 2.6-3m . Cutting work was done first
from two sides or one side of the middle
raise along the orebody strike
with slice-stope drifts being driven to
link up with the raises at the off and
port sides of the orebody to the
advantage of ventilation and transportation.

The slicing way was to utilize either
extraction drift or fan-shaped long face
on the basis of concrete circumstances.
Under the conditions that the length and
thickness of the orebody at higher and
lower slices were just the same or those
at higher slice were a bit too great, it
was more reasonable to utilize fan-shape
long face. If the higher part of the
orebody had been small and it had not
been in succession along the strike
(the area of higher slice mat had been
less than that of lower slice orebody)
the mat would have become a cantilever
beam or the whole mat would have
suspended in midair, thus leading to the
pressure acting upon it. For that reason
to stope with the extraction drifts was
safer under the circumstances. So stoping
with long extraction drifts along the
strike was carried out when the thickness
of the orebody at a few slices of its
higher part was less than 6m and the
orebody was not in succession, while
stoping with short extraction drifts
along skewed crossing strike was carried
out when the thickness of the orebody
was more than 6m. See Fig.1. about the
mining in the stope.

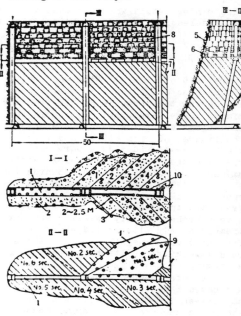

Owing to a little changes of the orebody at its a few lower slices, stoping with fan-shaped face was carried out. In stoping order each of slices was divided into several sub-sections, and the area of every sub-section was controlled within 100 sq. m. Separatedly stoping with extraction drifts, 1,3,5 should be stoped and then 2,4,6. The width of extraction drift was 2-2.5m. Fan-shaped face was advanced to 2m in every stoping circulation. Stoping with extraction drifts, it was difficult and inefficient to carry ore with scraper and face arrangement was simple.

In extraction drilling, 01-30 or YT-25 type rock drill was used. The depth of blast hole was 1.6-2m. The arrangement of holes in stoping with extraction drifts was the same for that in driving. The arrangement of holes in stoping with fan-shaped face was the same for that in stoping with long wall. The fire blasting cap and ammonium nitrate exposive were used for detonation. The freed ore was carried away by the scraper with power of 7 and 13kw. Fresh air was inletted into the stope through lower raise and contaminated air was exhausted out of the stope through higher rsise on the way. Stoping with extraction drifts, timber set posts were used for the stope support, their separation being 0.8-1m. In stoping with fan-shaped long face, timber posts with bars were used in rows, the distance between rows being 2m and the separation of posts being 0.8m. In order to economize timbers concrete brick posts were used for experiment at two slices instead of timbers. The specification of a concrete brick was 400mm 200mm 170mm. They were used to build brick posts with the cross section of 400mm 400mm, the distance between rows being 2m and the separation of posts being 1-1.5m. The supporting efficiency was very good and the timbers had been economized about 0 %.

.3 Caculation of back oressure

(1) the critical depth H_o of ground pressure

$$H_o = \frac{2\,s}{tg^2(45° - \frac{\alpha}{2})tg\alpha} = 50m$$

here s —— the half of stope span,
s = $\frac{1}{2}$ x 7 = 3.5m;
α —— the internal friction angle of rock, 56°.

Thus it can be seen that the overburden pressure of ore No 5 belonged in the pressure of shallow strata.

(2) the calculation of the back pressure P in stopes

$$P = G - 2F = 112 \text{ metric ton/m}$$
$$G = 2s\gamma H$$
$$F = 0.5\gamma H^2 tg^2(45° - \frac{\alpha}{2})\,tg\alpha$$

where H —— the distance between the orebody and the surface (about 40m),

γ —— the volume weight of rock (2 metric ton/m).

Thus it can be known from the calculation above that the pressure upon the unit area of stope back was

112/7 = 16 (metric ton/m).

(3) the checking calculation of the carrying capacity P of the timber posts in stope

$$p = \frac{\sqrt{KJ}\ F}{n} = 33 \text{ metric ton,}$$

$$\sqrt{KJ} = 249 - 1.65\frac{1}{J} = 157.6 \text{ metric ton/cm}$$

where F —— the cross-sectional area of pine post,
$$F = \frac{\pi d^2}{4} \text{ cm}^2 ;$$
d —— the diameter of the post (20 cm);

n —— the assurance factor of the pressive strength of the post (1.5);

l —— the length of the post (280 cm);

J —— the inertia moment of the post,
$$J = \frac{d}{4} ;$$

Obviously, every pine post sharing the

143

bearing area of 2 sq. m. was able to resist the back pressure of 16 metric ton per sq. m..

3.4 Cautions in the process of stoping

(1) The hanging & foot wall adjoining rocks could allow of no exposure. Planks and thatch ought to be timely used for maintaining so as to prevent the scaling-off and the cave-in of tailings-filled body at higher slice.

(2) In order to guard against the producing of cantilever beam, shearing force and break support ought to be strengthened in time to the edge of concrete mat.

(3) The breaking-dowd could not face towards the posts. If necessary, the posts should be reinforced prior to explosion.

(4) In the particular section without any concrete mat, stoping with extraction drifts should be made. After finishing the stoping of one extractic drift, fill should be carried out imediately.

4. FILL TECHNOLOGY

4.1 The making of fill material and the making installations

The mill tailings were utilized as fill material, whose chemical composition and size grading can be seen in table 1 and

Table 1. The chemical composition of tailings

Chemical composition	Pb	Zn	S	SiO_2	Al_2O_3	CaO	MgO	Sn
Content (%)	0.14	0.10	4.27	43.15	4.14	15.81	4.12	0.06

Table 2. The size grading of tailings

Grading (mm)		Crude tailings	Deslimed tailings	Spilling tailings
Mesh	Grain diameter (mm)	(%)	(%)	(%)
+65	0.20	3.19	4.77	0.13
+100	0.15	8.66	12.87	0.11
+200	0.075	23.30	32.99	0.29
+325	0.04	34.95	43.15	11.95
-325	-0.04	29.90	6.20	87.52
Tn total		100	100	100

The specific gravity of deslimed tailings was 2.86, its loose weight was 1.68 metric ton/ cubic m and its coefficient of permeability was 140 mm/ hour. The making process of tailings was that the tailings exhausted from the mill werw pumped by sand pump into hydraulic cyclone for desliming in the early stage, the spillage out of the cyclone was drained into tailings dam through outlet pipeline, and the deposit from the cyclone was transmitted into the storage pond for dewatering. In view of these facts that the mill was farther aprt from the stope (about 4 km) and the fill-tailings quantity

needed by the experimental stope was not too great, tailings were carried by truc to the tailing bin at the surface near the stope and then through mixing chambe were made into tailings mortar which was used as the fill material, after being deslimed and dewatered nearby the mill. About the fill system, see Figure 2.

The working capacity of the hydraulic cyclone was 2-3.2 cubic m (dry tailings The sand pump's lift was 20m and rate o; flow was 3.6 cubic m / hour. The two dewatering ponds were built up with blo stones. The volume of a pond was 50 cub;

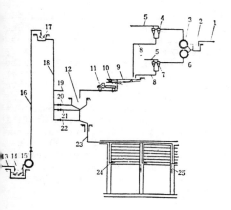

Fig.2. Tailings-fill flow diagram

1-- crude tailings
2-- balance pond
3-- 2.5 inch sand pump
4-- hydraulic cyclone (250mm)
5-- outlet pipeline of spillage
6-- reserve sand pump
7-- reserve hydraulic cyclone
8-- outlet pipeline of deposit
9-- dewatering ponds (50 cubic m)
10- loading scraper
11- carrying truck
12- tailing bin
13- discharge way for the seepage water
    from the stope
14- storage pond
15- water pump
16- force pipeline (4 inch in dia)
17- high-level storage pond (30 cubic m)
18- main flow pipeline (4 inch in dia)
19- sluice pipeline (1 inch in dia)
20- flow pipeline (3 inch in dia)
    within the tailing bin
21- flow pipeline (2 inch in dia)
    at the bin gate
22- flow pipeline (2 inch in dia)
    at the bell
23- main fill pipeline (3 inch in dia)
24- fill stope
25- discharge raise for the seepage water

metre. Scraper platform was set up at
one end of every pond for the purpose of
scraper's loading. The period of emptying
tailings into every pond was 10-12 hours.
After being dewatered in 1-2 days, the
tailings were going to be loaded and
transported.

The taling bin had the diameter of 7m,
height of 2.7m, floor slope of 15 degrees,
and volume of 115 cubic m. The water

pipeline 3 inches in dia which was circular arrangement in shape and that 1 inch
in dia which was radiative arrangement in
shape were laid at the bottom of the
tailing bin. On every circle of the water
pipeline (1 inch), there were 4 small
spouts drilled through, whose dia was
1.5mm. The separation of the circles
along the pipeline was 200mm. A tailings
chute 300mm in dia had been opened in the
centre at the bin bottom. The chute was
connected with a tailings pipeline 5
inches in dia,with a valve beneath the
chute. There was a cone cap 600mm in dia
on the chute for the use of preventing it
from being blocked up. When filling, as
long as the valve of the circular water
pipeline was turned on, the tailings
mortar could be drained away through the
chute by the density requested. About the
construction of the tailing bin, see
Figure 3.

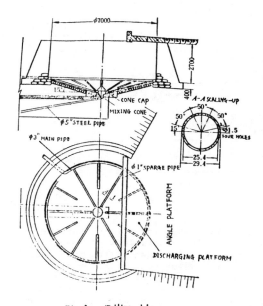

Fig.3.---Tailing bin

It was known through modelling that
when the dry tailings met with the scour
of pressure water they could be made into
tailings mortar, and that the density of
the mortar to the slope of the bin bottom
was in direct ratio. When water pressure

was 3kg/sq. cm, tne relationship between the density of tailings mortar and the slope(gradient) of the bin bottom is shown in Figure 4. When the slope of the bin bottom was 23 degrees, the relationship between water pressure and the density of tailings mortar(slurry concentration) is illustrated in Figure 5 and the relationship between water quantity and the density of tailings mortar is represented in Figure 6. The modelling result made it clear that the slope of the bin bottom was suited to the range of 15-20 degrees in order to obtain the requested mortar density of 65-68 %.

bell with the slant length of 1.25m, the slope of 25 degrees and channel section. Delivery tailings conduit was built abov the bell to make the surplus tailings mortar overflowing from the bell be able to discharge from it. The seamless steel tubes of 3-in-dia were used from beneath the bell to the raises. The ratio level-difference between the inlet α out of fill line to the total length of fill line was 2-4. Soft rubber tubes or large Nanzhu (bamboo, whose joints had been chiselled through) were erected quite close to the back in the middle of fill face. The fill line of stope may be seen in Figure 7.

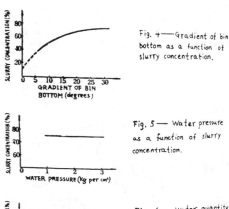

Fig. 4—Gradient of bin bottom as a function of slurry concentration.

Fig. 5— Water pressure as a function of slurry concentration.

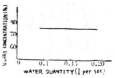

Fig. 6—Water quantity as a function of slurry concention.

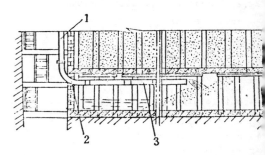

Fig.7. stope filling schematic diagram

1- fill steel tube (3-in-dia)
2- soft rubber tube (3-in-dia)
3- Nanzhu (bamboo)

The water used for filling was the internal mine water, flowing into the storage pond and then lifted into the high-level water pond by water pump. The high-level water pond was connected directly to the tailing bin by the water pipeline 4 inches in dia. The water pressure was 3kg/sq. cm.

The bin chute was closely connected with the seamless steel tube of 5-in-dia instead of mixed conduit. Its length was 28m and layout slope was 11 degrees. And its lower end was connected directly to the screen conduit covered by a wire screen with the mesh of 10 × 10mm. The conduit had a slant length of 1.3m, slope of 15 degrees and cross section of 0.5×0.4m. The lower end of tne screen conduit was connected directly to the

4.2 Fill preparation

Before filling, steel concrete mat ought to be poured. After finishing ore-carry, a cleanup of floor ought to be made, ste be paved, and then concrete be poured. The diameter of steel was 12mm. Its mesh was 200×200mm. The distribution ratio of concrete was 1:2:3 (cement-to-sand-to-crushed stone volume ratio). Having been made at the surface mixing-station, the concrete was transfered to the place abo stope and then put down to the stope through the ventilating tube (300-mm-dia in the raise-on-the-way. The concrete in the stope was raked level with the scrap pan turned up. See Figure 8. The thickness of steel concrete mat was 250-300mm.

At the worked-out sections near the or wall having need of filling, bamboo-mats were hung and nailed on the inner side o

146

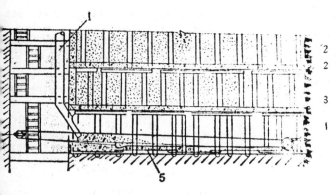

Fig.8. The pouring technology of steel concrete mat

1- the ventilating tube for dropping concrete (300-mm-dia)
2- the filled slice
3- steel concrete mat
4- scraper
5- steel (12-mm-dia, mesh 200x200mm)

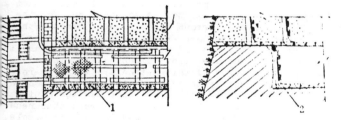

Fig.9. the hanging and nailing of bamboo-mat

1- long wood chip
2- bamboo-mat

the post. The bamboo-mats were used for separating the filled tailings and filtering water.

Finally, a concrete or red brick filtering-sand gate ought to be built near the extraction-drift mouth or raise mouth. Two or three thick-bamboo-tubes for watwer diversion were inserted in the place per 0.2m high on the gate wall. The construction of the sand gate can be seen in Figure 10.

.3 Cautions in filling

Before filling and in filling, the surface and fill spot ought to be in connection by telephone. At the beginning of filling, someone had to clean the fill pipeline with water for 2-3 minutes and check if it was unimpeded. In the process of filling, the quantity of supplying water had to be conctrolled. By the end of filling, the supplying water had to stopped, the surplus tailings mortar be drained away from delivery tailings conduit, and at last the bell & the fill pipeline be cleaned with water.

The fill order should be defined in accordance with the drainage direction. When that was in contrast to the direction of tailings injection, fill should be carried out backward. On the contrary, forward.

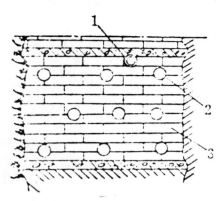

Fig.10. Sand gate

1- the rubber tube for filling
2- spill holes (120-mm-dia)
3- concrete pre-made brick

The accident of blocking pipeline usually happened at the bend of pipeline. When pipeline was blocked up, one end of the hard rubber tube at the bend ought to be opened and cleaned with water, and at the same time pipeline be stricken to make tailings flow out.

5 The organization of work

The stoping was made in 3 shifts per day. Each of shifts was 8 hours. The workers in the working face per shift were 10-12 persons. The producing personnel were 43, in whom 2 for managing, 2 for doing technique, 2 for tailings-making, 2 for maintaining equipment and 35 for mining. Specimen were in charge of filling.

6 Technical & economical indexes

(1) The stope-production-capability:
    82.7 metric ton/ day
(2) Fill capability:
    35-50 cubic m/ hour (dry tailings)
(3) Labour efficiency:
    5.1 metric ton/ man-shift (stoping)
    30 cubic m/ man-shift (filling)
    2.3 metric ton/ man-shift
       (working in the face)

(4) The working efficiency of an unit per shift:
    69 metric ton (drill)
    30-35 metric ton (scraper)
(5) Material consumption:

    0.0125 cubic m/ metric ton (wood)
    0.544 kg/ metric ton (explosive
    0.24 piece/ metric ton
        (blasting cap)
    0.62 m/ metric ton (fuse)
    0.544 kg/ metric ton (steel)
    0.039 cubic m/ metric ton (concre
    0.23 cubic m/ metric ton (tailing

(6) Direct cost for mining:

    4.12 RMB¥/ metric ton

(7) Fill cost:
    2.02 RMB¥/ metric ton
(8) Ore-loss rate:
    1.5 %
(9) Ore-dilution rate:
    3 %

7 Evaluation

The practice has made it clear that t mining method is workable one for the non-ferrous & noble metal mines with higher ore grade when the adjoining roc and ore are not extremely steady and th surface allows of no caving. The advan- tages of this method are that ore-reco- very rate is high, the ore-dilution rat is low. It is suitable for the mining o the orebodies with different thickness & slope angles. Development is simple. The ventilating condition is better tha that of slice caving. The operation under mat is safer. The pillars are not left and the stoping steps are simplifi

The existing problems about this meth are that the stoping technology is more compex, the producing capability is no high, the wood consumption is great; fi can not be contact with the mat (there is the space of 10-20cm without fill in every slice).

The following meassures are the deve- loping of this mining method in future. The mechanized level in stope must be raised and an attempt should be made at using loader to raise the capability of drawing. In order to economize high-tes cement we may utilize the cement replac ment, such as slag cement.
The tailings-cemented mat may be tried simplify the pouring technology of sand-crushedstone-cemented mat . Major efferts should be devoted to the study saving pit wood, such as the try of usi

concrete pre-made brick posts. In the
view of experimental results in two slices,
wood consumption was lowered from 0.0125
cubic m/ metric ton to 0.00618 cubic m/
mwtric ton. But the existing construction
is very difficulty. So the problem of
mechanized operation should be solved
with much attempt.

In brief, if we raise mining intensity
still further, speed filling, improve fill
material & technology, and save more wood,
this mining method will be sure to lead to
better technical & economical results.

Consolidated backfilling at Outokumpu Oy's Vihanti, Keretti and Vammala mines

V.A.KOSKELA
Outokumpu Oy, Mining Technology Group, Finland

ABSTRACT : Mining with backfilling must be justified by higher NPV of the mining project than in case of open stopes with ore pillars left for support.

The use of filling depends on the value of the ore especially when consolidated fill is considered. When consolidated fill is justified the question is to optimize the price of fill considering planning strength, capacity, filling materials and methods and the circulating speed of the stoping - filling phase.

Decisive factors in choosing the right filling method are the quality and cost of filling materials. If consentrator waste can be used the cost of the binding agent is the most important factor, always part of the consolidated fill can be replaced by cheaper material.

1 INTRODUCTION

For 30 years Outokumpu Oy has used consolidated backfill in its mines. This method was first tried out in Keretti mine using cemented backfill in 1952 and nowadays 4 other mines use consolidated filling as a part of their mining operation.

Consolidated filling at Outokumpu Oy´s mines can be divided into two groups according to the amount of water used in the filling process. Binding agent is added to conventional hydraulic waste fill from the concentrator and the mixture is transported to the stopes. In the optimum water content methods just enough water is added so that the binding agent can react. The consolidated filling method with optimum water content is used in the Vammala and Keretti mines and in Vihanti mine the hydraulic consolidated filling method is used. The mining with consolidated backfilling methods of the fore-mentioned mines will be described later in this paper.

The mines are small to medium sized. Vihanti and Keretti mine have been re-designed and modified for the use of mechanized trackless equipment. Vammala mine has been planned right from the beginning so that the most efficient use of modern equipment can be made in all of the stages of the mine operation.

2 VIHANTI MINE

The following separate orebodies are presently being exploited :
- Välisaari, mostly for zinc ore
- Ristonaho and Lampinsaari, pillar recovery of zinc ore
- E-orebody, Rämesaari, Eastern, for copper ore
- Isoaho, for zinc ore.

The mine comprises four shafts, an inclined ramp, main transport levels and two underground crushing stations. Ristonaho shaft is the main shaft for ore and personnel hoisting.

Sublevel stoping is applied either longitudinally or transversally to the strike. In the Ristonaho and Lampinsaari orebodies the stopes have been filled with classified tailings and pillar recovery is nowadays the only activity. When sublevel stoping is applied transversally consolidated fill is adopted as at Välisaari orebody.

Figure 3 shows the stoping method used in the Ristonaho and Lampinsaari orebodies. Difficulties in pillar recovery by undercut and fill method, waste dilution and low productivity of work were the main reasons which led Vihanti mine to consider consolidated filling. It started at Välisaari orebody by cemented filling. Stope width in the first phase is 30 m, as presented in

Table 1. Mine-Operation Data 1982

Mine	Vihanti	Keretti	Vammala
Start year	1954	1913	1976
Ore hoisted	963 000	405 000	321 000
Metals	Zn,Cu,Pb	Cu,Zn,Co	Ni
Electric power, MWh	15 900	12 200	6 100
Productivity ton/manshift (1)	38	12	28
Total m^3 of fill	207 000	109 000	68 000
Consolidated fill, m3	115 000	23 000	37 000
Water pumping, m3	870 000	1 600 000	760 000

(1) Tonnes of ore hoisted per total shifts including mine department, engineering services and contractors.

Figure 1. Outokumpu Oy's mines in Finland.

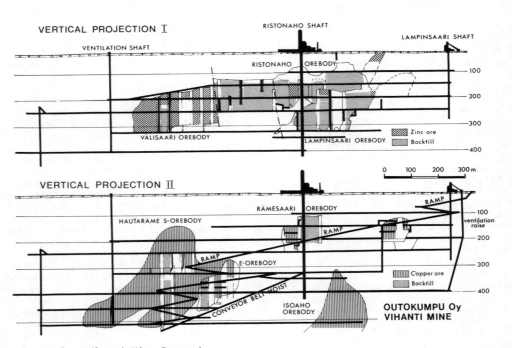

Figure 2. Vihanti Mine Lay-out.

152

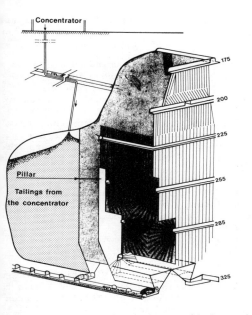

Figure 3. Sublevel Stoping Applied Longitudinally at Vihanti Mine.

figure 4, first phase stopes are filled with consolidated material. In the second phase 10 m pillars are left to support the fill while the 30 m wide stopes are empty. These pillars are blasted to the stope and loaded. The second phase stopes are filled with classified concentrator tailings.

Due to the big amount of water used in the filling process and the quality of

waste as filling material a lot of cement was needed to obtain a satisfactory pillar strength. The high price of cement made this method less economical and thus led to the research for cheaper binding agents to replace cement.

2.1 Consolidated Filling Method at Vihanti Mine

The precent consolidated filling method at Vihanti mine is based on the use of granulated blast furnace slag activated by $Ca(OH)_2$. The slag is brought from Raahe smelter to Vihanti mine by truck and the slag is stored in a silo. Between 50-60 tons of slag is used daily. Figure 5 is a flow circuit showing the slag grinding and mixing of the binding agent with the concentrator classified tailings.

Granulated slag is fed from the silo on to a belt conveyor and by a screw feeder into the ball mill. A feed maximum of 3.2 tons/h of slag can be obtained containing fineness of 2300 - 2500 cm^2/g. For every ton of slag 500 gms of grinding balls have to be added. Fine slag is pumped to the mixer where $Ca(OH)_2$ is added 1.5 % of the amount of the slag. The tailings from the concentrator are classified and conveyed to the mixer so that the slag-waste ratio is 1:11. Pulp density is controlled in the mixer at 40 % water per unit weight. The capacity of the process is 22.8 m^3/h of consolidated fill.

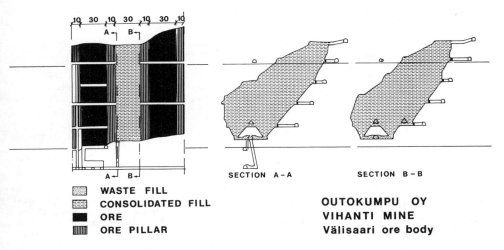

WASTE FILL
CONSOLIDATED FILL
ORE
ORE PILLAR

SECTION A-A   SECTION B-B

**OUTOKUMPU OY
VIHANTI MINE
Välisaari ore body**

Figure 4. Sublevel Stoping at Vihanti Välisaari Orebody.

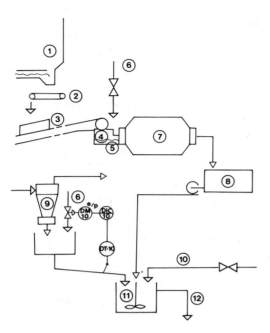

the middle of the stope. Because of classi
fication fine material and a larger amount
of slag flow against the stope walls leavi
the centre of the fill weaker.

Loading points and sublevels are built f
filling. Dams are erected at the lowest leve
to resist a pressure of 0.5 MPa and in sub
levels for 0.25 MPa pressure. Construction
of the dams is by shotcreting. For drainag
filtering lines are built from the bottom
up to each sublevel. Lines are of the same
type of plastic pipes used by farmers for
draining fields. The bottom of the stope i
filled with coarse tailings before the
consolidated fill is introduced.

At Vihanti 1 m^3 of consolidated fill
contains :
- 140 kg granulated slag
- 1.8 kg $Ca(OH)_2$
- 1 540 kg classified concentrator
  tailings
- 1 120 kg H_2O

The hardening time of Vihanti type conso
lidated fill is 3-5 years. After 1 year th
compressive strength of the fill has been
1.05 MPa in the stope. The area of fill to
opened at Välisaari orebody is 100 m x 60
(height x width)

Figure 5. Granulated Slag Grinding and
Mixing Station at Vihanti Mine.

1. Slag silo	8. Pump well
2. Belt feeder	9. Waste cyclone
3. Belt conveyor	10. Activator $Ca(OH)_2$
4. Box	11. Mixer
5. Screw feeder	12. Consolidated fill
6. Water	
7. Ball mill Ø 1800 x 3600, 100 kW, 27 r/min	

Table 2. Screen Analysis of Grinded Slag
and Classified Waste.

Sieve, mm	Slag % through	Waste % through
0210		93.0
0.149		82.0
0.105		52.0
0.074	88.7	32.0
0.040	58.0	
0.020	33.0	
0.010	23.0	

Table 3. Relative Filling Costs at Vihant
Mine 1982 at Välisaari Orebody (1).

	Consolidated fill	Wast fill
Relative amount of fill	0.6	1
Relative cost of fill	4.9	1
Slag, $Ca(OH)_2$, grinding %	84.0	–
Mixing, waste %	7.6	31.
Pipelines %	1.9	36.
Dams drainage ext. %	6.5	31.
	100.0	100.

(1) Capital is not included

Considering the Välisaari orebody the
total cost of fill used is 2.46. If cement
is used as a binding agent the total cost
would be 7.84.

3 KERETTI MINE

The orebody is an oblong plate cut by faul
It is 4 km long 300-400 m wide and usually
less than 10 m thick, although it attains
thickness of 30 to 40 m. The mine has been
in operation since 1913 in the eastern,
central and upper sectors. Nowadays the
principal mining area is located near the
Keretti shaft, which is the main shaft use

Mixed consolidated fill is taken from the
concentrator through Ø 90 mm polyethylene
pipelines 700 m to the stopes. Where the
pipeline is vertical it is placed in Ø
125 mm holes.

The fill is led from the hanging wall to

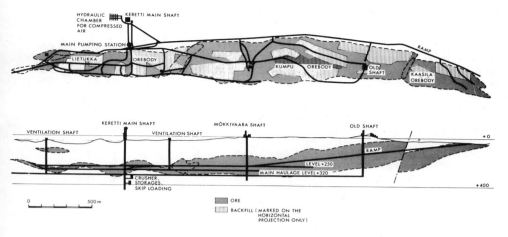

HYDRAULIC CHAMBER FOR COMPRESSED AIR — KERETTI MAIN SHAFT

MAIN PUMPING STATION

LIETUKKA OREBODY

KUMPU OREBODY

OLD SHAFT

KAASILA OREBODY

RAMP

VENTILATION SHAFT — KERETTI MAIN SHAFT — VENTILATION SHAFT — MÖKKIVAARA SHAFT — OLD SHAFT

+0

RAMP

LEVEL +250

MAIN HAULAGE LEVEL +320

CRUSHER, STORAGES, SKIP LOADING

+400

0   500 m

☐ ORE

▥ BACKFILL (MARKED ON THE HORIZONTAL PROJECTION ONLY)

**OUTOKUMPU Oy
KERETTI MINE**

Figure 6.   Keretti Mine Lay-out.

for personnel and ore hoisting. The mine has five shafts altogether, a ramp, three main levels and an underground crushing station.

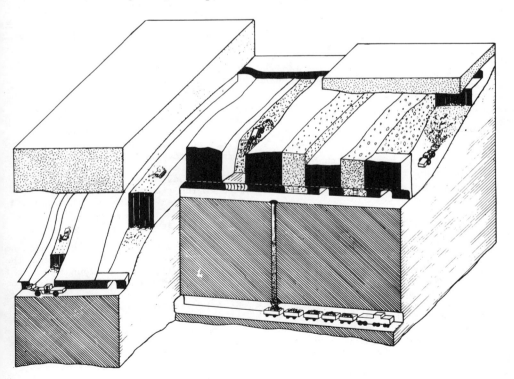

Figure 7.   Concrete Pillar Stoping at the Keretti Mine.

The main stoping method is a modification of cut and fill called 'Concrete Pillar Stoping'. In this system, as presented in figure 7, the mining area is divided into 6 and 8 m parallel rooms. In the first phase 6 m wide rooms are stoped and filled with concrete. In the second phase the 8 m wide ore pillars are stoped out and filled with classified tailings. For the concrete pillar method a mixing station was located at Mökkivaara from where the wet mixture was transported by trucks on the surface to the holes. Later on wet mixture was replaced by dry and water was added at the holes. Because of classification due to the amount of water used the concrete pillars proved to be too weak which caused dilution in the second phase.

Some stest were made to find out how strong the concrete fill could be made using sand as the aggregate with an optimum amount of water. The amount of cement used in these tests was 110 kg/m³ of fill and the strength of the fill was measured after 28 days.

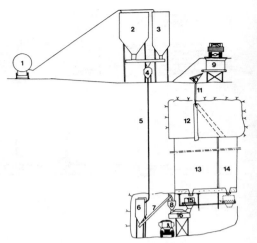

Figure 8. Underground Filling Station at Keretti Mine.
1. Railway wagon
2. Slag silo, 90 ton
3. Cement silo, 30 ton
4. Pneumatic sender
5. Ø 150 mm hole
6. Day-silos, 5 m³
7. Screw conveyor
8. Weighing for cement and slag
9. Feeder for gravel
10. Vibrating scre
11. Ø 250 mm hole
12. Pipe
13. Concrete fill gravel silo, 700 m³ (-50 mm
14. Shotcreting gravel silo, 300 m³ (-16 mm
15. Weighing for gravel
16. Mixer
17. Transport vehicle

Table 4.

	Compressive Strength MPa
Wet mixture with water	0.5
Dry mixture with water	0.4
Optimum water mixture by trucks	1.1
Optimum water mixture by slinger belt	2.3
Optimum water mixture by shotcreting	6.9

The result of the tests showed that the compressive strenght increased in compacted optimum water content backfill.

3.1 Consolidated Filling Method at Keretti Mine

Binding agent is brought to Keretti by rail and fed pneumatically to a silo. There are two silos, one for cement and the other for slag. In 1982 only cement was used. Cement or slag is pneumatically fed through 100 mm plastic pipe to respective day-silos at the concrete station. Screw feeders transport the material from the silos to a weighing scoop and mixer.

Gravel is fed into the mine through 250 mm diameter hole . A moveable pipe directs the gravel ment for shotcreting

to the 300 m³ silo while filling gravel drops straight down. The silos are positio ed above the mixer and the gravel is gravi fed into the weighing scoop and automatica ly dropped into the mixer. The mixed conc rete mass is emptied directly into waiting trucks by means of a vibrating chute, trans ported to the stoping area and dumped.

The whole system is fully automated and the composition of the concrete can easily be changed at any time. The capacity of t operation is 35 m³/h.

The optimum water content method does not need any drainage. The tunnels through the fill and the walls are of light constructic A wooden body of 'dams' is covered by Terra 1000 cloth.

The filling of the stopes takes place mainly in two stages.
First concrete is dumped into the stopes by trucks. In this phase the capacity is 150 m³/shift. When excessive height of the stope prevents the use of conventional trucks the slinger belt truck fills the remaining room with a capacity of 75 m³/shift.

Table 5. Screen Analysis of Gravel for Filling

Sieve, mm	% through
0.125	10.1
0.25	16.6
0.5	22.7
1	43.7
2	54.5
4	67.6
8	76.4
16	82.4
32	100

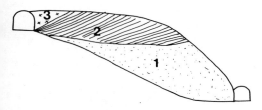

Figure 9. Different Stages of Concrete Fill.

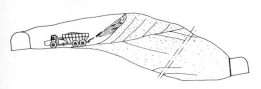

Figure 10. Concrete Fill Vehicle with Hydraulic Slinger Belt at Keretti Mine. Slinger belt truck :
- transport vehicle Volvo BM 7.5 m^3
- sliding frontwall
- stowing machine with slinger belt, 15 kW, 1500 r/min
- capacity ~75 m^3/shift

At Keretti mine 1 m^3 of consolidated fill contains :
100 kg cement
1 470 kg gravel
30-70 kg H_2O

The hardening time of the Keretti type consolidated fill is 1 month. After that the compressive strength of fill is 2.1 MPa containing 20-25 % air. The area of fill to be opened is 20 m x 60 m (height x length) respectively.

Table 6. Relative Filling Costs at Keretti Mine 1982 (1)

		Consolidated Fill	Waste Fill
Relative amount of fill		0.75	1
Relative cost of fill		3.78	1
Cement	%	39.3	–
Mixing, gravel/waste	%	43.8	10.0
Transportation	%	5.9	40.6
Dams ext.	%	11.0	49.4
		100.0	100.0

The total cost of fill related to local waste fill is 2.19.

4 VAMMALA MINE

The orebody is lens shaped about 500 m long, steep, dipping SSW and SW flattening at depth to the horizontal, plunging 40° NW. The average width varies from 5 to 30 m.

This is the most recent mine opened by the company. Ore hoisting is by trucks along the decline which has an inclination of 1:8. Full scale underground mining began in the autunm of 1978. The ore is exploited by transverse sublevel stoping. The standard stope width is 15 m and 15 m thick pillars are initially left between each pair of stopes. In the first phase stopes are filled with concrete. After the concrete pillars have hardened the second phase stopes are mined between the consolidated fill and filled with classified concentrator tailings.

The concentrator classified tailings are stored to be used later for consolidated fill or used straight away for waste filling. On the storing area exist 3 pools from where a contractor takes the classified waste by truck to the filling station about 2 km away.

The trucks dump the waste onto the screen above a little silo. From the silo a wagon feeder and a belt conveyor take the waste to a vibrating screen where lumps are broken up. Finally the waste is moved by a belt to a OKVA 200 mixer designed by Outokumpu.

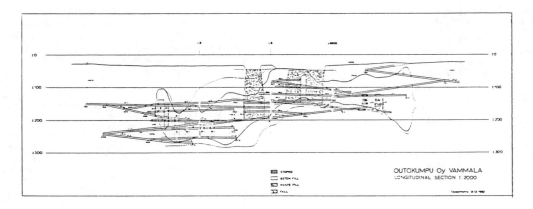

Figure 11. Outokumpu Oy Vammala Mine

4.1 Consolidated Filling Method at Vammala Mine

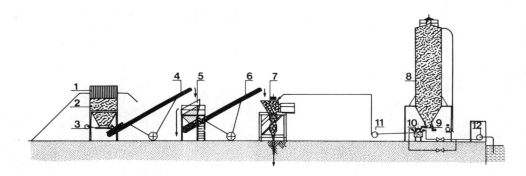

Figure 12. Filling Station at Vammala Mine.

1. Screen
2. Dim
3. Wagon feeder
4. Belt conveyor 800 x 1200
5. Vibrating screen
6. Belt conveyor
7. OKVA 200 mixer
8. Cement bin
9. Screw feeder
10. Grout mixer
11. Pump
12. Water pump

Cement has been used as a binding material since 1982 but slag is being used to re-place it and cement is only used as an activator. The binding agent is brought by truck to the mine site and emptied pneu-matically into a 125 m³ silo. From the silo a screw feeder takes cement to a mixer where water is added. The amount of water depends on the moisture content of the waste. Cement and water are pumped as grout to the concre-te mixing hopper OKVA200. The hopper is positioned directly over a 10" hole bored from the surface to the stope to be filled.

The capacity of the filling station is 45 m³/h when waste is used. If -20 mm gravel is added to the waste the capacity is 55-60 m³/h. The filling-station is operated by 1 man/shift.

Dams are constructed by shotcreting and are made to resist the same pressure as at Vihanti mine. Drainage lines are placed into the fill allthough water amount is small compared to that used in hydraulic methods. Nowadays research into the lightning of the dams and the drainage system is active.

One cubic meter of consolidated fill at Vammala mine contains :

100-105 kg   cement
200-220 kg   water
1600   kg   waste

Table 7. Sieve Analysis of Waste at Vammala Mine

Sieve, mm	% through
0.210	94.6
0.149	85.7
0.105	60.8
0.074	37.4
0.037	9.1
0.020	3.3
0.010	2.0

Compressive strength achieved with this mixture is 1.5 MPa. The hardnening time is 8 months. The area of fill to be opened at Vammala mine is 50 x 40-70 m (height x width) respectively.

Table 8. Relative Filling Cost at Vammala Mine

		Consolidated fill	Waste fill
Relative amount of fill		1	1
Relative cost of fill		13.5	1
Cement	%	62.6	–
Waste	%	23.2	32.0
Filling station	%	6.7	27.0
Pipelines, holes, dams ext.	%	7.5	41.0
		100.0	100.0

The total price of fill at Vammala mine compared to the price of local waste fill with quality ratio 1:1 is 7.25. When the present binding agent is replaced with slag, and cement is used as an activator the relative cost is 5.3.

5 COMPARISON OF THE PRESENTED METHODS

With the presented filling methods planned results have been achieved in practice. Comparing the hydraulic and optimum water content methods the following aspects can be noted :
- need for binding agent is 1.5 times greater in hydraulic method
- safety factor needs to be two times greater in hydraulic method
- capacity depends on the concentrator output in the hydraulic method
- hardening time (circulating speed) is 10-20 times longer in the hydraulic method
- drainage is essential in the hydraulic method
- the investment and running costs of the hydraulic method are lower.

Table 8. Filling at Vihanti, Keretti and Vammala

Item	Mine		
	Vihanti	Keretti	Vammala
Stoping method	Sub. stoping	Concrete pillar	Sub. stoping
Consolidated filling method	Hydraulic	Optimum W	Optimum W
Capacity of c. filling m^3/h	22.8	35.0	45
Operaging shifts/day	3	2	2 (1)
Binding agent	Slag	Cement	Cement
Aggregate	Waste	Gravel	Waste
Compressive strentgth MPa	1.0	2.1	1.5
Hardening time months	36-50	1	5-8
Water kg/m^3, consolidated fill	700-1100	70	200
Quality ratio waste fill/ cons. fill	0.6:1	0.75:1	1:1
Relative price of cons. fill (2)	4.9	56.3	21.3
Relative price of waste fill (2)	1	13.4	1.5
Relative price of total fill (2)	2.46	29.3	11.4

(1) not during winter months
(2) compared to Vihanti waste fill

6 CONCLUSION

To choose among consolidated filling methods
the one which gives optimum result for mines
purpose means that several factors must be
considered. After planning capacity and
strength of fill are desided must possible
aggregates and price be solved. The binding
agent and it´s amount for aggregates available can be tested out. When the stoping
method and the drying time of fill are in
balance the right method is the one with
minimum cost.

Mining with backfill at Kidd Creek No.2 Mine

D.L.MCKAY & J.D.DUKE
Kidd Creek Mines Ltd., Timmins, Canada

ABSTRACT. Mining the rich ore below 800 metres at the Kidd Creek Mine required a new approach to open stope blasthole mining. A backfill, made from available materials namely waste rock from the Open Pit, is crushed and sized into aggregate particles and mixed with a slurry of cement and water. This mixture, when fully cured, shows strengths in the range of 5 to 7 megapascals. In other words, we are dealing with a backfill which is essentially a weak concrete.

The mining method used at Kidd Creek is Blasthole Open Stoping. In the upper section (Kidd Creek No.1 Mine) a system of stopes and pillars are employed. Because of the depth at Kidd Creek No.2 Mine (800 metres plus), a totally new method of Blasthole Open Stoping was researched where no pillars are considered. Here mining commences at the centre of the orebody and as each stope is mined and filled (with consolidated fill), the adjacent stope is excavated. The mining against the consolidated fill has proven successful due mainly to the nature of the backfill, the size and shape of the stopes and the sequence in which they are mined.

INTRODUCTION

The Kidd Creek Mine is situated in Kidd Township forming part of the famous Porcupine Mining District. The mine itself is located 27 kilometres north of Timmins, Ontario which is approximately 700 kilometres north of Toronto, Ontario.

The orebody was discovered by diamond drilling in April 1964 after extensive airborne magnetometer surveys were carried out over the previous half dozen years. Following discovery, crews began metallurgical testing and construction of a mining plant so that mining and milling could start in earnest by the end of 1967. Since that time, a total of 53,500,000 tonnes of copper-zinc ore have been mined up to the end of 1982.

Mining at Kidd Creek Mine has followed three orderly stages. The first stage was an open pit to a depth of 220 metres which supplied the initial ore in 1966 until 1977. See Figure 1. A total of 28,300,000 tonnes were mined before the pit was exhausted. Prior to the completion of the open pit the second stage of mining or No.1

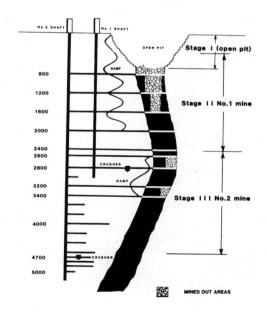

Fig.1 – Cross section showing three stages of mining.

Underground Mine, was developed. This mine replaced the open pit and now recovers that ore below the pit floor at 220 metres to a depth of 790 metres underground. The No.1 Mine reached full production of 273,000 tonnes monthly upon the completion of the open pit in 1977.

A second underground facility to mine ore below 790 metres to the 1400 metre horizon was started in 1974 with the first ore being hoisted in 1978. Full production was achieved late in 1981. The third stage of mining or No.2 Underground Mine was built in order to increase production by one third to a total monthly tonnage of 364,000 tonnes.

KIDD CREEK GEOLOGY

The Kidd Creek deposit of Archean Age, is one of the largest underground volcanogenic base metal sulphide deposits in the world. The deposit, which exhibits both complex structure and metal variation, attains a maximum width of 168 metres, a maximum length of 670 metres and is known to extend to a depth of 1,525 metres.

The deposit is divided into a north and south orebody divided by the middle shear or east-west fault. Immediately along this fault the orebody strikes east-west and is termed the middle orebody.

The Kidd Creek ores are composed of three major petrologic types, stringer ore, massive banded and bedded sulphides, and fragmental or sulphide breccia ore. The ores were formed syngenetically at the top of the Kidd Creek felsic pile, specifically at the mafic/felsic interface.

BACKFILL PREPARATION

Before underground mining began, it was known that backfill would be required if full recovery of the rich ore at Kidd Creek was to be realized. With both No.1 and No.2 Mines proceeding at an annual productive capacity of 4,500,000 tonnes of ore, it would require a total of 3,000,000 tonnes of backfill each year. It was estimated that approximately 75% of the backfill used would be consolidated fill made up of crushed rock and cement slurry. The remaining 25% of the fill would be run-of-mine waste rock.

In 1969 when No.1 Mine was under development, studies began on various types of backfill and methods for its placement. It

was agreed early in the planning that a competent backfill was necessary for support of walls and for reduction of dilution. These studies have been ongoing ever since and are continuously studied and monitored by the Rock Mechanics section of Mine Engineering.

REVIEW OF FILL TYPES

Prior to embarking on a fill type and distribution system, various types were studied from mines around the world. Visits were made to a number of mines in Canada from east to west coasts. Out-of-country visits were made to such mining countries as Sweden, Finland, South Africa, Rhodesia, Australia and the United States. Each mine visit contributed in one way or another to the final design of the present fill system.

The types of fill studied were:
1. gravel (sand)
2. waste rock
3. deslimed mill tailings
4. fill mixture (waste rock and mill tailings)
5. cemented mill tailings
6. consolidated fill
7. smelter slag

Although gravel was studied very closely, it was ruled out because it did not have the strength characteristics required, nor was there sufficient available gravel near the mine to meet the need. Waste rock was readily available from the 55,000,000 tonne dumps created from open pit mining. Unfortunately, waste by itself did not have the strength characteristics desired for the mining of pillars.

In deslimed mill tailings, which are a common source of backfill in use all over the world, we found two disadvantages in its use. First of all, the concentrator is 28 kilometres by railway from the Minesite. Therefore, fill would have to be either pumped via a 28-kilometre pipeline specially insulated for the Timmins winter, or hauled on the mine railway as a back-haul from the concentrator. The back-haul did look attractive even though a means had to be found to insulate the rail cars so that the wet fill would not freeze in the cars. However, the main reason for ruling out tailings was the fact that the ore has to be ground very finely in order to achieve liberation of the ore minerals for the flotation process. In other words, there were insufficient suitable mill tailings for backfilling.

The use of a mixture of rock and mill tailings as well as cemented mill tailings were ruled out for the same reason as above, namely the insufficient amount of mill tailings obtainable. Consolidated rockfill looked attractive because we did have in excess of 55,000,000 tonnes of run-of-mine waste rock on surface.

When these studies were being conducted, the use of smelter slag was not considered as it was not available at that time. With the start-up of the new copper smelter there are plans now underway to utilize some slag in the fill much in the same way that Mount Isa does in their filling operation.

A case for consolidated fill is made in a paper "The Development and Design of a Cemented Rock Filling System at the Mount Isa Mine, Australia", by K.E. Mathews and F.E. Kalsehagen. In that particular paper, the authors develop a good case for the reduction of dilution using consolidated fill. They go on to show the relationship between various ore grades and filling expenditures where a 15% return on investment is achieved.

To put the case more simply, consider the difference between $20 and $100 ore with a 25% dilution. With $20 ore and 25% dilution, we see a reduction in mine ore value of $5 to a net of $15 per tonne. Any effort or cost to reduce dilution therefore must be less than $5 or otherwise it is uneconomic. At this low value, it is next to impossible; therefore, it may be cheaper to accept the dilution.

On the other hand, when dealing with $100 ore at 25% dilution, we see a reduction in mine value of $25 to a net of $75 per tonne. With $25 to work with, one can afford to spend money on fill to reduce dilution and thereby increase the value of the ore mined.

The ore at Kidd Creek Mine is closer to the example of the higher grade. Consequently, it was decided early to strive for 100% recovery of all ore with minimum dilution. In other words, the mining and filling methods would be designed to maximize profits.

PRELIMINARY STUDIES

In order to achieve minimum dilution, it was necessary to obtain a fill that could stand unsupported for a height of 120 metres. To achieve this, we needed a fill with a compressive strength of 3.5 megapascals. We arbitrarily doubled this value to 7 megapascals to give ourselves a safety factor of 2.

The early thinking back in 1969 and 1970 was to crush waste from pit waste dumps on surface to minus one centimetre and pump this underground with a cement slurry to the stopes. Various crushing and pumping tests were conducted by outside laboratories on pulp samples containing 60, 65 and 70% by weight of crushed solids. Although a number of problems were encountered with the pumping of these pulps, the main reason for abandoning this approach was the cost of crushing and sizing the stone and the cost of cement required to achieve the required strength.

DESIGN MIX

In 1970, the idea of pumping a mixture of finely crushed stone and cement slurry was abandoned. In the meantime, other methods were studied particularly those of Geco Mines (Manitouwadge, Canada) and Mount Isa (Australia). At Geco, rock is quarried on surface and channeled down over broken ore in the stope to be filled. The ore is pulled evenly to prevent the waste rock placed above from rat-holing through the ore. They then pour cemented hydraulic fill (tailings to cement ratio at 30:1) on top of the quarried rock in the stope which fills the voids and knits the mass together. It is reported that heights of 60 metres for free standing walls are achieved.

At Mount Isa, the cemented rockfill is produced by introducing rockfill and cemented hydraulic fill into the same pass concurrently at an average ratio of 2:1. The rockfill is mined from a siltstone quarry and is crushed to a size of 100% passing 30 centimetres and 5% passing 3 centimetres. The cemented hydraulic fill consists of Portland cement, ground copper furnace slag, and tailings in the ratio of one part cement to two parts slag to make a pulp density of 68% to 72 % solids by weight. Strengths of one megapascal for unconfined compression were reported. Also, several large fill exposures in the order of 90 metres wide and 60 metres high were achieved.

Laboratory and field tests were conducted in 1973 to obtain a design mix for the fill required at Kidd Creek. After a number of trials the mix which consistently gave good results was that of:

163

75% - 15 + 1 centimetre
25% - 1 centimetre + 100 mesh
with a 5% weight of cement mixed with an
equal weight of water. This mixture is a
variation of a mixture used by Hydro Quebec
in one of their earth dams.

After casting a number of cylinders, the
engineers and technicians conducted various
drop tests both on surface and underground.
The drop tests were performed to determine
the effect of segregation and degradation.
Also at the same time experiments were conducted on the methods of mixing.

DESIGN AND CONSTRUCTION OF AGGREGATE CRUSHING PLANT

Design and engineering for an aggregate
crushing and screening plant commenced in
the spring of 1974 after the design mix had
been approved. Crushing and screening
tests were made by various manufacturers of
crushers and screens. After a series of
tests were made, we adopted a flow sheet
which consisted of three stages of crushing
and two stages of screening.

The existing 54-74 primary gyratory crusher used for open pit mining is included
in the flow sheet to crush the run-of-mine
waste to minus 15 centimetres. From the
primary crusher the material is conveyed
to a 400 tonne surge bin to be fed into a
48 S Telsmith crusher which also existed as
part of the plant when the open pit was
operating. In order to increase the
throughput of the Telsmith, a scalping
screen was installed before the Telsmith to
remove all the minus 3 centimetre fraction.
From the Telsmith, the material is routed
to a double decked screen where all the
minus one centimetre is screened off and
sent to the fine bin while part of the
coarse fraction is sent onto a gyra disc
(tertiary) crusher. The product from this
crusher is again screened with part of the
coarse directed back to the tertiary crusher while the fines are mixed with the
first fine product. The two products
obtained from the plant are (-15 + 1
centimetre) aggregate and (-1 centimetre
+ 100 mesh) sand in the ratio of 3 to 1,
aggregate to sand.

The rated capacity of the existing plant
is 5,500 tonnes per day. The plant was
doubled in size recently to 11,000 tonnes
per day with the addition of another Telsmith and gyra disc crusher each, along
with screens.

After the flow sheet was adopted, Kidd
Creek engineers and designers commenced
detailed work on site. The surface and
underground plants to be designed and constructed consists of:
a) An aggregate preparation plant i.e.,
a plant to take run-of-mine rock and reduce
this material into aggregate and sand.
b) A cement handling plant i.e., a load-
in facility that could receive bulk cement
from either rail hopper cars or tanker and
transport this to surface cement storage
silos.
c) A slurry mixing plant i.e., a mixing
station where water and cement are mixed in
equal proportions in measured batches which
can be sent underground on demand.
d) An aggregate and sand transport system i.e., a means of taking the aggregate
and sand underground, mixing these to their
proper proportions, and transporting these
to the orebody.
e) A slurry transport system i.e., a
means of pipelining the slurry from the
slurry plant to receiving tanks underground
where the slurry is mixed with the fill.

Construction commenced in the summer of
1974 and continued all that year. By snow-
fall, all civil work was completed and the
building was enclosed so that the mechanical and electrical work could continue over
the winter. The plant was ready for trial
early in the summer of 1975.

METHOD OF MIXING

Along with the design of the surface preparation and the underground transport
plants, it was necessary to determine the
best method to mix the aggregate, sand and
slurry.

The mixing of the aggregate with the sand
is done by feeding these onto the same conveyor underground. The feeder below the
aggregate bin feeds three times more material than the feeder below the sand bin,
thereby achieving the 3 to 1 ratio required
in the design mix.

The mixing of the aggregate and sand with
the slurry required some brain storming.
The method adopted initially was to spray
the fill mixture into a baffled raise. The
slurry coated the fill mixture as it cas-
caded down the raise falling from one baf-
fle to another. The mixing tests conducted
on surface showed a 20 metre raise was
needed with baffles every three metres to
obtain good coating of the particles. It

164

was later proven underground that 20 metres of raise was not necessary as the material became well mixed through handling as it was being transported to the stopes. It was found good mixing could be obtained by bypassing the mixing raise altogether and discharging the fill and slurry through a two metre diameter culvert pipe three metres long with two baffles in it mounted directly above the discharge point in the stope to be filled.

Another method of mixing was by slushing. In this method, a load of fill was dumped into a slusher trench; with slurry sprayed on top.

The action of the scraper dragging the backfill from one end of the trench to the discharge point over the stope mixes the fill and slurry.

A third method was to spray a measured amount of slurry over a truck load of fill. The truck in turn transported the fill to the stope and dumped the mixtrue into a three metre culvert mounted vertically over the discharge point in the stope to be filled. The action of the materials cascading through the culvert into the stope below coated the dry fill particles with slurry.

The most recent method used is similar to the commercial concrete batch plant. A measured amount of fill and slurry are placed in a drum mixer which revolves to mix these ingredients. The mixed product is then dumped into a truck which in turn transports the mixed backfill to the stope and discharges it there.

HAULAGE

The original method of hauling backfill was with five-yard scooptrams. In both No.1 and No.2 Mines the scooptrams picked up the mixed backfill at the bottom of the baffled raise (mixing raise) and transported it to the stope. This method was equipment intensive and at the same time messy. No matter how much effort was made to keep the haulage-ways clean, the end result was still a mess. This method of transport was soon abandoned in No.1 Mine in favour of conveyors.

In 1975, a system of conveyors were installed out in the footwall of No.1 Mine about fifty feet above the 800 level. The conveyor system runs parallel to the ore-body, and at various points along the ore-body, conveyor crosscuts are developed in order to convey fill into the particular stope to be filled. All fill is sprayed with slurry prior to it being discharged through a three-metre long, two-metre diameter baffled culvert pipe described previously.

At the No.2 Mine, where the stopes are only a quarter of the size of those found in No.1 Mine, a system other than conveyors had to be found because of economics. Rather than spend large sums of money on conveyor moving and installation, it was decided instead to use ten and thirteen-ton Jarco trucks. These are as flexible as scooptrams but without the disadvantage of spilling fill onto the roadways. Also, their carrying capacity is much greater than the scooptram. To date there are eight trucks in use at No.2 Mine.

MINING METHOD EVOLVEMENT

At No.1 Mine, the method used is sublevel blasthole stoping. The stopes are generally 18 to 24 metres wide, 30 metres long and 90 metres high. These dimensions are varied somewhat to meet the geological conditions. The pillars between the stopes are 24 metres wide. The sill pillars separating each mining horizon are 30 metres thick.

The main level development consists of a footwall and hangingwall fringe drifts with crosscuts driven down the centre of each pillar. A system of drawpoints are driven from each crosscut at 7 metre centres under each stope. Undercut drill crosscuts are driven at the junction of the system of drawpoints. Fan drills, drilling 54 millimetre holes are used to drill off the undercut.

Sublevels 60 metres and 90 metres above the main draw drifts are driven to conveniently drill the stope with 20 centimetre holes using a Robbins 11D rotary drill. See Figure 2. Slot raises are bored using a Robbins 61-R borer reaming 1.2 metre diameter holes are generally positioned in the hangingwall of the stopes.

Wagner ST-8 scooptrams equipped with spade-type bucket lips are used to muck from the drawpoints to the orepasses.

All stopes upon completion are backfilled with consolidated fill consisting of 75% -15 + 1 centimetre and 25% -1 centimetre + 100 mesh aggregate material mixed with 6% by weight of 50:50 water-cement slurry.

165

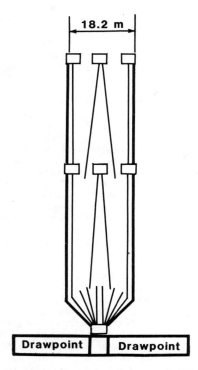

Fig.2 - Typical Sublevel Stope Drilling
        Pattern

Lockerby Mine operated at Falconbridge Nickel Mines Ltd. At this time we had the opportunity to see: 1) the Avoca Method used at a depth of 670 metres, 2) Cut and Fill Method at a depth of 1,525 metres and 3) Open Stope Blasthole Method at a depth of 1,280 metres. After visiting the Lockerby Mine and seeing open stoping being successful at these depths, we decided to try the Open Stope Blasthole Method on the 2800 level which is at a depth of 850 metres. It was decided that the stope dimensions would be held to 15 metres wide, 30 metres long, and 60 metres in height. Also after the stope was mined out, it was to be filled immediately with consolidated fill.

The first open stope to be designed was the 28691 Stope located in the centre of the North Orebody adjacent to the hanging-wall. See Figure 3. The approximate

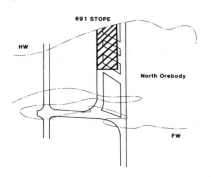

Fig.3 - 2800 Level North End Undercut

When No.2 Mine was in the planning stage, it was recognized that the method used in No.1 Mine could not apply because of the mining depths involved. A standard flat back cut and fill mining method was investigated but was turned down because it was labour intensive and therefore too costly. Mechanized cut and fill methods utilizing ramp development both upwards and downwards from a footwall fringe drift was studied. In fact, one stoping block was developed with this method in mind.

This method was later held in abeyance in favour of trying the rill type cut and fill method developed at Avoca, Ireland. One transverse stope with dimensions of 9 metres wide, 45 metres long and 60 metres high was mined using the Avoca Method utilizing consolidated fill. Although this method is viable, it did have the disadvantage of being a small 'batch' system. What we were looking for was a system similar to that used in the Kidd Creek No.1 Mine which would work at our ground pressures.

At the time when we were investigating the Avoca Method, we made a visit to the

dimensions were 15 x 30 metres in plan and 60 metres in height. Three drawpoints 4.3 metres wide by 3.4 metres high were excavated at the bottom of this stope to facilitate the mucking with ST-5 scooptrams. The main extraction crosscut was driven down grade towards the hangingwall as it was planned to excavate a return airway and water collection drift along the hanging-wall. This drift heading never came about.

Some 45 metres above the extraction level, there were two drill crosscuts driven along the North and South boundaries of the stope. See Figure 4. These two cross-cuts were joined at the hangingwall and again with a slot drift at the other end of the stope.

The development at the very top or over-cut horizon of the stope was a single

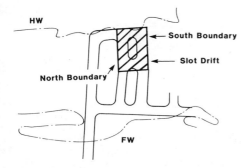

Fig.4 - Plan of 28691 Stope Middle Cut
Horizon - 28/3 Level

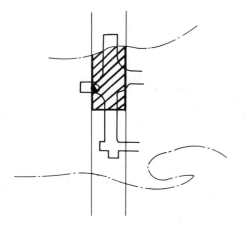

Fig.5 - Plan of 28691 Stope Overcut Horizon
2600 Level

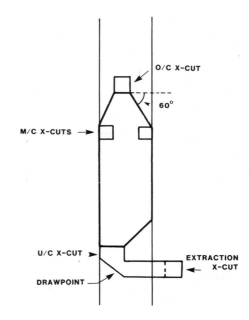

Fig.6 - 691 Stope

crosscut driven down the centre from foot
to hangingwall. See Figure 5.

To provide for a tight filling situation
and minimize the openings on the overcut
horizon, it was decided to fan out the
overcut blasthole rings. The resulting
wedge slope of the stope back enabled the
excavated stope to be backfilled tightly,
thereby increasing the ground stability.
All holes drilled from the overcut horizon
were drilled with Gardner-Denver DH123
drills mounted on a small self-propelled
carrier. The average length of these holes
were 15 metres and were 54 millimetres in
diameter. See Figure 6.

The middle cut was drilled with a Robbins
11D rotary machine drilling holes of 200
millimeter diameter to a depth of 31 metres
down to the undercut horizon. The slot
raise for the stope was drilled with a

61-R raiseborer at a diameter of 1.8
metres. Because we were uncertain of the
competency of the hangingwall, we placed
the slot on the footwall side of the stope.
Although it caused an access problem, we
were prepared to live with this and therefore excavated access at the hangingwall.

After the slot was excavated, the blasting sequence was roughly to: a) blast one-
third of the undercut from the slot to the
hangingwall, b) blast the remaining undercut all the way to the hangingwall, c)
blast two rings at a time upward in a stepped fashion towards the middle cut horizon,
d) the remainder of the middle cut and
overcut were taken in one blast later. See
Figure 7.

When the stope was emptied of all broken
material, a mixture of crushed rock (75%
-15 + 1 centimetre and 25% -1 centimetre +
100 mesh) and 5% by weight water-cement
slurry (1:1) was introduced into the stope.

From mining the first stope (691) we
found the hangingwall to be relatively competent. Also, we experienced little, if
any, sloughing of ore on either side of the
stope.

On the next stope (661), we decided that
we would leave a 15 metre pillar between

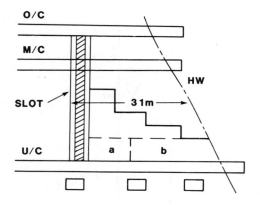

Fig.7 - 691 Stope

On the 2600 level or overcut horizon, the development for this stope was similar to that of the previously mined stope. That is, a crosscut was driven down the stoping panel. This provided a drill crosscut whereby the top portion of this stope could be drilled.

At a distance 15 metres below this horizon two parallel crosscuts were driven along the boundaries of this stope to provide drill horizons for the middle cut. A second sublevel was opened 15 metres above the undercut horizon in order to provide access for slot drilling and blasting. A cross-section of this stope shows the four (overcut, middle cut, undercut slot and undercut) horizons excavated for the drilling and blasting. See Figure 9. On the

it (661) and the previously mined stope. This allowed us to mine this particular stope at the same time while filling the 691 stope. At this point of our mining we still thought that we would use transverse pillars between each stoping panel.

The undercut development differed from 691 stope in this way. The main extraction crosscut was driven down the centre of the stope whereas in the former it was driven down in the adjacent pillar. Also two drawpoints were driven from the pillar at the South side. The main extraction crosscut served both as an undercut drilling horizon and as a drawpoint. See Figure 8.

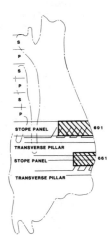

Fig.8 - Plan of Undercut Horizon

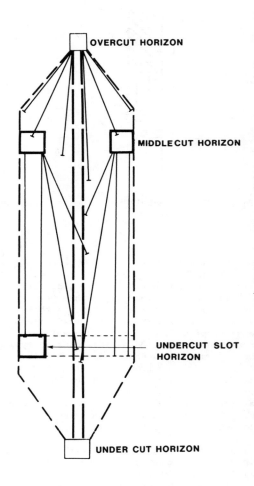

Fig.9 - Typical Stope Cross-Section

168

hangingwall side of the stope we drilled a
1.8 metre diameter raise bore hole from the
overcut to undercut horizon as a slot
raise. The overcut and undercut was long-
hole drilled with 54 millimetre diameter
holes while the middle cut and slot undercut were drilled with an In-the-hole drill
drilling 114 millimetre diameter holes.

The blasting sequence utilized a combination of horizontal and vertical slicing.
However, prior to blasting any of the upper
portion of the stope, we blasted and
removed all of the undercut. The middle
cut blasting consisted of opening the slot,
slicing the bottom 15 metres of each ring
and blasting the remainder of each ring
separately. The overcut was extracted by
vertically slicing each ring. This particular blasting sequence method generated
large blocks resulting in much more time
spent on secondary blasting than that of
the previous stope. Again on completing
the mucking, we filled the stope with consolidated fill.

What we had done so far was to mine two
hangingwall stopes in virgin ground at a
depth of 850 metres. These stopes were
separated from one another by a pillar of
15 metres. The next experiment was to mine
against a stope that had been previously
mined and filled. The stope selected was
beside the first one mined. Here, we would
mine against the footwall side of the
stope.

This new stope, called 692 Stope, had
dimensions in plan of 15 by 15 metres by
61 metres in height. See Figure 10.

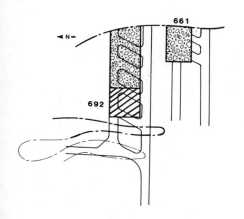

Fig.10 - 692 Stope Adjacent to Previously
Mined and Filled Stope

The undercut was developed similar to that
of the first stope (691) and was drilled off
with 54 millimetre diameter holes. All
other horizons were drilled with 114 millimetre diameter holes using an In-the-hole
drill.

In order to achieve as near to 100% recovery, we mined the undercut using similar
methods as the Avoca rill method of mining.
The sequence followed was to blast two
undercut rings which amounted to 3 metres
of advance and muck out the broken ore. A
fill fence was next constructed and consolidated fill was dumped down the slot raise
at the hangingwall side of the stope. The
fill would form a rill approaching 45°.
See Figure 11. After it had set for eight

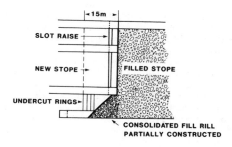

Fig.11 - Cross-Section 692 Stope Showing
Consolidated Fill Rill

hours, the miners blasted another two rings
of undercut holes and the sequence was
again repeated. This procedure created a
45° rill which allowed the operators to
achieve nearly 100% recovery of the broken
ore.

The rest of the stope was blasted such
that the slot and several rings were blasted from the sublevels above in order to
create a 20% void. After this was achieved
the remainder of the stope was blasted,
mucked and filled with consolidated fill on
completion of the mucking cycle.

The mining of the stope adjacent to consolidated was successful. Visual observation from the various horizons up the stope
height showed the fill walls were standing
admirably. Dilution was 3%. From this
experiment, we were assured that the concept of mining without pillars was feasible.

The next experiment was to mine the pillar between two previously filled stopes.

Having successfully mined against the fill at the end of a previously mined stope, we believed this experiment (mining the pillar) whereby two stope walls adjacent to fill would be the final test to prove whether or not our method was sound. Also, this pillar stope was located alongside of the hangingwall which adds to its critical nature. The exposed fill along the sides was 21 metres long and 61 metres high, whereas, the end adjacent to the hangingwall was 15 metres wide and 61 metres high.

Again, the undercut was drilled with 54 millimetre diameter holes while all the other horizons were drilled with 114 millimeter diameter holes. Also, so as to achieve 100% recovery of all blasted material it was the practice to excavate and fill the undercut void in the same manner as was done in the previously mined stope. The rill formed allowed the operators to completely empty the stope without going into the stope and exposing themselves or their machines. Since this stope was mined, we were able to purchase a remote controlled scooptram which can move right out into the open stope for cleaning up the remnant ore, thus eliminating the necessity of building a rill described in the previous paragraph. To date this vehicle has paid for itself many times over both saving time in constructing a rill and recovery of the remnant ore in the stope.

The mining of this particular stope was successful as the walls (two fill walls bounding the stope) stood up extremely well. It was estimated that 41,690 tonnes of ore were blasted and 42,910 tonnes were mucked. This works out to a dilution factor of 3%. As a result of this, it proved the method was successful and its success was due mainly to the quality of the consolidated backfill used.

THE FINAL METHOD USED

The orebody is divided into stoping panels 15 metres wide at right angles to the strike. Ideally, the mining sequence begins with mining the panel at the centre of the orebody and progressing from hanging to footwall in one direction and from the centre panel of the orebody out to both ends in the other direction. It has been described as an oval area with a T-shape superimposed on it. The first stope to be mined is the one at the intersection of the upright and cross-bar of the 'T' which is adjacent to the hangingwall. After this stope has been mined and filled, and after

allowing one month for the backfill to cure a second stope immediately behind the first stope and in the same panel is mined. After the second stope is mined and backfilled, the two hangingwall stopes beside the original stope are mined along with the third stope (in the same original panel) behind the second stope.

The exposure to highly stressed ground is limited. Mining initially those stopes adjacent to the hangingwall relieves the particular panel to be mined from the high horizontal stresses. See Figure 12.

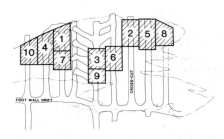

Fig.12 - No.2 Mine Stoping Block Pattern

Consequently, all horizontal stress is directed out to the ends of the orebody. (There has been some evidence that the consolidated fill does withstand stress but this is only a fraction of the total.)

While mining and filling operations are being conducted in the active stopes, other stopes are being prepared for later mining. Slot raises and longhole drill patterns are being drilled in subsequent stopes to that mining will progress in a continuous manner. See Figure 13.

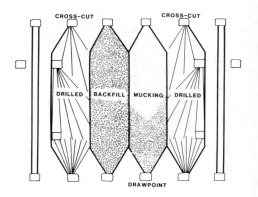

Fig.13 - Stopes in Various Stages of Development and Mining

170

Mining at Kidd Creek No.2 Mine is carried over a vertical height of 180 metres before silling out for the next block of ore. The sequence of mining in a vertical plane is similar to that described when mining in plan. The centre panel is mined from hanging to footwall, then the adjacent panels are mined. The panel immediately above the first panel is mined next followed by the panel adjacent to the first panel on the lower horizon. The sequence used to mine a 175 metre vertical block of ore is shown in the diagram. See Figure 14.

After each stope is mined it is filled with consolidated fill.

The plan for mining the sill pillar is to mine the top portion of the stope up under the previously mined out block of stopes. A sublevel 7 metres below the main level will be established with crosscuts driven between and below the previous stope extraction crosscuts. See Figure 15.

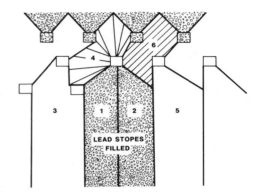

Fig.15 - Mining Sill Pillars

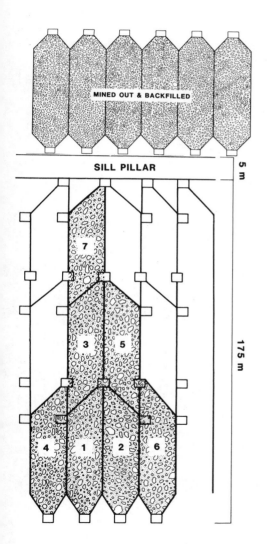

Fig. 14 - Sequence of Mining a Block of Stopes

These sublevel crosscuts act as the drill crosscuts to drill off the overcut and the sill pillar. The top section of the overcut (top 7 metres) and the sill pillar are blasted after all the broken ore from the stope has been removed. After the ore from this latter blast has been removed the stope is filled. All sill pillars and overcut sections of the stopes are mined together progressing out from the centre and expanding to each side. As before only 15 metres of ground along the panel is mined at any one time.

Since these early experiments in 1979, a total of over 24 stopes representing 1.0 million tonnes of ore has been mined and approximately 0.6 million tonnes of consolidated backfill has been placed. Roughly speaking, the average stope contains 49,000 tonnes and requires 800 manshifts to mine and fill for an efficiency of 60 tonnes per manshift. The statistics for a typical stope are shown in the appendix.

The development of the open stope mining method without pillars covered a period of approximately 3½ years. The success of the method is due of course to the consolidated fill used. However, without the cooperation and coordination of all parties involved, the operators, the maintenance people and those of engineering, the success of this method could not have been achieved.

Belford, J.E. 1981, Sublevel Stoping at Kidd Creek Mines. Symposium Design and Operation of Caving and Sublevel Stoping Mines: SME-A.I.M.E.

Pabst, M. 1981, Kidd Creek No.2 Mine Boost Production to 20,000 TPD. Canadian Mining Journal.

APPENDIX

STOPE ANALYSIS FOR 2800 NORTH
To End of October 1982

	Total (All Stopes)	Average/ Stope
No. of Stopes	23	
Total Tonnes of Ore	1,122,958 MT	48,824 MT
Tonnes of Backfill	665,227 MT	28,923 MT

Manshifts

Drilling	4,309	187
Blasting	2,161	94
Haulage	4,382	191
Ground Support	404	18
Filling	4,576	199
Secondary Blasting	2,337	102
Ventilation	92	4
Electrical	50	2
Other	306	13
TOTAL MANSHIFTS	18,617	802
TONNES ORE/MANSHIFT	60.32	60.88
TONNES BACKFILL/ MANSHIFT	145.37	145.34

ACKNOWLEDGEMENT

The writers wish to thank Kidd Creek for their permission to write this paper. Appreciation is given to the Engineering Department for the help in preparing the illustrations and in compiling data and to the Geology Department for their very concise description of the mine geology.

BIBLIOGRAPHY

McLeod, P.C. and Schwarts, A. 1970, Consolidated Fill at Noranda Mines Limited (Geco Division). CIMM Bulletin.

Mathews, K.E. and Kalsehagen, F.E. 1973, The Development and Design of a Cemented Rock Filling System at the Mount Isa Mine, Australia. Mount Isa: Jubilee Symposium on Mine Filling.

Yu, T.R. 1978, Rock Mechanics Symposium and Underground Blasting Meeting. Timmins: Internal Memo.

A review of the backfill practices in the mines
of the Noranda Group

J.H.NANTEL
Noranda Mines Ltd., Montréal, Canada

ABSTRACT: Noranda Mines Limited is a large Canadian mining, concentrating and metallurgical company; it is also involved in manufacturing and forest products. Total revenues in 1981 amounted to over U.S.$2.5 billion, of which the mining and metallurgical sectors accounted for 40%. In 1981, Noranda was operating 37 mines, 23 were underground operations of which 15 were making use of backfilling materials; an additional 4 mines are considering the use of fill in the near future. The total output extracted from all mines totalled 84 million tonnes in 1981, of that tonnage 19.5 million tonnes were obtained from underground mines, some 60% of the tonnage mined from underground operations necessitated the use of backfill materials. The backfill materials used include: non-consolidated granulated slag, rock fill, rock fill consolidated with cemented classified mill tailings, dewatered classified mill tailings, alluvial sand, cemented and non-cemented classified mill tailings transported hydraulically.

1 INTRODUCTION

Backfilling of underground stoping areas has been practiced in the mines of the Noranda Group for nearly fifty years. In 1981, there were 15 mines in the Noranda Group utilizing some sort of backfill material to extract underground ores. The percentage of the ore mined with the use of fill accounts for nearly 60% of all ore mined by underground methods at Noranda.

In 1981, the Noranda Group operating mines produced 226,208 tonnes of copper, 530,393 tonnes of zinc, 118,708 tonnes of lead, 7,470,500 grams of gold and substantial quantities of silver, molybdenum, potash and other commodities.

Backfilling, in the Noranda Group, was first started at the Horne mine, located in Noranda, Québec, Canada. Although production at the mine commenced in 1927, experimental use of backfill materials was not started until 1933.

Observations on the cementing action of fine sulphides on surface dumps, and on floors underground, raised the possibility that pyrrhotite tailings could be used to obtain a consolidated fill underground (Patton 1952). Moreover, since the overburden on the surrounding countryside was primarily clay, with sand or gravel not being readily obtainable, slag was a most obvious contender as a fill material. The combination of the chemically active pyrrhotite tailings from the mill and the reverberatory furnance slag was found to be well suited to meet the needs of ground support at the mine.

This was the first known use of reverberatory slag and pyrrhotite tailings as fill and proved to be eminently successful but exclusive to the Horne mine.

At Kerr-Addison Gold Mines, Limited, of Virginiatown, Ontario, a program of investigation was started in 1947 to find a method of making a suitable underground backfill material out of the cyanidation tailings. After considerable experimentation, hydraulic backfilling became routine around 1952 (Hawkes 1955).

In the 1960's, Geco Mines developed another unique backfilling system. Soon after start-up of the mine in 1957, it became apparent that large tonnages of sand would be required for backfill purposes (Barnett 1969).

During the first 4 years of operation (1957-61) open blasthole stoping was practised at the western end of the orebody. This method consisted of primary stope development and pillars. Favourable economics led to a continuation of this method but with the addition of rock fill and final stabilization of the fill by the introduction of cemented hydraulic fill.

Initial experimental work on the preparation of sand fill from flotation tailings was begun in late 1959. In 1962, a research program on cemented fill was initiated and in 1964 a cemented-fill preparation plant was installed.

With hydraulic tailings being successfully used at the Geco and Kerr Addison mines, normal Noranda mining practice now includes a large percentage of its mining with backfill. Backfill is especially important in Noranda's operations since most of the orebodies are lenticular, and have vertical or near vertical dispositions in host rock of varying degrees of competence.

At the present time the following mines are involved in backfilling operations: Brunswick Mining and Smelting, Geco Division, Matagami mine, Chadbourne mine, the mines of the Pamour organization, namely: Pamour No. 1, Pamour No. 3 (formerly Aunor), Schumacher mine, Timmins Underground Project (formerly Hollinger), Mattabi mine, Lyon Lake Division, Kerr Addison, Goldstream Division and Tara. Orchan mine ceased operations in 1982.

The following mines are embarking on, or are considering the use of backfill: Heath Steele mine, Gaspé mine and Norita mine. The possibility of using backfill at Central Canada Potash mine is also being assessed.

The types of fill material used by the mines of the Noranda Group cover, undoubtedly, the most diversified array of fill materials used by any one mining organization in the world and the experience acquired by the Noranda personnel is unmatched.

This paper will attempt to make a brief review of the backfill practices of the mines of the Noranda Group and the future course of action will be discussed.

2 REVIEW OF BACKFILL PRACTICES

Only a limited number of mines, the most representative operations, will be described in this paper. For each operation reviewed, the author will attempt to cover the following elements: type and sources of fill, role of fill, properties of fill, mining methods and method of fill placement.

2.1 Chadbourne mine

The type of backfilling practiced at the Horne mine (closed since 1976), is now being considered for a small gold mine, located in the centre of the Town of Noranda, Québec. Considering the proximit of houses and the need to recover the pillars, the backfill material needs to be competent and the backfill program must be carefully laid out. No subsidence can be tolerated on surface. A study was conducted in 1982 (Nantel 1983), to determine the properties of slag-pyrrhotit tailings mixtures similar to the fill used at the Horne mine. The test results indicated that it was still possible to create a backfill material similar to the Horne mine with the constituents still available on refuse dumps around the Horne mine.

1. Type, sources of fill and method of placement.
The granulated slag is available in large quantities from the Horne site, a few miles from the Chadbourne mine. The pyrrhotite tailings are available from an old tailings pond and will need to be loaded, trucked and classified in the Horn mill before it can be mixed with the slag at the Chadbourne mine. At present, only granulated slag is used at Chadbourne to fill the mined out lower portion of the ore body. The fill is introduced and brought to its final destination through a series of fill passes originating on surface. A form of pneumatic stowing is being considered to place the fill in hard to-reach areas.

2. Role of fill.
The functions of the fill material at Chadbourne are: i) ground support to permi the recovery of ore pillars and ii) in the long-term to prevent surface subsidence. The fill material needs to be competent enough to remain stable when exposed to one face for limited periods of time. The fill material also needs to be competent enough to accept some of the load previously supported by the ore. To avoid any surface ground movement, the settlement in the fill plus any void between the top of the fill and the crown pillar shoul be kept below 5% of the thickness of the crown pillar.

3. Fill properties.
The following properties were determined for a mixture of 90% slag and 10% pyrrhotite tailings by weight, after a 112-day curing period; void ratio: 0.41, bulk density: 2.25 g/cm^3, uniaxial compressive strength: 0.331 MPa, modulus of compressio 1.3 MPa/% strain, cohesion: 0.083 MPa,

internal angle of friction: 37° and a percolation rate of approximately 40 cm/h.

4. Mining methods.
The ore is mined with a modified vertical crater retreat method and the voids left by the mining activities are filled after the ore has been removed (Gagnon 1981).

2.2 Brunswick Mining and Smelting

The type of backfilling method practiced at Brunswick Mining and Smelting No. 12 mine is characterized by the placement of dry fill.

1. Type, sources of fill and methods of placement.
Because of the extensive grinding required to liberate the valuable minerals of the ore and the high content of pyrite (80%) and pyrrhotite (5%), tailings were rejected as a potential fill material (Gignac 1978). Open pit waste was selected as fill material instead, because it was available in sufficient quantities and the costs appeared relatively low. A secondary source of fill is the waste rock produced from development rock.

The fill material is run through a gyratory crusher to yield a minus 15 cm product; it is then dumped into bored fill raises or stockpiled on surface. The pathways followed by the fill material underground are somewhat tortuous; the principle is to bring the fill above a mining block by gravity and then to rehandle it to a given borehole, through which the fill will be finally passed to the desired stope. The final fill placement in the stoping area is done by the mucking equipment.

2. Role of fill.
At Brunswick Mining No. 12 mine, where the ore is removed in a series of horizontal slices in highly mechanized cut and fill stopes, the fill assumes first and foremost the role of a working platform. The second role of the fill is to provide ground support, i.e. minimize wall closure, absorb a proportion of the lateral stresses and consequently help reduce the stresses on the remaining pillars. The fill is, of course, used to permit maximum ore recovery.

3. Fill properties.
The sequence of passing and rehandling of the fill results in size reduction and degradation. The size of the largest particles decreases from 100 mm on surface to 29 mm on the 2300 foot level, the distribution modulus from 0.7 to 0.45 and the coefficient of uniformity increases from 13 to 24. The bulk density, as measured in the top layer, is 2.45 g/cm³.

The relative density (Dr) was calculated to vary from 35 to 50% depending on the depth of the sample in the lift. Due to the difficulties in assessing the strength properties of rock fill, these have not been determined at Brunswick.

4. Mining methods.
The mining method used almost exclusively at Brunswick Mining is a highly mechanized variety of longitudinal cut and fill method. The method is described in a number of internal documents and papers (Gignac 1978, Moerman 1980).

2.3 Geco mine

The backfilling practices of Geco mine are probably the best known throughout the world, along with those of the Horne mine.

1. Mining methods.
Since the beginning of operation in 1957, approximately 90% of the ore production has originated from sub-level blasthole, with the remainder coming from development work and cut and fill stoping. At the time of writing, all primary mining is essentially completed; all future ore production will originate from the ore remaining in the pillars.

A primary transverse blasthole stope was generally mined in two 122 m lifts with the top lift being removed first. The ore was drilled from perimeter sub-levels spaced at 30 m vertical distance. The sub-levels were connected by one or more slot raises. After approximately 25% of the ore had been removed to create a void, the remainder of the shell was blasted in one shot. As soon as the final shot was blasted, waste rock from the surface quarry was introduced into the stope. This addition of waste rock was necessary to control ground movement both in the waste and pillar walls. The ore-waste interface was maintained horizontal by means of a positive draw control program. The bottom lift was mined in the same fashion as the top lift. When the panel was completely mined out, bulkheads and drainage outlets were installed in all openings and a mixture of mill tailings and cement at a ratio of 30:1 was introduced into the stope. The pillars were mined in the same method as the primary stopes except that the pillars are generally mined out in three lifts for a better draw control of the broken ore between the two fill walls.

Lately (1975), 11.5 cm dia. holes are used to drill the pillars; much less hole rehabilitation is required than with the customary 5.4 cm dia. holes. With the advent of the modified drilling practice, the blasting practice has been changed. The new blasting sequence proceeds level by level coupled with corresponding removal of the broken ore to provide the required void for the following blast: the undercut is blasted first, then the slot raise from the undercut to the first sub-level, then sufficient ore is blasted to accommodate the remaining ore from this layer of ore. The procedure is repeated for the following sub-levels. The broken ore is left as close to the unblasted back as possible to minimize movement of the fill walls (Weeks 1981).

2. Properties and role of fill.
At Geco, the fill assumes good ground support properties and permits the recovery of ore pillars. Although the fill material is not normally required to remain stable when exposed on one face, such fill exposure was made over the years; in one instance, the consolidated fill was exposed for a height of 60 m (Schwartz 1978). A secondary role of fill is to provide an adequate control on the dilution of the ore by waste materials.

3. Type and sources of fill.
The rock fill originates from a surface quarry operated especially for mine fill. The rock is transported to a surface dump area, the fill then follows a series of fill raises to finally fall into the desired mining block.

The hydraulic fill is transported underground via a bore hole-pipeline system and it consists of classified flotation tailings transported at a pulp density 65-67% solids by weight. For the bulk filling operations in the stopes, cement is added in the ratio of 1 part to 30 parts of tailings by weight.

2.4 Kerr Addison

At the Kerr Addison mine, located in northeastern Ontario, 11.2 million tonnes of hydraulic backfill have been used to replace 26.1 million tonnes of ore mined out since 1953.

1. Type, sources of fill and method of placement.
The fill material is prepared by recovering 42% of the mill tailings through hydro-cyclones. The material is introduced underground through 10-cm bore holes at 50% solids by weight.

2. Mining methods
The fill material is used in conjunction with several mining methods at Kerr Addison. Conventional cut and fill accounts for approximately 30% of the tonnage mined. A variety of square set and fill is used at depths where ground

pressures are higher. Undercut and fill methods are also practiced in zones of difficult ground. Shrinkage stoping with delayed fill is also employed.

In cut and fill stopes, the last 50 cm of each lift is poured with fill mixed with Portland cement in the ratio of 7:1; no cement is added in the first section of the lift. Similarly, no cement is added to the fill in the square set and shrinkage applications.

3. Properties and role of fill.
All fill material is cycloned to yield a percolation rate of 10 cm per hour.

In the cut and fill stopes the fill is used mainly as a working platform and for ground support.

2.5 Norita mine

Norita mine, located near Matagami in northwestern Québec, is in the process of converting from a sub-level retreat mining method, used on the upper levels, to an inverted bench long hole method, where a mixture of rock fill and cemented tailings will be used to fill the mined out stopes. Once the cemented fill has consolidated, the pillars will be extracted and filled with rock fill without the addition of cemented tailings.

1. Mining method
The ore zone, 550 m below surface, has been layed out with stopes and pillars 11 m in width, 65 m in height and averaging 25 m in length. The drilling will be accomplished with in-the-hole equipment drilling blastholes 16.5 cm in diameter. The stopes and pillars will be mined out in a carefully laid out sequence Once a stope has been pulled empty, waste rock, mined in a nearby waste stope, will be introduced in the stope and filled. At this period in time two alternatives are envisaged, i) the cemented tailings fill will be introduced simultaneously with the rock fill to form a homogeneous mass, or ii) the cemented fill will be introduced once the entire column has been filled with waste rock. The final decision will depend on the conditions of the wall rock during the active mining period.

2. Sources of fill and method of placement.
The rock fill will be mined from a waste zone located a distance of 150 m from the active mining zones. The mining methods used for the rock fill will be similar to the methods used for the ore. The reasons for obtaining the rock fill underground are depth and the bad ground conditions near the ore body, rendering

prohibitive the driving of a waste pass, and the lack of suitable fill material on surface.

The tailings fill will be fabricated at the Matagami mine milling facilities (where the Norita ore is processed). The classified tailings fill will be trucked to Norita and stored on surface. Cement and water will be admixed to the required ratios and the slurry will be introduced into the mine via a pipeline in the shaft and at appropriate levels.

3. Properties and role of fill. Various mixtures of cemented fill were tested and the strength parameters obtained. The following properties were determined for mixtures of 10:1, tailings to cement ratio: a) after a 28-day curing period; bulk density: 1.90 g/cm^3, uniaxial compressive strength: 0.85 MPa, cohesion: 0.15 MPa, internal angle of friction: 51°, b) after a 112-day curing period; bulk density: 1.93 g/cm^3, uniaxial compressive strength: 2.19 MPa, cohesion: 0.28 MPa, internal angle of friction: 61°.

The fill will be used to permit total recovery of the ore and to offer ground support during the active mining period. The fill will also need to have adequate strength to remain free-standing over a height of 60 m for a limited time period. Stability calculations were performed and will be used to select the quantities of cement required in the cemented fill mixtures.

2.6 Gaspé mine

Gaspé is proposing to change the mining method of room and pillar it has used since the beginning to a bulk filling mining method. It is estimated that the method using fill will be safer and cheaper to use.

1. Mining method.
The proposed mining method (Sauriol 1982) is a long hole mining method with delayed non-cemented backfill. Each stope, 150 m in length, 23 m in width and 30 m in height will be drilled off from one upper sub-level with 16.5 cm diameter drill holes. Once completely mined out, the stope will be filled with classified mill tailings. After a suitable drainage period, the adjacent stope will be mined by following the same procedure. The fill deposited in the first stope will be standing at an angle of 60°, due to the configuration of opening blasted in the stope. It is expected that the fill will remain stable in the first stope while the ore is being pulled in the second stope. Once the second stope is pulled empty, it will be filled with classified tailings fill.

2. Sources of fill and method of placement.

The mill tailings of the No. 2 Gaspé concentrator will be classified to obtain a backfill material having a percolation rate of 18.0 cm/h. Sufficient quantities can be produced to supply the bulk filling requirements. The fill slurry at 70% by weight solids will be introduced underground through a series of bore holes and pipelines.

3. Properties and role of fill.
The following properties were determined on the Gaspé fill material (Archibald 1982); granulometry: 25% plus 75 microns, coefficient of uniformity: 5.0, specific gravity of solid particles: 3.1 g/cm^3, in-situ density (15% water): 2.16 g/cm^3, porosity: 0.41, pulp density at 10% solids by weight: 1.88 g/cm^3, percolation rate: 18.7 cm/h, apparent cohesion at 15% water: 26 kPa, internal angle of friction: 34°, modulus of compressibility: 1.6 MPa/% displacement.

At Gaspé, the fill material will have to be capable of supporting the back and walls of the mined out zones without allowing excessive deformations during the mining period and also after a lengthy period to insure the general stability of the mine. The fill needs to be stable at an angle of at least 60°.

3 FUTURE COURSE OF ACTION AND CONCLUSIONS

With the establishment of a Mining Technology Division at the Noranda Research Centre of Pointe Claire, Québec, backfill studies will be given even greater emphasis. It is intended to devote more attention to the determination of the various backfill properties before a new mine is opened and plans are underway to study the fill properties at existing mines with the object of improving the role played by the fill material in the overall mining cycle.

The future approach to the design of fill systems at Noranda will include the analyses of the following components:
a) functions of fill;
b) criteria for the satisfactory behaviour of fill;
c) specification of required fill properties;
d) methods of obtaining desired fill.

As a matter for standardization, all fill materials at Noranda will be tested for the following: granulometry, percolation rate, liquefaction characteristics, densities, porosity, void ratio, transportation characteristics, attrition of particles, pyrite and pyrrhotite contents, structural components: shear resistance, cohesion, friction angle, uniaxial compressive strength, triaxial strengths, modulus of deformation and rate of recovery from mill tailings. The self-heating properties of fill materials is also made on potential fill materials.

Once the functions of fill have been carefully assessed and the criteria for the satisfactory performance been defined, stability, bearing capacity, and subsidence calculations could be made using the results of the tests performed. Optimization of fill properties is then possible.

The above determinations and calculations influence greatly the selection or rejection of potential fill materials at a mine; the tests will often indicate the nature of the modifications needed before a material can be used for underground applications. In the final analysis, the mine operator wishes to develop a fill material with the required properties in the most economical manner, taking into consideration the availability of materials and other constraints.

REFERENCES

Archibald, J.F. and Nantel, J.H. 1982. Evaluation préliminaire des résidus du concentrateur comme source de remblai hydraulique à la Division Mines Gaspé – Murdochville. Test work performed at Queen's University, Kingston, Ontario, Canada.

Archibald, J.F. and Nantel, J.H. 1982. Results of Tests and recommendations for the utilization of sand materials from the Matagami area as underground backfill material at Norita. Test work performed at Queen's University, Kingston, Ontario, Canada.

Barnett, C. 1968. Preparation of hydraulic sand and cement backfill at Noranda Mines Limited, Geco Division. Canadian Mining Journal, December 1968: 47-48.

Gagnon, S.G., Robertson, R. and Lecuyer, N. 1981. Blasting controls at the Chadbourne Project. Fifth Annual Underground Operators Conference, February 1981.

Gignac, L. 1978. Filling practices at Brunswick Mining – No. 12 Mine. Mining with Backfill, CIM Special Volume 19, 1978: 30-36.

Hawkes, G.J. 1955. The Preparation of hydraulic backfill at Kerr Addison Gold Mines, Limited, Virginiatown, Ontario. Canadian Mining Journal, January 1955: 58-64.

Moerman, A. 1980. Continuous search for lower cost mining methods ... looks to hydraulic jumbos. The Northern Miner, October 1980.

Nantel, J. and Lecuyer, N. 1983. Assessment of slag backfill properties for the Noranda Chadbourne Project. CIM Bulletin, January 1983: 57-60.

Patton, F.E. 1952. Backfilling at Noranda. CIM Transactions, Vol. LV: 137-143.

Sauriol, G. 1982. Projet de minage de la zone C Nord-Ouest par Chantier à longs trous avec remblayage retardé, Revision 1. Mines Noranda Limitée. division Mines Gaspé, internal report.

Schwartz, A. 1978. Pillar recoveries using consolidated fill at Noranda Mines Limited, Manitouwadge, Ontario, Mining with Backfill, CIM Special Volume 19, 1978: 56-68.

Weeks, R.M. 1981. Down-the-hole drilling at Geco. Fifth Annual Underground Operators Conference, February 1981.

ACKNOWLEDGEMENTS

The author wishes to express his appreciation to all the persons of the Noranda companies and affiliates for their participation in assembling the information contained in this paper.

Fill operating practices at Mount Isa Mines

L.B.NEINDORF
Mount Isa Mines Ltd., Australia

ABSTRACT: Different types of fill are used to backfill voids created by underground mining at Mount Isa Mines Limited. These require different reticulation and handling systems.

Some fill operating practices are outlined for the various stoping methods used at Isa Mine.

Finally future fill developments and their applications are discussed.

INTRODUCTION

The Isa Mine of Mount Isa Mines Limited in North West Queensland, Australia, currently produces 5.4 million tonnes of copper ore and 3.7 million tonnes of silver, lead and zinc ore annually. Of this, 5.1 million tonnes of copper ore and 2.5 million tonnes of silver, lead and zinc ore are produced by sublevel open stoping and 1.2 million tonnes of silver, lead and zinc ore are produced by cut and fill (MICAF) mining.

Various types of fill and filling systems are used to backfill the approximate 3.0 million cubic metres of void created annually. The success of the mining methods used is very dependent on the fill properties and the fill operating practices used.

This paper discusses the fill types, filling systems and fill operating practices used to backfill the voids created.

WET FILL STATION (WFS)

Tailings from the Copper Concentrator and the Lead/Zinc Concentrator are pumped to separate tailings collection boxes at the wet fill station. The copper tailings are directed to a dual bank of primary cyclopacs and the lead tailings directed to a similar dual bank of secondary cyclopacs. Cyclone underflow sand is then directed to either a sand collection tank or a cement mixing tank. From these tanks it is pumped underground as hydraulic fill (HF) or, after the addition of a cement and slag slurry, as cemented hydraulic fill (CHF).

To maintain a high quality fill, manual sampling for sizing is carried out daily in addition to continuous density monitoring. Pulp density is maintained at 64-70% solids (Ashby and Hunter 1982).

Given WFS and stope availability, Copper and Lead/Zinc Concentrator tailings combined, will result in the WFS producing two streams of fill running at approximately 200 tonnes per hour. Production of fill and placement is continuous and in line with the operations of the two concentrators.

Approximately 2.2 million tonnes of hydraulic sand fill is produced annually by the WFS.

2 FILL TYPES

2.1 Hydraulic fill (HF)

Hydraulic fill (HF) consists of uncemented deslimed mill tailings and is particularly suited to cut and fill mining. It has no strength or free standing ability and drains freely, which allows the next mining lift to begin with minimum delay. The HF helps stabilise the hanging wall and footwall of the narrower steeply dipping lead orebodies, and provides a level, stable floor from which to work. To achieve this, current sand size specifications are less than 9% passing 10 microns. See Figure 1.

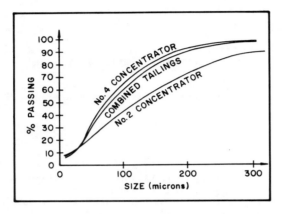

Fig.1. Deslimed tailings size distribution

HF is also placed in some lead and copper open stopes where regional ground stability is required and where there is no likelihood of future fill exposures.

2.2 Cemented hydraulic fill (CHF)

This is produced by adding 3% Portland Cement and 6% ground Copper Reverberatory Furnace Slag (CRFS) by weight to HF. Research (Thomas 1971) concluded that this ratio gives optimum strength characteristics at minimal cost.

The grind of the CRFS dramatically affects the cementing ability of the slag. The specification to obtain maximum benefit is for 90% of the slag to pass 400 mesh.

CHF is used in both lead and copper open stopes where the fill is required to have strength and free standing ability so that, when exposed by adjacent pillar extraction, fill dilution is minimal.

2.3 Rockfill (RF)

A shortfall in available deslimed tailings was recognised (Mathews and Kaesehagen 1973), and reduced by introducing a graded rockfill underground. It also offered the opportunity to:
1. reduce Portland Cement usage,
2. fill stopes more quickly,
3. reduce introduced water underground,
4. attempt different pillar mining methods.

Uncemented rockfill is used to fill pillar voids in the massive 1100 copper Orebody where there is no planned future exposure of fill. It is used to give regional ground stability. It consists of crushed and screened Kennedy Siltstone between 25 mm and 300 mm in size.

Currently 1.7 million tonnes of RF is mined annually in a surface open pit.

2.4 Cemented rockfill (CRF)

RF and CHF are mixed in various ratios to give CRF. It is used in 1100 copper Orebody open stopes where future pillar mining will expose fill.

3 FILLING SYSTEMS

3.1 Hydraulic fill system

From the WFS one of four centrifugal pumps is used to pump two fill lines of either HF or CHF horizontally to one of three vertical holes. Basically, each hole is up to 300 mm in diameter with a minimum of the top 60 metres lined with basalt pipe. The collar lining is installed because of high wear experienced in previous holes in the weaker upper oxidised zone of these holes. Vertical drops from the surface are up to 385 metres. See Figure 2.

The fill on reaching a major level underground is reticulated via a 150 mm flanged rubber lined pipes and 150 mm diameter vertical bore holes. The introduction of larger hole drill rigs to Isa Mine has led to longer, more accurate and more efficient underground fill holes. This is has resulted in less pipework and breast plate installation and maintenance.

At the collar and bottom of each vertical hole, a breast plate is

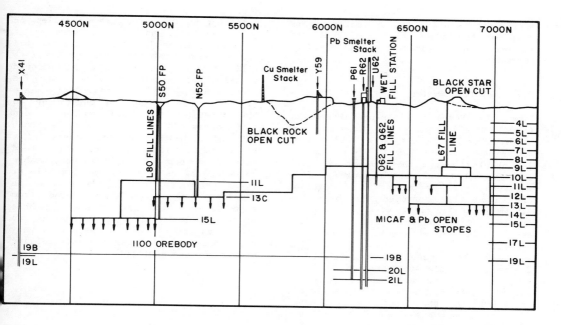

Fig.2. Surface to underground hydraulic fill system

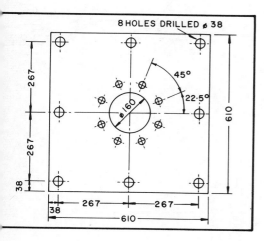

Fig.3. Plan of breast plate

installed. A breast plate consists of a 32 mm thick 610 mm square mild steel plate rockbolted in position and then grouted with a quick setting cement between the plate and the country. See Figure 3.

Due to line velocities in excess of 3.5 m/s a highly resistant fill line was desirable. 3 metre lengths of rubber lined pipe has proven to be reliable. It is installed by three man pipe fitting crews by placing pipes on pipe hangers resin grouted to the back or roof of headings. If possible, horizontal pipes runs are kept to a minimum of 50 metres to minimise pipe wear.

To keep pipe wear to a minimum, long sweeping bends rather than sharp corners are desirable. However, tee-pieces rather than 90° elbows have proven successful. Valves are generally not installed in the underground fill system. Line changes are affected by dropping out a pipe from one line and reconnecting to an alternate line.

Fill pipes, fittings and breast plates are inspected visually for defects at least twice a shift. A fill line monitoring system is currently being examined to reduce fill spillages and improve the hydraulic fill system performance.

3.2 Rock fill system

From the Kennedy Siltstone Open Cut (KSOC), siltstone is hauled by Caterpillar 773 trucks to a 2.13 x 1.52 jaw crusher which has a 300 mm bottom jaw opening. From the crusher the rock is conveyed to a screen house. After screening off the minus 25 mm material the rock is conveyed approximately 2.5 km on a 1.2 metre wide belt to one of two fill passes. The rock fill system is shown schematically in Figure 4.

Both RF passes were raisebored 2.4 metres in diameter most of the way. N52, the most northerly pass goes to 13C sublevel (580 metres below the surface), and S50 to the south to 15 Level (728 metres below the surface). N52 and S50 have surface pits developed around them and these hold approximately 20 000 and 30 000 tonnes respectively. To assist the RF control and chute

maintenance arc gates are installed on S50 at 11 Level and N52 at 12 Level.

These rock fill passes are choke fed and discharge onto the underground rock fill conveyor system. The underground systems are shown in Figure 5. The underground fill conveyors are 1.2 metres wide and can convey up to 800 tonnes/hour on 13C sublevel and 1200 tonnes per hour on 15 Level. Both 13C and 15 Level conveyors consist of a heavy major trunk conveyor which feed onto a number of lighter demountable conveyors by using a series of mobile or fixed trippers. Using this system the massive 1100 Orebody is covered from east to west and as far south as has presently been mined. Extensions to these existing systems are currently being examined to cater for rock filling of future more southerly stopes.

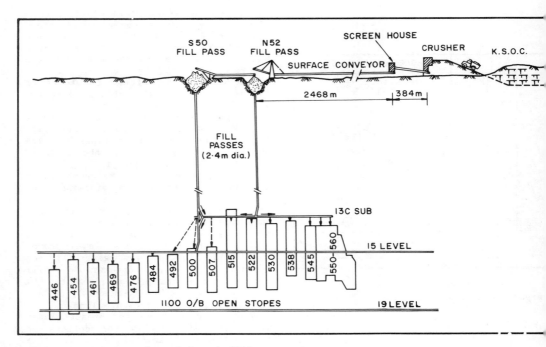

Fig.4. Long section of graded rock fill system

182

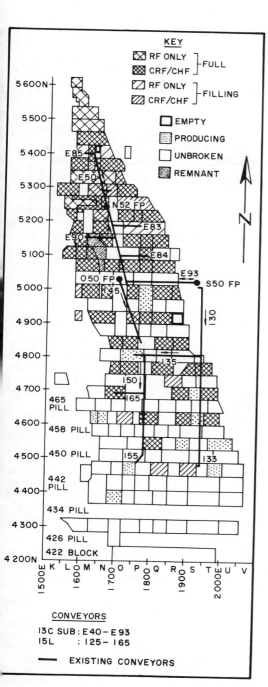

4.1 MICAF (cut and fill stoping) filling

Cut and fill stoping is used in the narrow tabular silver, lead and zinc orebodies in the northern section of the mine. Generally the width is restricted to a maximum of 11 metres and a minimum of 4 metres (Goddard and Bridges 1977).

Currently there are two MICAF areas operating between 9 and 11 Level and between 13 and 14 level. Once the initial sill horizon has been mined, slotted 100 mm diameter galvanised pipe or polythene agricultural pipe is laid along the entire strike length on the footwall and hanging wall of the orebody (Ashby and Hunter 1982). Hessian or a coarse cotton sleeve is wrapped around to either pipe type.

Drain towers are built on brick and concrete foundations at either end of an orebody, or at the filling limit for each section being filled in longer orebodies. The towers consist of tightly rolled lengths of 150 mm square wire mesh and are wrapped with four layers of hessian to act as a filter. Four towers approximately 3.2 metres high are built close to each drain raise which are located at strategic points along each orebody. See Figure 6. Drain raises are kept open by bolting pre-cut and drilled crib timbers together. The raise is then wrapped with chicken wire, hessian and polythene. Fill water can then only escape through the drain raises via the drainage towers and connecting 100 mm diameter drain pipes. The water is directed to sumps located on the level below.

Before each new lift, drain pipes from the drain towers are blanked off in the drain raise.

Prior to filling all ore passes, accesses and crosscuts are closed off with 460 mm thick brick bulkheads constructed on concrete pads and pinned at the walls. For ore passes two 460 mm thick brick bulkheads are built approximately 1.5 metres apart. Wire mesh is placed in the void between the bulkheads and concrete is then blown in. This is done to prevent fill collapsing into the ore pass during future lifts.

In the longer orebodies sand fill ramps and barricades are built to separate cut and fill cycle activities. Plastic sheeting is laid down the inner

Fig.5. 1100 Orebody rock fill conveyor system

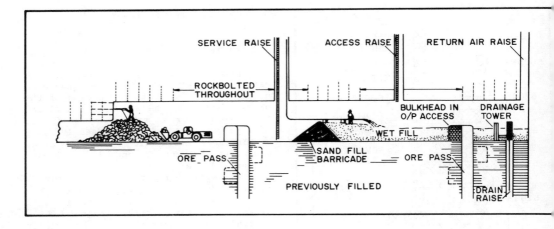

Fig.6. Typical layout of a MICAF flatback stope

slope to prevent excessive seepage. Hydraulic fill placement commences at this barricade with 6 metre long x 100 mm diameter victaulic piping being suspended from existing rock bolts and supported by wire rope. Tee pieces are installed in sections of the line to allow for back filling as the pipe is extended along strike and the fill builds up to a height of 2.6 to 2.8 metres from the back.

Generally the fill has drained sufficiently within three days of placement to allow flat back drilling for the next 3 metre lift to commence.

4.2 Silver, Lead and Zinc sublevel open stope (SLOS) filling

Some of the wider silver, lead and zinc orebodies are mined using sublevel open stoping (Goddard 1981).

Once the open stope has been mucked empty and filling is required, bulkheads are built on each sublevel. The standard thickness is 460 mm. However due to the narrower troughs at the bottom of some stopes, resulting in fill levels rising faster, double thickness bulkheads may be installed. See Figure 7.

At each bulkhead location, which is a minimum of 3 metres from the stope edge,

the floor is blown off to solid. A concrete pad is then poured as the foundation for the bulkhead. Pins are installed in the side of the drive and the straight bulkhead is constructed with a man access hole. Initially this man access hole allows access to enable cement grouting of the space between the brick work and the rock from inside the bulkhead. Later it allows access for installation of percolation pipes and for stope inspections before the fill level reaches that horizon. Rubber flaps are usually installed over the man access holes for ventilation control. Bulkheads are also grouted and sealed from the outside. Sometimes it becomes necessary to shotcrete badly cracked ground in the vicinity of bulkheads to prevent fill leaking from the stopes.

When the CHF or HF approaches the level the man access hole is bricked in.

Two percolation pipes are installed through each bulkhead. They consist of slotted 100 mm galvanised pipe wrapped in hessian and fixed to valves outside the bulkhead. Recently polythene agricultural pipe has been trialled successfully as an improved percolation pipe. This pipe has also given improved stope drainage by suspending it from one sublevel to another. See Figure 8.

Percolation water is either pumped into the sublevel drainage system or

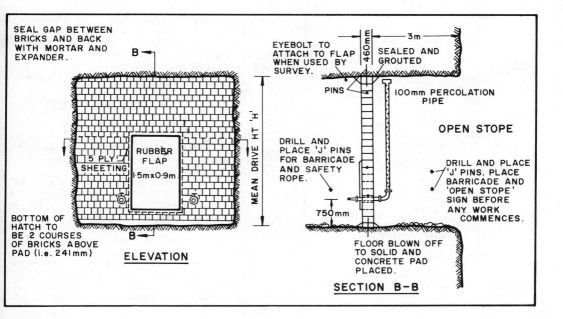

Fig.7. Typical bulkhead layout

SEAL GAP BETWEEN BRICKS AND BACK WITH MORTAR AND EXPANDER.

B →

RUBBER FLAP
1·5m×0·9m

5 PLY SHEETING

MEAN DRIVE HT 'H'

BOTTOM OF HATCH TO BE 2 COURSES OF BRICKS ABOVE PAD (i.e. 241mm)

B →

ELEVATION

EYEBOLT TO ATTACH TO FLAP WHEN USED BY SURVEY.

3m
SEALED AND GROUTED

PINS

100mm PERCOLATION PIPE

OPEN STOPE

DRILL AND PLACE 'J' PINS FOR BARRICADE AND SAFETY ROPE.

DRILL AND PLACE 'J' PINS. PLACE BARRICADE AND 'OPEN STOPE' SIGN BEFORE ANY WORK COMMENCES.

750mm

FLOOR BLOWN OFF TO SOLID AND CONCRETE PAD PLACED.

SECTION B–B

allowed to gravitate via drain holes to the main level drainage.

Filling is usually done through a specific fill hole into the back of each stope or through the return air raise/cut-off raise, into the back of the stope. Stopes are generally filled continuously and would only be stopped due to drainage problems or fill unavailability.

CHF filled stopes, after filling, are allowed to drain and the fill gains sufficient strength to allow stoping of adjacent pillars to commence within four months of completion of filling of the adjacent stope.

4.3 Copper SLOS filling

The egg-crate stoping technique (Mathews 1972) and the development and deviation from this technique (Alexander and Fabjanczyk 1981) results in rib and transverse pillars between primary stopes in the 1100 Orebody. See Figure 9. The mining of the

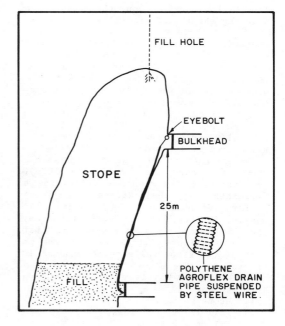

FILL HOLE

EYEBOLT

BULKHEAD

STOPE

25m

POLYTHENE AGROFLEX DRAIN PIPE SUSPENDED BY STEEL WIRE.

FILL

Fig.8. Level to level stope drainage system

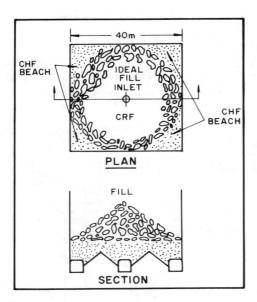

Fig.9. Ideal 1100 Orebody fill practice

1100 Orebody relies on the good free standing ability of the fill.

After a primary open stope has been emptied, bulkheads, identical to those used in the silver, lead and zinc SLOS, are built. Initially CHF is introduced to cover the top of the bulkheads in the draw points at the bottom of the stope. CHF and RF are then added together to form varying fill mass types (Cowling and Gonana 1976). CHF and RF are generally introduced through the same raisebored fill pass.

Investigations of the mining method used (Hornsby and Sullivan 1978), indicated the importance of obtaining a fill mass that, when exposed by adjacent pillar mining would be stable. To ensure the stability of fill exposures, the RF/CHF ratio is determined so that the more uniform CHF forms beaches on each of the stope walls to be exposed in the future. See Figure 10.

The RF/CHF ratio is determined taking into account:
1. stope geometry,
2. fill inlet positions,
3. fill trajectory.

It is of economic advantage to maximise the amount of rock and generally this is achieved by locating the fill pass centrally.

Depending on the heights and shapes of adjacent pillars and the status of surrounding stopes RF may be introduced by itself to provide regional fill and ground stability.

RF introduction to the 1100 Orebody has:
1. reduced the use of more expensive CHF.
2. made up for shortfalls in hydraulic fill.
3. permitted faster filling rates and consequently quicker pillar extraction.
4. reduced drainage problems.

Some isolated stopes or parts of stopes may be filled only with HF.

It has become increasingly important to schedule stope filling to minimise interference with stope extraction. Stope drainage, adjacent ore passes and faults and stope filling locations in relation to other producing stopes and other exposed fill masses must all be considered. These factors must also be considered in MICAF and lead SLOS extraction.

Generally a minimum of four months is allowed for filled stopes to drain and cure before mining adjacent stopes which would expose CHF or CRF.

To ensure maximum benefit is gained from CRF, regular stope inspections are carried out to monitor stope filling. Photographs and stope surveys are carried out on each major sublevel. This information is used to design future adjacent pillar stopes.

5   FUTURE FILL DEVELOPMENTS

In August 1982, a silver-lead-zinc ore pre-concentration plant was commissioned. The process involves heavy medium separation with a reject material between 10-14 mm being produced. Approximately 1.2 million tonnes of reject, also known as aggregate, will be produced annually.

In March 1983 an aggregate fill mixing plant will be commissioned. Aggregate will be conveyed from the Heavy Medium Plant (HMP) stockpile to launders, where up to 35% by weight of the aggregate will be combined with CHF or HF from the WFS to form Cemented Aggregate Fill (CAF) or Hydraulic Aggregate Fill (HAF). The mix will be directed into a new lined fill hole to 6 Level. Via a new fill pipe-line network, the HAF and CAF will be directed initially to lead SLOS.

The new fill pipe-line system consists of 3 metre lengths of flanged 150 mm NB polyurethane lined mild steel pipe. The polyurethane line is 6 mm thick. Trials of this pipe have proven it to be superior in this type of application to rubber lined pipe.

A trial established the suitability of CAF as a fill material (Davidson 1978). Tests included pumping pipe wear, in-situ placement and in-situ property testing.

Due to the large amount of aggregate being produced by the HMP, other means of utilising the material as backfill have also been investigated. Initial trials have indicated that the free running aggregate can be handled by existing surface and underground fill conveyors and fill passes. Consequently it will be introduced as aggregate fill underground as a replacement for RF in voids where it will not be exposed by future mining. Investigations into its use as a replacement for CRF will be examined later.

There is still much to learn about operating practices of handling HAF, CAF and AF underground. However, with the present operating and technical skills in handling CHF, HF, RF and CRF, Mount Isa Mines Limited is confident that the placement of the new types of fill will be successful and offer additional flexibility and more economic fill practices.

ACKNOWLEDGEMENTS

The author would like to thank the management of Mount Isa Mines Limited for permission to publish this paper and also those people of the Mining Division who assisted in its preparation.

REFERENCES

Alexander, E.G. & Fabjanczyk, M.W. 1981. Extraction design using open stopes for pillar recovery in the 1100 Orebody at Mount Isa. International Conference on Caving and Sublevel Stoping. Denver.

Ashby, I.R. & Hunter, G.W. 1982. Filling operations at Mount Isa. underground operators conference. Queenstown.

Barret, J.R. & Cowling, R. 1980. Investigations of cemented fill stability in 1100 Orebody, Mount Isa Mines Limited. Queensland, Australia. Institution of Mining and Metallurgy. 89:A118-A128.

Cowling, R & Gonano, L.P. 1976. Cemented rockfill practice and research at Mount Isa Mines. AMIRA, Excavation Design. Wollongong.

Davidson, C.W. 1978. Cemented aggregate Fill – underground placement, Mount Isa Mines Limited. Technical Report No. 53, Nov. 1978.

Goddard, I.A. 1981. The development of open stoping in lead orebodies at Mount Isa Mines Limited. International Conference in Caving and Sublevel Stoping. Denver.

Goddard, I.A. & Bridges, M.C. 1977. Development in Lead-Zinc mining methods at Mount Isa, Australia. Lead-Zinc update, S.M.E., New York. p.89-109.

Hornsby, B. & Sullivan, B.J.K. 1978. Excavation design and mining methods in the 1100 Orebody, Mount Isa Mine, Australia. Aus I.M.M. Conference. North Queensland. p.171-181.

Mathews, K.E. 1972. Excavation design in hard and fractured rock at Mount Isa Mine, Australia. 8th Canadian Symposium on Rock Mechanics. Ottawa. pp.211-230.

Mathews, K.E. & Kaesehagen, F.E. 1973. The development and design of a cemented rock filling system at the Mount Isa Mine, Australia. Jubilee Symposium on Mine Filling. Mount Isa. p.13-23.

Thomas, E.G. 1971. Cemented fill practice and research at Mount Isa. Proc Australian Institute of Min. Met. No. 240.

Proceedings of the International Symposium on Mining with Backfill / Luleå / 7-9 June 1983

Slinger belt stowing technique for cemented backfill at the Meggen mine

I.ROHLFING
Sachtleben Bergbau GmbH., Lennestadt, Germany

SYNOPSIS: Due to varying dipping conditions different mining methods are in use in the Meggen mine: sublevel-shrinkage mining, room-and-pillar mining or cross-cut-mining, the two latter methods combined with cemented backfill. All development and stoping faces are equipped with LHD-units. The backfill work started with pneumatic stowing and changed over to slinger belt stowing because of higher performance and lower overall costs derived from this backfill method. A cost comparison between rockfill, pneumatic fill and slinger belt fill is presented.

GEOLOGY,ORE RESERVES AND OVERALL PRODUCTION FIGURES

The mineral deposit at Meggen is a complex mixed sulphide orebody which contains iron pyrites, sphalerite and galena. As a result of volcanic activities in Devonian times, the ore was deposited on the sea-bed between limestone as hanging and sandy shales as footwall country rock. Subsequently these marine sediments were violently tilted and folded against a strong barrier formed by calcareous reef beds.

The strike length of the orebody is about 3 km (see Fig.1) and the deposit has a strong dip variation from vertical to horizontal (see Fig.2). The width varies between 1 m and 6 m, average 3.5 m.

Mining activities in the Meggen district began in the late 19th century. Today there are some 12 mio. tons of proven reserves left. The actual production is 1 mio. tons of run-of-mine ore per year and about 80.000 tons of waste rock out of development drifts.

Since 1972 trackless diesel-driven equipment has been used underground for loading, hauling and dumping (LHD), for ore transportation, drilling and other service work. Today we have road access for these mobile vehicles from the surface to every point of the mine through the Walter Ramp (illustrated in Fig. 1).

MINING METHODS

Due to varying dipping conditions of the deposit different mining methods are in use. In steep dipping districts a combined sublevel-shrinkage stoping method is practiced (see Fig.3). Flat lying parts are mined by room and pillar (see Fig.4a) or cross-cut stoping methods (Fig.4b), both systems combined with backfill.Semi-steep parts are extracted by a cut-and-fill method.

Due to the long striking length of the orebody, the variations in metal content and the necessary production of 1 mio.tons of run-of-mine ore per year, some 40 stoping faces have to be in production. Today more than 25% of these stoping faces are situated in flat lying or semi-steep dipping parts and must be backfilled. Furthermore, these faces are distributed over the whole striking length of the orebody and over levels 8, 10, 11 and 12. During the next year a backfill stope will be installed on level 9 as well.

SELECTION OF THE BACKFILL METHOD

When starting with backfill mining six years ago naturally the application of the most common backfill method, the hydraulic filling was discussed. In general the following requirements for economical

backfilling with hydraulic material must be met:
- the economical disposition of preparation plant tailings or sand fill material with a high enough dewatering ratio
- horizontal and vertical tubes which are able to provide an efficient fill transport to various stoping points; in the case of the Meggen mine more than one level (see Sect."Mining methods") must be supplied over a distance of some 3.000 m adaptable for different mining methods
- drifts below the backfilled rooms to pump off the percolated water.

Since the above mentioned conditions do not exist in the Meggen mine, their realisation would mean such a high investment that hydraulic fill cannot be applied economically in the mine. Only dry placement methods are in use: rockfill out of development drifts (5%) and slinger belt stowing with cleaned rock fill (95%). The latter method is a newly developed one, after having started with pneumatic stowing at rather high overall backfill costs.

BACKFILL QUALITY

Backfill quality depends on the characteristics of the fill material and backfill procedure.

Fill material characteristics

Backfill material from development drifts is limestone or sandy shales with no definite grain size distribution. Here and especially when backfilling with cleaned waste from our heavy media plant cement is added so that we generally build up a consolidated backfill.

The above mentioned cleaned waste consists of more than 50% limestone with compressive strength values of about 1000 daN/cm^2. The rest of the material has strength values of about 500 daN/cm^2. Fig. 5 illustrates the grain size distribution of our material and beside this a derivation of an optimum size distribution with minimum pore volume. (Fuller distribution). It can be seen that the cleaned waste rock material has no optimum but a usable characteristic as to the grain size. Essential for the use of slinger belt stowing machines is a grain size of not more than 50 mm.

The cement additive is about 50 kg per m^3 backfill material. Based on laboratory tests and practical knowledge this amount of cement is sufficient for a support

strong enough to allow a complete extraction of the ore as well as a good enough safety margin for the miners and machines doing stoping work. We do not yet know the effects of our confined backfill mining upon possible surface damage. We hope however that a newly started research programme will deliver computable values to answer these questions.

Backfill procedure

The backfill procedure should guarantee a sufficient mechanical compaction of the fill material and a complete filling of the mined out rooms even in the flat parts of the mineral deposit.

The compaction of fill material by the slinger belt system is comparative to pneumatic stowing and provides a surplus of compressive strength compared e.g. with rockfill performed by LHD-units.

The complete filling of a mined out room in a flat lying deposit can only be achieved with pneumatic or slinger belt stowing.

The early support over a relatively large area by backfill material ensures sufficient safety for men and machines, a reduction of roof support (e.g. timbering, bolts or concrete) and of dilution by rockfall especially from the hanging wall.

SLINGER BELT STOWING

The slinger belt stowing machine is illustrated in Fig.6. It consists of a feed cone from which the cleaned waste fill flows over a proportioning bucket wheel to the fast running slinger belt. This belt is driven by a 15 kW electric motor at speeds up to 20 m/sec. The material can be thrown about 8 m high and over a distance of 14 m. The whole machine is generally installed near the stoping points and can be transported to the faces by 2 cubic-yard LHD-units. The slinger belt may as well be fed by these LHD-units.

A more effective slinger belt stowing system is shown in Fig.7. The diesel-mobile vehicle operates as follows:

The truck is equipped with articulated steering. The truck body is shaped like a trough with a sliding blade being moved by a hydraulic cylinder. The electrically driven slinger belt machine (see Fig.6) is attached under the rear end of the truck body. In order to adapt it to this type of mounting, the feed cone has been shortened and the foundation structure eliminated.

There is an opening in the bottom of the truck body allowing the waste fill to be fed at an even rate onto the slinger belt by the sliding blade pushed forward by the hydraulic cylinder.

Another possibility of discharging the truck is by closing the above mentioned opening automatically and by unblocking the slewable rear port. Now the fill material will be pushed out by means of the sliding shield without passing the slinger belt.

A photograph of the $6m^3$-capacity slinger-belt truck is shown in Fig.8. The vehicle was built by Messrs. Hermann Paus, Ems-bühren, based on drawings of Sachtleben Bergbau GmbH, both in West Germany. Today five of these machines are employed underground in Meggen, the latest one of $10m^3$ capacity.

Charging of the slinger belt trucks is shown in Fig.9. The fill material flows from a chute over a vibrating trough to the truck after the proper amount of cement has automatically been mixed into the waste fill. The cement is stored in a $6m^3$ bin near the chute and is trans-ported to this point pneumatically out of a bigger bin on the surface.

BACKFILL COSTS

As mentioned above the backfill material consists mainly of cleaned waste rock from the heavy media plant on the surface. Today the total amount of waste rock pro-duced in the separator is transported into the backfill chute by means of a 30m-long conveyor belt. Otherwise this material would have to be trucked to a waste deposit at the respective cost of 3.20 DM per m^3, so that with backfill this amount can be entered on the credit side.

The backfill distribution for a con-sumption of 100.000 m^3 per year is shown in Fig.10. On levels 8 and 9 the waste rock flows directly from the main waste chute to the slinger belt trucks (after having automatically been mixed with cement); on level 10 the material is transported by railway over 600 m and 1500 m to shorter chutes, where the back-fill is dumped and then charged to the slinger belt trucks. The average one-way transport distance from the chutes to the stoping faces is about 350 m.

On the one hand the personnel cost for backfilling depends on the time consump-tion for the filling procedure. Here you can determine fixed times for the slinger-work (1 min/m^3), the positioning of the truck (5 min/cycle) and filling of the truck (4 min/cycle). The transport time is determined by the hauling distance and quality of the road. In the Meggen mine we calculate with 1 min/100 m on the average. Based on these figures it takes 22 min/cycle for a $6m^3$-slinger belt truck to backfill a room over a distance of 350 m. Furthermore, the capacity of one $6m^3$-truck is 106m^3 per drift (6.5 pro-ductive hours), with one driver.

On the other hand personnel cost must in-clude maintenance work for the diesel-mobile vehicles and other service work at the faces (backfill) and at the filling stations.

Including this, on an average there will be a performance of 60m^3 backfill per man-shift based on $6m^3$-trucks and thereof per-sonnel cost of 4.- DM per m^3.

Expenses for material are mainly for the slinger belt machine, the acticulated trucks and cement (50kg/m^3 backfill) at an amount of 8,- DM per m^3.

The amortization is derived from the investment for the slinger belt truck and for the filling station. With an average filling volume of 60.000 m^3 material being supplied by one filling station and a life-time of 10.000 working hours of the back-fill truck one can calculate with 3.- DM per m^3 investment cost.

In Fig.11 the above derived figures are summarized and illustrated in a column dia-gram (procedure No.9). In Fig.11 altogether nine backfill procedures are illustrated. Nos. 1 and 2 are pneumatic stowing methods of 30m^3/h capacity machines and 150m^3/h units respectively. On the very right hand side of the Figure the performance in m^3 per manshift is written down as a basis for personnel cost calculation. The total cost can be seen on the left hand side of the Figure in the column-diagram.

The rockfill methods (Nos.3, 4, and 5) are calculated on the basis of employing 6t, 10t and 15t normal trucks, hauling the material from the filling station into the neighbourhood of the stopes to be filled. From there 1.5 m^3-capacity LHD units back-fill the rooms.

The slinger belt stowing procedures Nos. 6 - 8 are based on the employment of trucks (see rockfill methods) and LHD-units charg-ing the slinger belt illustrated in Fig.6. The No.9 procedure is the slinger belt truck as described above of a capacity of $6m^3$ per cycle. Performance calculated in m^3 backfill per manshift can be seen at the right hand side of Fig.11 and the total costs are presented once more in the column diagram.

191

The lowest costs can be attributed to the slinger belt truck backfill procedure with roughly 15.- DM/m³. This is mainly due to low personnel costs because of the relatively high performance of 60 m³ backfill per manshift. The costs of material are comparable to those of other backfill procedures whereas the investment costs are much lower.

The low investment costs are due to the fact that the slinger belt truck backfill method can be integrated into the existing trackless mining system without difficulties. For example there are no additional expenses for the preparation of backfill stopes contrary to the pneumatic stowing systems. Experiences with this system have shown that e.g. additio-

nal development drifts were necessary even in the host rock to ensure a straight installation and the shortest distances possible of the pneumatic stowing tubes.

Furthermore, the slinger belt truck system allows to react with great flexibility to influences created by other stoping work. This criterion of a backfill method is extremely important when a small amount of backfill material per stoping point has to be stowed and the number of stoping faces is low because of the nature of the mineral deposit. Especially in this case the backfill has to push rather than lag behind the remaining work sequences, otherwise the mining costs will be too high to reasonably implement a stoping method based on backfill.

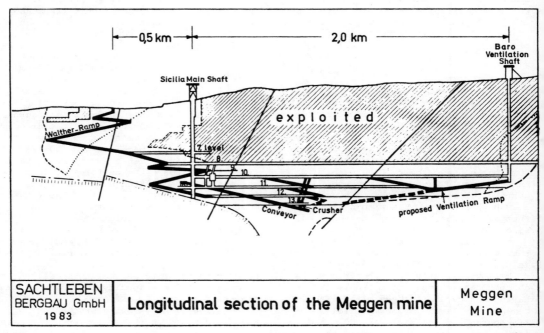

| SACHTLEBEN BERGBAU GmbH 1983 | Longitudinal section of the Meggen mine | Meggen Mine |

fig. 1

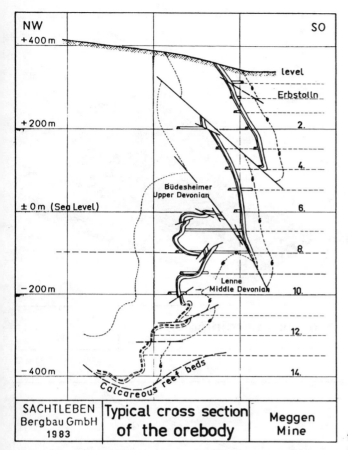

NW
+400m

NW
SO

+400m

+200m

± 0 m (Sea Level)

−200m

−400m

level

Erbstolln

2.

4.

Büdesheimer
Upper Devonian

6.

8.

Lenne
Middle Devonian   10.

12.

14.

Calcareous reef beds

| SACHTLEBEN Bergbau GmbH 1983 | Typical cross section of the orebody | Meggen Mine |

fig. 2

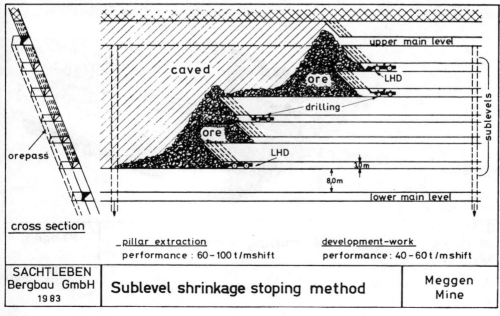

cross section

caved

ore

LHD

drilling

ore

LHD

orepass

upper main level

sublevels

10m

8,0m

lower main level

pillar extraction
performance: 60–100 t/mshift

development-work
performance: 40–60 t/mshift

| SACHTLEBEN Bergbau GmbH 1983 | Sublevel shrinkage stoping method | Meggen Mine |

fig.3

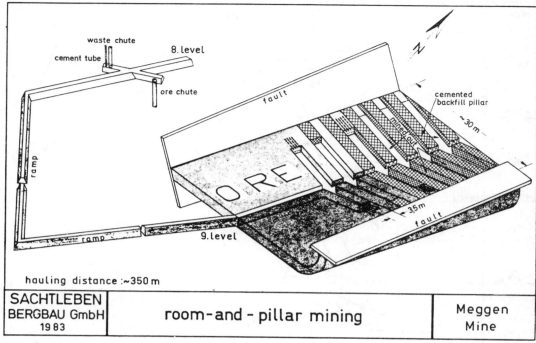

SACHTLEBEN BERGBAU GmbH 1983	room-and-pillar mining	Meggen Mine

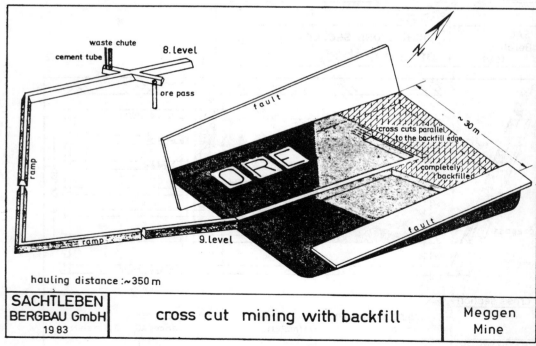

SACHTLEBEN BERGBAU GmbH 1983	cross cut mining with backfill	Meggen Mine

fig. 4b

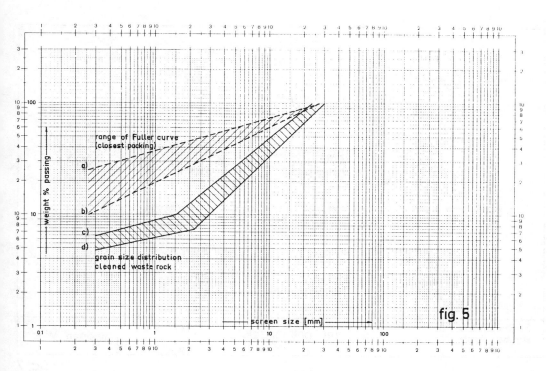

range of Fuller curve
(closest packing)

a)

weight % passing

b)

c)

d)

grain size distribution
cleaned waste rock

screen size [mm]

fig. 5

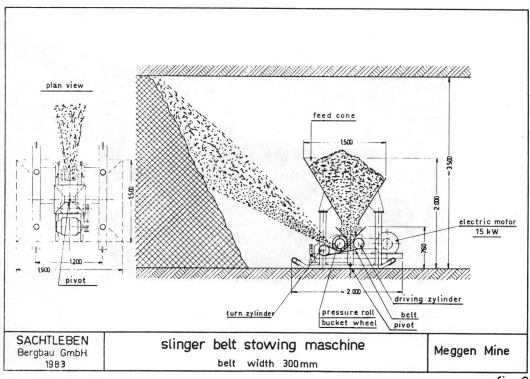

plan view

feed cone

1500

~3500

2.000

electric motor
15 kW

1500

760

1200

1900

pivot

~2.000

turn zylinder

pressure roll
bucket wheel

driving zylinder

belt
pivot

SACHTLEBEN Bergbau GmbH. 1983	slinger belt stowing maschine belt width 300mm	Meggen Mine

fig. 6

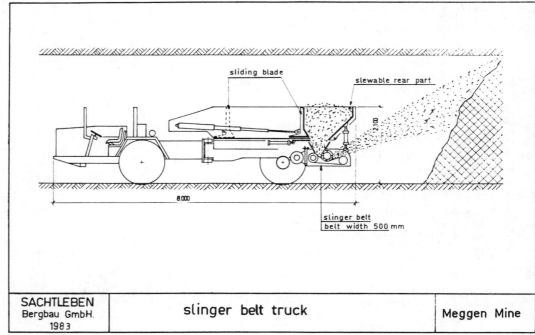

| SACHTLEBEN Bergbau GmbH. 1983 | slinger belt truck | Meggen Mine |

fig. 7

fig. 8

6m³-slinger belt truck

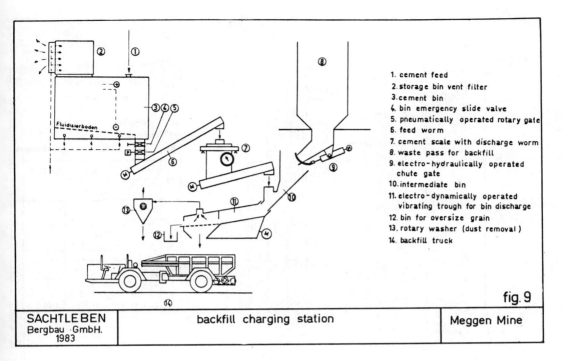

1. cement feed
2. storage bin vent filter
3. cement bin
4. bin emergency slide valve
5. pneumatically operated rotary gate
6. feed worm
7. cement scale with discharge worm
8. waste pass for backfill
9. electro-hydraulically operated chute gate
10. intermediate bin
11. electro-dynamically operated vibrating trough for bin discharge
12. bin for oversize grain
13. rotary washer (dust removal)
14. backfill truck

fig.9

| SACHTLEBEN Bergbau GmbH. 1983 | backfill charging station | Meggen Mine |

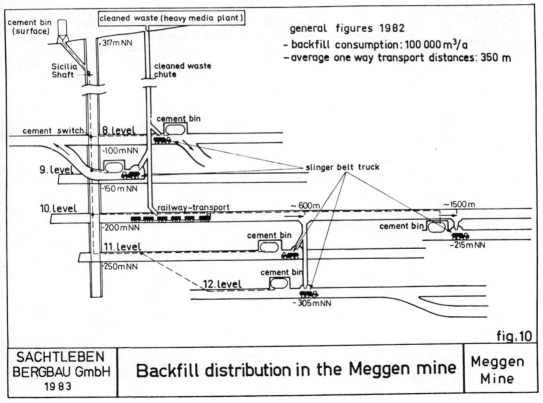

cement bin (surface)

cleaned waste (heavy media plant)

Sicilia Shaft

317m NN

cleaned waste chute

cement bin

general figures 1982
- backfill consumption: 100 000 m³/a
- average one way transport distances: 350 m

cement switch.   8. level

-100m NN

9. level

-150 m NN

slinger belt truck

10. level   railway-transport   ~600m   ~1500 m

-200mNN

11. level   cement bin   cement bin

-250mNN   -215m NN

cement bin

12. level   cement bin

-305mNN

fig.10

| SACHTLEBEN BERGBAU GmbH 1983 | Backfill distribution in the Meggen mine | Meggen Mine |

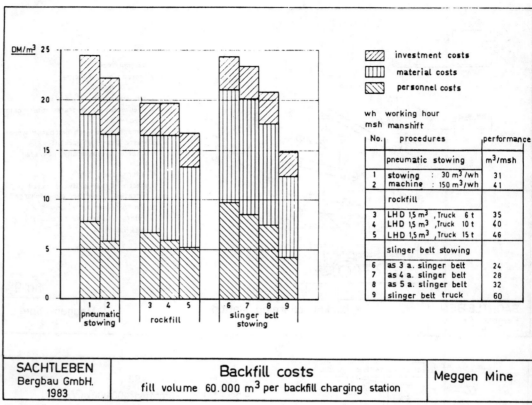

No.	procedures	performance
		m^3/msh
	pneumatic stowing	
1	stowing : 30 m^3/wh	31
2	machine : 150 m^3/wh	41
	rockfill	
3	LHD 1,5 m^3 ,Truck 6 t	35
4	LHD 1,5 m^3 ,Truck 10 t	40
5	LHD 1,5 m^3 ,Truck 15 t	46
	slinger belt stowing	
6	as 3 a. slinger belt	24
7	as 4 a. slinger belt	28
8	as 5 a. slinger belt	32
9	slinger belt truck	60

investment costs

material costs

personnel costs

wh working hour
msh manshift

SACHTLEBEN Bergbau GmbH. 1983	Backfill costs fill volume 60.000 m^3 per backfill charging station	Meggen Mine

fig. 11

Initial experience in the extraction of blasthole pillars
between backfilled blasthole stopes

D.A.J.ROSS-WATT
Black Mountain Mineral Development Co. (Pty) Ltd., Johannesburg, South Africa

ABSTRACT: A previous paper by C.L. de Jongh, "Design Parameters Used and Backfill Materials Selected for a New Base Metal Mine in the Republic of South Africa" was presented at the symposium, "Application of Rock Mechanics to Cut and Fill Mining", held in Sweden in June 1980. This paper covered the technical parameters for blasthole stope and pillar layouts and for the selection of backfill material at Black Mountain Mineral Development Company (Pty) Limited's Broken Hill Orebody.

At Black Mountain, the mining of ore containing copper, lead and zinc sulphides as well as silver takes place. Copper, lead and zinc concentrates are produced, with the major portion of the revenue being derived from lead and silver.

The total ore reserve is some 35 million tons of ore grading about 7,5% lead and 95 grams/ton silver.

The area amenble to blasthole stoping, dipping at greater than 50°, contains some 3,5 million tons of ore grading over 11% lead and between 175 and 200 grams/ton silver. Due to the high grade of the ore, a blasthole stope and pillar method is employed which aims at 100% extraction. The balance of the orebody, which is flatter dipping is being mined by cut and fill methods.

To date over 1,9 million tons have been extracted from blasthole stopes. In order that the blasthole pillars may be extracted the stopes are backfilled with a material containing desert sand, mill tailings with appropriate cement addition. Total backfill placed in blasthole stopes is over 750 000 tons and total backfill placed including cut and fill stopes is over 1 million tons. The extraction of the first pillar is completed and extraction of the following pillars is in progress.

This comprehensive paper describes the mining methods employed in blasthole stopes and pillars, the operation of the backfill plant, the preparation of stopes for filling, backfill reticulation, filling and drainage of stopes and the results obtained in extraction of the initial blasthole pillar stopes.

1 MINING METHODS IN GENERAL AT BLACK MOUNTAIN

At Black Mountain, the mining of ore containing copper, lead and zinc sulphides as well as silver takes place. Copper, lead and zinc concentrates are produced, with the major portion of the revenue being derived from lead and silver.

The Broken Hill ore reserves are contained in two orebodies extending over a total strike length of approximately 1 600 metres and from a surface outcrop in the western area down to approximately 700

metres below surface in the eatern extremity. The two orebodies run conformably where both are present and are separated by a barren schist material varying between five and thirty metres thick. The orebody as a whole plunges below surface at 23° in a direction N64°E. The dip of the orebody varies from approximately 55° to virtually flat in parts of the lower eatern area. (Figure 1) The area in which blasthole stoping is employed has been folded upwards so

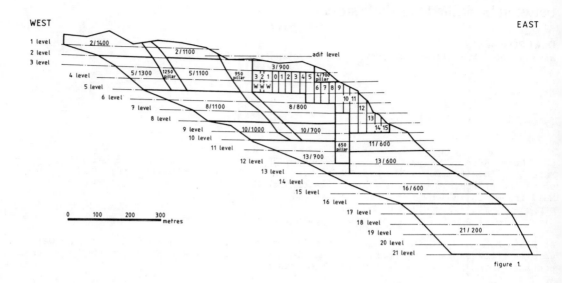

VERTICAL PROJECTION OF THE BROKEN HILL OREBODY SHOWING BLASTHOLE STOPES
AND PILLARS, CUT–AND–FILL STOPES AND MAJOR STABILITY PILLARS

figure 1

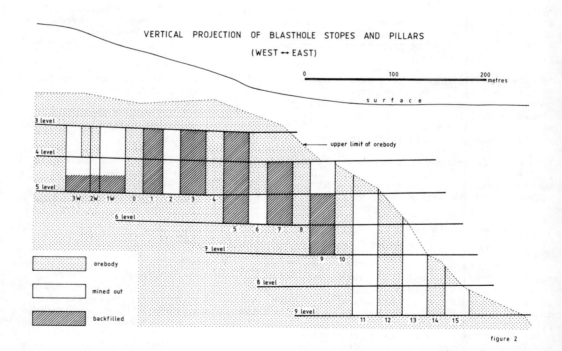

VERTICAL PROJECTION OF BLASTHOLE STOPES AND PILLARS
(WEST ↔ EAST)

figure 2

that the dip is approximately 55° and the strike approximately EW. This area contains some 3,5 million tons of high grade ore with average lead values of over 11 percent and average silver values between 175 and 200 grams per ton. The steep dip and the competent hanging wall make this area amenable to high productivity, open stoping blasthole methods. This area is the most accessible part of the orebody. The horizontal width of the orebody varies from five to forty metres. Due to the high grade of the ore, it is essential that as close as possible to a hundred percent extraction be achieved in this area.

The balance of the orebody strikes on average at N60°W with the dip ranging from 50° to virtually flat. In general, the grade decreases with distance from the blasthole stoping area. Average lead values are approximately 6,5 percent and average silver values approximately 65 grams per ton. The flatter dip precludes the use of open stoping methods and various cut and fill stoping methods will be used throughout this part of the orebody. Horizontal width of the orebody varies from five to ten metres up to possibly seventy or eighty metres in the flat dipping sections. A system of cut and fill stopes nominally 200 metres on strike, separated by major dip stability pillars 50 metres wide, has been provisionally laid out. The major pillars will be extracted at the end of the life of the mine. It is expected that high levels of extraction will be achieved in this way.

2 MINING METHOD - BLASTHOLE STOPES

The blasthole stoping area is divided into blasthole stopes (3W, 1W, 1,3,5,7,9,11, 13,15) each 28 metres wide on strike, and blasthole pillars (2W, 0,2,4,6,8,10,12,14) variously 10,20 and 28 metres wide on strike. As the blasthole stopes are extracted and filled, the blasthole pillars will be extracted and filled in their turn. (Figure 2)

Levels are positioned 35 metres apart vertically, and, at these elevations, footwall drives are driven parallel to the orebody approximately 20 metres into the footwall. Access crosscuts are mined to the orebody in the pillar positions. Drill drives are mined in the orebody at 6 metre centres leaving small drive pillars. At the east and west of each stope, a cut out and an access crosscut, respectively, are mined. On the lowest drawpoint level, two drawpoint cross-cuts are mined for each stope and connected by a trough drive. A slot raise of 1,8 metre diameter is bored from level to level at the intersection of the footwall contact and the pillar line. (Figure 3)

The blastholes are 165mm in diameter and up to 45 metres long, being drilled from level to level down the dip of the stope. Three rows of holes are drilled in the slot with burdens of 3 metres and spacing of 1,5 metres. In the remainder of the stope, a burden of 3 metres and a spacing of 4,6 metres is used. In the undercut, 51mm blasthole rings are drilled. The 165mm blastholes are drilled using crawler rigs with DTH pneumatic hammers.

The undercut is charged with 560mm x 38mm x 60 percent Dynagel and Cordtex detonating fuse. The 165mm blastholes are charged with up to three decks of Anfex, each deck separated by a one metres plug of sand. Each metre of blasthole carries 17kg of Anfex so that 13 metre decks consist of 221kg of Anfex and each hole contains about 660kg. Primer charges are lowered to the centre of each deck. The primer consists of two electric detonators taped to Cordtex detonating fuse which is, in turn, threaded through a 400g Pentolite booster. Heavy lead wires are connected and used for lowering the assembly down the hole. The top three metres of every hole is tamped with sand to minimise brow damage and fly dirt. Short period delay (S.P.D.) electric detonators are used exclusively and all sets are connected in series. Since each deck has a different delay period, the maximum quantity of explosives being detonated simultaneously is of the order of 220kg. This keeps blast damage down to a minimum.

The first blast taken is the slot undercut. This is followed by removal of the slot in three or four blasts. Undercut rings and main stope ring blasting then progresses. Normally two or three stope rings are blasted at a time, yielding between 5 000 and 25 000 tons of ore per blast. (Figure 3)

The status of blasthole stopes at the end of August 1982 was as follows: -

Stope 3W   56 000 tons blasted In Progress
Stope 1W   72 500 tons blasted Depleted
Stope 1   137 800 tons blasted Depleted
Stope 3   235 200 tons blasted Depleted
Stope 5   363 300 tons blasted Depleted
Stope 7   196 200 tons blasted Depleted
Stope 9   302 100 tons blasted Depleted
Stope 11 292 300 tons blasted Depleted
Stope 13 182 500 tons blasted In Progress
Stope 15   31 700 tons blasted In Progress

3 BACKFILLING AND PILLAR STOPING IN GENERAL

After depletion, the blasthole stopes must

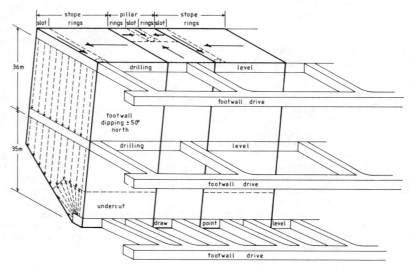

figure 3

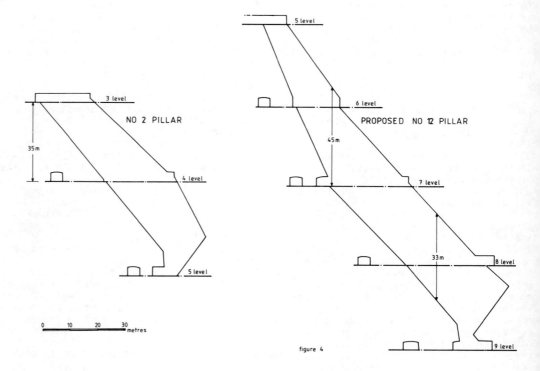

figure 4

be backfilled in order that the adjacent pillars may be extracted. Once the pillars have been extracted, these voids will, in turn, be backfilled with a suitable backfill material.

When the appropriate blasthole stopes have been backfilled, drilling access is developed at the 35 metre footwall intervals and the pillar blocks are drilled off using 165mm diameter blastholes. In the case of the pillars, the 1,8 metre slot raise is bored on the hangingwall in the centre of the pillar. A slot is then blasted across the hangingwall between the filled stopes. If a side slot is used, concentrated slot blasting will take place against the backfill wall. If a central slot is used, backbreak may deny access to blastholes in the narrow side remnants. The undercut is created by means of blastholes from the level above. On the lowest level, a single drawpoint is mined.

Charging up procedures are as for the blasthole stopes. The EW slot is initially extracted followed by rings of blastholes blasted northwards into the slot. Blasting is carried out from the lowest level upwards.

The status of blasthole pillars at the end of August 1982 was as follows: -
Pillar 2   126 800 tons blasted   Depleted
Pillar 6   32 500 tons blasted   In Progress
Pillar 2W   13 100 tons blasted   In Progress

4 THE BACKFILLING OPERATION

4.1 The required strength of the backfill

For the blasthole stopes, the criteria for the strength of backfill are as follows: -

a. If pillars are extracted one level at a time with each block being backfilled before continuing, then the backfill must be freestanding for at least 35 metres vertically. It is preferable however to extract pillars to their full extent before backfilling, thus avoiding the necessity of having to load ore on the backfill surface with a remote loader. The extent of exposed backfill in 2 pillar is for 70 metres vertically. The maximum extent of exposed face will be in the case of the 11 stope 12 pillar interface, where backfill will be exposed for approximately 140 metres vertically, i.e. 4 levels. (Figure 4) The maximum effective vertical height, from the footwall to the hanging wall, will be a function of the dip of the orebody and the horizontal width from footwall to hangingwall. A 50° dip and 40 metre horizontal width for example implies a maximum effective vertical height of 40 tan 50° = 47,7

metres.

For this criterion, the failure of the backfill under its own weight alone is considered and the strength requirement is a function of height, width and density. The mechanics of failure and the shape of the failure surface are important. The orientation of the failure, whether vertical, along the direction of dip, or a combination of the two, is important in considering whether maximum effective vertical height or some increased value of effective height must be considered. Strength requirements for the backfill were initially fixed using a combination of first principle calculations (vertically orientated failures) and reference to literature covering backfilling experience on other mines. Based on the above, an unconfined compressive strength (U.C.S.) of 700kPa was recommended for the backfill in the blasthole stopes. This strength would allow for a freestanding height of approximately 60 metres. The recommendation was considered conservative, but nevertheless realistic, in view of the high ore grades in the blasthole area and the uncertainty surrounding both the required strength for the backfill and, in the early stages, the strength of material achieved.

In fact, approximately 50 percent of the blasthole stopes had been filled before extraction of the first pillar was completed, giving full scale experience of the backfill. The required strength criterion was reduced during this period, first marginally, based on further discussions with overseas operations and consultants, and then to a U.C.S. value of 400kPa based on extensive large scale model testing carried out at Black Mountain.

b. The backfill must be sufficiently resilient to withstand the effect of the 165mm blasthole blasting in the pillars. The resilience of the backfill will decrease with higher cement contents. The strength of the backfill based on the 700 kPa value fixed by criterion (a) does, in fact, require a somewhat higher cement content than that recommended to allow the required resilience. (1:13 compared with 1:20 percentage of solids by weight.) The risk associated with inadequate strength was, however, considered to be greater than the risk associated with blast damage. As pillar blasting experience is gained, the burden and spacing of blastholes adjacent to the backfill can be adjusted accordingly.

c. The backfill in the base of the stope must be sufficiently strong to allow the crown pillar of the cut and fill stope below to be stoped out completely.

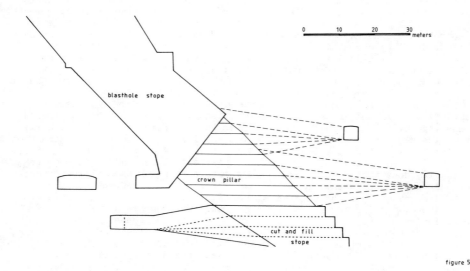

figure 5

TYPICAL GRAIN SIZE DISTRIBUTIONS OF MILL TAILINGS
CLASSIFIED MILL TAILINGS AND DESERT DUNE SAND

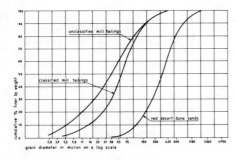

figure 6

(Figure 5) Conversely, the cement content in this portion of the stope must not be so high as to inhibit percolation. Information on backfill properties for undermining is scarce. After consultation with overseas consultants and consideration of procedures overseas, a decision was made that the strength of backfill required by criterion (a) would be suitable for this purpose. (When the strength criterion for a freestanding face was reduced, the strength of the backfill material in the trough portion of the stope was maintained. If necessary, the undermining will be carried out on a remnant cut and fill basis with the widths of single cuts being kept minimal (3 to 4 metres).

The placement of reinforcement on the footwall is not practical due to lack of access, but every effort is made to completely clean the footwall of the trough so as to eliminate non homogenous backfill.

d. The backfill must be sufficiently strong to allow for overall stability of the mine as extraction progresses. Due to the width of the orebody, 10 to 40 metres in the blasthole stoping area, it is unlikely that any redistribution of stress will, in fact, take place through the backfill. As mining progresses, stresses will be distributed to the major dip stability pillars, the unmined portion of the orebody and the host rock on the boundaries. The backfill will, however, maintain the integrity of the hangingwall by preventing progressive scaling. In this instance, it is predominantly the bulk of the material that is required. The backfill specified by the other criteria will be sufficiently dense.

4.2 The backfill material

The strength required of the backfill (700 kPa, approximately 100psi) is relatively low compared, say, with that required of a normal concrete mix (21mPa, approximately 3 000psi). The cost, availability and means of mixing and transport of the components of the backfill thus become the most important criteria in designing a backfill to meet the required strength specifications.

Potential backfill materials available on site are mill tailings, wind-blown dune sand and rock which could be quarried for this purpose. Significant characteristics of these materials are as follows: -

a. The tailings are very fine, have a high magnetite content and a specific gravity of over 3,5.

b. The dune sand is relatively coarse and well graded, but has poor viscous properties. Dune sand requires to be loaded and transported for several kilometres.

c. Most mines using quarry rock as a backfill have large waste dumps either as a result of opencast operations or underground waste produced over an extended period. No such dumps exist at Black Mountain, thus rock would have to be quarried and would constitute an expensive backfill component.

Extensive tests were carried out on the following backfill materials and combinations of backfill materials: -
dune sand;
milled dune sand;
unclassified tailings; (+38μ);
classified tailings : dune sand, 50:50;
classified tailings : dune sand, 75:25;
unclassified tailings without magnetite;
unclassified tailings : dune sand, 50:50
plus a limited number of tests on various mixes with screened and unscreened rock. Typical grain size distributions of mill tailings, classified tailings and dune sand are shown in figure 6.

Availability of mill tailings

A typical metallurgical balance during high grade blasthole stoping is as follows: -
94 000 tons ore milled (all dry tons);
1 500 tons copper concentrate produced;
1 000 tons lead concentrate produced;
3 500 tons zinc concentrate produced;
78 000 tons mill tailings available for milling.

If unclassified tailings are placed in the stopes, the full 78 000 tons of mill tailings will be available. If the tailings are classified by taking out most of the -38μ (or similar cut-off) material a loss of some 30 percent of the mass will occur, i.e. from 78 000 to 55 000. The mass of material required to fill the void created by the extraction of 94 000 tons of ore depends on the specific gravity of the ore (4,0) and on the dry specific gravity of the backfill material once placed. At 2,0, 47 000 tons of backfill material will be required and at 2,5, 58 750 tons of backfill material will be required. It is considered impractical to return tailings from the tailings dams for backfilling and thus tailings can only be stored to the extent that storage and retrieval facilities are available at the backfill plant itself. At a specific gravity of 2,0, unclassified tailings would have to be placed for 60 percent and classified tailings for 75 percent of the hours during which milling takes place. This would not have allowed for initial high backfilling rates to stopes already mined before the backfill plant startup, for backfill plant breakdowns or for backfill stoppages due to scheduling problems underground. Should the backfill material prove to have an even higher specific gravity, the situation will obviously be worse.

If classified tailings only are used, problems will arise in building tailings dams consisting of fines alone. The decision was that, at least in the early stages during the backfilling of the blasthole stopes, some additional backfill would be required from the point of view of availability alone.

Viscosity - pulp density relationship

This property is important in determining the method of transport of the backfill. For backfill materials having a low viscosity, or alternatively, for excessively high water contents, the solids will settle causing a blockage. For backfill materials with a high viscosity or alternatively, for very low water contents, the frictional resistance will become excessive causing a blockage.

All underground backfill transport and placement is by gravity, the most extreme instance in the blasthole stopes to date

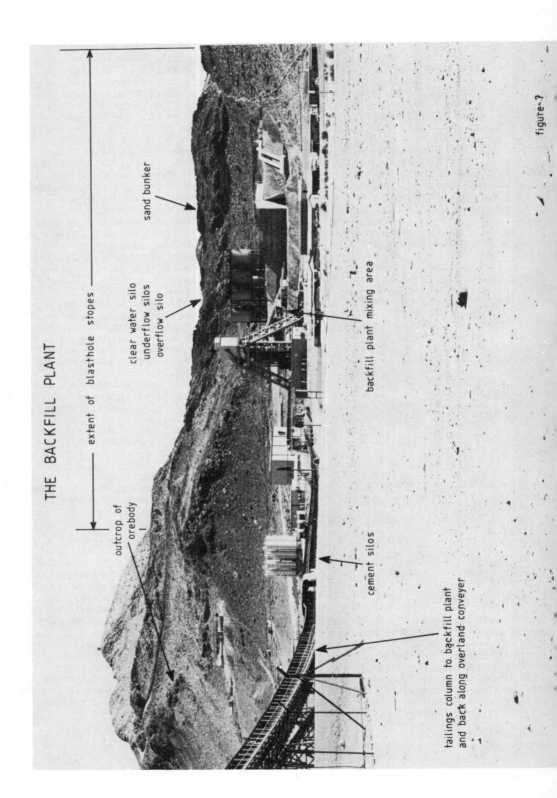

THE BACKFILL PLANT

extent of blasthole stopes

outcrop of orebody

sand bunker

clear water silo
underflow silos
overflow silo

backfill plant mixing area

cement silos

tailings column to backfill plant
and back along overland conveyer

figure 7

206

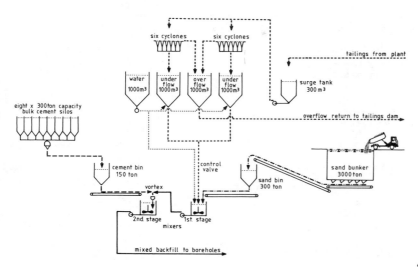

figure 8

being transport over 220 metres horizontally with 75 metres of available head.

Bench scale tests indicated that dune sand alone would not become sufficiently viscous for transportation. Conversely, classified and unclassified tailings and mixtures with dune sand indicated a rapid increase in viscosity at around 76 percent to 78 percent pulp density. Due to the high specific gravity of the fill material, these pulp densities represent large volumes of water, excessive for cement hydration and which create problems in properly draining the backfill. The backfill has been placed at pump densities of around 73 percent to date. It is considered essential however that full scale tests be carried out to attempt to increase the pulp density. These are planned as soon as a surface borehole can be placed at risk.

Permeability

In view of the low pulp densities at which the backfill is placed, it was considered essential that the backfill material be sufficiently permeable to allow percolation of excess water through the fill to the drainage points.

Percolation rates for classified tailings and dune sand mixes were shown to be around 100mm per hour in bench scale tests whereas classified tailings alone had a percolation rate of less than 40mm per hour. At higher cement content the percolation rate decreases to the extent that, at 1:10 cement to solids by weight, percolation rates reduced to around 10mm per hour. As is described later, drainage by percolation has been insignificant in the filling of the blasthole stopes. most of the drainage being through decantation.

Unconfined compressive strength

Unconfined compressive strength gives a measure of the shear strength of the material, this being a combination of frictional resistance and cohesion. Resistance to failure of the freestanding wall of backfill material depends on the shear strength (cohesion and frictional resistance) present along potential failure surfaces. Unconfined compressive strength was used as a criterion for the design of the strength of the backfill material and is used to control backfill quality by taking U.C.S. tests on a daily basis.

U.C.S. tests on the various backfill materials were carried out, with cement ratios varying from 1:8 to 1:40 cement to solids by weight.

Material	U.S. Values (kPA)		
	7 days	14 days	28 days
Unclassified tailings			
1:10 cement by mass	714	984	1 197
1:25 cement by mass	137	303	337
1:40 cement by mass	113	208	369
Classified tailings			
1:10 cement by mass	603	839	951
1:25 cement by mass	145	185	203
1:40 cement by mass	78	92	102
Classified tailings : dune sand 75:25			
1:10 cement by mass	749	658	1 107
1:25 cement by mass	195	203	203
1:40 cement by mass	103	108	109
Classified tailings : dune sand 50:50			
1:10 cement by mass	727	850	1 231
1:25 cement by mass	191	225	251
1:40 cement by mass	92	128	167

In the blasthole stopes, early strength is not important, but is was considered important to maintain adequate permeability. A backfill material consisting of 50 percent classified tailings, 50 percent dune sand and 1:13 cement all by weight was chosen so as to give a U.C.S. of 700kPa.

4.3 The backfill plant

The specially designed mixing plant is situated in the shaft area so as to be in a position to discharge directly to the near vertical boreholes leading from surface to the various levels underground. This is so that no significant pumping of the final backfill material is required. (Figure 7) The design of the plant was based on an initial requirement of 2 500 tons of solids per day. The additional capacity was incorporated to allow for peaks in filling which may be required for production scheduling. Backfill production of up to 3 000 tons of solids in an 8 hour shift has been achieved. The capital cost of the backfill plant was approximately R5 million. Figure 8 illustrates the general arrangement adopted for the processing of the backfill material.

Tailings from the concentrator are pumped over a distance of approximately 1 000 metres through a 250mm internal diameter class 10 high density polyethylene (HDPE) pipeline to the backfill plant. All tailings are pumped via this route w drawoff of material as required and with the balance being returned to the tailings dam. This is to prevent a stop-start operation with resultant settling and blockages in the pipeline. Some 5 000m³ of tailings are pumped through this route daily, these containing appro mately 2 800 tons of solids.

The tailings are received in a 300m³ capacity surge tank from which they are pumped to either one of two banks of six 250mm cyclones where they are classified or direct to the overflow silo for retur to the tailings dam if not required. Th underflow is passed to two conical-bottomed, 1 000m³ capacity underflow sil The overflow passes across to the conica bottomed, 1 000m³ capacity, overflow sil before being allowed to settle and the w is allowed to decant off. A fourth 1 00 capacity tank is used for the storage of settled underground water. Water is dra from this tank and injected at high pressure (1 000kPa) into the base of the cone of the underflow backfill tailings silo in order to agitate the solids duri discharge. The use of water as an agitating medium is effective and is preferred to mechanical agitators or compressed air. If not added here, water would in any case have had to be added later in the circuit. Compressed air is available in any case of emergency. The discharge tailings then gravitate through a control valve by way of a 100mm internal diameter class 10 HDPE pipeline to the first of the two mixing tanks.

Sand is brought to the backfill plant area by truck and stored in a 3 000 tons capacity surge bunker before being transferred by belt conveyor to a 300 ton capacity sand bin. It is then drawn from the bin at a similar controlled rate and transferred by belt conveyor to the first stage mixing tank mentioned above.

The quantity of underflow tailings entering the first mixing tank by way of the gravity pipeline is measured by means of a densitometer and the sand feed is automatically controlled to match this rate so as to maintain a 50:50 or other appropriate mix. Make-up water from the clear water tanks is introduced to the mixing tank via the gravity pipeline. The water feed is automatically controlled so as to achieve the required pulp density (approximately 73 percent solids by weight). The mixing tank capacity of 28 cubic metres allows a retention time of 10 minutes for the tailings and sand so as to ensure a homogenous mixture of the two at the correct density. The first stage mix is then pumped to another, identically sized tank, through a vortex mixer, being added by a variable speed feeder at the appropriate rate. The final mixed backfill is then pumped at a specific gravity of approximately 2,0 to the boreholes through which it enters the mine. Continuous plots of all appropriate quantities appear in the control room.

In addition to the standard backfill, backfill of classified tailings and cement only and of sand only have been produced for the cut and fill stopes.

4.4 Underground backfill reticulation

The backfill placement in blasthole stopes to date has been via 172mm diameter boreholes drilled directly from surface to the level where backfill is required. Holes were drilled to 3,4,5, and 6 level from surface at dips varying from 64° to 83°. The vertical heads of the holes varied from 40 metres to 148 metres and the hole lengths from 41 metres to 163

metres. Subsequently, holes were drilled from 6 to 7 and 8 levels and from 9 to 10 levels to extend the reticulation to the lower blasthole stope blocks and the cut and fill stopes. The latter holes are 292m in diameter being a function of the availability of the type of boring machine.

The backfill holes from surface pass through a zone of weathered schist over the first 20 to 30 metres. Initially it was felt that excessive wear may cause scaling and hence collapse and blockage of the holes. In two of the holes to 4 level, two and three 13 metre lengths of 140mm class 6,3mm HDPE piping respectively were butt-welded together and lowered down the holes to protect the upper weathered zone. A decision was made to use HDPE pipes for lining and not steel lining so that the holes can be redrilled if a blockage occurs. The lining in one of the holes wore through and had to be removed but no serious blockage took place. Experience has shown however that no lining was, in fact, necessary and over 800 000 tons of solids have passed through these holes to the end of August, 1982. A single hole has, to date, taken over 100 000 tons of solids.

The backfill pumped is from the second stage mixing tank to the holes via a 150mm flexible rubber hose. A 1,5m length of steel pipe is fixed to the end of the flexible hose. This is merely inserted into a stand pipe which has been grouted into all the borehole columns. No special receiving tank or "breathing" arrangements have been installed.

At the base of the hole a rubber-lined casing pipe is installed. This telescopic type arrangement has a rubber gasket which seals the casing to the sidewall. The casing is secured using eyebolts. For the initial boreholes, the casing pipe was extended to the footwall for support with a takeoff for backfill about half way down. Subsequently, 150mm flexible rubber hoses were connected directlly to the casing pipes and no serious problems occurred. The 150mm flexible hose connects to the main backfill column, 150mm internal diameter class 10 HDPE piping. This piping is suspended on the same "parrot swings" as the compressed air and water service pipes. Additional support is provided at sharp bends but movement of the pipeline during filling is slight and no problems have arisen.

To date the greatest length of backfill transport is 310 metres horizontally with a 12m increase in vertical head. Indications are that the head of fill which

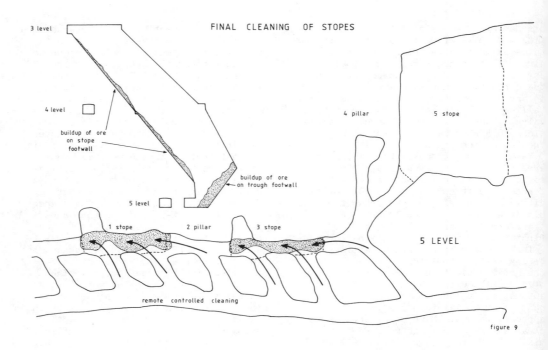

FINAL CLEANING OF STOPES

3 level

4 level

buildup of ore
on stope
footwall

buildup of ore
on trough footwall

5 level

4 pillar

5 stope

1 stope

2 pillar

3 stope

5 LEVEL

remote controlled cleaning

figure 9

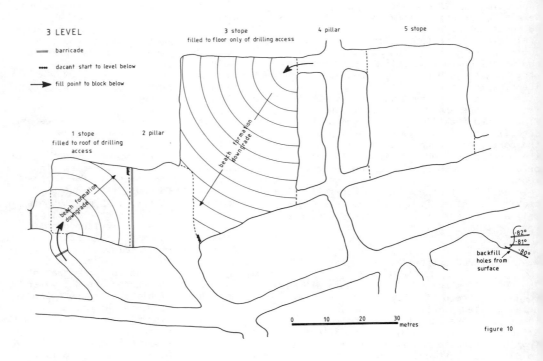

3 LEVEL

barricade

decant start to level below

fill point to block below

3 stope
filled to floor only of drilling access

4 pillar

5 stope

1 stope
filled to roof of drilling
access

2 pillar

beam formation
downgrade

beam formation
downgrade

-82°
-81°
-80°

backfill
holes from
surface

0   10   20   30
                    metres

figure 10

built up in the borehole was between 35 and 70 metres. This was in the case of a cut and fill stope on the lower levels.

No significant wear has been noticed on the pipelines to date, even at sharp curves. Experiments were carried out with butterfly valves in the line for particular reasons but these valves were worn away within a single shift. Experiments were also carried out with Y pieces in order to fill two stopes simultaneously (or to spread backfill in cut and fill stoping). The flow tended to favour one leg of the Y leg blocking the other in the process.

During initial filling, 150mm flexible hoses were connected at the end of the backfill pipeline for discharge into the stope. This resulted in severe movement of the flexible hose which was almost impossible to control. Almost no movement is however present when the pipeline discharges directly into the stope.

4.5 Final cleaning of blasthole stopes

Once blasting in the stope is completed and the ore has been loaded from the drawpoints to the extent which loading can be carried out safely, final cleaning of the stope commences. Cleaning of the stope must be carried out for two reasons: -
  a. To empty the stope completely of value-bearing ore, and
  b. To clear the footwall of the trough so that a homogeneous fill will be available for the mining of the crown pillar from below.

Initially the entire footwall of the stope is checked and cleaned as thoroughly as possible. The footwall of the stope is not completely smooth and some buildup of ore does occur. This is cleaned using water either through a normal hose or a jet of some sort. The major problem in this regard is access to the stope.

Once the footwall cleaning is complete, the base of the stope must be cleaned using a remote controlled loader. Ore is left between the drawpoints after normal loading and this causes a buildup of ore on the footwall of the trough drive. (Figure 9) The remote control unit is a standard Wagner ST 5B Four Wheel Drive Scooptram. Tramming capacity is 6,8 tons and bucket capacity is 3,03m³. The unit is fitted with remote control facilities in the form of a cable spool, a trailing cable and a control panel which is identical to the operating panel on the scooptram itself. The control consists of a steering lever with a micro-switch for brakes, a boom and bucket control lever, a start button, a park brake and

engine switch-off control, a throttle control and a forward-reverse control. The control panel connects to 28 wires in the cable which connect via 28 brushes to a slip ring inside the spool. Control of the scoop is then via a bank of solenoid switches. Problems encountered are damage to the cable and various electrical problems. The system has, however, proved entirely suitable for the required application at Black Mountain.

The operator keeps the loader within sight and normally within 30 metres of himself. Once the bucket is loaded and the loader is clear of the stope, the operator climbs onto the loader and drives to the tipping point. To date stopes 1,3,5,7 and 9 have been cleaned and the cleaning of stope 11 is in progress. Tonnage trammed by the remote loader varies between 2 000 and 10 000 tons per stope with minimal ore being left in the trough. The cost of the remote loader was covered by the additional revenue of the first two stopes cleaned.

4.6 Backfill barricades

Positioning of barricades

Backfill barricades are placed in all accesses to the stopes being backfilled. The positions of barricades for stope 1 and 3 are shown in figures 10,11 and 12.

Strength of barricades

The vertical height of the backfill placed behind the barricades will vary from a few metres on the uppermost levels up to 140 metres in the case of pillar 12. The first barricades designed were for the drawpoint levels of stopes 1 and 3. The vertical height of backfill to be considered was 70 metres in both cases. If a 70 metre height of fluid backfill material of specific gravity 2,0 is considered, the pressure at the barricade would be: -
70 x 2,0 x 10kPa = 1 400kPa (200psi)

In order to withstand this level of pressure, a major bulkhead construction would be required. The backfill material does not however remain fluid, and, provided that most of the excess water is removed by decantation and percolation, it is unlikely to return to a fluid state for any reason. Once the material is placed the solids settle, the bulk of the excess water rises to the surface and stabilisation of the backfill material by the action of the cement proceeds. The decant water is removed and percolation

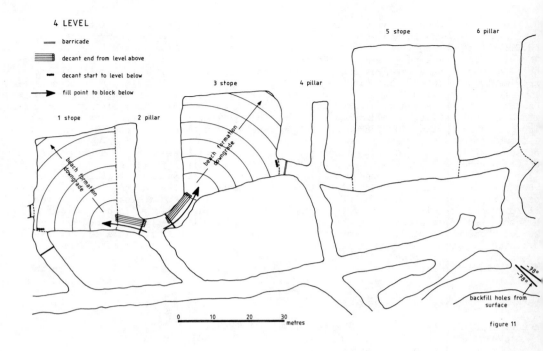

4 LEVEL

- ▬ barricade
- ▤ decant end from level above
- ▬▪ decant start to level below
- ➤ fill point to block below

1 stope 2 pillar 3 stope 4 pillar 5 stope 6 pillar

bench formation downgrade

bench formation downgrade

-78°
-78°

backfill holes from surface

0   10   20   30 metres

figure 11

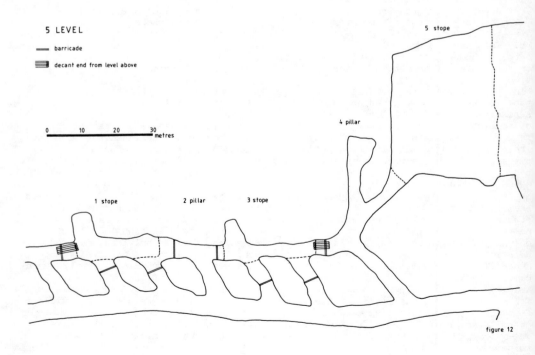

5 LEVEL

- ▬ barricade
- ▤ decant end from level above

0   10   20   30 metres

5 stope

4 pillar

1 stope   2 pillar   3 stope

figure 12

of further excess water proceeds slowly. Once the backfill material has settled it, in fact, tends to become self-supporting. The main function of the backfill barricade is thus to contain the backfill material as the backfill rises past the elevation of the barricade. This provided that stabilisation of the backfill is quick enough and that drainage of the backfill material is adequate.

The barricades for stopes 1 and 3 were designed to take a pressure of 70kPa (10psi). This is equivalent to: -

$70 \div 20 = 3,5$ metres

The significance of this pressure is that the barricade will retain a head of 3,5 metres of fluid backfill above the centre point of the barricade. Backfill is, in fact, only poured in 1,5 metre lifts to beyond the top of the barricade thus providing an adequate factor of safety. The design height of 3,5 metres above the centre of the barricade will provide for sufficient stabilised material to prevent a possible washaway penetrating to the barricade and exposing the barricade to excessive pressure from fluid backfill.

It is possible that, over time, water from the backfill may connect, from much higher elevations, along the footwall of the stope or along fissures and expose the barricade to excessive pressure; i.e. for a height of 70 metres the pressure would be 700kPa (100psi). Two factors, however, must be considered. The first is that failure of the barricade will occur but will not be associated with large volumes of runaway material. The second is that, provided the barricade is of a permeable design, the pressure will be dissipated.

Various barricade designs were considered before finalisation was reached. Amongst the factors considered were: -
a. Adequate strength;
b. Permeability;
c. Method of sealing, particularly on hanging wall;
d. Cost of materials;
e. Ease of construction;
f. Availability of required materials;
g. Re-use of required materials.

The initial barricade used is shows in figures 13 and 14. In addition to the main construction shown, a brick wall 1,5 metres high was constructed in front of the barricade to collect decant and percolation water. A V notch was installed in the wall and drainage measurements taken. (Figure 15) The main members of the barricade are two or three vertical universal beams fixed to the hanging wall and footwall by means of heavy duty, fabricated gussets and pinning steel. The fabric of the barricade consists

of horizontal minepoles covered first with wire weld reinforcement or 10gg diamond mesh and then poly-propelene hessian as a filter cloth. Decant pipes and percolation pipes pass through the barricade. The poly-propelene hessian is sealed up against the sidewall by means of a cement grout. The backfill is poured to 1,5m and allowed to stand for the next 16 hours before further filling takes place. Once the third lift is completed, a sealing wall is built between the backfill and the hanging wall and the backfill is carried up past the top of the barricade. An alternative method of sealing at the hanging wall, as well as of supporting the upper end of the main columns, is shown in figure 16. Efficient sealing of the filter cloth to the sidewalls and of the sealing wall against the hanging, and around the decant and percolation pipes, is critical, as major losses of backfill material can occur. The barricades are placed as far from the open stope as possible for reasons of safety of construction and of obtaining as long as possible a plug of stabilised backfill material in the access drive.

Various changes are being made to the barricade as more experience is gained. It is however essential that changes in barricade design are carefully monitored and it is standard procedure to request written permission from the Assistant Manager in a record book before backfill is placed behind any barricade.

4.7 Drainage and drainage reticulation

When running at full capacity of around 300 tons per hour of solids and at a pulp density of between 70 percent and 73 percent, over a 100 tons per hour of water is, in addition, being placed into the stope. Of this water, approximately 15 percent will be used in the hydration of the cement and the balance must be removed from the stope by drainage or remain in the backfill.

Once the backfill is placed, the coarser solids settle rapidly, a suspension of fines occurs above the coarse layer which becomes progressively less dense towards to surface. Once filling is stopped, the fines settle and the water rapidly becomes clear. This forms the characteristic fines layers, not normally more than about 150mm thick, on top of each layer of backfill. Where filling is over a large expanse, the backfill material runs away from the point of pouring at an angle to form a beach of between 5° and 10°.

INITIAL BARRICADE DESIGN

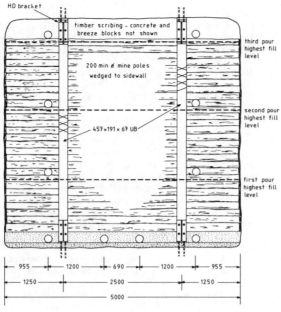

HD bracket

timber scribing – concrete and breeze blocks not shown

200 min ⌀ mine poles wedged to sidewall

457×191× 67 UB

third pour highest fill level

second pour highest fill level

first pour highest fill level

955   1200   690   1200   955

1250   2500   1250

5000

FRONT ELEV

figure 13

INITIAL BARRICADE DESIGN

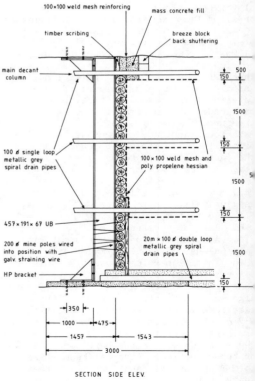

100×100 weld mesh reinforcing

mass concrete fill

timber scribing

breeze block back shuttering

main decant column

100 ⌀ single loop metallic grey spiral drain pipes

100×100 weld mesh and poly propelene hessian

457×191× 67 UB

200 ⌀ mine poles wired into position with galv. straining wire

20m × 100 ⌀ double loop metallic grey spiral drain pipes

HP bracket

500
150

1500

150

1500

150

1500

150

350

1000   475

1457   1543

3000

SECTION SIDE ELEV.

figure

214

ALTERNATE BARRICADE DESIGN

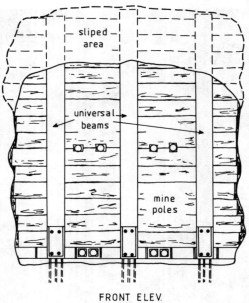

FRONT ELEV.

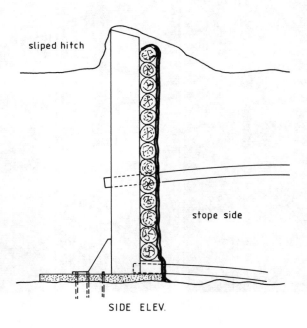

SIDE ELEV.

figure 16

Initially, the entire surface is covered with water but, as the water is drained, the beach become dry at the filling point. As the backfill passed the main levels, experience was that, immediately the surface of the backfill was dry, it hardened rapidly to support at least the weight of a person.

Decant drainage is achieved by decant pipes lowered from the level where filling is taking place to one of the drawpoints below. Due to the high percentage of water which is removed by decantation, the proper installation and protection of these pipes are critical. Once the backfill has risen above the level of the barricades, damaged decant pipes cannot be replaced. Each decant pipe unit consists of two 100mm diameter corrugated plastic pipes wrapped in filter cloth. The plastic pipes themselves are perforated by slots 10mm x 20mm and at a density of approximately 40 per metre. Each unit is approximately 50 metres long. The units are wrapped around a steel member on the backfilling level in order to secure them adequately so that there is no chance of a pipe slipping down into the stope. Initially a duty of 300m² per unit was considered but it has become standard practice to install four units in all stopes as an insurance against a damaged unit. (Figures 10,11 and 12) The decant pipes are placed as far as possible from the filling point so as to prevent blinding of the filter cloth by splashing and in order to get the decant pipes to the lowest possible portion of the beach. Some blinding of the pipes does occur due to the suspension of fines, but experience has been that decant water is adequately drained on an 8 hour pouring, 16 hour standing basis. This means that no progressive buildup of water occurs as filling proceeds. A certain amount of water will always be present in the backfill due to inability to place the decant pipes at the lowest point of the beach and to blinding of the pipes. The guide used to ensure that no progressive buildup of water is occurring is merely to check before filling commences that at least a portion of the beach has risen above the water. If this is not done, a dangerous accumulation of water can occur. In practice, up to 40 percent of the total water is removed by decanting, although much less was experienced where fissures drained water to adjacent open stopes or haulages (Figure 15)

Initially, percolation pipe units similar to the decant units were placed through the barricades. These looped out over the footwall or over the last layer of backfill

placed. No French Drain facilities were provided and the backfill material quickly blocked the filter cloth. Percolation drainage through the barricades themselves, through fissures and through percolation pipes, has been very slow. Diamond drill holes drilled into the backfill from the barricade also yielded very little water. Flow from the barricades rapidly slowed down to the extent that the barricades merely remained moist. Once the barricades were removed no drainage by percolation was apparent. Experience has shown that, while decant drainage is vital and requires carefull attention, percolation drainage is much less important and no special arrangements are now made other than ensuring that the barricades are permeable.

Drainage through the mine is important. The large quantities of water involved must be kept away from tramming operations in particular and other mining operations in general. A system of 165mm diameter boreholes carries all backfill water to the underground settler and pump station which returns the water to the clear water tank in the backfill plant.

4.8 The backfilling operation

Communications are on three levels. The Assistant Manager will communicate changes in standard backfill mixes or pulp densities. The Underground Manager communicates the scheduling of backfilling and the mixes for the various stopes. The underground operating personnel, under control of the backfill Shiftboss, communicate directly with the backfill plant control room personnel during pouring of this backfill. This latter communication is vital and is carried out via a "Genephone" telephone system with a normal underground telephone system as a backup.

Once the underground personnel have inspected barricades and reticulation, the control room operator is contacted. The control room operator confirms and sends down a water flush. Once a steady flow of clear water flows from the correct backfill pipe, backfill material is called for. The period of flushing will vary between one and three minutes. As soon as filling is completed, a flush is again called for and the plant is only stopped once clear water indicates that the reticulation is clear. Flushing always precedes and follows filling, including cases of emergency shutdowns.

Filling past the initial barricades in the bottom of the trough required careful

ROOF CONDITIONS OF STOPES 1 AND 3
AND PILLAR 2

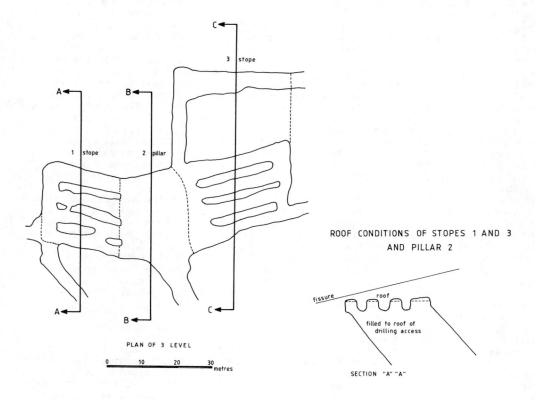

PLAN OF 3 LEVEL

0   10   20   30 metres

figure 17

ROOF CONDITIONS OF STOPES 1 AND 3
AND PILLAR 2

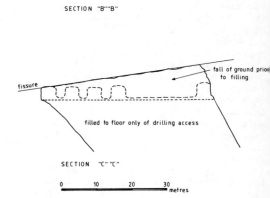

fissure   roof

filled to roof of
drilling access

SECTION "A" "A"

fissure   fall of ground during blasting

slot

SECTION "B""B"

fissure   fall of ground prior
to filling

filled to floor only of drilling access

SECTION "C" "C"

0   10   20   30 metres

figure 18

attention. In this area, the rate of rise of the backfill, and hence the rate of increase in head, is greatest. Leakages cause expensive losses of backfill material and the spillage causes at least inconvenience, if not severe damage. The backfill lifts are confined to 1,5 metres. A major advantage of the type of barricade installed is that construction can proceed with backfilling, allowing continuous access to backfill material. This provided important information on the backfill material during early backfills. Once the top of the barricade has been sealed, care must be taken not to build up an excessive head of fluid backfill material above the barricade by placing too large a lift.

4.9 Instrumentation

Flatjacks were installed in both vertical and horizontal orientations in stopes 1 and 3. These were to measure the buildup of pressure due to the backfill material. The maximum increase in pressure in any one flatjack was 65kPa (9,3psi). The other flatjacks however showed much smaller increases and the pressures dissipated with time rather than increased with the height of the backfill. Piezometeres were installed to measure pore water pressure. This instrumentation was not successful, with few instruments operating satisfactorily. The maximum pressure obtained was 25kPa (3,5psi) but this again dissipated. Results indicated that after initial loading due to backfilling directly past the barricades, very little additional laoding developes. The installation of instrumentation at barricades has now been dispensed with.

4.10 Control samples

The most important control tests carried out are unconfined compressive strengths tests. During each pour four samples are taken on surface from a valve just downstream from the second stage mixing tank. Sample bags are made up of a plastic cylinder cut from a roll supported by a wooden ring at the top and a wooden disc at the bottom. Drainage holes are drilled in the disc and a filter paper prevents losses of fines. Once filled, the sample bags are hung from a frame in the sample preparation room and allowed to cure for 7 days, 14 days, 28 days and 3 months respectively. Initially the samples were allowed to dry but subsequently arrangements have been made to keep the

samples saturated by means of a drip feed. This more accurately represents underground conditions. On completion of the curing period, the sample is tested in a press with a 55kN proving ring. The U.C.S. value for each sample is obtained and a sketch made of the mode of failure. The specific gravity of the sample and moisture content are then measured. From time to time, percolation tests and grain size analyses are carried out for experimental purposes. Research and development work has been carried out on floculants, slagment and fly ash. Samples are sent to the concentrator laboratory from time to time for cement content determinations.

4.11 Control test values obtained

Naturally four small samples per day taken from a pour of 2 500 to 3 000 tons of solids causes some scatter in the results obtained. This is due both to variations in cement content and in the tailings – sand ratio. Representative average U.C.S. results in kPa are as follows: –

	7 Days	14 Days	28 Days	3 Months
6:1	1 320	1 380	1 700	2 050
13:1	510	560	720	920
15:1	370	500	560	730
20:1	360	440	500	640
30:1	210	250	260	290

Moisture contents vary between 14 percent and 20 percent by weight. These moisture contents represent approximately 50 percent and 71 percent of the original water content in the backfill material. Limited tests on samples for underground backfill show similar moisture contents.

5.0 PILLAR EXTRACTION

The first pillar extracted on Black Mountain was 2 pillar situated between 1 and 2 stopes and extending from 3 to 5 level. Currently, extraction of 6 pillar is in progress. This section describes the extraction of 2 pillar

5.1 Support of the drilling access in the stopes

When the stope drilling access is originally mined, pillars are left to support the roof of the access while drilling is in progress. Figure 17 depicts the support in stopes 1 and 3. The narrow

3 LEVEL

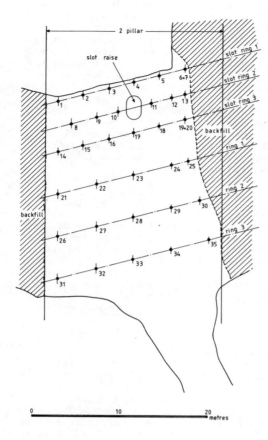

figure 19

4 LEVEL

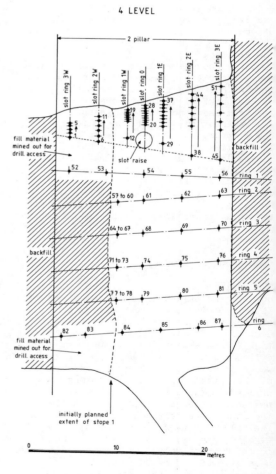

figure 20

220

pillars are typical of drilling access support.

The large pillar in stope 3 was due to some non-standard drilling required to remove an additional block of ore made payable by high metal prices. Naturally, support pillars must be compatible with the drilling layout.

Once the ore is extracted, both the hanging wall itself and the roof of the uppermost drilling access are left unsupported. The rock of the orebody and hanging wall is competent, but is traversed by shallow southerly dipping fissures. These combined with fault planes give an extremely blocky nature to the rock. The major fissures cross the hanging wall plane at near vertical angles and, apart from minor falls, have not created a problem. Falls of ground from the hanging wall result in large, low-grade blocks in the stope which required considerable secondary blasting, but have not to date represented a stability problem. The major fissures, however, cross the roof of the drilling access at a low angle and cause a severe support problem. (Figure 17 and 18) Support must be installed prior to extraction of the uppermost block of the stope. In stopes 1 and 3 the severity of the problem was not foreseen and no support was installed in the roof. No fall of ground occurred in stope 1, but a major fall of ground occurred in stope 3.

The drilling accesses for pillars were initially totally mined out for ease of drilling the somewhat more complex slot pattern, and with regard to the smaller width of the pillars. The access of pillar 2 was almost completely mined out prior to the extraction of the uppermost blocks of stopes 1 and 3. The need for support in pillar 2 was recognised and a pattern of 15 to 20m long cables 19mm in diameter was installed at 5 metre centres. Subsequently, a pattern of 10 metre long cables 25mm in diameter was installed at 2,5 metre centres. While these cables had a strength of 40 tons and would only support some 2 to 3 metres in tension, the additional length was considered necessary to maintain support if initial falls of ground took place, and to assist in maintaining the intergrity of the roof beam. The roof above the pillar remained stable during drilling, but, during blasting, with the slot complete and with three rings to be drilled, the fall of ground in 3 stope extended into pillar 2, causing very large slabs of rock to cover the remaining holes. Blasting and cleaning of these rocks had to take place prior to blasting the three rings.

The roof in stope 1 remained in place during blasting which enabled the backfill to be brought up to support the roof. This was achieved by building a backfill barricade across the western edge of pillar 2.

The support of drilling access areas is now recognised as an important consideration and all pillar accesses are now being laid out with support pillars as for stopes. Adjustments are being made to drilling layouts where necessary. The 4 level drilling access of pillar 2 and both drilling accesses of pillar 6 were completely mined out, with 2,4m x 12m grouted rebar, installed at approximately 2 metre centres as development progressed. In all cases, the condition of the roof deteriorated and extensive 25mm cable bolting had to be installed. In the lower, and in some of the upper pillar drilling accesses, the hanging wall is discontinuous along strike as the adjacent stopes have been mined out past the elevation of the access. In addition, the blasting of the stopes creates additional fracturing in the pillars, and the extraction of the adjacent stopes will tend to cause some opening of existing fractures and fissures.

5.2 Blasthole drilling layouts

Figures 19,20 and 21, plans of 3 ,4 and 5 levels respectively depict the drilling layouts for pillar 2. A typical section is shown in figure 22 and figures 23 and 24 show the layouts for typical slot rings.

For the blasthole stopes, the slot raise is drilled on the intersection of the footwall and the pillar line and the slot is blasted from footwall to hanging wall. The rings are then blasted towards the slot in a strike direction. In the case of pillars, a transverse slot did not appear to be feasible. Should the slot be placed against the backfill, the concentration of blasting and the limited breaking face would be likely to damage the backfill. In addition, rings would be blasted in a strike direction against the backfill wall. Should the slot be placed centrally in the stope, back damage to the remaining remnants may cause difficulty in access to blastholes and a resultant loss of ore.

In the case of pillar 2, the slot raise was positioned centrally towards the hanging wall of the stope. The slot is then blasted along strike to expose the full width of the hanging wall. The main blasthole rings are blasted northwards

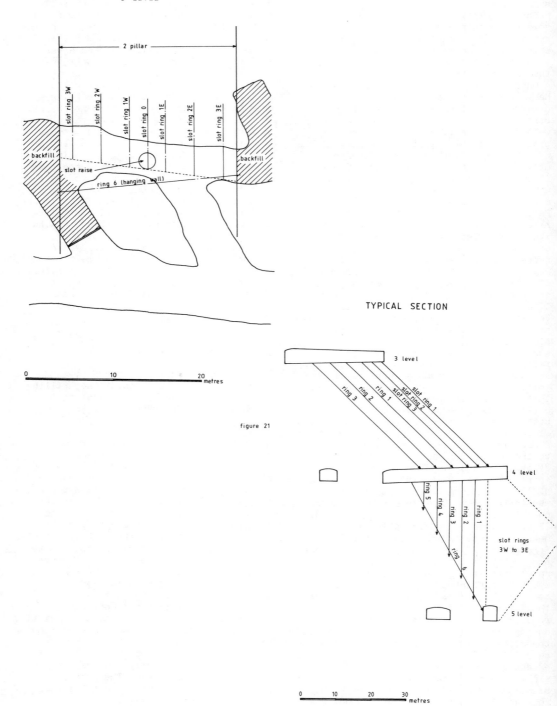

5 LEVEL

2 pillar

slot ring 3W
slot ring 2 W
slot ring 1W
slot ring 0
slot ring 1E
slot ring 2E
slot ring 3E

backfill

slot raise

ring 6 (hanging wall)

backfill

0   10   20 metres

figure 21

TYPICAL SECTION

3 level

ring 3
ring 2
ring 1
slot ring 3
slot ring 2
slot ring 1

4 level

ring 5
ring 4
ring 3
ring 2
ring 1

ring 6

slot rings
3W to 3E

5 level

0   10   20   30 metres

figure 22

222

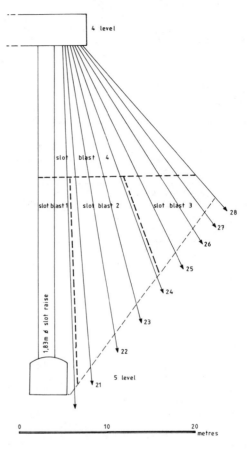

figure 24

TYPICAL SECTION THROUGH SLOT 3 TO 4 LEVEL
(SLOT RING 2)

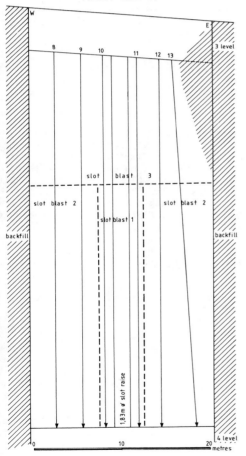

figure 23

223

towards the hanging wall. Holes are laid out as for blasthole stopes at between 90° and about 45°. The burden and spacing in the slot is nominally 3m x 3m and the burden and spacing in the main rings nominally 4m x 4m. For pillar 2, the trough itself was also blasted by means of 165mm blastholes. The outer blastholes are placed 1,5m from the backfill, a parameter which was later changed.

The drilling and blasting parameters for 2 pillar are as follows: -

	tons	Drilled	Charged	Explosives	kg/ ton	tons/
Slot 3 - 4 Level	17 700	871	811	14 425	0,81	20,3
4 - 5 Level	38 500	1 675	1 241	22 150	0,58	23,0
Total Slot	56 200	2 546	2 052	36 575	0,65	22,1
Rings 3 - 4 Level	36 700	661	616	11 000	0,30	55,5
4 - 5 Level	33 900	885	753	13 475	0,40	38,3
Total Rings	70 600	1 546	1 369	24 475	0,35	45,7
Total Stope	126 800	4 092	3 421	61 050	0,48	31,0

On 4 level the drilling access for stope 1 originally extended eastwards to the position shown, with a much narrower pillar 2 envisaged. This area was backfilled together with stope 1, and some mining of backfill was required during the development of pillar 2, 4 level access. Due to bad roof conditions, not all the backfill could be mined out and holes had to be angled westwards to extract the portion of 2 pillar under this filled area. This backfill subsequently mixed with the ore when the 4 to 5 block was extracted.

All blastholes are positioned by means of survey lines and all blastholes are surveyed and plotted when completed. Holes with unacceptable deviation are redrilled.

To date, no difficulties have been encountered in drilling the pillar blocks, which could have been expected due to more fracturing than the stope blocks. Great care is, however, taken to identify and re-drill any holes which penetrate in-to backfill material. This is done by means of watching penetration speed and drill cuttings. Blastholes may penetrate into the backfill material, due either to deviation of the blasthole or to excessive overbreak during the mining of the adjacent stope. Blasting of holes which have penetrated the backfill may cause severe damage to the backfill wall.

5.3 Blasting procedures

With reference to figures 19 and 20, the sequence of blasting of pillar 2 was as follows: -

Block 4 to 5

Blast No.	Holes	Number of millisecond delays	Kg of Anfex
1 (slot)	20,12,29,30,13	12	1 875
2 (slot)	15,16,22,23,24,31,32,33,A1,A2,A3,38,39, 40	21	4 400
3 (slot)	25,26,27,28,12,14A,17,18,19,29,34,35, 36,37	18	2 350
4 (slot)	12,13,14,15,16,17,18,19,20,21,22,23, 24,25,26,27,28,29,30,31,32,33,34,35, 36,37	21	3 513
5 (slot)	38,39,40,41,42,43,44,A1,45,46,47,48, 49,50,51,29,34,37	21	4 100
6 (slot)	6,7,8,9,10,11,1,2,3,4,5	17	2 650
7 (slot & rings)	6,9,11,1,3,5,52,54,55,56,57,58,59,60, 61,62,63,82,84B,85,86,87	21	6 100
8 (rings)	64,65,66,67,68,69,70,71,72,73,74,75,76, 82,83,84,85,86,87	18	3 538
9 (rings)	78,79,80,81,82,83,84,85,86,87	10	1 500

Block 3 to 4

Blast No.	Holes	Number of millisecond delays	Kg of Anfex
1 (slot)	10,11,17	6	1 125
2 (slot)	2,3,4,5,6,8,9,12,15,16,18,19,9A	17	2 275
3 (slot)	1,2,3,4,5,6,8,9,10,11,12,13,14,15,16, 17,18,19	11	3 175
4 (rings)	26,27,28,29,30,31,32,33,34,35	17	5 100

Holes marked A and B are redrilled holes.

As for blasthole stoping, holes are decked, but individual decks for pillar blasting are limited to 7 or 8 metres. As a rule, not more than two individual decks would be initiated simultaneously and, where three decks are initiated simultaneously, the total explosives will be limited to 300kg of Anfex.

As shown in figures 23 and 24, blasting of the slot takes place from the bottom up as the slot advances from the slot raise outwards across the full extent of the hanging wall. In the 4 to 5 level block, the slot in fact advances first in a transverse direction from the slot raise to the hanging wall, and then in a longitudinal direction across the width of the hanging wall. The initial opening up of the slot is critical as bad breaking will cause major delays. The appropriate holes are all measured before each blast takes place and each blasting layout is issued as a detailed schedule from the Planning Department.

One major problem occurred in the blasting of the 3 to 4 level block of pillar 2. After the extraction of the slot and during the time of preparing to blast the last three rings, a major fall of ground occurred from the roof of the 3 level

225

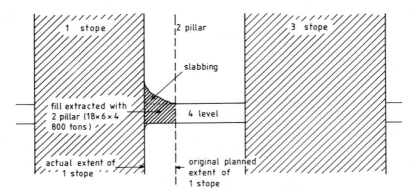

E - W VERTICAL SECTION STOPES 1 AND 3 AND PILLAR 2

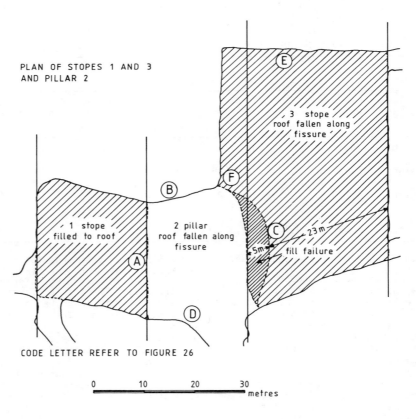

PLAN OF STOPES 1 AND 3
AND PILLAR 2

CODE LETTER REFER TO FIGURE 26

figure 25

drilling access. This fall was associated with the southerly dipping fracture running just above 3 level and was in fact merely an extension of the fall of ground in 3 stope. (Figure 18) The three rows of holes were covered by large slabs of rock and blasting and then cleaning of the rock had to take place. It was not safe to clear the rock right to the third row of holes due to the proximity to the edge of the slot and the possibility of under-cutting or bad fracturing at the edge. The result was that only the back two rows of holes could be blasted with the fore-most of these two rows having double the normal burden. The blast broke out success-fully, but it appears that a greater than normal proportion of large rocks occurred. In addition, a failure of the backfill wall in 3 stope some 5m deep, and extending from 3 level to 4 level, occurred. It is likely that the excess burden on the rings contributed to this failure.

5.4 The exposure of the backfill walls

Experience to date has been that closely spaced vertical fractures or discontinui-ties tend to form in the backfill. These run NS or parallel to the backfill walls and result in slabbing or scaling of the backfill walls on a vertical plane surface or a slightly curved surface. These fractures or discontinuities may be the result of shrinkage or vertical drainage channels and are probably aggravated by penetration of the blast waves into the backfill.

Were no scaling to take place, the back-fill wall would take on the rough appear-ance of the original pillar sidewall. This is not the case and the smooth sidewalls give an indication of at least some sca-ling. This scaling is not easily measured, but estimates to date are that the loss of backfill is not significant being about one metre thickness.

Two major backfill failures have, however, occurred. (Figure 25) The first was due to the change in size of pillar 2 from 14 metres wide to 20 metres wide at a time when the drilling access had already been mined. The remaining portion of the drilling access was backfilled together with stope 1 and the result was that, during extraction of pillar 2, some 800 tons of backfill mixed with the ore.

The second failure was in the stope 3 backfill wall between 3 and 4 levels. This is depicted in figures 25 and 26. The failure extended back for approximately 5 metres at the deepest point on 3 level

and was estimated to be about half this extent on 4 level. A rough estimate of the backfill to ore is about 3 000 tons. The blasting of the last two rings in the 3 to 4 level block likely contributed to this failure, but the failure could also be associated with the possible pene-tration of one of the slot holes into the backfill.

Neither of these failures represent a stability problem. At the date of writing, the broken ore in the stope was not yet depleted, but backfill walls had been exposed to over 40 metre heights.

5.5 Dilution problems

The major problem in diluting the ore with backfill is the effect on metallurgy. The pH of the ore entering the concentra-tor has risen from a natural pH of just below 7 up to levels of 8 and 9 and occasionally 10. This however is the com-bined effect of dilution in blasthole pillars and from cut and fill cappings After initial severe disruption of the metallurgical process, changes have been made to handle moderately higher pH ore values.

The dilution of ore with backfill has a secondary effect of increasing the moisture content of the ore. This causes problems in the underground orepass sys-tem, the ore transfer system and in the crushing plant.

Changes made both to the drilling lay-outs in the blasthole stopes and to the cappings in the cut and fill stopes are currently reducing dilution to manageable levels.

5.6 Proposed adjustments to blasting

As the blasthole pillar blocks are gene-rally more fractured than the stope blocks, it is likely that the burden and spacing of the blastholes can be increased. Initially, however, the only parameter being changed is the spacing of the blast-holes from the backfill sidewalls. This distance has been increased from 1,5 metres to 2,5 metres with success.

6.0 PROPOSED FILLING OF PILLAR STOPES

Once the particular pillar stope has been emptied of ore, backfilling will start as soon as possible. The trough of the stope will be filled with the same 1:13 mix as for the blasthole stope in order

fall of ground along fissure

FAILURE OF 3 STOPE BACKFILL
WALL BETWEEN 3 AND 4 LEVELS

figure 26

to allow for mining of the cut and fill crown pillar below. The backfill to the next level can be placed with no cement and as little sand as possible because the plug of 1:13 backfill has been provided. The criteria of the backfill from this point on are the following: -

a. Sufficient bulk to give overall stability.

b. Sufficient strength and adequate drainage to form a plug at the next level of barricades and so that no possibility of liquifying results. Tests will be done at the first level of barricades. The likely final backfill will have little or no cement content and a low dune sand content. Decant drainage facilities as for blasthole stope filling will be provided.

Once the trough area has been filled, the opportunity exists of tipping develop-ment waste rock and any other scrap mate-rial into the pillar excavation during backfilling.

ACKNOWLEDGEMENT

I wish to thank the Consulting Engineer and Management of Black Mountain Mineral Development Company (Pty) Limited, for permission to publish this paper. I wish also to pay tribute to all those persons involved in the design and construction of the backfill plant, those involved in the designing and layout of the blasthole stopes and those involved in day-to-day operation of the system.

This paper was originally presented to the Association of Mine Managers of South Africa. I wish to thank the Association for permission to republish this paper.

Mining with backfill at Indian Copper Complex

P.A.K.SHETTIGAR
Hindustan Copper Ltd., Bihar, India

ABSTRACT: An attempt has been made to highlight various efforts made in a developing country to exploit a chalcopyrite orebody from greater depths by phased mine mechanisation against a background of traditional manual operations at five underground mines on eastern part of India, operated by Indian Copper Complex, Hindustan Copper Limited. The reasons for the adoption of mining with backfill are outlined at the outset. It deals with the description of current stoping practices, problems encountered and some of the solutions already accomplished. Introduction of small capacity electric LHDs, electro-hydraulic rock breakers, rubber tyred self-propelled drill wagons, cable bolting etc have contributed for significant improvement in productivity and reduction of mining costs. There has been substantial reduction in accident rate owing to improved ground conditions and better working environment. The paper concludes with a brief description of future plans.

1 INTRODUCTION

Indian Copper Complex has been operating Mosaboni mine since 1924 and it is situated at a distance of 214 KM south-west of Calcutta. It is the deepest mine (1230M below surface) in Singhbhum Copper Belt. Presently there are five underground copper mines operating on a strike length of 20 KM (refer to lease plan at Figure 1) with the following production capacities:

Mosaboni mine	–	3200	tpd
Pathargora mine	–	400	"
Surda mine	–	1300	"
Kendadih mine	–	200	"
Rakha mine	–	1000	"
Total		6100	tpd

2 BRIEF GEOLOGY

Copper deposits occur in a well known Singhbhum Copper Belt extending over more than 160 KM in an arcuate fashion from Durapuram in the west to Bahargora in the east of Singhbhum district of Bihar State. Regionally metamorphosed pelitic sediments and volcanogenic rocks of Precambrian age constitute the main rock types. The area suffered intense tectonic disturbances between 2200 to 600 million years in four orogenies giving rise to structures which control the mineralisation; but the area has been a stable mass since then. The tectonic disturbances gave rise to a series of geo-anticlines and geo-synclines with isoclined overturned folds. A major deep-seated zone of shearing and thrusting developed and contains all the copper deposits of the belt. The rock within the shear zone is highly brecciated, mylonitised and hydrothermally altered to various degrees.

Modern mining started at the copper belt in the early part of 20th century and is confined to the south-eastern 20 KM stretch of the belt with five operating mines. Copper mineralisation is confined to the shear zone rock like quartz-chlorite-biotite-schist, sheared granite and silicified schists and quartzites. The mineralisation is controlled by various opening resulting from shearing and to some extent pore space filling. These shears are mostly parallel to subparallel to regional foliation and occur as closely spaced parallel fissures. The mineralisation has a flat dip of 25° to 35° towards NEE and a general strike of N 15° – 25° W. Ore minerals consist mostly of chalcopyrite, pyrite, pyrrhotite and subordinate amounts of arsenides and tellurides in addition to apatite and magnetite. Small quantities

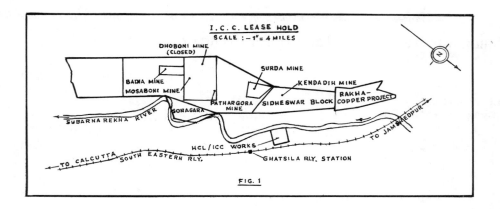

FIG. 1

of gold, silver, tellurium, selenium, cobalt, nickel etc. are associated in the chalocopyrite ore and some of them are recovered as by-products.

3 NEED FOR MINE MECHANISATION

A number of operational parameters were evolved during the several decades of operations with more emphasis for the improvement of mining operations. Keeping in tune with the time, the efforts were directed towards mining the cream out of the mineralised zones even to narrow widths till late 60s. The manual operations such as hand-shovelling in moderately dipping stopes, filling of mine cars in drives, hand-tramming over long distances, break-ing of large boulders on grizzlies etc. imposed serious restrictions in increasing the productivity of workpersons. As a result of the above deficiencies, there was spiralling effect on the operating costs. Consequently, the reliance on muscle power had to give way for the commitment for mine mechanisation. In this context, it was considered appropriate to change the earlier philosophy for computation of ore reserves based on mining the rich zones only. A change in concept of mining to assay contacts with lower cut-off and pay limit values permitted the formation of wider orebodies by including lean shear zones in the vicinity of richer ones which made it possible to consider the possibility of deployment of various types of mining equipment. Therefore, the strategy was to cut down operating costs through judicious adoption of bulk mining methods to the extent possible as this high productive and superior technology have been already established in the developed countries with remarkable success. Mine mechanisation

was, therefore, the only solution to ensure economic exploitation of the mineral under adverse environments of increased wages and salaries and higher costs of energy, steel, equipment, spares, consummables, timber etc. However, two serious limita-tions were faced — one in the deployment of larger equipment in old underground mines with narrow openings from surface and also in underground, and the other in the orebody itself which has an average dip of $30°$ over a strike length of 20 KM. It is recognised all over the mining world that introduction of an efficient type of mine mechanisation is rather difficult under such a situation. This is a fact of life for this mining belt where most of the country's known copper ore reserves occur. With a view to increasing indigenous production of copper, a strategy was evolved by the Government to consider small scale expansion of the operating mines and also the major expansion of the adjoining deposits in Singhbhum Copper Belt. This will permit the conservation of valuable foreign exchange which is rather vital for the growth and development of the country. However, substantial amount of capital investment will be necessary during the next decade to achieve the above objective.

4 CURRENT STOPING PRACTICES

Since over a decade substantial changes have been brought about in underground mining practices of Mosaboni group of mines with an emphasis on the establishment of safer, faster, highly productive and econo-mic mining methods. Incidentally, it invo-lved the use of steel for roof support in place of conventional timbers. A brief description of various stoping methods

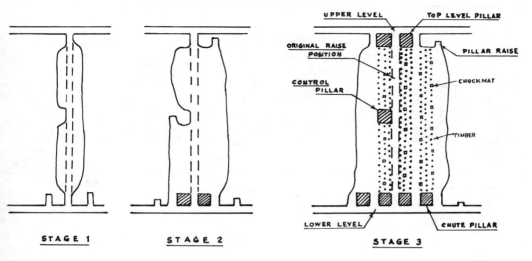

LINE DIAGRAMS SHOWING BREAST STOPING (NOT TO SCALE)

STAGE 1   STAGE 2   STAGE 3

FIG. 2

currently followed is given in the following paragraphs.

4.1 Breast stoping

For over four decades, the entire mine
production was obtained from Breast stopes
by mining the best zones to the minimum
possible width and thus achieving a mill-
head grade exceeding 2.2% Cu. By this
method of stoping, the lodes exceeding 4M
in width could not be mined and handling
of long timbers all the way from surface
to scattered underground workings was found
to be extremely laborious. In addition,
immediate support of the roof was becoming
increasingly difficult. However, as a
result of new mining practices introduced
since over a decade, only about 3% of the
mine production is presently contributed
by Breast stoping. It is now practised in
areas with an ore width less than 1.5M.

To start with, a raise is put up along
hangingwall of the orebody dipping at
about 30° between two successive levels
with a vertical interval of 37.5M. 2M long
and 32mm dia holes are drilled on each
face at an angle of 45° − 50° using Holman
Silver 3/Atlas Copco RH 659 rock drills
mounted on airlegs and about 60 − 80 Nos.
of drill holes are blasted electrically
at a time. Both faces of the raise are
advanced in the strike direction by leaving

systematic crown and chute pillars for the
protection of the levels. Random stope
pillars are left in geologically disturbed
areas. The excavated area is supported
systematically by erecting 20cm dia timbers
at intervals of 1.5M along dip and 2.4M
along strike (as shown in Figure 2). In
addition, 60cm chockmats are erected at
4.5M interval as an additional support of
the roof. An electrically operated 22.5 KW
double drum slusher installed on the lower
horizon, is used to haul the broken ore
down a stope into timber chutes from where
it is filled into 1.4 tonne side-tipping
mine cars. After the extraction of ore is
completed in a block, strong barricades
are fixed in the chute raises and backfilling is done. About 800 − 1000 tonnes per
month of ore is broken from one such stope.
Recent trials with two lengths of 0.75M
long tor-steel grouted bolts, suitably
jointed with a coupling sleeve, have demonstrated the possibility of eliminating
the use of timbers for roof support in
such stopes and mining of narrow widths
by "Room and Pillar" method.

4.2 Room and Pillar stoping

This method of stoping is generally used
in ore zones not exceeding 6M in width and
not more than 80M in strike. Initially a
raise is mined along the hangingwall

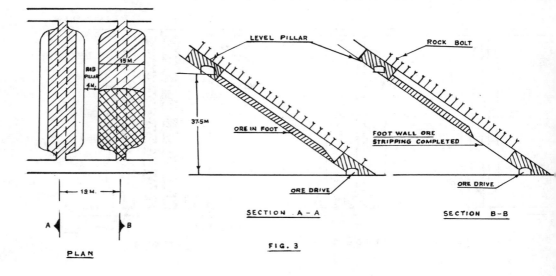

FIG. 3

contact of the orebody from lower horizon and both the faces of raise are advanced to a span of 15M leaving 4M wide rib pillars between such rooms as shown in Figure 3. Drilling of holes, their blasting and the scraping of ore etc. are done as explained earlier. Timely support of the roof soon after blasting and dressing of loose, is done by systematic rock bolting of hangingwall strata with 1.5M long and 20mm dia grouted tor-steel bolts at a spacing of 1.2M x 1.2M. After the completion of hangingwall cut for a height of 2M, the ore left in the floor is stripped out by drilling 2M - 2.4M long holes. In case of wider orebody, the floor stripping and scraping is repeated. Systematic chute pillars and level pillars are left to protect the levels. After the completion of mining, strong barricade walls are constructed at the bottom and backfilling of the room is done. Adjoining blocks are systematically extracted by adhering to a pre-determined stoping sequence. Presently about 1000 - 2200 tonnes per month of ore is extracted from a Room and Pillar stope. On some occasions upto 3000 tonnes of ore per month has also been mined from such stopes. Presently about 27% of the mine production is obtained by Room and Pillar stoping.

4.3 Horizontal Cut and Fill stoping

Horizontal Cut and Fill method of stoping has been successfully used to mine ore zones exceeding 4M in width and a minimum strike length of 80M. Initially a sill drive in ore is developed 8M (inclined distance) above the ore drive and full width of the orebody is exposed for a maximum height of 4.8M as shown in Figure 4. Generally 2.4M high horizontal cuts are taken by drilling 2.4M long and 38mm dia horizontal holes using jack hammers mounted on airlegs. The drill holes are blasted with ANFO to achieve reasonably satisfactory fragmentation. On the footwall side of the ore drive, a haulage drive is driven from where orepasses are excavated in waste rock to hole into the stope at intervals of 40M especially where Cavo 310 have been deployed. Generally a stope is worked in two panels so that a loading equipment can be effectively deployed in both the panels alternatively. The hangingwall is supported by rock bolts systematically at 1.5M x 1.5M pattern. As the drilling and blasting progress, a 310 Cavo or 0.76M^3 electric LHD is deployed to load and haul all the ore into the orepass. 1.5M dia orepass rings made out of 10mm thick steel plates are welded inside the stope to serve as manway and orepass through the backfill. Subsequently the panel is backfilled leaving about 2.4M of head room. As the preparation for

LAYOUT OF CUT & FILL STOPE

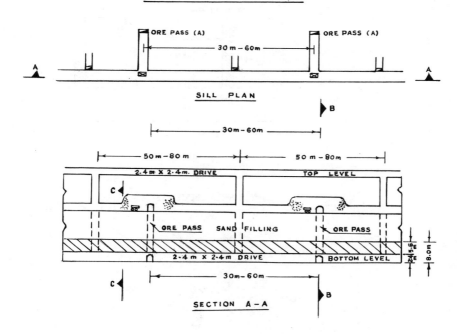

ORE PASS (A)   ORE PASS (A)

30 m - 60m

A   A

SILL PLAN

▶B

30m - 60m

50m - 80 m   50 m - 80m

2·4 m X 2·4 m. DRIVE   TOP LEVEL

C◄

ORE PASS   SAND FILLING   ORE PASS

2·4 m X 2·4 m DRIVE   BOTTOM LEVEL   8·0m

C   30m - 60m   ▶B

SECTION A - A

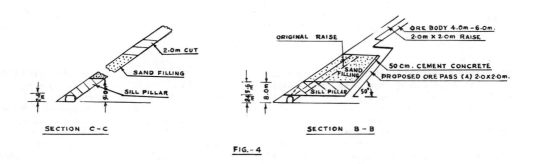

2·0m CUT

SAND FILLING

SILL PILLAR

SECTION C-C

ORIGINAL RAISE   ORE BODY 4·0m - 6·0m.
                        2·0m X 2·0m RAISE

SAND FILLING   50 Cm. CEMENT CONCRETE
                        PROPOSED ORE PASS (A) 2·0x2·0m.

SILL PILLAR   50°

8·0m

SECTION B - B

FIG.- 4

backfilling is made in one panel, the ore is extracted from the other by adhering to rigid cycle of operations. Nearly 70% of the present mine production is obtained from Cut and Fill stopes i.e. Horizontal Cut & Fill and Post Pillar stopes. A production of 2500 - 3500 tonnes per month is generally achieved from a Horizontal Cut & Fill stope by the use of a Cavo 310 loader or 0.76M^3 electric LHD of Wagner and Jarvis Clark make.

4.4 Post Pillar stoping

Post Pillar stoping has been successfully adopted in orebodies exceeding 6M in width and a minimum strike length of 80M. It is almost identical to Horizontal Cut & Fill stoping as explained above except for the formation of 4M x 4M insitu vertical pillars ("Post Pillars") which give additional stability to the roof for breaking wider spans of excavation. The post pillars are spaced at intervals of 13M along strike and 9M across it as shown in Figure 5.

233

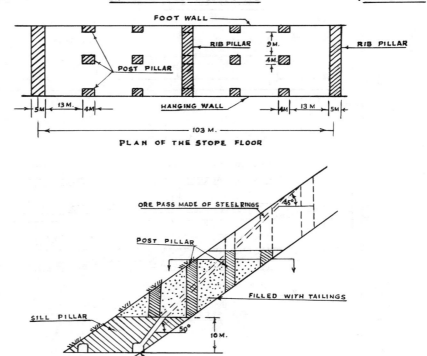

LAYOUT OF POST PILLAR STOPE   (NOT TO SCALE)

PLAN OF THE STOPE FLOOR

SECTION

FIG. 5

After the completion of sill development and full exposure of the lode from hanging-wall to footwall contact, a free face is created near the hangingwall contact so that drilling can proceed systematically towards footwall side. A small stretch of the cut on footwall will have to be separately stripped as the last stage of drilling cycle. Rock bolting of both the hangingwall and the back (in orebody) closely follows the drilling and blasting cycles and it is done by using 1.5M long and 20mm dia tor-steel grouted bolts at 1.5M x 1.5M pattern. Normally 2.2M high cuts are taken by drilling 2.3M long uppers at 70° to horizontal using two boom rubber-tyred self-propelled stope wagons mounted with 2 nos of Joy AL 60 drills. The height of the excavation is limited to a maximum of 4.8M above the backfill. The broken ore is loaded by a LHD (0.76 - 1.67M³ capacity) and it is dumped into nearby orepasses. Soon after the entire ore is cleared from the stope, the orepasses are raised by

using 1.5M dia steel rings and backfilling is completed in a panel leaving a head room of 2.4M. In the meanwhile, the LHD is deployed in the adjoining panel so that continuous stope production is maintained. A minimum of 2 nos of orepasses in each panel are considered essential to ensure efficient loading. Presently 3000 - 4500 tonnes per month of ore is produced from a Post Pillar stope depending upon the width of orebody and the availability of loading equipment. On some occasions high production rates upto 6400 tonnes per month has been achieved from such stopes by the deployment of a 0.76M³ electric LHD. When lean zones of mineralisation are encountered within a wide Post Pillar stope, it is possible to resort to selective mining. Similarly low grade ore can be conveniently left out on hangingwall and footwall contacts. Thus Cut & Fill stopes permit the adoption of selective mining when the necessity arises; however, it will impede the smooth and efficient movement

234

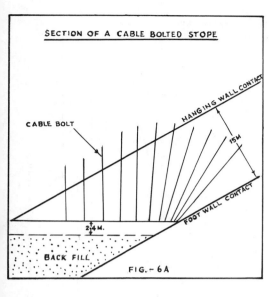

SECTION OF A CABLE BOLTED STOPE

FIG. - 6A

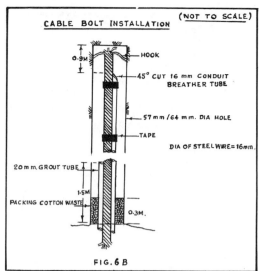

CABLE BOLT INSTALLATION   (NOT TO SCALE)

FIG. 6 B

of the LHD to some extent, thus marginally affecting its output.

5 PROBLEMS OF MINING IN DEPTH

As the stoping operations are extended to deeper levels (around 900M below surface), more difficulties have been encountered with regard to ground control, ventilation condition as a result of increased quantity of heat radiating from freshly exposed rock surfaces and increased cost of services. These factors will have the impact of higher operating costs and therefore they need to be considered while assessing the overall economics of mining from depth.

5.1 Increased strata pressure

Timely installation of rock bolts, restriction of the area excavated at any one time, early backfilling of voids before excavating the adjoining areas, adherence to planned sequence of stoping etc. have contributed significantly for the improvement of ground conditions at Mosaboni mine. It is to be remembered that unfavourable ground conditions would demand more systematic mining with better ground control techniques so that larger tonnages can be produced from limited number of working faces which would ultimately permit the economic exploitation of orebody from greater depths.

5.2 Adverse ventilation

With gradual downward extension of workings, the wet bulb temperature will rise steadily in spite of strict enforcement of humidity control measures. When improved environment at the working faces cannot be established within acceptable limits by circulating larger quantity of air, mine cooling is generally resorted to. At Mosaboni mine, chilled service water will be used shortly by commissioning a 600 RT refrigeration plant. It is thus obvious that mine ventilation costs will increase substantially while exploiting the mineral from greater depths and therefore maximisation of production is essential to offset such increased ventilation cost.

5.3 Increased cost of services

Extension of various services to deeper levels - such as compressed air, water, electricity, ventilation, supply of materials, pumping, man winding and establishment of additional facilities such as sub-stations, pump sumps, ore transfers, new shfts etc. will call for increased investment which will ultimately increase the operating costs. It is, therefore, imperative that maximisation of production and adoption of improved techniques are essential to ensure the overall economics of mining, particularly while exploiting low grade deposits.

235

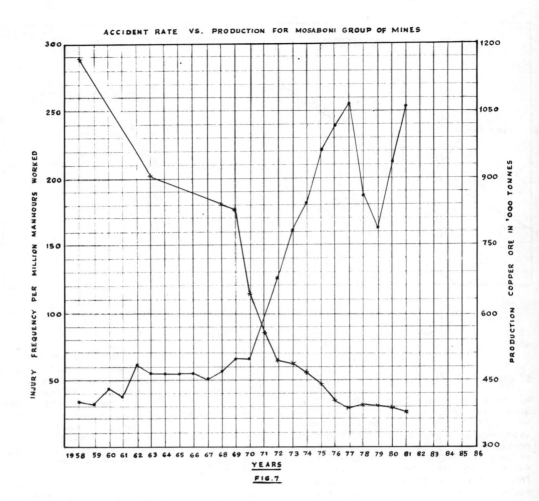

ACCIDENT RATE VS. PRODUCTION FOR MOSABONI GROUP OF MINES

FIG.7

6 BACKFILLING

The tailings from the concentrator is deslimed in single stage through 375mm dia hydrocyclones. A typical analysis of the tailing feed from the concentrator, overflow and underflow from the cyclones, is given below :

Mesh size	Cyclone Feed	Cyclone O/F	Cyclone U/F
+ 45	0.6	0.0	5.7
+100	22.6	1.6	42.2
+200	34.6	16.8	30.5
+325	18.7	29.2	11.4
-325	23.5	52.4	10.2

The underflow from the cyclones is taken down the mine by mixing with adequate quantity of water through 75mm dia bore holes suitably cased in weathered zone near the surface. Subsequently, it is distributed to various stopes through 110mm dia (OD) HDPE pipes in shallow mines and seamless steel pipes in deeper mines. More number of boreholes are being drilled to convey backfill to deeper levels to effect better economy in its transportation. The backfill requirement of small mines located far away from the concentrators, is catered for by the return trips of the ore-carrying dumpers. For this purpose the underflow from the cylones is taken to a paddock where the water is allowed to be drained out and reasonably dry classified tailing is scraped into

the dumpers by means of a triple drum
slusher located outside the paddock. The
backfill taken by dumpers to the shallow
mines is dumped into the bunker at the
respective mine from where it is led into
a hopper and mixed with minimum quantity
of water. It is then sent down the mine
through boreholes/pipelines.

Nearly 50% of the tonnage milled in the
concentrators is recovered for use as
backfill at the respective mines. At
Mosaboni mine where a 2700 tpd concentrator is situated, the backfill is sent down
the mine with a bulk density of 55% - 60%
(i.e. solids by weight) at the rate of 60
tonnes per hour. The percolation rate of
water through the backfill in the stopes
has been found to be 15cm per hour when
the percentage of fines (i.e. below 325
mesh) is within 10% by weight. As a result
of introduction of backfill in the mines,
following gains have been observed:

(a) Efficient and safer mining of wider
orebody has been made possible. (This
was not possible by Breast stoping practised earlier.)
(b) Easier mechanisation of drilling,
loading and hauling operations.
(c) Better strata condition has been
achieved through timely rock bolting and
stabilisation of the ground with backfilling of voids.
(d) Improved percentage of extraction
of ore.
(e) Reduction in leakage of ventilating
air through worked-out stopes which has
significantly contributed for improved
and efficient ventilation.
(f) Generation of larger tonnage at
individual stopes with higher productivity and consequently reduced mining
costs.
(g) Easier supervision.

7 ROCK MECHANICS STUDIES

A study was carried out to find out the
deformation of post pillars subsequent to
backfilling for the optimisation of their
size. For this purpose a strain gauge
borehole extensometer with a remote
read-out facility was developed by National Geophysical Research Institute,
Hyderabad. The extensometer was installed
in a horizontal hole of 33mm dia and 2.5M
long in a post pillar at a depth of 300M
below surface and periodic readings were
taken by a strain indicator to monitor the
deformation of the post pillar. During the
above installation, arrangements were being
made to place third fill in the stope. The
data so obtained was plotted on a graph -

Time Vs Deformation. The above study
revealed that a systematic lateral expansion of the post pillar at a rate of
0.046mm per day upto 130 days i.e. till
the commencement of placement of fourth
fill. While backfilling for fourth cut
was in progress, the pillar was found to
contract rapidly by about 4.3mm. This
phenomenon could be explained by using
the concept of effective stresses as a
result of water pressure generated in the
pores and fractures within the pillar
decreasing the vertical stress in the
pillar leading to its contraction. A couple
of days after the completion of backfilling, the pillar started expanding rapidly
in the lateral direction till it recovered
to its original level. This might be attributed to natural dewatering of the backfill
with the passage of time. Similar phenomenon was observed while placing fifth
fill in the stope. Thus it was evident
that the post pillar above the fill
deformed considerably during mining and
it was stabilised again by the lateral
confinement provided by the backfill.

With a view to evaluating optimum size
of rib pillars in Room and Pillar stopes,
measurement of change of stress in such
pillars using borehole inclusion solid
stressmeter was considered at a depth of
722M below surface. The instrument was
fixed in a 90cm deep and 40mm dia borehole
drilled along the foliation of the orebody
in the middle of a rib pillar of a room
before its filling. The room was then
backfilled and mining of the adjacent room
was continued till the 4M wide rib pillar
was reached. During this period, a sharp
increase of stress upto 40 Kg/cm^2 and
subsequently a gradual rise upto 45 Kg/cm^2
was observed. Another instrument installed
in the same pillar became inoperative and
therefore further studies will be carried
out by using vibrating wire stressmeters
to determine the principal stress and its
direction.

8 PROBLEMS ENCOUNTERED

Some of the problems encountered while
mining Cut and Fill stopes have been
briefly discussed in the following
paragraphs :

8.1 Drainage of water from the fill

The percolation rate of water in backfill
in Cut and Fill stopes was found to be
reducing after the extraction of 3-4 cuts
and backfilling, which resulted in the

accumulation of large quantity of water inside the stope thereby delaying faster cycles of operation. This problem has been overcome by drilling NX size boreholes through sill pillar and extension of these passages to the stopes by fixing old perforated pipes suitably wrapped with wire mesh and jute before pouring fresh backfill into the stope. Similar perforated pipes are laid after every 2-3 cuts on the floor along the strike direction and also across it (in case of wide orebody) to ensure efficient drainage of water from all parts of the fill.

8.2 Cable bolting

Occasional fall of ground from the planes of weakness like joints, slips and fractures at the back and hangingwall have resulted in complete dislocation of the normal mining activities in some of the Cut and Fill stopes. Although a few of the joint planes were prominently visible in the stopes, some others did not exhibit any signs of unstability which ultimately fell down in wedge-shaped blocks. However, recent introduction of cable bolting by drilling 57mm dia and 18M long holes at a spacing of 3M x 2.4M and cement grouting after placing a 16.2mm dia high tensile steel cable in them has substantially stabilised the ground conditions in such stopes (Figure 6). This has prompted the management to extend this pre-support system to more number of stopes where criss-cross joint-planes are known to exist. It has also resulted in the achievement of better fragmented rock which in turn permits efficient loading and faster rate of extraction from such stopes.

8.3 Electric LHDs

The use of electric LHDs to load, haul and dump ore in place of pneumatically operated 310 Cavos has resulted in improved underground environment (reduced noise level, more comfort to the operators, improved visibility due to the absence of fog) and low energy requirement. The problem of low compressed air pressure at the workings, already experienced in different areas of Mosaboni mine, has been partially overcome by the use of more number of electric LHDs. Thus, it would be very beneficial for the management to consider the introduction of more electric LHDs on account of their substantially low consumption of electric power as compared to compressed air operated machines.

8.4 Steel chutes

Recent development of manually operated steel chutes in place of conventional wooden chutes to handle larger tonnage from high productive stopes has speeded up the filling of mine cars. It has also eliminated high incidence of injuries to workpersons while removing the stopper boards in wooden chutes to fill the mine cars.

8.5 Electro-hydraulic rock breakers

Manual breaking of boulders at the grizzlies by hammering had been the practice at this mine. This being a very arduous task it has been replaced by mechanised breaking where concentration of output could be achieved. For this purpose electro-hydraulic rock breakers of 50 - 100 tonnes per hour capacity are being used. In addition to relieving the physical exertions of manual breaking of boulders, it has reduced the accident rate considerably.

8.6 Re-circulation of backfill

The problem of re-circulation of backfill along with the ore is a serious problem which has considerable impact on the achievement of projected mill-head grade. Although more care has been taken during the operation of LHDs in Cut and Fill stopes to avoid too much of digging into the backfill, no tangible improvements have yet been seen. Presently there is no arrangement in the mines to use cemented fill as top layer of the backfill in a stope. In this context, it is to be appreciated that cost of portland cement in India is rather high which will make the justification for the use of cemented fill more difficult especially while mining low grade chalcopyrite ore with about 1% -1.2% Cu. Although efforts were made to use wooden slabs/planks to prevent LHD bucket digging into the backfill, the whole system was found to be very cumbersome and hence it was given up.

8.7 Feeble movement of Air

One of the serious problems in all the above methods of stoping is to provide effective ventilation. Although 600 - 800 M³/min of air is circulated through high productive Post Pillar stopes, the velocity of the ventilating air cannot be maintained at a satisfactory level owing to a large

238

cross-sectional area. The possibility of
solving the above problem by the use of
brattice cloth for coursing the air is
rather cumbersome. It is also inconvenient
to use flexible ventilation ducts with
small capacity electric blower inside such
wide stopes. However, no effective solution
has yet been found to overcome the above
problem.

9 IMPROVED SAFETY STANDARDS

As a result of timely rock bolting of the
hangingwall and the back in the stopes and
systematic backfilling of the voids created
by stoping, there is a general improvement
in the ground conditions leading to improved
standard of safety at the workings. This is
evident from the production and accident
statistics as given at Figure 7, especially during the last 10 years. This can be
attributed primarily to the introduction
of new mining methods. The exposure of
workpersons to adverse roof conditions,
rolling stones on inclined planes, unfavourable ventilation conditions etc. has been
gradually eliminated with major proportion
of the mine production coming from Cut and
Fill stopes. As a result of the mine mechanisation, considerably reduced physical
exertion is required by the workpersons in
many of the operations and this will keep
them mentally alert for long periods. These
factors, no doubt, have substantially
contributed for the reduction of accidents.

10 FUTURE PLANS

Some of the future plans to achieve further
improvement in safety, productivity and
overall economics are briefly highlighted
below :

(a) Possibility of drilling 25M long and
57mm dia holes from a sub-level in Room
and Pillar stopes so as to minimise the
repetitive cycles of various mining
operations and to obtain faster rate of
extraction.
(b) Possibility of use of 165mm dia
blastholes in wide and steeper orebodies
of Rakha mine has been under consideration. On its successful introduction,
substantial reduction in operating costs
and establishment of safer working environment in the mine will be possible.
(c) Possible use of cemented fill in
richer blocks of different mines is under
the active consideration of the management. In this connection, laboratory
scale tests have established that cemented
fill with a composition of 81.5%
(by weight) of deslimed tailings, 12.5%
boiler ash and 6% portland cement can

give a compressive strength of 11 Kg/cm^2
after a period of 28 days.
(d) Deployment of electro-hydraulic
jumbos for faster development, down-the-
hole drills for drop raising, more
electro-hydraulic rock breakers for the
grizzlies at critical areas and underground crushers to facilitate proper
sizing and faster movement of ore have
been contemplated.

11 CONCLUSIONS

With growing demand for copper, it is vital
for India to maximise indigenous production
of the metal by exploiting larger deposits
in promising areas of Singhbhum Copper Belt.
Any such venture will have to be necessarily well-mechanised by adopting the latest
mining technology so that bulk mining of
low grade copper ore at low operating costs
is possible. The confidence gained in
Mosaboni group of mines in the establishment of Cut and Fill stoping will, no doubt,
contribute significantly in achieving the
above objective. Relatively cheaper and
more productive methods of stoping like
blast-hole mining may have to be adopted
in some form or the other to maintain the
cost-effectiveness of the operations.

12 ACKNOWLEDGEMENT

The author is thankful to the management
of Hindustan Copper Limited for according
permission for the publication of this
paper. He is grateful to Mr M V N R S Rao,
Chairman-cum-Managing Director, Hindustan
Copper Limited, Mr M A Khan, Director
(Operations), Hindustan Copper Limited and
Mr R C Bahree, General Manager, Indian
Copper Complex for their valuable guidance
from time to time. Thanks are also due to
Mr J Chattopadhyay for going through the
paper critically and for offering sugge-
stions. The author wishes to thank the
staff members of both operating and service
departments of Mosaboni group of mines for
their timely assistance in the preparation
of this paper.

REFERENCES

(a) David G.Rasmussen and Gorden M.Pugh,
"Ramp-in-Stope access" - Engineering and
Mining Journal, August 1979.
(b) Stand Dayton, "Mount Isa mixes multiple concepts of mining with advanced fill
technology" - Engineering and Mining
Journal, June 1978.
(c) S J Patchet, "Fill support system
for deep level gold mines".

The development and introduction of LHD undercut-and-fill mining at Homestake's Bulldog Mountain Mine

R.A.SIMONSON & G.C.LOGSDON
Homestake Mining Co., Creede, USA

ABSTRACT: Homestake Mining Company's Bulldog Mountain Mine is located in Creede, Colorado, USA. Three hundred tons per day of silver and lead ore is produced from a narrow (2'-8'), steeply dipping (60°-90°) vein system. Through the development and introduction of an LHD (load-haul-dump) undercut-and-fill mining method along with improvements in the quality of hydraulically-placed, cemented backfill; better grade control and increased productivity have been achieved.

1 INTRODUCTION

Homestake's Bulldog Mountain Mine is located in Creede, Colorado, USA. Ore production is 300 tons per mine day at a grade of 16 ounces silver per ton and 1.5 percent lead. Total production in 1982 was 1,358,000 ounces silver and 1,720,000 pounds lead. The operation employs 100 people.

Homestake began producing silver and lead concentrates at the Bulldog in 1969. The mine consists of 5 levels, one of which is inactive (FIG 1). Currently all ore is produced below the 9360 haulage level and must be hoisted at the #1 winze. The present known ore zone has a strike of approximately 6000' and a vertical dimension of 800'. The ore zone's limits have been defined vertically and to the south, but are open to the north. The ore body consists of a complex system of narrow (2'-8') veins, dipping steeply from 60° to 90°. The vein structures pinch and swell irregularly along strike and down dip. Mineralization is also very irregular along strike and dip. Stope walls vary from strong to highly fractured. The ore is generally weak and unconsolidated. The host rock is a rhyolitic ash flow tuff. Major economic ore minerals are native silver and galena.

2 ORE DEVELOPMENT

Diamond drilling is rarely used in deter-

mining stoping blocks due to the spotty, irregular nature of mineralization and poor core recovery. Instead, cross-cutting to the vein and scramming on-vein is the delineation method. Ideally an ore block would be scrammed its entire length on top and bottom levels. Often this is not possible and the ore block is scrammed on top and perhaps cross-cut at regular intervals (300'-400') on the bottom.

Until 1981, all scramming at the Bulldog had been done with electric slushers. These slusher scrams were driven 8' to 10' above track level. A wooden bridge with a muck chute was built over the track and the ore slushed into muck cars below. In January 1981, in an effort to raise productivity through increased mechanization, a 1 yard LHD was put into service scramming on-vein at track level. Scram mining widths proved acceptable after a brief learning period using the 48" wide LHD. Productivity and flexibility proved to be higher as compared to conventional scramming with electric slushers. This led to the purchase of 2 more LHDs in 1981 to increase the scramming rate. With the success of LHD scramming, an effort began to develop a mining method using LHDs.

3 MINING METHODS

The original method at the Bulldog was ascending cut-and-fill which required miners to work under an unstable back. During the early 1970's the method was changed to

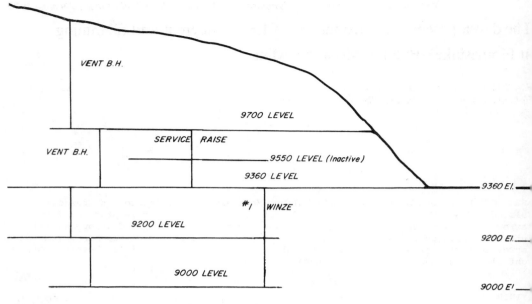

Figure 1.　Idealized long section schematic looking east.

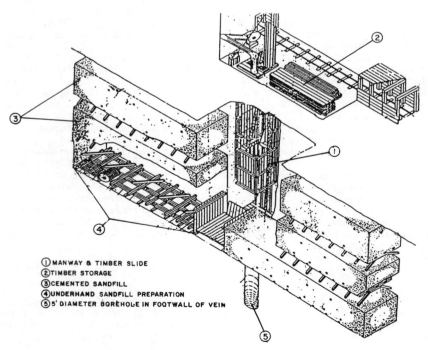

① MANWAY & TIMBER SLIDE
② TIMBER STORAGE
③ CEMENTED SANDFILL
④ UNDERHAND SANDFILL PREPARATION
⑤ 5' DIAMETER BOREHOLE IN FOOTWALL OF VEIN

Figure 2.　Idealized conventional-UCF.

undercut-and-fill due to the unconsolidated nature of the vein. The method is currently evolving from undercut-and-fill using 30 h.p. electric slushers and bored ore passes (conventional-UCF) to LHD undercut-and-fill (LHD-UCF) using 1 yard, rubber-tired LHDs.

3.1 Conventional-UCF

Conventional-UCF (FIG 2) uses slushers mucking to a 5' diameter borehole in the footwall. Due to the irregularity of dip in the 200' vertical ore blocks, the borehole can range from being in the vein to 25' in the footwall. Sinking on the borehole to the vein is completed before mining can be started. The sinking makes room for installation of a manway and supply chute. When these service facilities are completed to the floor of the cut to be mined (approximately 14' below previous cut), the rock is mined from the borehole to the hanging-wall below the floor level. This provides a slot for the ore to be gravity fed to the borehole. Once the slot is completed, a work deck (bridge) is installed over the borehole and slot. This deck supports the slushers and grizzlies.

With the borehole sunk, the manway completed, and the slushers installed; mining along strike away from the borehole begins. Mining proceeds 150' each direction from the borehole to an arbitrary cutoff point halfway between the 300' crosscut intervals. The mined out cut is then

backfilled and the cycle returns to the sinking stage in preparation for the next stope cut.

The conventional-UCF method is labor and supply intensive. This leads to a low in-stope productivity of only 12.5 tons per man-shift. Coupled with the low productivity is the problem of stope production cycles. Conventional-UCF stopes average over 50 percent in the non-productive, or "down", cycle. The "down" cycle includes sinking, installing manway and supply chutes, installing the floor and slushers, preparing the stope for backfill, and backfilling the stope. The actual mining of ore is the shortest lived aspect of the total conventional-UCF stope cycle.

3.2 LHD-UCF

Stoping with LHDs began below the 9000 level in 1982. Earlier a diamond drilling program determined the structural limits of the ore body below the 9000 level. Currently, economic mineralization is thought to extend only 50' to 80' below the 9000 level. This shallow extension of the ore below the 9000 level could not justify sinking the #1 winze and creating a lower level; therefore, a method using LHDs that could create their own access to the ore had to be developed.

Declining on-vein for access was developed for two reasons; to minimize waste produced per ton of ore developed, and to begin ore production immediately. Prior to stoping,

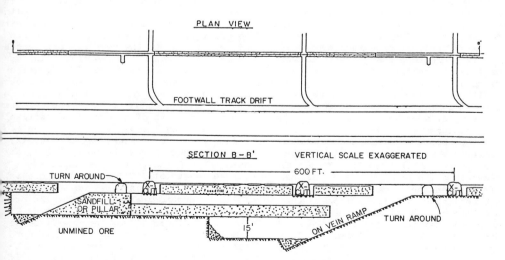

Figure 3.   Idealized on-vein LHD-UCF (no level below).

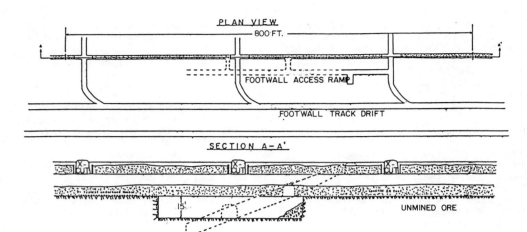

Figure 4. Idealized off-vein LHD-UCF (no level below).

2500' of vein strike length had been scrammed at track level on the 9000 level. Scramming was done from 9 cross-cuts, each approximately 300' apart.

Each LHD stope block was designed to mine the ore between every other cross-cut, or 600'. In preparation for stoping, the scrams were backfilled to within about 80' north of the cross-cuts from which the declines would start. This 80' allows the necessary room for loading, a turnaround doghole for the LHD, and the required length to get the decline under the above sandfill (FIG 3). The 7'x8'x15' turnaround is the only off-vein waste development necessary to begin stope production. Since the decline is on-vein, ore production begins immediately if ore-grade mineralization is present in that stretch of vein. The decline is designed to have 8' of clearance as it goes under the above sandfill. It is continued until the desired stope-cut height is reached. Stope-cut heights vary between 12' and 16' depending on the competency of walls.

As a stope block progresses down and to the north, the stope end lines stairstep down and to the north (FIG 3). If ore is not present at the end line, a pillar can be left between the stope to the south and the decline of the next stope to the north. When ore is present, the pillar is mined in order to maximize recovery. The stope must then be completely filled with a cemented-sand backfill in the decline area. The decline for the next stope may then be established in the

cemented-sand with no loss of ore. Keepi the southern stope one cut below the northern stope is necessary in most cases

Three on-vein LHD-UCF stopes covering 1800' are presently producing at the Bulldog. At least two more stopes to the north are anticipated, bringing the total strike length to 3000'. Some of the stop have and will have multiple headings due to vein splits. Because the ore is expected to end about 50' to 80' below th 9000 level, these stopes are expected to end after 4 or 5 cuts.

In one case, a major vein split is bein mined with access from an off-vein declin (FIG 4). This ore split is 700' long wit no possibility of extension to the north or south. Due to the vein's short length on-vein access was not feasible. In this case, the decline is driven 20' from the footwall of the vein. The vein is access by cross-cutting from the decline when th desired elevation has been reached. Mini proceeds in both directions along strike after reaching the vein.

To mine new ore blocks between levels, (as opposed to below the 9000 level), off vein LHD-UCF can be used if sufficient strike length of 300' to 1500' of ore is delineated. At this time, no LHD-UCF sto is in production between levels. An ore block is being readied for production however. Figure 5 shows an idealized pla for this ore block. All newly developed ore blocks are being considered for LHD-UCF mining as the first choice method if possible.

Production drilling is done with hand-held jacklegs. Non-electric delays and 1" by 16" water-gel explosives are used for blasting. The broken ore is mucked with the 1 yard LHDs and loaded into 3.5 ton ore cars. The loaded train is then hauled to the shaft station and dumped. Presently, 3 to 4 men are working the 4 LHD-UCF stopes below the 9000 level. They have up to 7 on-vein headings and 1 off-vein decline heading available for production. Production, grade, and cycle requirements determine how many and which headings are active on a daily basis. This crew is responsible for drilling, blasting, mucking, and hauling the rock to the shaft. They also do all preparation for backfilling, the actual backfilling of the stopes, and hauling of supplies to the stopes from a central supply station.

After a stope cut has been mined to its end line, it is prepared for sand backfilling. The stope floor is leveled with 6" to 12" of broken ore. This amount of ore is generally left as mining proceeds to maintain a smooth road bed. In sandfill preparation, it is left to act as a cushion for the next lower stope cut. Round timbers (6" to 8" diameter) are then laid from wall to wall on 7' centers. These timbers are set in timber hitches that have been installed in the walls and/or secured with 3/8" cable attached to an eye pin installed 4' up on the wall. Often it is only necessary to hitch or cable tie the timber on the hanging wall and simply wedge the timber to the footwall. Unmilled, round timbers have recently replaced more expensive 8"x8" milled timber in the sandfill preparation.

Chain link wire is laid over the timber for the length of the stope. The wire is nailed to the timber and stretched along the floor between the timbers. Care is taken to get the timbers and wire on the floor. This allows the cemented-sand backfill to be poured with no danger of slabs hanging below the wire and timber on the next cut below. If the wire was stretched across the tops of the timber, as was previously the method, uncemented sand was poured first to fill this 8" gap. The sand then harmlessly fell out below the wire on the next lower cut. However, substantial ore dilution resulted from this 8" layer of sand. The wire should be on the floor for improved grade control and safety.

A simple burlap covered bulkhead is constructed at the stope end. The bulkhead is designed to be as water tight as possible in order to avoid losing cement and sand through it. Clear water is decanted over the bulkhead and through the burlap. The hanging of 2" plastic pipe with distribution valves and hoses every 20' completes the stope preparation. Backfilling begins with a 3' layer of cemented-sand followed by 4' to 5' of sand. The desired cemented-sand mix is 15 percent cement by weight. Through experience and testing 15 percent cement has been determined to safely provide enough strength to the mix. When the cement drops below about 10 percent by weight, its addition does not add appreciable strength to the sand and therefore is wasted. The 4' to 5' layer of sand is added to prevent wall movement and add weight to the cemented-sand layer to prevent lifting from blasting.

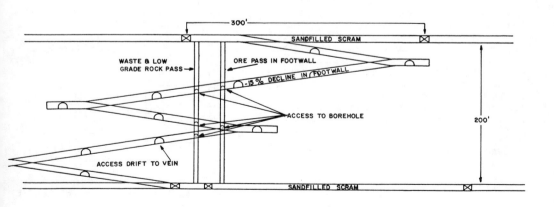

Figure 5. Idealized off-vein LHD-UCF (level to level).

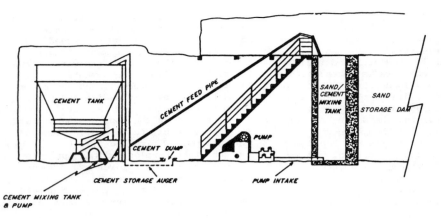

Figure 6. 9700 level sandplant.

5 SAND PLANT

When the amount of cement in the backfill drops below the desired 15 percent by weight, failures in the stope backfills begin to occur. Backfill failures cause safety problems, production losses, and grade dilution. The control and consistency of cement addition at the sand plant is most critical to the overall quality of the stope backfill. When failures began happening regularly, problems at the sand plant were apparent. The former auger system for conveying the dry cement to the mixing chambers for mixing with the sand slurry was inadequate. It was inconsistent due to variabilities in the cement moisture and temperature, which caused extreme variations in the feed rate. A steady, adequate flow of cement could not be maintained. To eliminate this problem, the auger was eliminated and replaced with a cement slurry pump. The dry cement is fed directly from the storage tank to the pump. A small amount of water is added to slurry the cement. This slurry is then pumped to the mixing chambers for addition to the sand slurry (FIG 6). Cement flow control to the pump is from a variable speed auger in the bottom of the cement storage tank. The cement pump then maintains a steady, correct flow of cement to the sand mix.

6 ADVANTAGES OF LHD-UCF

6.1 Grade control

Mineralization along the strike of the vein is very irregular. It often varies from waste, to low-grade, to high-grade in dis-

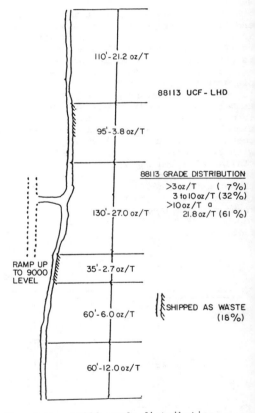

Figure 7. 88113 grade distribution.

tances of tens of feet. Segregation of grades along strike is very beneficial to grade control. Mucking with LHDs allows round by round or even bucket by bucket

246

segregation. Loading rock from the face directly into muck cars makes possible this segregation. This segregation is not practically achievable with conventional-UCF because the mining and mucking process in the stope is necessarily independant of the loading and haulage from the bore-hole.

Figure 7 shows the grade distribution in the LHD-UCF stope 88113. This ratio of ore, low-grade, and waste is typical of the entire vein system.

6.2 Productivity

In-stope productivity is 25 tons per man-shift for LHD-UCF. Conventional-UCF productivity is 12.5 tons per man-shift. The actual mining rate along vein in an LHD-UCF stope is not any faster; its gain in overall productivity comes from lower down time and increased ease of supply handling. There is no time spent sinking, building work decks, or constructing man-ways with the new method. The only down time in LHD-UCF mining is for sand back-fill preparation and backfilling. This time is necessary in both methods.

6.3 Flexibility

The ease of stope set-up and the use of non-captive LHDs provides tremendous flexibility. The miners can move easily and quickly among available headings as production and grade requirements dictate.

7 DISADVANTAGES

The mechanized method has the disadvantages of requiring more ventilation and mainte-nance. Ventilation is being handled with 20 h.p. vane-axial fans and 20" vent bag. Two diesel mechanics service the fleet of 5 LHDs.

8 CONCLUSION

The LHD-UCF method has helped reduce unit mining costs through increased productivity and improved grade control. The method accounts for 40 percent of mine production after one year and its share will increase in 1983. All newly developed ore blocks are being evaluated for LHD-UCF stoping as the first choice method.

Evolution of newer techniques of mining thick coal seams and wide orebodies with filling in India

R.D.SINGH
University of Jodhpur, India

D.P.SINGH
Banaras Hindu University, Varanasi, India

ABSTRACT: In India, mining of coal and ore with filling has been practised for several decades primarily for safety and conservation. Stowing is insisted upon for extracting pillars in thick coal seams (more than 4.8 m thick). The pillars are extracted in lifts of 2.4-3m height in ascending order in conjunction with hydraulic sand stowing and this technique has been applied to extracting pillars in seams as thick as 28m, though the technique has not been successful in seams thicker than 8-9m. A combination of partial extraction and full extraction of pillars in thicker seams developed in multi-sections or in contiguous seams has shown promising results. In recent years experimentations have been done to work flat and steep thick seams by slicing on longwall method with stowing in ascending order. Longwalling in slices parallel to roof and floor has been successful in seams 9-10 m thick and for working thicker seams cross-inclined slicing has shown promising results in laboratory studies with equivalent material models. For mining, wide orebodies filling has established its merits. Recent trials have established the technique of post pillar stoping with filling for mining a copper deposit with a true width of more than 8m and strike continuity of not less than 60m. In another part of the country for mining a 15-65m wide lead and zinc deposit post pillar stoping with filling has been started in recent months. In an uranium mine to mine 8-10 m wide ore section cut and fill stoping with filling has been successfully done. Mining of wide orebodies with filling has given good strata control, improved percentage recovery of ore and high production and productivity. 'R & D' projects are in progress to optimise the mining techniques.

1 INTRODUCTION

In India, coal seams more than 4.8 m thick are considered as thick seams. Such seams contain more than 75 per cent of India's total coal reserves of 85444 million tonnes of all types (proved, indicated and inferred). The majority of thick seams are in the range of 6-10m and one exceptionally thick coal seam attains a thickness of 138m.

These seams occur at all depths: some occur close to the surface and some at great depths (600-1200m). They are generally banded; the bands varying from a few milli-metres to several centimetres in thickness. Pyritic intrusions are also common and all seams are prone to spontaneous heating.

Gradients of the seams vary from a few degrees to almost vertical. Cleats are not well defined and the coals are usually hard. These seams are overlain by massive sandstones or hard shales and the roofs are difficult to cave. Mining of such seams by underground methods has always been beset with problems of strata control and spontaneous heating: this has necessitated experimentation and search for appropriate technology of mining.

Similarly, for mining wide ore-bodies of copper and lead and zinc experimentation is under way to establish suitable techniques of mining for maximum conservation and safety.

Experiences thus far gained indicate that mining with backfill could be successful for mining thick coal seams as also wide orebodies in many situations. In this paper a brief review of the experimentation and evolution of newer techniques of mining thick coal seams and wide orebodies with filling have been presented. Also characteristics of filling materials and likely developments in mining with filling have been indicated.

2 MINING OF THICK COAL SEAMS WITH FILLING

Abundance of thick coal seams at shallow depths encouraged adoption of bord and pillar mining in the past and many thick seams were developed on bord and pillar method; some in multi-sections and today over 2000 million tonnes of coal are locked up in these pillars. Extraction of these pillars has become problematical because of difficulty in strata control and many workings have witnessed premature collapses, fires, etc. This has necessitated search for appropriate technology to extract pillars in thick seams.

Hither to extraction of pillars in thick seams has been done in conjunction with hydraulic sand stowing(Singh 1969). On the average about 1.37 tons of sand were stowed for each ton of coal extracted. Generally, pillars in thick seams are extracted in lifts of 2.4-3m height in ascending order with stowing. After the bottom lift has been extracted the process is repeated as in the first lift till the full thickness of the seam is liquidated. For example, at colliery 'V' in the Raniganj coalfield pillars in Laikdih seam (14.63m thick) were extracted in four lifts in ascending order in conjunction with stowing. The seam has 'Jhama'(naturally burnt coal) at the floor and sandstone at the roof and has a dip of 1 in 6.

The seam was developed on bord and pillar system in two sections: (i) along the floor; and (ii)along the roof with 1.82 m coal left in the roof. The pillars were 24.38m x 24.38m centres. In a depillaring district, diagonal line of face was maintained, the working faces in the dip levels being kept in advance by half a pillar. A level split was driven in the pillar to be extracted: dip and rise slices approximately 2.4m high x 4.8m wide were then taken from the original level up to half the distance of the length of the pillar. After extraction of the slices, the void was stowed solid with sand, leaving a rib of 1.8 m. The next slice was then taken and so on. After the first lift of all the pillars in a panel was extracted and coal replaced with sand up to a height of 2.4 m, the second lift was developed over the stowed sand followed by third lift. The fourth lift was extracted from the top section already developed initially below which 2 m of coal was left to form a solid floor.

This technique has proved successful, if the seam could be extracted in three lifts i.e. 7 - 8 m thick seam. After the third lift, roof troubles are invariably experienced and extraction operation is difficult.

2.1 Extraction of pillar in thick and contiguous seams

As stated earlier, in India coal seams occur in quick succession, especially in the Jharia coalfied. If the parting between the seams is small, the workings of one may affect the workings in the other seam. In Donets Coalfield (USSR) 'undermining' takes place if the parting between the seams is less than $12m \times 3.5 \ m^2$ where, m is the working thickness of the bottom seam (Shevyakov, 1965). In British Coalfields pillars left in the top seam have affected workings up to 274 m below (Scurfield 1970). The partings between seams in Indian coalfields are usually too small and therefore design of safe pillar extraction systems is often difficult and rather a tricky exercise. Examples of some trials are given below:

At colliery 'S' in the Jharia Coalfield XIV seam 8.8m thick and XIII seam 5.94m thick occur close together with a parting of only 1.5m - 1.8m. Development was done on bord and pillar system along the floor of XIII seam and along the roof of XIV seam. The size of the pillars centre to centre was 30.5m x 30.5m. The pillars and galleries in XIV seam were vertically above those in the XIII seam. Only partial extraction was done in XIII seam. After development two rise to dip splits and two level splits 6.09m wide x4.5m high were driven in each pillar, giving an extraction of 60 per cent. Stooks 6.09m x 6.09 m were left in the goaf and stowed with sand. The XIV seam was worked in two lifts on the longwall method. The bottom face was leading the top face by 30.5 m. Both the top and bottom goaves were stowed solid with sand. This technique, however, did not become popular.

At colliery 'K' XIII seam (6.6 m thick) and XIV seam (8m thick) with a parting of 1.5m occur at a depth of 167.6m and were developed on bord and pillar system. XIII seam was developed along the floor leaving 0.9m inferior coal. The width and height of galleries was 3.6m and 2.6m respectively and the pillars were 25.5 m x 25.5m centres. XIV seam was developed along the roof in the same manner. The pillars and galleries in XIV seam were vertically above those in the XIII seam. During depillaring, the pillars in XIII seam were splitted and the splits were stowed solid with sand emplaced hydraulically leaving stooks 7.5m x 7.5m. This operation was done in two lifts. 1m above the stowed stooks of XIII seam, XIV seam pillars were splitted and the splits stowed with sand in the bottom lifts. Thereafter the top 4.8m was extracted fully with stowing. In another variant the top 3.6m of coal was extracted by caving. The stooks were extracted in slices such that the exposed roof did not exceed 90 sq.m. This method, however, did not work well. The above technique of extraction of pillars gave the following percentage extraction:

1.1. Percentage extraction in XIII seam - 50 per cent.

2. Percentage extraction in XIV seam where full top 4.8 m of coal was extracted with stowing - 65 per cent.

3. Percentage extraction in XIV seam where top 3.6m was extracted with caving - 60 per cent.

Support in the dipillaring area consisted of wooden props and chocks and the average consumption of timber was 8.49 m^3/1000 tonnes of coal. The above method gave good strata control. Ground subsidence varied between 5.48 and 16.45 cm.

In the past some seams were developed in three to four sections. Extraction of pillars in such seams developed in multi-sections has defied all ingenuity so far. In most of the situations partial extraction in the bottom section with stowing and full extraction to a height of 4.8m in the top section are done. The overall percentage extraction comes to around 60 per cent where caving can be done in the top section and to about 50 per cent where surface has to be protected.

The need for the protection of surface has necessitated the use of bord and pillar partial extraction system. In this system some 30-40 per cent of coal is left in situ. It is recommended that for areas where surface features are to be protected at depths not exceeding 150m, a factor of safety of 1 and for areas free of surface features a factor of safety of 0.5 may be used in designing the pillars left in situ provided the strength$_2$ of coal is not less than 400 kg/cm^2 (GhoseCommittee report 1973).

2.2 Slice mining by longwall methods

The first experiment to introduce longwall method to work thick seams in conjunction with stowing was taken up at colliery 'P' in the Raniganj coalfield to work Dishergarh seam 4.2m thick at a depth of over 500m primarily to control bumps. The 2.1-m bottom section was worked by single unit or double unit longwall face laid dip and rise and advanced on the strike. The goaf was stowed solid by hydraulic sand stowing in strips of 3.6 m. About 2 years after the extraction and stowing of the bottom section 2.1-m top section was developed on bord and pillar method over the stowed goaf of the first lift and the pillars were extracted in conjunction with stowing. The longwall face in the bottom section was quite successful. The convergence in the face was 25 mm every 3.6m of face advance. The gate roads stood well; the convergence in the middle of a double unit face stabilized 140 m behind the face (Figure 1) and it was 14% of the first slice height; some 11% of the convergence took place in the first 30m advance of the face (Guin 1962). The incidents of bumps were reduced and the roof control was generally good. But the top action working by bord and pillar method did not give satisfactory results.

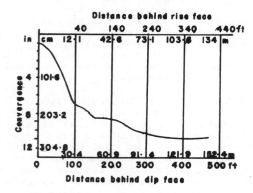

Figure 1. Convergence in the middle gate of a double unit long-wall face.

At another colliery at a depth of 745.84m the same seam was worked in two slices both by longwall method in ascending order in conjunction with solid hydraulic sand stowing. There were occasional bumps in the bottom section but the workings in the top section were free from bumps. The technique of slice mining by longwall methods was established and its application was extended to thicker seams also.

At colliery 'J' in the Jharia coalfield XIV seam 8.5 - 9.1m thick dipping at 1 in 6.5 was worked in three ascending slices in conjunction with hydraulic sand stowing at a depth of 441 m. Two variants have been used. In one pannel the bottom face was laid along dip and rise and advanced on the strike. After the bottom section was extracted and stowed, two strike faces were opened for the second slice, one on each side of a central gallery staggered by 22.5m and after the second slice faces had advanced 15m, the third slice faces were opened one on each slide of the central gallery. Thus in a panel there were 4 faces in all; 2 to work second slice and 2 to work the third slice; the faces advanced to the rise. In the second variant three strike faces were opened; first the bottom slice was carried up the dip with stowing; then the second slice face some 13.5 - 22.8m behind was opened over the stowed sand and lastly the third slice face was opened 13.5 - 22.8m behind the second slice face.

2.3 Slice mining of thick and steep seams.

Slice mining of thick and steep seams with stowing in ascending order has been tried out at two collieries. Two variants were experimented: (i) slices parallel to the roof and floor and (ii) horizontal slices between roof and floor across the seam.

1. Slices parallel to the roof and floor

A seam 7.31 m thick inclined at 35° was opened out between two horizons 100 m apart at a depth of 260 m for the application of inclined slicing (slices parallel to the roof and floor) in ascending order with stowing. First a slice 3.5m along the floor was taken by a longwall face 100 m long laid along the strike and advanced to the rise. In a panel two such faces were opened out, one on each side of a central gallery. The faces were staggered by 25m. While coaling was done at one face, stowing was done at the

other face. The face was supported
by props and bars; the interval
between the adjacent props was kept
1.2m. As the face advanced, stowing of the goaf was done every 6 m
with sand emplaced hydraulically and
the props were stowed in the goaf.
The second lift was started over the
stowed goaf of the first lift and
the faces were carried forward as in
the first lift. This method has
proved successful and is now adopted
even for working thicker seams.

2. Horizontal slicing

A trial to work a seam 16-22m thick
dipping at 29° by horizontal slicing
in conjunction with stowing in
ascending order was taken up at a
colliery in the Jharia coalfield.
Two faces,one one each side of a
central gallery were opened out, the
length of the block being 200m and
the height of the block 100m. The
faces were laid out between roof and
floor and retreated towards the
central gallery. The working height
of the face was 3m and the face was
supported by timber props and bars.
As the face retreated to the central
gallery, the void was stowed solid
with sand emplaced hydraulically.
After the first slice was extracted,
the extraction of the second slice
was commenced and so on. With this
method 3 slices on the left hand
side of the central gallery and 6
slices on the right hand side of
the central gallery could be extracted. The workings experienced
roof and side falls; fractures
developed in the back and also in
the face wall. Convergence on the
face was erratic in different
slices (Figure 2) and also
spontaneous heating occurred in the
panel. The experiment was abandoned.

3 MINING OF WIDE OREBODIES

Wide orebodies of copper, lead and
zinc and uranium occur in certain
locations in India and for the
exploitation of all these wide
orebodies the stoping method
adopted is dependent on backfilling.
As for copper, in order to augment
the ore reserves, the working width
of the orebody in Singhbhum copper
belt has been increased so as to
include even leaner ores. This has
necessitated search for appropriate
stoping technology which could be

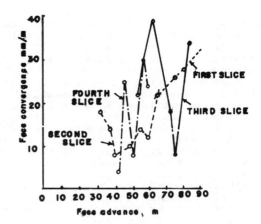

Figure 2. Convergence in different
slices of a thick seam
worked by horizontal
slicing.

productive and economic. Post
pillar stoping has been adopted
with this objective in view since
1973 and has been established as
a standard method for mining
orebodies with a true width of
more than 8 m and strike continuity
of not less than 60m. Table 1
gives the main design parameters
of a post pillar stope,(Khan 1976).

Table 1. Main design parameters of
post pillar stope.

Length of stope along strike	60-190 m
Width of stope	8-30 m
Size of post pillars	4-6 m along strike * 4-7 m across strike
Size of rib pillars	3-4 m along strike X full width of the orebody across the strike.

---

* Pillars more than 4 m x 4 m are
left only if necessary to eliminate
unpayable ground or to support
incompetent ground.

Development of the sill floor
starts with a pillar of 8-10m being
left over the haulage level. Back

stripping is carried out in two stages and maximum height before filling is 4.7m The first layer of classified tailings 2.2m is then placed. This leaves 2.5m of clearance between the ore back and the tailings floor. The back is then rock bolted on a 1.5 m x 1.5m pattern. Next another slice of 2.2 m is taken from the back and the void created is filled leaving a clearance of 2.5m above the fill. The process is repeated until the full height of the orbody has been extracted. Mill tailings are used as fill material. The bearing capacity of the fill varies between 3 and 5 kg/cm^2 depending upon the slime content of the fill and the time allowed for consolidation . Post pillar stoping has established its superiority over other forms of stoping e.g. breast stoping and room and pillar stoping (Table 2).

Table 2. Results obtained by different methods of stoping.

	Breast stoping	Room and pillar stoping	Post pillar stoping
Output per man shift, tonnes	1.5	3.5-4.0	3.5-5.5
Production per month, tonnes	800	1500-2000	3500-400
Cost/tonne of ore, Rupees	44	37	22

The ground control in the stope is good. Investigations are in progress to study the movements/deformations of pillars so as to optimise the design parameters. The measurements of deformation of post pillars taken over a period of about ten months have shown little deformation (less than 1mm): after about 2 months of the installation of instruments there was a tendency to slight relaxation which was followed by compression.

In another part of the country a lead and zinc deposit of maximum width of 65m occurs. Post pillar stoping with filling has been introduced in a part of the mine with 24m wide orebody. The size of the pill-

ars are 6m x 6m placed 20m centre to centre both in strike direction and across the strike. This gives a maximum span of 14m. The height of the slices is 3m. Mill tailings are used as filling material which are deslimed in a hydrocyclone. The underflow is used as fill material to which 10% cement is added In the second lift 5% cement will be added and thereafter no cement will be added. The water percolation rate of 7 cm per hour is obtained.

An uraniferous ore body 10-12m wide in Singhbhum belt is being worked by cut and fill method of stoping. Deslimed mill tailings with a water percolation rate of 10 cm per hour are used for filling.

4 "R & D" LABORATORY STUDIES

Laboratory studies using equivalent material models made with sand and wax mixture was done to study the applicability of horizontal slicing with filling to work thick and steep seams. The workings of IX/X seam 20-22m thick inclined at 45° were simulated on a scale of 1:100. The experiment showed that up to 3 slice could be worked satisfactorily. Thereafter cracks developed in the back at the hangingwall end and triangular wedges were formed which were hanging dangerously ready to fall.

Singh (1979) did extensive study of the different methods of slice mining of thick coal seams using equivalent material models. The models were constructed with a mixture of gypsum and sand in suitable proportions on a scale of 1:100. His studies showed that horizontal slicing of thick seams with filling in ascending order caused large scale strata movement after fourth slice. The face conditions deteriorated due to frequent falls near the hangingwall. Incline slicing also did not succeed after the third slice. During the working of the fourth slice back loosened and bed separation occurred up to a height of 23 cm (23 m in the field). Field trials confirmed these finding In another study faces laid at a gradient of 20°-30° opposite to the direction of the dip of the seam with a length of 20-25m could be worked up to 11 slices in ascending order with stowing without any strata control problems and surface subsidence was only 2.5 per cent of the seam height. The model studies showed that in the Jharia coalfield at Sudamdih and Chasnalla, cross-inclined slicing with filling in ascending order was a suitable method of mining thick and steep coal seams.

5 CHARACTERISTICS OF FILLING MATERIALS

In coal mining river sand is widely used as a filling material. Indian river sand comprises mostly quartz (over 80-90 per cent) and small percentages of garnet, ilmenite, tourmaline, feldspars and shales. The percentage of clay is negligible: it varies from nil to 1.5 per cent in different river beds. Table 3 gives some physical characteristics of Barakar river sand (Chatterji 1963):

Table 3. Physical characteristics of Barakar river sand

Microscopic Examination	Specific gravity	Hardness (Moh's scale)	Size analysis B.S. mesh	Percentage
Comprises mainly quartz with feldspar, garnet, tourmaline, etc.occurring in minor quantities	2.65	7	+10	6.48
			10-30	46.60
			30-60	42.80
			60-120	3.50
			120-200	0.30
			-200	-

The percolation rate of different fractions of this sand has been found to be from 0.45 to 6.29 metres per hour. The in-place compression of sand stowed layer is around 5-10 per cent. In recent years projects have been taken up to use crushed stone from quarry dumps as the fill material in coal mines. These crushed stones give a compaction of up to 30 per cent.

In the metalliferous mines e.g. copper, lead and zinc and uranium mines mill tailings are used as fill material. Screen analysis of mill tailings used as filling material in copper mines is given in Table 4 (Singh, 1982).

Table 4. Sieve analysis of fill material used in Copper mines.

Size	%
+ 45 mesh	4.9
+ 100 mesh	41.7
+ 200 mesh	31.6
+ 325 mesh	11.8
- 325 mesh	10.0

The percolation rate of the above fill material is of the order of 15 cm per hour. The fill material when placed in the stope consolidates sufficiently in 5 days time to permit the operation of machines over it.

The screen analysis of fill material used in an Uranium mine is given in tables 5 and 6 (Batra 1971). The mill tailings are deslimed in hydrocyclones in two stages to obtain the desired quality of the fill material:

Table 5. Screen analysis of deslimed tailings (Stage 1).

Mesh	Feed	Overflow	Underflow
+ 48	3.0		4.3
+ 65	6.5		8.9
+100	9.5		13.2
+150	12.0		16.6
+200	14.0	0.8	19.1
+270	13.0	1.6	17.5
+325	7.0	4.0	8.3
-325	35.0	93.6	12.1
	100.0	100.0	100.0

The percolation rate of the deslimed tailings used is more than 10cm per hour.

For the first time in India cemented fills have been introduced at a mine to work a thick lead and zinc deposit by cut and fill post pillar stoping method. The mill tailings are used as fill material with 10% cement in the first slice and with 5% in the second slice. Thereafter only mill tailings with-

out cement will be used. Percolation
rate of this fill is 7 cm per hour.
Full results have not yet been
established.

Table 6. Screen analysis of
deslimed tailings (stage 2)

Mesh	Feed	Overflow	Underflow
+ 48	4.3		4.7
+ 65	8.9		9.8
+100	13.2		14.6
+150	16.6		18.3
+200	19.1		21.0
+270	17.5		19.3
+325	8.3	3.3	8.8
-325	12.1	96.7	3.5
	100.0	100.0	100.0

6 CONCLUSIONS

Continued practice of mining with
filling for exploitation of thick
coal seams is foreseen in situations
where surface mining is not possible
or strata has to be maintained intact. River sand as in the past will
be used as filling material, though
there will be increasing use of
crushed stone obtained from surface
mining. For mining wide orebodies
extended use of filling with mill
tailings will be done. As for the
technology of mining, the experiences to date have shown that:

1. In coal mining, slice mining
with filling could be successful if
the seam can be liquidated in three
slices. For working thicker and
steep seams cross-inclined slicing
has been found be a suitable method
in model studies. Longwall method
of mining is the correct choice.
But where pillars have been already
formed on bord and pillar system,
extraction of pillars could be done
in lifts in seams of 8-9 m thickness.
In thicker seams or in contiguous
seams a combination of partial
extraction and full extraction with
full stowing of the goaf is the
suitable method.

2. In the mining of wide ore -
bodies, post pillar stoping with
fillings has established its efficacy
In some situations cut and fill
stoping may yield desired results.

REFERENCES

Guin, M.N. 1962 Mining Dishergarh
seam at depth. IMMA Review. 1 :4:
111-20.

Chaterji, B.N. 1963. Scientific
tests and studies on stowing
materials. CMRS research paper
No.9, Reference CMRS-M3/9.

Shevyakov, L.S. 1965. Mining of
mineral deposits. Moscow, Foreign
language publishing house, p.563.

Singh, R.D. 1970. Techniques of
mining thick coal seams in India,
In Mining and Petroleum technology
Jones M.J. ed. (London IMM 1970),
87-109, (Proc. 9th Common W.Min.
Metall. Congr; 1969, Vol.1).

Scurfield, R.W. 1970. Staffordshire
mining layout for mid 1970s.
The Mining Engineer. 130. 122:
73-83.

Batra, M.K. 1971. Private communications.

Report of (Ghose) committee on
splitting of pillars (1973).
Directorate general of mines
safety.

Khan, M.A. 1976. Development of
copper deposits-strategy for
future. unpublished lecture
delivered at the Indian School
of Mines, Dhanbad, 15 Nov. 1976.

Singh, T.N. 1980. A study of the
behaviour of ground due to mining
thick coal seams by equivalent
material models. Ph.D. thesis
submitted to Indian School of
Mines, Dhanbad.

Singh, B. 1982. Private communications.

Proceedings of the International Symposium on Mining with Backfill / Luleå / 7-9 June 1983

Mixed order slicing of 20m thick moderately dipping coal seam in conjuction with stowing

T.N.SINGH & B.SINGH
Central Mining Research Station, Dhanbad, India

R.D.SINGH
MBM Engineering College, Jodhpur, India

ABSTRACT: Inclined slicing in ascending order with hydraulic stowing caused difficult strata control problems after 3-4 slice workings due to highly pulverised nature of coal seam. Taking 3 slice working feasible in weak seams, a thick seams could be divided into equivalent sections and extracted separately adopting ascending order slicing while the sections were worked in descending order. The system of mixed order slicing was found to be problematic and wasteful due to very low strength of IX/X seam of Jharia coalfield. The coal along the hanging wall parted after two slice working itself while the lower section became unsafe even under 4m thick coal band.

INTRODUCTION

Extraction of coal seams underneath water bodies and surface features required hydraulic stowing of goaf. Recovery from 20m thick geologically disturbed coal seams of Jharia coalfield was poor by horizontal and inclined slicing even with effective stowing in the moderately dipping coal seam. The strata condition after III slice working by inclined slicing became unsafe and hence the remaining 9-10 section could not be worked. Inclined slicing in mixed order was thereafter thought as an alternative method of mining under the same geo-mining condition. A thick seam under this programme was divided into two sections of 9-10m and each one of them were worked by inclined slicing in ascending order in conjunction with stowing while the sections were worked in descending order. The experiences with the method - mixed order slicing in equivalent material model under the geo-mining conditions of Sudamdih block is covered here in.

LOCALE OF THE EXPERIMENTAL SITE

The Sudamdih area lies on the north bank of river Damodar and coincides with the great boundary fault and the Patherdih thrust. The coal bearing formation in the block is Barakar measure, exposed in the northern part of the property in which almost all the coal seams outcrop. The typical cross section of the seams in the area is shown in Figure 1. The area is infested with a series of faults varying in throw, inclination and structure. These faults frequently change the inclination of the seams in different sectors from 28° to 65°. These faults seam to be sympathetic to the great boundary fault. The mica peridotite dykes and sills have intensively damaged and distorted the seams. The thickest(12-27m) and the most valuable IX/X seam of the area is burnt to jhama in Chasnalla side and partly burnt on the outcrop side. The XV seam (7.5-16.5m) is also burnt throughout the outcrop zone and also in the deeper portion towards south. The IX/X seam contains over 30 million tonnes mineable coking coal reserve of the area. A major portion of this reserve lies underneath river Damodar which is the main perennial source of water. A part of the leasehold is under the railwayline. Inspite of huge coking coal reserve in the area, extraction was abandoned due to these surface constraints and the geological disturbances. The mining activities were resumed later in 60's but the success is still remote in the absence of suitable method of mining.

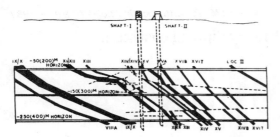

Fig.1. Seam section of Sudamdih area

COAL MEASURE FORMATION

Exploratory drilling at a rate of 5
boreholes per sq.km. was done to prove
and estimate the reserve. A representative section of the strata is shown
in Figure 2, according to which, the
formation contains 78% shaly sandstone.
The physico-mechanical property of the
formation is also shown in the same
figure. The immediate roof of IX/X
seam appeared to be well laminated and
incompetent. The coal seam was highly
disturbed and crushed along the hanging
wall over 10m while the lower section
was well cleated and friable. Slight
burning of the top section in patches
was noticed. The disturbance of the
seam increased in the deeper side where faulting was very prominent. The
floor of the seam was strong and
stable shaly sandstone.

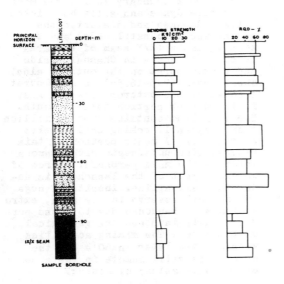

Fig.2. Strata section near the panel

METHOD OF EXTRACTION

Thick moderately dipping IX/X seam of
Sudamdih mine with surface features
like river Damodar, railwayline and
colliery offices was worked by horizontal and inclined slicing in conjun
-ction with hydraulic stowing. These
trials were unsuccessful due to
active roof movement during V slice
working with horizontal slicing and
III slice by inclined slicing.

Mixed order slicing was concieved as
an alternative variant according to
which, 2 separate sections of 8-10m
could be worked in 3 slices in ascending order. The separate sections
were to be worked in descending order
one after the other. This proposition
was simulated in an equivalent material model before undertaking the
field trial.

MIXED ORDER SLICING

Mixed order slicing sequence envisaged
mining of 19-20m thick seam in two
separate sections (Figure 3), 9m thick
top section in 3 slices each of 2.5m
to 3m thickness inconjunction with
stowing. The bottom section was to be
worked thereafter in ascending order
to the extent possible under the
stowed goaf.

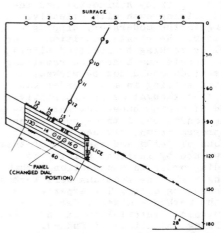

Fig.3. Mixed order slicing sequence
and instrumentation.

As 8-9m thick section has been worked
without active strata movement, it was
expected to extract the top section
without roof control problems. Replacement of the weak fragile coal section by sand pack was likely to avert,

258

a) the heating danger due to cavity formation along the weak contact pla -ne, and
b) the lamination of the lower coal beds and reduce the coal band deterioration due to its cushioning effect.

The lower 10m thick section could subsequently be worked in ascending order under sand pack. It was expected that under stowed top section, 2-3 slices from the bottom section could also be worked.

EQUIVALENT MATERIAL MODEL STUDY

This proposition was tried in an equivalent material model constructed on a geometrical scale of 1:100, for which, the geometrical dimensions and the strength parameters were reduced by dimensional analysis. The mass strength of the formation was deduced from the sample strength by the use of the weakening coefficient based on the rock quality, joint characteristic and the geological discontinuities. The factor of 0.3 appeared to be adequate in view of its faulting, intrusion and the rock quality designation. For the coal seam - top section, it was taken as 0.1 while for other sections, it was 0.2.

Gypsum based artificial materials were used to simulate the formation with mica as the bedding plane contact. The geo-mining boundary conditions simulated in the model is resolved as follows:

Seam thickness   - 18 to 20m
Average depth of the panel - 120m
Thickness of the slices - 2.5 - 3m
Inclination of the seam - 30⁰

MODEL INSTRUMENTATION

The experiment was aimed at to see the strata movement failure pattern and subsidence aspect while adopting the mixed order slicing. The dial gauges were mouted around the workings according to the programme shown in Figure-3. After the extraction of the top section, the gauges 13-16 were replaced in the coal band immediately under the stowed goaf. The other gauges 9-12 were set vertically above the panel.

Visual observation of the cracks, fissures and strata dislocation was regularly made. Photographs were taken

to depict the strata condition at different stages.

MODEL WORKINGS

The first slice of 3m thickness was worked immediately over the bottom section. The strike face was advanced from dip to rise at a rate of 2m/cycle and stowed at 6m interval. The second and the third slices of 2.5m each were to be worked in ascending order one after the other, leaving 0.5m thick coal band in the floor. Stowing at closer interval of 3-4m was done later to offer early resistance to the fragile coal roof, which tended to be unstable during the upper slice workings.

OBSERVATIONS - TOP SECTION WORKING

No bed separation or dislocation of the roof rock mass was observed upto 50m advance of the first slice. Face advance to 54m resulted fine parting along the hanging wall contact plane, which progressively extended upward to 27m height with the completion of the panel (Figure 4).

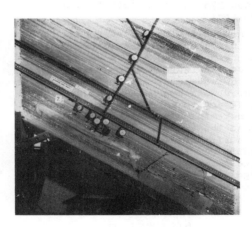

Fig.4. Bed separation due to first slice working.

The second slice working further de-established the coal roof and the superincumbent rock mass. The bed separation travelled to 34m height with 14m advance of the second slice face. The bed separation changed to a lenticular cavity (Figure 5), after 40m advance when the workings got connected to the cavity through cracks. These cracks were making 65 to 70⁰

259

angle to the plane of stratification. With 52m advance, the coal was crumbled to fragments, making it difficult to work further.

Fig.5. Lenticular cavity due to second slicing working.

Strata Movement

The movement of the surrounding rock mass became clear even with 45m advance of the first slice face. All the gauges up to 12, i.e., 12-18 showed appreciable movement after 60m advance. This was the stage when bed separation 27m above the workings was observed.

The gauges 13-18 mounted around the workings showed slight movement up to 45m advance, the maximum value being 11.5cm over gauge 18. Subsequently, with 60m advance, all the gauges showed abrupt increase in reading, indicating active movement. The trend of the movement of point 11 lying just beyond the cavity, 9 lying close to the surface is shown in Figure 6. The location of the cavity in between the gauge 11 and 12 is obvious from the gauge reading which increased from 8.2cm in gauge 11 to 37.5cm in gauge 12.

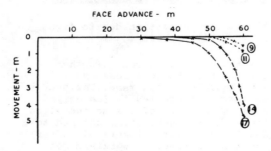

Fig.6. Movement due to first slice working.

The second slice working induced a new wave of movement from the very beginning over all the gauges. The trend of movement of some of the points is shown in Figure 7, which shows that the gauge 11 remained stable, showing very little movement while the lower points 14 and 17 moved much faster.

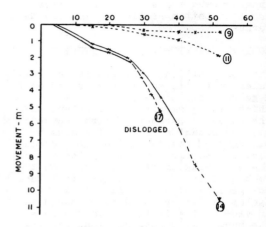

Fig.7. Movement after second slice working.

Gentle surface subsidence was indicated with the first slice face advance to 60m, representing smooth settlement of the superincumbent rock mass. The subsidence factor with 60m advance was just 0.7% of the working height.

The second slice working induced surface subsidence much earlier so much so that after 15m advance, it reached up to 1% of the working height. The overall subsidence due to this working remained within 1.3% of the working height. The width/depth ratio for this subcriticed panel was approximately 0.6, resulting in low surface subsidence.

Results

An effort to mine 18-20m thick seam with top 8-10m fragile coal band under the condition of Sudamdih mine was made by mixed order slicing with stowing. The seam for this proposition was divided into two sections - top and bottom and each slices of the indi-vidual section were worked in ascending order while the sections were worked in descending order. The fin-

260

dings of the top section working are as follows:

a) A slice 3m thick located in the middle of the seam could be worked without any strata control problem, damaging deformation of the surronding rock mass or surface subsidence.
b) The panel width of 60m introduced bed separation and loosening up to the height of 27m above the seam.
c) The second slice working aggravated the bed separation and created roof control problem from the very beginning when the parted roof exerted dead load over the coal roof. The coal band was traversed by a number of cracks and fissures even in advance of the face.
d) The bed separation extended upward to 34m horizon where a lenticular cavity formed. The rock mass in between the cavity and the workings was traversed by fractures and cracks.
e) The fracture planes were making 65° and 70° angle to the bedding planes against the advancing and the stagnant ribs respectively.
f) The third slice coal was highly disturbed due to sagging and inherent weakness. The support of the flowing coal mass was practically difficult and hence the slice could not be worked.

BOTTOM SECTION WORKINGS

The bottom section slicing was started after the completion of the top section workings. The fourth slice of 3m thick -ness laid along the floor was advanced from dip to rise. The V slice of 2.5m thickness was started over the stowed goaf after 20m advance of the fourth slices face and both were worked simultaneously with the staggered faces. Active strata movement was observed over the fifth slice face where a crack traversed the face and interrupted the workings(after 32m advance of the fourth slice face). The faces in the fourth and the fifth slices were restored and further advanced to 34m and 14m respectively when heavy strata movement was again observed. The lenticular cavity formed during the top section workings widened to over 4m (Figure 8), along with the widening of the rise side fracture plane. It appeared that the total mass within the arch tended to flow

down towards the dip.

Fig.8. Disturbed strata condition due to mixed order slicing.

The fifth slice face was stopped and only the fourth slice face was advanced with regular and tight stowing after every 2 cycles of advance. The face was advanced up to 58m, during which, cracks in the coal roof across the total 7m band were frequent. Subsequently, the extraction of the fifth slice even under 4m thick coal band became very difficult as the roof was heavily fissured through which sand started flowing.

OBSERVATIONS - BOTTOM SECTION WORKING

Strata Movement

The fourth slice face advance to 24m caused no deformation of the coal roof or of the broken rock mass. The gauge 12 showed some movement probably because of the compression and consolidation of the debris. Major movement was noticed after 30m advance, when the fifth slice face advanced to 10m. At this juncture, only 4m thick coal band was left over in between the stowed goaf and the workings.

Heavy movement continued throughout the fourth slice working even when it was worked alone after 34m advance. The roof coal was badly crushed by this time. The extraction induced erratic/heavy movement during which the gauges 13-16, set in the roof collapsed.

The gauges 12-16 showed slow progressive deformation during the fourth slice working. The upper strata beyond 9 remained uneffected till 36m advance followed by slow movement up to the surface (Figure 9).

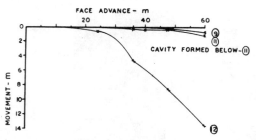

Fig.9. Anchor movement due to IV
slice working.

The extraction of the fifth slice
induced movement from the very beginning. The movement beyond gauge 12
was, however, low. It showed the
existence of stable lenticular cavity
below gauge 11(Figure 10).

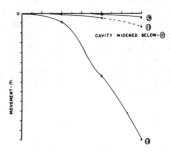

Fig.10. Anchor movement due to V
slice working.

The workings of fourth and fifth
slices were resumed after complete
settlement of the strata and no subsidence was recorded up to 36m advance of the fourth slice face. Subsequently, the surface subsided slowly
to 2.8% and 7% of the fourth and
fifth slice working height respective
-ly.

Results

The IV slice was worked in conjunction
with stowing, leaving 7m thick coal
band as a roof under the stowed goaf.
The thickness of the coal band decrea
-sed to 4m over the V slice which was
worked simultaneously with staggared
face. The salient features during this
and the upper slice workings were as
follows:

a) A 3m thick slice under 7m thick
weak coal band was safe to be work

in conjunction with stowing, without any strata control problem.
b) When the coal band thickness was
reduced to 4m during the second
slice working, the roof coal sagged
and developed cracks.
c) Intensive fissures developed all
along the roof coal during the
second slice working; making it
difficult to work. Sand frequently
flowed down through these cracks
to the face.
d) The lenticular cavity became
very wide and got connected to the
working level through cracks, the
rise side crack being comparatively
more prominent and active.

During this process, the dip side
cracks tended to close down due to
en masse sliding of the total broken
mass to the dip side. The front side
crack opened even when the face was
30-40m away; indicating that by this
time the whole mass within the active
deformation zone developed the flowing
tendency.

DISCUSSION OF RESULTS

It was planned to work 19m thick seam
in mixed order slicing when the seam
was divided into two sections. The top
9m section was worked in ascending
order in conjunction with stowing,
followed by the bottom section working
also in ascending order in slices. In
this way it was expected to work up to
5 slices. The results however proved
that the mixed order slicing was comparatively more wasteful, troublesome
and caused higher degree of surface
subsidence under the geo-mining condition of Sudamdih. The tensile strength
of the seam was practically zero, and
of very low stability index (50).

The mixed order slicing became difficult as the third slice working caused
cracks ahead of the face, large scale
movement of the hanging-wall and diffi
-cult face condition. This appeared to
be because of the weak pulverised coal
parting which was located along the
hangingwall. The working in the top
section induced heavy roof convergence
even before the stowing. This reduced
the efficacy of the sand pack as a
roof support and induced bed separation along the coal - shale contact
plane and resulted in cavity 27m above
the hanging wall.

The second section working under the fragile and well cleated coal roof also of low stability index(approximately 75) resulted in frequent roof fall and heavy strata movement during fourth and fifth slice workings. Roof fall and fractures in advance of the face became so frequent that the workings had to be abandoned after the second slice working.

CONCLUSION

The mixed order slicing under the geo-mining condition of Sudamdih failed to deliver the desired recovery because of poor bond along the hanging wall and well cleated and pulverised nature of coal seam. The pertinent factors responsible for its failure are summarised as follows:

Cleat frequency	- 20/m
R.Q.D. of the Seam	- 30%
Top sector RQD	- 10%
Bending strength	- 10 kg/sq.cm
Top 3m section	- 5 kg/sq.cm
Stability Index	- 50
Safe span of coal band of 3m thickness	- 3m

Under this condition, the coal roof tended to part and converge before the stowed sand could offer any effective resistance. This resulted in opening of the cleats formation of cavity with connecting cracks and other strata control problems during the top section workings.

During the lower section workings, the coal band of IV and V slices under 6m thick sand bed converged heavily, resulting in open cracks and reduced safe span. As a result, different strata control problems and flowing in of sand during V slice workings were observed. The sand pack normally offered effective resistance to the roof with maximum open span of 10m. The value for IX/X seam was invariably below this limit.

The mixed order slicing might be feasible if the stability index of the coal seam/band was 300 or more. This is possible with poorly cleated(2-5 cleats/m), and strong (bending stren -gth - 20 kg/sq.cm) coal seam of RQD more than 50%. Under this condition, the limiting span would be of the order of 10m even for 3-4m thick coal band, enabling slicing with low con-vergence and less strata control problem.

263

Development of stope backfilling at Bor copper mine

RADIVOJE STANKOVIĆ
Technical Faculty of Bor, Yugoslavia

ANTE GLUŠČEVIĆ & ZORAN PETKOVIĆ
Faculty of Mining & Geology, Beograd, Yugoslavia

ABSTRACT: Copper ore deposit in Bor is being permanently exploited since 19o3. In the stoping technology the backfilling methods hold a dominant place. Until 1961 stopes were dry filled with waste rock, the material being obtained as overburden during the open pit mining. Owing to the aggravated mining-technological conditions, i.e. owing to a lower parcentage of metal in ore, it became necessary to replace this expensive mode of filling. Taking advantage of the experience from all over the world, as well as the results of the proper researches, a system, the filling with cyclonized flotation tailing was projected and run-in. During recent years the mine was faced with a new problem, i.e. it was necessary to mine subsequently found ore bodies containing a high percentage of metal, located in the zone of the safety pillar of the main shaft and some other important surface objects. To solve the problem a system of filling the stopes with flotation tailing and building of concrete safety pillars has been projected.

1 INTRODUCTION

Exploitation of the copper ore deposit in Bor is being done on a permanent basis since 19o3. Since 1928 the deposit is being mined simultaneously by open pit and underground method.

The capacity of the underground ore production ranged from several hundred thousand tons to the maximum of a million tons a year.

Descending with exploitation work into greater depths, at the same time increasing costs and decreasing the content of metal in the ore, mining-technical and economical conditions of exploitation as a whole were aggravated.

For the mining of the ore deposit, depending on geological and other conditions, in the various parts of the deposit (ore bodies), several different mining methods such as cut-and-fill, sublevel stoping, shrinkage, square-set mining and others were applied, however the dominant part belonged to mining methods with filling.

In order to maintain the economy of mining, because of aggravated mining conditions, the need for constant improvement of the mining methods, especially methods with filling was imposed.

A visible progress in the improvement of mining methods was realized in the filling technology of the mined spaces by replacing waste rock filling with cyclonized flotation tailing.

The economy of the mining was significantly improved, also the intensity of the mining was increased by applying the new filling method. This fact, besides the geological conditions, was decisive for the elimination of other mining methods, and helped to decide only on mining methods with the filling, which are exclusively applied at present.

Starting from the belief that for scientific and expert public the results that were achieved by way of applying mining methods with stope filling in this particular ore deposit will be interesting, we will proceed in presenting the most significant mining methods and the achieved results.

2 CUT-AND-FILL - BOR MINING METHOD

Over a long period of years, this method has been applied for the mining of large ore bodies featuring area of several thousand square metres, steep inclination and hard ore.

The height between the two levels is 6o m, stopes are arranged transversally and 12-16 m wide ; safety pillars between the stopes are 7-1o m wide.

The preparation for the mining includes making of main haulage drift in the hanging and in the footwall of the ore body, cross-cuts through every second safety pillar, short drifts from cross-cuts, ore passes and manways, filling drifts in the upper horizontal pillar and drillings for the transportation of the filling.

The mining is started from the main level, from short drifts. The first slice is mined from the height of 5.5-6 m. Having finished the mining of the first slice cribbed timber ore passes and manways are made and the stope is filled up to the height of 2.5-3 m. The mining height of the subsequent slices is about 2.5 m and this mining cycle alternating with filling is repeated up to the height of 8 m under the upper main level. The remaining height of 8 m is left to feature as the horizontal pillar.

The mining technology consists of the making of horizontal drillings, and the blasted ore is transported by LHD units to the ore passes.

The stope supporting is carried out by chokes or roof bolting. As a rule, chokes are removed before filling except in cases where there is danger of roof collapsing.

As was mentioned earlier, the filling is alternating with mining. The hydraulic flotation tailing is brought from the station on the surface to the distributing level in the mine, by drillings of Ø 1oo-15o mm. Distribution all over the mine is carried out by steel pipes of Ø 5o-14o mm, and the filling of the stope itself from the filling drift is carried out trough drillings of Ø 75 mm.

The filling of a stope is preceeded by preparation consisting of timber cribbing of the ore passes and manways up to the filling height and wrapping them with burlap, as well as making bases for flexible filling pipe lines when dealing with stopes of greater lengths. Before the filling is begun, the checking of passability of the filling pipelines is carried out by letting the clean water through for about 15 to 2o minutes.

Drainage of the stopes is carried out through corresponding raises wrapped in burlap. On the levels, the water is drained by canals leading to settling ponds, and then to the pump station sump.

Percolation speed of the filling material in the conditions of the Bor mine, is about 3o-5o N/cm^2, and that makes possible further mining, 24 hours after

the filling has been carried out.

The Bor experience has shown that in order to secure load-bearing capacity, the filling must not contain clay particles, but it is indispensable that it contains 2o% sand particles smaller than 2oo meshes.

By replacing the waste rock filling with the hydraulic filling, significant improvements in technical-economical results have been achieved, and here the most significant ones are shown:

Table 1.

I n d e x	Waste rock filling	Hydraulic filling
Mining intensity (t/m^2/year)	16-21	33-43
Filling output (m^3/manshift)	3,48	37,8
Mining output (t/manshift)	4-5	12-15
C o s t s (%)	1oo	32,2

The layout of this method is shown in Figure 1

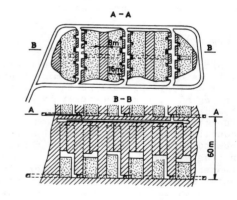

Fig.1. Bor mining method

3 CROSS-CUT WITH FILL

This method, until today, has been applied on two locations, and that: for the mining of horizontal safety pillars of the ore body "Čoka Dulkan" and for mining of the ore body "G". On both locations physical-mechanical properties of the ore, which contained up to 15% of copper where such, that the earlier described Bor method could not be applied.

Dimensions of horizontal safety pillars mined by this method were:

length loo m, width 16 m, height 8 m. Preparation for the mining consisted of making a tramming drift in the footwall on the level of the lower limit of the pillar and connecting this drift via ore passes and manways to the low mine level.

The first cross-cut is made beside the vertical safety pillar and its dimensions are 2.5x3 m. When this cross-cut is made in the whole length of approx. loo m, that being the stope length, the hydraulic filling of the cross-cut is carried out. Preliminary, at the beginning of the drift, a bulkhead is placed and wrapped with burlap. Filling pipelines are inserted into the drift, and the filling is carried out up to the roof. After filling and draining of the stope, a new cross-cut (stope) is made beside the filled one.

This tehnology is repeated until the complete stoping of the first slice has been caried out, and then the whole cycle of the preparation and stope making is repeated on the next slice. The complete horizontal pillar 8 m thick is mined in the described manner, in three slices.

In a similar manner the ore body "G" was mined, except that the mining height was 6o m.

By this method the average outputs of 15 t/manshift were ashieved, with minimal ore losses of 5% and delution of 2-3%.

The layout of the described manner of the mining is shown in Figure 2.

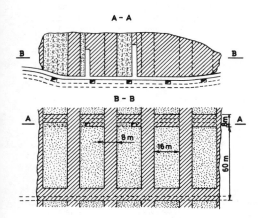

A - A

B - B

8 m

16 m

60 m

Fig.2. Cross-cut method.

4 "POST" PILLAR METHOD

This is a new method for the conditions of the Bor mine, just begun to be applied in the ore body "Brezanik" (a similar method is applied in the mine Falconbridge). "Brezanik" is a low procentage ore body and taking this fact into consideration it would have been logical to apply sublevel caving or block caving, however as the dislocation of the objects on the surface would be costly it was indispensable to chose again one of the methods with filling.

The mining of this ore body by the existing Bor method was not to be taken into consideration, as to achieve economy, a high capacity rate of production was to be ashieved, and that asked for the application of the biggest mechanization in all phases of the technological process (not to be achieved by the existing method). However the chosen method fulfills the demanded conditions and makes possible a considerably greater coefficient of ore recovery ratio in the primary phase of exploitation.

As with previous methods, the ore body is divided into levels 6o m high, however with difference, every second level is used for haulage.

Vertically, the ore body is divided into blocks 46 m wide. The blocks are separated by rib pillars 4 m wide. Inside the block, square pillars 8x8 m are formed by mining, and mining fileds between the row of pillars are lo m wide (Figure 3).

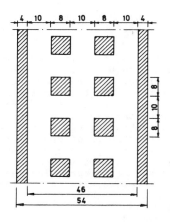

4 10 8 10 8 10 4

8

10

8

46

54

Fig.3. Arrangement of "post" pillars in the mining field.

Preparation for the mining consists of: making a ramp by which the main levels are connected. From this ramp, every 9 m of the height in the ore body footwall, tramming drifts are made and through them the ore is transported to ore passes

267

that are connected with the haulage level.

From tramming drifts, two cross-cuts are made to each ore block, leading to the ore body from which the mining is begun. These cross-cuts are used for the mining of three slices, as shown in Figure 4.

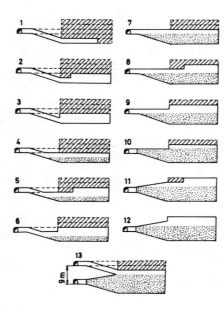

Fig.4. Mining phases from one of the cross-cut levels.

By the project, the blasting down of the ore was to be carried out by vertical drillings, and that demanded leaving empty space of 3 m between the filling and the roof. The acquired experience when mining the first two slices has shown that when having an empty space of 3 m bigger lumps of ore get dislodged from the roof as a consequence of clay layers, therefore as a definite solution horizontal drilling and filling of each mined slice up to the roof, was taken up. By changing the manner of drilling, the mining capacity is somewhat decreased, but it can be compenstated by a greater number of drilling points.

The filling of stopes is done by pipelines placed through the already prepared ramp, tramming drifts and crosscuts. Water is drained through drainage water passes which follow the progress of the stope in height.

5 SUPPLYING THE MINE WITH FILLING

Supplying the mine with filling (hydro) that is flotation tailing, is being done from two sources: existing mine dressing plant and a newly built dressing plant of the mine "Veliki Krivelj".

The mine dressing plant secures filling for all ore bodies, except ore body "Brzanik", which gets the filling from the open pit dressing plant "Veliki Krivelj" located near Bor.

5.1 Supplying the mine whit filling from the mine dressing plant

The dressing plant is located at the distance of 17oo m from the mine cyclone station. By recovery of the flotation tailing up to 43%, existing, as well as future needs of the mine for hydraulic fill tailing are completely satisfied.

The filling, supplied from pyrite flotation cycles, is used for the filling of stopes in the mine.

The tailing supplied by the flotation cycle is brought to the pump station thickener, located in the dressing plant itself.

From the dressing plant to the cyclone station the tailing is hydraulically transported by two pumps which are connected to work in series.

The basic characteristics of the transporting system dressing plant - cyclone station are the following:

- pumping capacity   82.0 l/sec
- pulp density   15oo.o kg/m³
- water - solid ratio   1 : 1
- solids in pulp   5o.o %
- pipeline diameter   254.o mm
- pipeline pulp speed   1.6 m/sec
- geod. pumping hight   28.1 m
- calculated hydraulic resistance   46.9 m
- overall manometric pumping height   75.o m

The plant for the preparation of the filling (hydrocyclonic station) is built near the mine, that is, on the surface above the ore deposit.

The pulp for the ore dressing plant, which is transported from the dressing plant to the hydrocyclonic station by pipes whose diameter is 254 mm, enters the tank for the primary pulp, from the tank it goes to the plunger pump, from the pump to hydrocyclones, and from there two products get out: the filling for the mine (sand) and the overflow which through the return pulp tank and

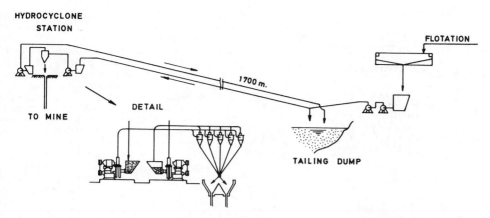

Fig.5. The layout of the preparation of the filling in the existing dressing plant.

return pump gets back to the tailing dump.

The filling for the stopes goes, from the cyclones to the collecting basin of the sand pulp, which in turn is connected to the drillings used for the transportation of the filling to the mine.

The layout of the equipment and the preparation technology of the filling are shown in Figure 5.

The grain size distribution of the primary pulp shows variations depending on the content of aluminate, and here are the average values :

Meshes		Min.(%)	Max.(%)
	+ 48	o.5	2.5
- 48	+ 65	5.o	1o.o
- 65	+ 1oo	8.o	13.o
- 1oo	+ 15o	6.o	12.o
- 15o	+ 2oo	5.o	13.o
	- 2oo	6o.o	75.o

The high content of the class minus 2oo meshes is a consequence of the demanded grade of milling because of the flotation of the copper and the content of clay in the ore.

The grain size distribution of the filling after cyclonizing in average amounts to :

Meshes		Participation (%)
	+ 48	8.35
- 48	+ 65	17.o5
- 65	+ 1oo	23.63
- 1oo	+ 15o	12.62
- 15o	+ 2oo	13.oo
	- 2oo	25.35

The shown grain size distribution of the filling material makes possible percolation speed 2o-6o mm/h, an average speed of 4o mm/h, and the load bearing capacity of the drained filling 3o-5o N/cm^2, which makes possible use of the LHD mechanization on it.

Accomplished percolation speed makes possible more intensive filling, as well as mining, sufficient for securing the production capacity.

The prepared filling material is mixed with water in the hydrocyclonic station, to achieve the ratio solid : water of 66 : 34. Transportation of so prepared pulp, from the hydrocyclonic station to the exploitation levels, is done by vertical drillings whose diameter is 15o mm. In order to secure safe transport, a spare drilling is always made.

Distribution stations are installed on the main levels and they serve for the directing of the filling to horizontal pipelines.

On the levels, the filling is distributed through steel pipes 14o mm in diameter, and from levels to stopes by drillings or vertically placed steel pipes through filling raises.

5.2 Supplying the mine with the filling from ore dressing plant "Veliki Krivelj"

Location of the open pit dressing plant belonging to the mine "Veliki Krivelj" is very favourable for the supply of stopes of the ore body "Brezanik" with filling.

The hydrocyclone station for the preparation of the filling was built in the

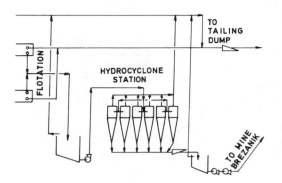

Fig.6 The layout of the preparation of the filling in the dressing plant Veliki Krivelj.

dressing plant itself. For the preparation of the hydrofilling, the overflow of one section of the pyrite dressing plant is used.

The preparation technology of the filling consists of the following:

The overflow, by way of gravitation, enters the tank of the mud pump, the pump then supplies the battery of six hydrocyclones with the pulp. In the hydrocyclones, the classification between the sand and the overflow is carried out. The overflow is collected and by gravitation led to the canal for the tailing, and thereby led to the tailing dump of the mine "Veliki Krivelj", while the sand, whose density is about 60% solid and with about 25% size class - o.o74 + + O mm is collected and through the sampler taken to the tank of the pumps connected in series.

Sand from the hydrocyclone, whit previously added water, froms pulp in ratio 56 : 44 = solid : liquid, it is pumped to the elevation point 476.5 m through a plastic pipeline ∅ 15o mm and further, by indentical pipelines, by gravitation, it is taken to fill drillings.

The hydrocyclone station equipment consists of : primary pulp tank, sand (filling) tank, mud pumps, two pumps, six hydrocyclones ∅ 5oo mm and sampler.

The layout of equipment and preparation technology of the filling are shown in the Figure 6.

The stope filling of the ore body "Brezanik", from this hydrocyclone station, should begin in the second half of this year, therefore, for the present there is no practical experience how whole system works.

In connection with supplying the mine with hydro-filling, the following shoul be mentioned :

1. Supplying the mine with filling fr the existing dressing plant, will be ca red out until 199o, when the work on th open pit of Bor will cease.

2. Considering a very long exploitati life of the open pit "Veliki Krivelj", after 199o, the overall needs for the filling will be secured from the dressing plant of this mine.

3. Cyclonized tailing from open pit "Veliki Krivelj" is of a considerably better quality than the tailing obtaine from the existing dressing plant, owing to more favourable participation of lar ger fractions in the filling. Better qu lity of this filling will secure greate draining speed in the stopes. That will increase the intensity of mining.

6 MINING OF THE ORE BODY "NOVO OKNO"

The ore body "Novo okno" has been disco vered during recent exploration work. As has already been mentioned, it lies in the zone of safety pillar where the haulage main shaft, dressing plant, fac tory of liquid oxygen and tailing dump are located.

Because of the mentioned objects whos displacing was not to be considered, for the mining of this ore body a method ha to be chosen, which would completely make impossible deformations of the sur ounding rocks , and via them of the sur face.

Besides the above mentioned, the chos method had to have maximal recovery rat as the ore contained a high percent of copper, gold, silver and other rare metals.

Taking into consideration all relevan facts, and after detailed exploration work, a modified Bor method was chosen for the mining of this ore body.

The esential features of this method will be :

1. The mining will be done transversally that is, across the strike of the ore body.

2. The ore body is divided into stopes lo m wide, and into pillars 4 m wide. Th height of pillars and stopes suits the thikness of the ore body and ranges from 4 to 2o m. The length of stopes and pillars is equal to the width of the ore body and that is 6o-12o m. Arrangement of stopes and pillars in the ore body is shown in Figure 7.

3. The safety pillars will be mined,in the primary phase, and then the empty space obtained by the mining will be filled by concrete, hardness 3ooo N/cm^2 which will form artificial safety pillars.

4. After hardening of concrete pillars, the process that lasts 18o days, we proceed to the mining of the rooms and their filling with flotation tailing.

At present, in the ore body, preparation for the mining of the pillars is carried out, and on the surface a concrete plant and hydrocyclone station are being finished.

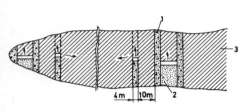

1 - CONCRETE PILLAR
2 - HYDRAULIC FILLING
3 - ORE

Fig.7. Vertical cross section of the ore body with arrangement of stopes and pillars.

6.1 Technology used for the making of concrete pillars

The technology itself,used for the making of cast concrete on the surface and making of concrete pillars in the ore body, is divided in three phases (Figure 8).

1. Preparation of the concrete mixture on the surface.

2. Transporting the concrete to the mine and transporting it to the place of building-in.

3. Building-in of the concrete into the mined space.

The preparation of the concrete mixture is carried out in the concrete plant located in the immediate vicinity of the drilling which is used for transportation of the concrete into the mine. Beside the plant, there is sand and gravel storage place, the aggregate bin and cement silo.

The weight ratio of the concrete components is water : cement : aggregate = 1 : 2 : 9.

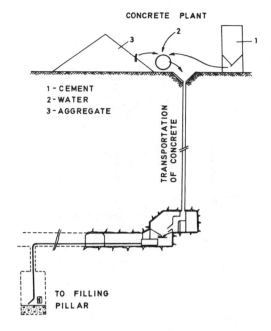

CONCRETE PLANT

1 - CEMENT
2 - WATER
3 - AGGREGATE

TRANSPORTATION OF CONCRETE

TO FILLING PILLAR

Fig.8. Making, transporting and building in of concrete.

Maximal capacity of the concrete plant is 18 m^3/h. The overall quantity of the concrete to be built into pillars is ~ 1oo,ooo m^3.

In order to decrease the price of concrete, there is a possibility to use,instead of the aggregate, crushed dross obtained from the metalurgical procesing the copper concentrate in the air furnaces.

Transportation of the ready concrete into the mine and transportation to the place of building-in is done in the following way:

The concrete mixture is dosed to the drilling funnel by which it is transported to elevation point + 73.2 m where a shock absorber is placed (Figure 9).From the shock absorber the concrete mixture is drawn into the intermediate bunker, from which it is poured into the concrete pump funnel, from there the concrete mixture is transported, trough pipelines, to the place where the building-in is carred out.

Building the concrete into the mined space of the safety pillar is carried out after the mining of the definite height of the safety pillar, if the pillars are of lower height, or after the

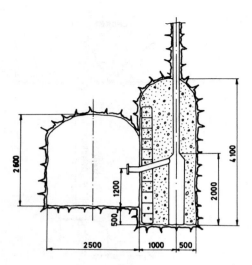

Fig.9. The shock absorber for concrete

Cleland,S.R. & Singh,K.H. Development
    "post" pillar mining of Falckonbridg
    nickel mines Ltd, CIM Bulletin vol.
Society of Mining ENgineers. Mining En
    neering Handbook. Volume 1.
    p. 12-233 - 12-253, New York
Technical documentation of copper mine
    Bor.

mining of one slice high 3 m, if the
pillars are higher.
   The concrite filling is carried out by
moving the pipeline over the filling
area. The built-in concrete is vibrated
in order to achieve the necessary hardness.
   The expected concrete filling capacity,
when making the pillars, will be about
11 m^3/h.

7 CONCLUSIONS

From the presented paper it can be concluded that the mine of Bor, during 8o
years of its existance, owing to specific conditions, had to apply the methods
with the filling. In order to increase
the production and to keep up the economy, new technologies which offered
satisfying effects had to be constantly
applied. Confirmation of this statement
are two new methods, which after deta-
-iled research work are now intoduced
(Brezanik, Novo okno) and from which,
through practical work the confirmation
of the projected parameters is expected.

REFERENCES

Gluščević,A.B. 1974. Opening and under-
   graund mining methods for ore deposits,
   p. 158-2o6. Beograd, Minerva

Development of stoping methods with the application of back fills at Kolar Gold Field

R.R.TATIYA
University of Jodhpur, India

ABSTRACT: Mining of Gold in India at Kolar Gold Field has completed a century. After six decades of active life of Kolar Gold Field the workings underground were very much extended and these mines were considered amongst the deepest mines of the world. The deposits here have also been known to possess very high inherent stresses. These factors resulted in the problems of rock burst and ground control.

Initially open stoping and timber supported stoping methods were in vogue. With the increase in frequency and severity of rock burst occurence the use of granite masonry pack walls as a backfill in the stopes once brought a revolution in K G F mines but it was only an interim relief against the occurence of rock burst. Through field studies and laboratory investigations the closure beteen walls in the stopes were studied and the practice of supporting the stopes back with deslimed mill tailings came into operation but its use at great depths due to increase in humidity has been a limiting factor. Use of concrete fill with stope drive method of stoping which was started about 25 years ago ultimately helped in minimising the occurence of rock bursts and the method gave satisfactory results in mining many ore shoots, shaft pillars and remnants.

INTRODUCTION

The world famous centre of hard rock mining and Asia's richest Gold area the Koler Gold Field (KGF) of M/S Bharat Gold Mines Limited is located in the Karnataka State of South India, KGF has completed a hundred years of its mining activities in the year 1980. The presence of old workings evidences that the mining here had been even 2,000 years old.

Many enterpreneurs ventured to exploit this deposit including M/S John Taylor of U.K. who did a commandable job in systemising the mining activities in this field. Initially the field has been worked by five mines namely the Balaghat, Oorgaum, Champion Reef, Mysore and Nundydroog. The first two mines were closed due to exhaustion of ore deposits and the other three mines are presently working. After six decades of active life these mines were considered amongst the deepest mines of the world. Presently the field is confined to an area messuring 8 Km long, 2 Km wide and to a depth of workings of more than 3 Kms from the surface.

The KGF proved to be the world's most difficult gold deposits ever exploited due to very high inherent stresses in the rocks leading to violent ground movement and a high geothermic gradient. This is regarded as pioneer development in the mining technology.

GEOLOGY

The KGF is situated in the belt of highly metamorphosed volcanic rocks having a general north south strike and dipping toards the west at 40° to 45° at shallow depth but changing to nearly vertical at depth. The major rock types in the area are dark hornblende schist or lower Dharwar age which are surrounded by

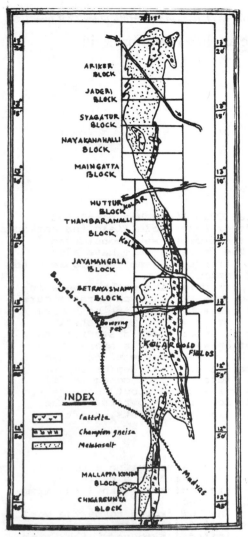

Fig.1. Schist belt of Kolar Gold
Field.

granites and gneisses of Archean age.
In the present working area there
are atleast 26 known lodes out of
which only two gold bearing quartz
lodes which are in the form of veins
and lenses, namely the Champion and
Oriental lodes are of great economic
importance. The Oriental lode which
is on the west of Champion lode, is
sulphidic in nature and accounts
for the bulk of reserves and the
sources of ore minod in Nundy droog
Mine. The schists are folded, faulted and intruded by dolerite, dykes

and pegmetities.

Most of the gold output sofar
has bome from Champion lode on
which all the mines have been
worked. A very significant break-
through was made in the recent
past in the Central part of KGF
within the Champion Reef Mine
when presence of gold in the
amphibolites apart from earlier
known quartz veins was recognised.

Presently the grade of payable
ore is 6 gm/tonne and that of the
low grade ore is 4 gms/tonne.

As per the laboratory investigation the KGF host rock schist is
very strong in compression and
possesses a high young's modulus
of elasticity. The average uniaxia
compressive strength is about 2670
kg/sq.cm and 3440 kg/sq.cm in the
direction parallel and perpendicular to schistocity and corresponding young's moduli of elasticity
are 8.79×10^5 kg/sq.cm and $7.03
\times 10^5$ kg/sq.cm.

The quartz which is a lode
matter is brittle and has high
breaking strength under compression as 4,350 kg/sq.cm.

MINE DEVELOPMENT

In order to develop the deposit
in the field about 65 km of shafts
of various sizes and shapes have
been sunk so far. All the three
mines of KGF are having major
shafts for winding ore from deeper
levels followed by main haulages
from where several secondary
shafts have been sunk. In addition
the mines have quite a few tertiary shafts to reach the bottom
levels.

The Edgar's shaft, which goes
to a depth of 1121 m to the 51st
level, serves the Mysore Mine,
Gifford's shaft, which goes to a
depth of 1996 m to 70th level,
serves Champion Reef Mine and is
one of the world's deepest shafts.
The Henry's shaft, which goes to
a depth of 1242 m to 48th level,
serves the Nundydroog Mine. The
shafts are either circular or
elliptical in shape and are
concrete and brick lined throuth-
out.
Besides shaft sinking operation,
the field has been developed by
driving a net work of mine openings in the form of levels drives,
cross-cuts, winzes and rises.

All the mines of KGF are inter-connected. Presently all the drives and raises, are supported by steel arches with sufficient lagging of timber and granite blocks.

THE STOPING PRACTICES

The stoping practices at KGF can be grouped as follow:

1. Earlier stoping methods
2. Stoping methods with use of granite masonry packwalls
3. Stoping with the use of deslimed mill tailings
4. Stoping with the use of concrete fills

1. Earlier stoping methods-Looking into the history of KGF mines we find that the mining here has been labour oriented since its inception, till 1950 the average daily employment had been about 25,000 with the strength of underground labour force as high as 20,000 for a considerable period. All the five mines progressed at a satisfactory level both laterally as well as along the dip

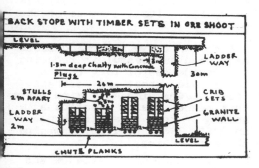

Fig.2. One of the earlier stoping methods.

The open stoping methods were in use to a depth of 150 m or so, No pillar used to be left in stopes except in few cases where there used to be lean patches of ore in between the rich ore blocks. Within twenty years of mining and after occurence of first rock burst in 1898 at their Oorgaum mine at a depth of about 320 m, the use of timber for the purpose of supporting stopes came into operation. Heavy timber stulls, timber stulls with

head boards of softer timber as cushion between the rigid stulls and the ground, crib sets both empty as well as filled etc. were in use with overhead system of stoping.

With the increase in depth and rock pressure the consumption of timber increased abnormally. After thirty years of mining the problems of working at depth, heat and humidity, began.

Shrinkage stoping as practicised in low grade reefs which had to be finally abandoned due to pyritic nature of ore which necessiated secondary blasting to draw the stuff . Slabbing had also taken place complicating the ore drawing further and diluting the already low grade ore. Rock bursts were quite frequent.

2. Stoping methods with the use of Granite Masonry Packwalls-

With the Passage of 3-4 decades of mining at KGF it was noticed that too rigid supports used to fail due to taking more load than they could withstand and the yielding supports resulted into wall closures, failure of surrounding rocks and inturn failure of support itself. Thus it was felt necessary to change system of support. In 1927 the experiment on the use of granite packwalls at Oorgaum mine proved be a success. The use of granite blocks which were obtained from granite quarries at the surface and subsequently transported to working stopes were started for building masonry walls in the working stopes of almost all the mines. The waste rock was also used to build the masonry packwalls. The use of MASONRY PACK WALLS with the use of granite and cement mortar REVOLUTIONISED the stoping practices at KGF.

The granite Pack walls were initialy used with Back stoping (over hand stoping). These Pack walls were built 2.4 m long and 2.4 m high with 1.5 m gaps on strike dip but at greater depths the continuous pack walls were built on level to level with two

275

setted inter gaps. But the ore used
to fall on workers perticularly in
stopes with very weak ore, the incidences of rock burst were also there.
This method was, therefore,
discontinued.

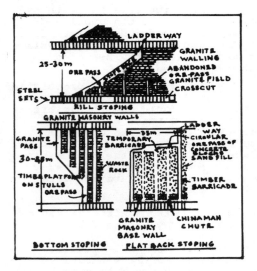

Fig.3. Stoping methods with the
use of granite pack walls.

The Bottom stoping(under hand
stoping) was then started with
granite pack walls, which used to
collapse as the wall rocks could
not provide enough support for them
to settle down and resist the
pressure. The pack walls used to
collapse mainly as the bed timber
was collapsing due to hanging wall
and foot wall. As even this could
not improve the conditions, timber
was placed on rails, supported by
plug bars fixed in the holes drilled
into the walls but the collapsing
of timber still continued as both
walls were very week.

In 1954, a major rock burst took
place in 43 ore shoot area of
Champion Reef Mine with this method
in use. This practice, therefore,
was discontinued. Meanwhile a technical committee set up by M/s John
Tayler & Sons Ltd which was working
on the problems of rock burst and
ground control gave its report and
recommended Rill Stoping with the
use of granite packwalls with rigid

'V' sequence. Some of the
recommendations with regard to
working in the stopes of the said
committee were as follows:

Development : Drives and levels
liable to rock burst to be steel
setted and lagged; holling to be
made from larger excavations
to smaller, and not vice versa;
when approaching a fault, fissure,
dyke or drive at an oblique angle
development to be turned so that
the approach is at right angle.

Stopes : No pillar or remnants to
be left within an oreshoot stope
preparation by stripping either
back or the bottom of the drives
to be kept minimum, winzes to be
kept to the minimum number o for
adequate ventilation, with
parallel reefs, one to be stoped
ahead of other, with flatter dips
hanging wall reef to be taken firs

Stoping Sequence: stope faces to
confirm to a longwall face as far
as possible.

Support and wall closure:Permanent
support(granite packwall) to be
installed as rapidly as possible
to resist initial closure of the
walls.

Although these recommendations
have been successful in reducing
the incidences of rock bursts
but in Glen shoot and Northern
folds of Champion Reef Mine where
this method was in practice a
series of rock bursts took place
in 1962, 1963 and 1966, damaging
the entire area.

It was through experience and
field investigations established
that solid granite supports used
at depth and slow rate of ore
extraction from stopes did nothing
to assist in providing controlled
relief to wall rocks and also it
helped in throwing high stresses
on the abutments in advance of the
face and created a situation which
was beyond control.

The closures between the stope
walls behind the rill stope face
with granite packwalls were
generally errecti. (The maximum
measured closure did not exceed
12-15 % of stoping width).

3. Stoping with the use of deslimed mill tailing :

Besides the limitations of the granite pack walling as outlined above the cost of stoping with its use was alarming and it was in every body's mind to go for an alternative media of stopes filling and in 1953 a plant to feed deslimed mill tailings was ultimately installed at their Nundy droog mine which was streamlined within a short duration

By this time with the installation of air cooling plant at this mine the environmental conditions in underground workings were improved and the use of hydraulic stoping did not effect them adversely.

The deslimed tailing feed (consists of quartz hornblende with variable content of suplphide minerals) gets oxidised and cement the sand and provides a firm fill in the stopes without reducing its water drainage capability. The deslimed mill tailing was thus fairly suited their presently working stopes which lies between

Henry's and Golconda shafts of their Nundydroog mine. The deslimed feed before being sent down is repulped to a consistency of 40% sand and 60% water. The filling rate is in the order of 120,000 to 150,000 t/year.

The stoping practice followed with this fill is HORIZONTAL CUT AND FILL.

The stope blocks are prepared keeping the level interval varying from 25 to 100 m and length of 22 m along strike.Levels are connected either by winzes or raises. The work starts from one of the winzes connecting two levels and it includes:

1. Preparation of chetty stopes
2. Heightening and equipping the drive (level)
3. Main stoping–Breaking of ground upward and preparation of orepasses simultaneously
4. Filling the excavated area

The chetty stopping of the level consists of excavating the ground to a depth of 1.8 to 2.4 m (depending upon width of reef) and

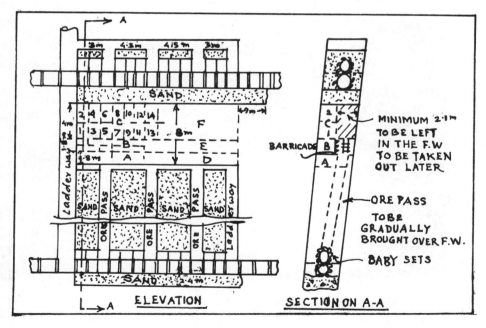

Fig.4. Cut & Fill Stoping with the use of deslimed mill tailings

filling the same either with concrete or sand. Once the level is chetty stoped it is supported with the use of steel arches lagged by timber or granite blocks. The track is then laid in the drive (level). The drive is then heightened by taking a slice of 1.8 m and baby sets 1.2 m in diameter (small steel arches) are put over the main steel sets. A second slice of 2.1 m is further taken, then top and the sides of the arches are packed with cloth and then over them the sand is filled, leaving the space over the baby sets for the construction of ore passes. The ore passes are built up by the use of prefabricated concrete blocks in successive layers.

The process of stoping advances by taking the slices of 1.8 m (for reef greater than 3 m width) or 2.4 m (for reef less than 3 m width) till it reaches within 5.4 m from the bottom of chetty stope above. The ground in this vicinity is taken in various stages and temporarily supporting the stope face as per the systematic timbering rules.

In the upper horizons of KGF this method is working satisfactorily but at greater depth it could not be adopted as water increases humidity and wet bulb beyond the limits.

STOPING WITH THE USE OF CONCRETE FILL

A brief account of various stoping practices as outlined above reveals that each practice had its limitations mainly due to any one or combination of more than one of the following reasons:
1) Inadequacy of support work
2) Delay in supporting
3) Large open area before supporting
4) Increase in wet bulb temperature beyond permissible limit

After the rockbrust in the year 1952 in 43 ore shoot area a team of KGF engineers under the leadership of Late Mr. J.K. Lindsay suggested a stoping practice called SUB LEVEL DRIFT or STOPE DRIVE METHOD with the use of concrete fill.

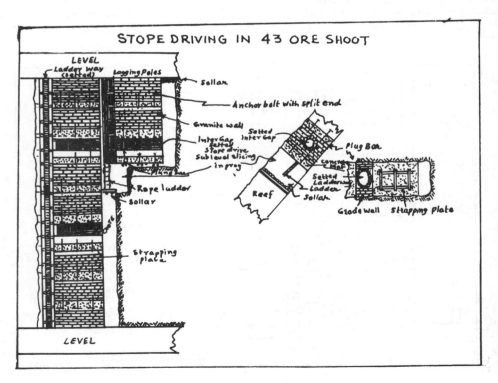

Fig.5. Stope drive stoping starting from top level.

The method suggested was the modification of conventional bottom stoping and it was decided to build packwalls on a bed of reinforced concrete. This concrete was held in position by rock bolts. The reinforced concrete was laid on the stope face itself. Anchor bolts with split ends were used to ensure proper bonding between the reinforced concrete and granite packwall. The ground was extracted in slices not exceeding 2.4 m in height only two intersetted gaps were left in this stope between the levels and all ladder ways were setted (fig.5). The method proved a success and the entire 43 ore shoot area, measuring 118 m x 260 m, was successfully stoped out. The success of this practice was mainly due to:

1) Exposing very little area at a time and that too for a shorter duration prior to supporting.
2) Making the support closer to working face by concreting on the face itself.
3) Ensuring the proper coherence and placement of supports by the use of rockbolts, anchor-bolts and reinforcement.
4) Providing safe comunication with setted inter gaps and ladderways.

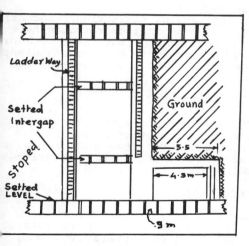

Fig.6. Stope Drive Stoping Starting From Bottom Level.

The method was further modified n the recommendations of expert's ommittee in 1967 which suggested the se of concrete fill in preference o granite pack walls on the concrete mat.

It was further pointed out that Voidage in a fill is an important factor which determines the load yield characteristic of the materials in the fills. The voidage for different back fills used here is as follows:

Type	% Voids	Remarks
sand fill	38	measured
Dry granite walling	34	Measured
Granite and Mortar	11	Estimated minimum voidage
Concrete mix	5-17	Depend upon the mix chosen

The reason given by the said committee for the use of concrete mix is as follows:

"When closure takes place, the load on the granite is not accepted as uniform stress throughout the granite because of irregular surfaces of the granite blocks. Local high spot in the intimate contact are subjected to a very high stress even though the average load on granite fill may be small. Failures of the blocks occurs until the load is distributed evenly through out the support. Due to this reason, the immediate load that can be resisted by granitesupport tends to be lowand corresponding initial closure rate high. If a concrete mix with a similar voidage is used, this would not be so because of its uniform nature and while the amount of closure at the centre of a large spot would be similar to both types of support, the closure in the immediate vicinity of the stope face should be less in the case of concrete fill. This should tend to reduce localised rock burst".

The experience in KGF mines so far indicates that concrete is a more reliable form of support which meets the requirement of a good support.

Stope drive method can be applied either by starting form top level and advancing towards bottom or vice versa. Starting from top level gives the advantage of working the workers under the safe roof but the ventilation is sluggish whereas starting from bottom ensures better ventilation, ease in ground breaking but workers have to work under exposed ore which is not very safe.

Stoping is done invariably in a sequence with three to five stopes in each sequence. Usually the top stope is one slice ahead of the stope below.

The practice of Stope Drive has thus proved to be a success in mining of many ore shoots, shaft pillars and remant pillars. The areas which have been badly damaged as a result of earlier rock bursts are under extraction by the use of this method.

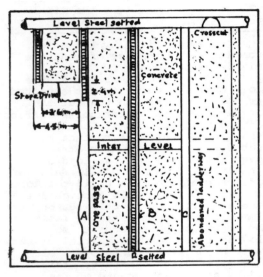

Fig. 7. Stope drive method with concrete fill.

The cost of extraction by this method is high but since its adoption from the last 25 years there is a definite decline in frequency and severity in the occurence of rock bursts, as illustrated in the table given below:

Year	Number of Reported Rock Bursts
1957-61	162
1962-66	124
1967-71	29
1972-76	59
1977-78	24

The problems of rock burst and ground control at KGF are still not under control. Efforts are on hand to understand mechanism of rock burst for its prediction and prevention.

CONCLUSIONS

The history of 100 years of mining at KGF reveals that the governing factor for changing from earlier methods of stoping to the prevalent stoping practices is the occurance of rock burst. Through field investigations, research and experience this has been established that frequency and sevirity of rock brust is not directly related to depth. Rock bursts have also occured even at shallow depths. The important factors for occurance of rock burst are the physical properties of rocks, very high inherent stresses, size and shape of excavation and the in homogeneit' of rock due to the presence of geological disturbances which invol' planes of weakness either in themselves or at their contact.

Use of granite pack walls as backfill once brought a revolution in KGF mines but ultimately deslimed mill tailings at moderate depths and concrete fill at greater depths have proved to be most relia ble back fills.

The rate of extraction, producti vity of the stoping practices here low, cost is high but these methods have proved to be a success against the problems of ground control and rock bursts.

ACKNOWLEDGEMENT

The willing assistance and useful advice of the staff at mining Engineering Department of Jodhpur University and KGF mines is gratefully acknowledged.

REFERENCES

Gupta, P.D. 1982, The Kolar Gold Field - Past, Present and future. Jnl of Mines, Metals and Fuels, March 1982, 79-86
Krishnamurthy, R.1968, Ground Contr and Rock burst at KGF, IM & EJ special September 1968, P-70-80
Rao, Sambasiva.S. 1980, Stop Drive methods, Bhart Gold Mines Ltd, Centenary Souvenir, P-158-171

Proceedings of the International Symposium on Mining with Backfill / Luleå / 7-9 June 1983

The state of development of pneumatic stowing in the hard coal industry of the Federal Republic of Germany

K.H.VOSS
Bergbau-Forschung GmbH., Essen, Germany

SYNOPSIS: Following a critical comparison of solid stowing and caving, this paper describes the current state of pneumatic stowing in the Federal Republic's hard coal mining industry.

Besides, some operational results achieved on current pneumatic stowing faces in Germany are presented which demonstrate that, apart from a high face output, remarkable output figures per manshift can be reached on solid stowing faces equipped with modern pneumatic stowing equipment.

One of the essential pre-conditions of efficient pneumatic stowing is the availability of high-capacity waste haulage facilities, apart from a favourable infrastructure for stowing. Some information will also be given on this aspect.

Underneath objects that are particularly susceptible to subsidence, the Federal Republic's hard coal industry intends to use compact early-bearing stowing packs in the future. The means of obtaining such stowing packs are illustrated.

Finally, some purposeful and desirable developments in the pneumatic stowing sector with regard to the coming years are presented.

1 TRENDS IN THE PERCENTAGE USE OF PNEUMATIC STOWING IN THE FEDERAL REPUBLIC'S HARD COAL INDUSTRY

Over the last 20 years, the trend in the percentage use of caving and solid stowing in the Federal Republic's hard coal industry was characterized by an essential increase in the use of caving. This was partly due to the drastic reduction of the hard coal industry's output potential and to the resulting choice of the most economical mining methods. A second reason was the fast advance of powered supports which initially could not be provided with an appropriate pneumatic stowing system that would allow economical operation of solid stowing faces.

The share of pneumatic stowing, which accounted for more than 30 % of the Federal Republic's hard coal output in the 50's declined to approximately 3 % in the 70's. It has, however, again reached about 6 % and is further increasing (Fig. 1).

The main reasons for the renewed and intensified use of stowing systems must be seen in the ever more restricted areas available for waste tips and in the high costs of subsidence as the Ruhr mining industry is to a large extent working below densely populated areas, some of which with buildings that are particularly sensitive to surface subsidence.

2 ADVANTAGES AND DISADVANTAGES OF PNEUMATIC STOWING

The advantages of solid stowing in the Federal Republic's hard coal industry are the following:
- Reduction of surface subsidence. As a result of the general cost increase, the subsidence costs of the mines are drastically rising, particularly with the increasing extraction of thick seams by caving.
- Reduction of internal mine damage, i.e. damage to the mine structure. This internal mine damage may be several times as severe as surface subsidence.
- Improvement of the face climate. The rock temperature is constantly rising with continued progression of face workings into ever greater depths. Through the tight isolation of the coaled-out cavity from the face, rock

Year	Caving		Gravity Stowing		Hydraulic Stowing		Flow Stowing		Pneumatic Stowing		Solid Stowing total		DMS-underground	
	Ruhr %	SAAR %	Ruhr %	SAAR %	Ruhr %	SAAR %	Ruhr %	SAAR %	Ruhr %	SAAR %	Ruhr %	SAAR %	Ruhr %	SAAR %
1960	47,5	72,6	18,4	2,2	-		7,9	1,6	-		32,5	17,5	52,5	27,4	2.102	2.013
1965	64,4	49,9	16,0	-		16,1	3,2	-	16,4	34,0	35,6	50,1	2.766	2.740
1970	68,8	47,7	20,8	-		21,5	1,9	-	8,5	30,8	31,2	52,3	3.843	3.632
1975	89,2	65,6	6,1	-		12,5	0,8	-	3,9	21,9	10,8	34,4	3.855	4.060
1980	90,2	85,2	5,6	-	-		1,0	-	5,2	14,8	9,8	14,8	3.943	4.645

1983	Share of different stowing methods in the german coal mining industry	Stbv TB 13076

Fig. 1

heat can hardly penetrate into the face area in case of pneumatic stowing as compared with caving. Furthermore, caving will entail air leakages through the insufficiently packed goaf. As a result, air temperatures are 4° - 6° lower today on pneumatic stowing faces as compared with similar caving faces.
- Better control of methane emissions. With solid stowing, peak emission values at dropping atmospheric pressure can in particular be prevented, which occur as a consequence of large gas volumes emitted from worked-out cavities that are frequently present on caving faces.
- Reduction or elimination of the need for waste tips. Regarding the dense population of the Ruhr district in particular, there is a lack of suitable areas for waste tips in acceptable distance to the mines. Besides, there is growing public resistance to waste tips, especially due to the annoying effects of waste transport by trucks.

However, application of pneumatic stowing has disadvantages, too:
- The need for the installation of a second haulage system to transport stowing material from the surface to the faces.
- The organizational difficulties of a stowing face and the relatively low capacities of pneumatic stowing equipment compared with that of modern winning machines, resulting in a slow-down of the daily rate of face advance.
- The additional operational costs consisting of shift expenditure for stowing material transport and application, compressed air costs and machine rental.

3 DEVELOPMENT OF TECHNICAL EQUIPMENT FOR PNEUMATIC STOWING

Up to the end of the 60's, pneumatic stowing was exclusively used in connection with individual prop supports. The corresponding technique consisted in setting timber props into the area to be stowed before removing the metal face supports. In individual cases, characterized by favourable roof conditions, the goaf could be kept open by rear-cantilevered articulated bars (Fig. 2). After stowing of

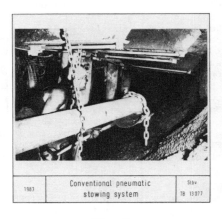

1983	Conventional pneumatic stowing system	Stbv TB 13077

Fig. 2

6 - 8 m wide sections each, the pneumatic stowing pipeline was manually shifted into the new goaf. The goaf width was normally 1.50 to 2.00 m und was separated from the remaining open face area by means of a paper wire mesh or a jute cloth before stowing. Some mines also used a "walking"-type wall (Fig. 3).

Combining the described pneumatic stowing system with powered supports raised severe problems. On individual faces, attempts were made to set timber props behind the powered supports in order to keep the goaf open, but this method did not yield satisfactory economical results.

However, the pneumatic stowing machines used for these conventional stowing methods had already considerable through-put capacities which, in connection with a 225 mm diameter pipeline, reached 200 - 250 m³ per hour under favourable conditions, as can be seen from diagram 4.

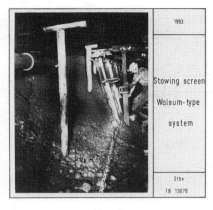

Fig. 3

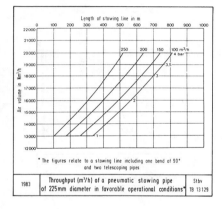

Fig. 4

3.1 Pneumatic stowing with lateral discharge

Starting from the basic idea that the pneumatic stowing pipeline could be shifted most conveniently with powered supports in a closed string, a pneumatic stowing system was developed from 1968 onwards which no longer required the stowing pipeline to be decoupled for shifting. In spacings of 4 - 5 m each, the stowing pipeline was provided with mechanized units for lateral discharge of the stowing material. As shown in Fig. 5, the lateral discharge units are of cage-type construction featuring a straight pipe length for material through-flow and a blow-out blade for material discharge, both of which are hydraulically

coupled with the stowing pipeline. Swinging-on and out of these two elements is hydraulical, too.

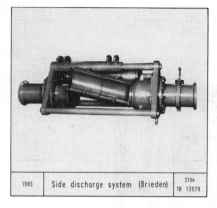

Fig. 5

Initially, the stowing pipeline to be shifted in a closed string was positioned on hydraulically hinged columns attached to the skids of the powered supports. Later on, the pipeline was suspended, free to swing, at the rear-cantilevered canopies of the powered supports allowing vertical adjustment (Fig. 6). This solution proved to be the most appropriate one as it allows the stowing pipelines to be shifted in a straight line without being damaged.

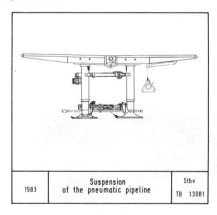

Fig. 6

Pneumatic stowing with pipelines to be shifted in a closed string and lateral discharge units is feasible down to a lower seam thickness of approximately

1.80 m. In connection with this system, a simple stowing screen consisting of sheet metal plates and rubber belt loops mounted at the goaf-side legs of the powered supports has proved to be quite useful (Fig. 7).

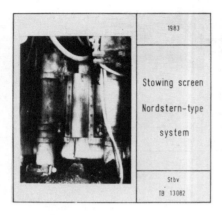

Fig. 7

Telescopic pipes (Fig. 8) and ball joints movable under pressure (Fig. 9) have been developed as useful accessories to this new stowing method. The telescopic pipes allow for a continuous extension and shortening, respectively, of the gateroad pipeline so that the pipes have to be mounted or demounted only once a day (Fig. 10). The ball joints movable under pressure allow for a deflection of the pipeline up to 15° and are preferably used on faces with sudden dip changes.

Fig. 8

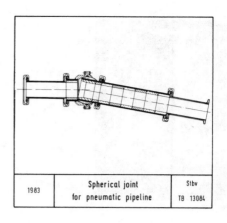

Fig. 9

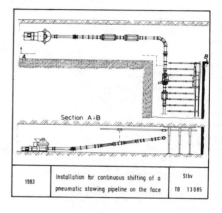

Fig. 10

3.2 Pneumatic stowing with stowing screen

In combination with powered supports, a stowing system still largely based on conventional pneumatic stowing systems was developed at Hugo mine of Ruhrkohle AG in the 70's. The stowing material is section-wise blown from the maingate to the tailgate and is discharged at the front end of the pipeline to form narrow packs. The roof is supported by rigid rear-cantilevered canopies of the powered supports. The props and wire mesh commonly used in conventional pneumatic stowing are replaced by a movable partition wall (Fig. 11). It consists of 7.5 or 9 m long elements, corresponding with the length of the stowing section worked.

The stowing pipeline is mounted on the face
side of the partition wall; it is vertically
and horizontally adjustable, and is shifted
in sections together with the individual
elements of the wall, which are connected
neither with the face supports nor among
themselves. Over a longer period of time,
they thus effectively support the stowed
goaf until they are advanced to the
supports by two block-and-tackle gears
immediately before filling of the next
stowing section. In this way, the old
vertical stowing slope remains unsupported
for only 8 - 10 minutes. Immediately behind the goaf-side legs, rubber mats are
attached to the canopies of the powered
supports to seal off the stowing elements
from the roof. The wall elements are
anchored in the goaf by means of wire ropes
to avoid their being shifted towards the
coal face as a result of the pressure
exerted by the fresh stowing material.

Since the Hugo-type system works with
narrow stowing sections between the old
stowing front and the partition wall, with
the material being discharged at the front
end of the pipeline, the stowing material
is highly compressed and has thus a good
supporting effect on the roof behind the
face area.

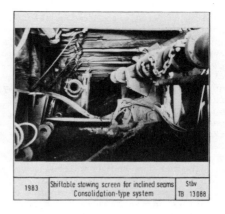

Fig. 12

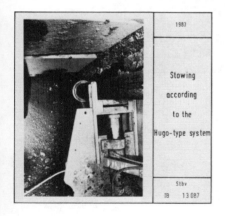

Fig. 11

A modification of the Hugo-type partition
wall for inclined seams was developed at
Consolidation colliery of Ruhrkohle AG
(Fig. 12). This partition wall used also
in connection with four-leg chocks is
hydraulically height-adjustable to prevent
the stowed material from flushing at the
waste to produce a slope of highest stability in inclined seams, in particular. The
individual wall elements are drawn up by
long hydraulic jacks attached to the

supports. It must be noted that both the
wall elements and the stowing pipelines
are separately and hydraulically anchored
in the upward direction of the dip.

A combination of supports and partition
wall for inclined seams was also developed
at Erin colliery of Eschweiler Bergwerks-
verein (Fig. 13). In this system, the
stowing pipeline is integrated in a lem-
niscate-stabilized three-leg chock with
the support area screened off on the
goaf side by a telescopic flushing shield.
After shifting of several support units
at a time, the stowing material is discharged in upward direction through the
openings between the support chocks resulting from the support lead.

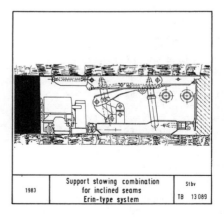

Fig. 13

3.3 Pneumatic stowing with stowing shield chocks

A further development of pneumatic stowing, initially with regard to thick seams in particular, was made in the Saar district. Using novel stowing shield chocks, a partly mechanized stowing system with front-end discharge of the stowing material was developed at Luisenthal colliery of Saarbergwerke AG. The goaf is kept open by a rigid bar hinged to the shield canopy which can further be provided with a 3 m long trailing bar the end of which always remains in the goaf (Fig. 14). The stowing pipeline is suspended at the rear-cantilevered canopies and can be adjusted hydraulically in horizontal and vertical direction. The shield itself is designed as a tight and rugged stowing screen which advances together with the supports.

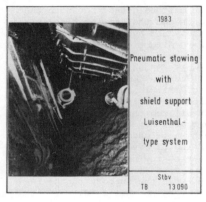

Fig. 14

Starting from the tailgate, 7 - 9 m long sections are filled with the stowing material. The pipeline situated in the section to be stowed is manually decoupled and placed close to the flushing shields of the supports. During stowing of the next section, the pipe becomes immersed in the stowing material. The area to be stowed can be entered from the travelling track through a hydraulically opening trap door in the flushing shield.

Provided the stowing sections are not too wide (only 2 support advance steps, if possible), a very compact waste can be obtained with this system since the support resistance of the four-leg chock shields is very high so that the convergence remains low and the roof in the waste is prevented from caving by the trailing bar in particular.

3.4 Pneumatic stowing with mechanized front-end discharge

Coupling and decoupling of the pipe strings is still rather time -consuming and often difficult with the various pneumatic stowing systems developed at Hugo and Luisenthal collieries. Therefore, it was inescapable to develop a pneumatic stowing pipeline, shiftable in a closed string, for mechanized front-end discharge of the stowing material.

Such mechanized front-end discharge units have been developed since 1980 (Fig. 15). They feature a box-shaped construction with an integrated hydraulic coupling and two thrust pistons. After decoupling of the two pipe ends, the discharging pipe is laterally displaced in a link by the two thrust pistons, positioned below and above the pipe end, so that the stowing material is axially discharged. By tilting of the link, the jet of stowing material can practically assume any projection parabula desired.

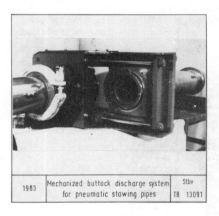

Fig. 15

The mechanized front-end discharge units are installed in spacings of 7 - 9 each, with the stowing pipeline conventionally suspended below the rear-cantilevered canopies of the supports (Fig. 16).

At Nordstern colliery of Ruhrkohle AG, the first face completely equipped with such mechanized front-end discharge systems has already yielded remarkable operational results, as will be described in the next section. Tests with such systems are already under way at other collieries in the Federal Republic and in France.

A corresponding front-end discharge system was developed for seam thicknesses below 2 m (Fig. 17). It is still being

286

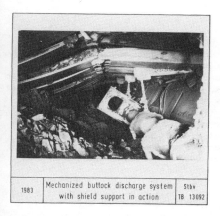

| 1983 | Mechanized buttock discharge system with shield support in action | Stbv TB 13092 |

Fig. 16

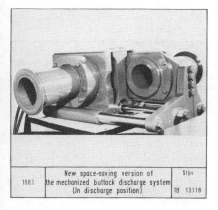

| 1983 | New space-saving version of the mechanized buttock discharge system (In discharge position) | Stbv TB 13118 |

Fig. 17

tested so that there are no operational results as yet.

4 OPERATIONAL RESULTS AND EXPERIENCE WITH MODERN PNEUMATIC STOWING FACES

The above-mentioned newly developed pneumatic stowing methods used in combination with powered supports are already quite efficient and are characterized by a low shift requirement.

It has not been possible so far to design pneumatic stowing systems which would allow a saleable face output of more than 5,000 t per day as is known from caving faces; nevertheless, a saleable output in the range of 3,000 - 4,000 t per day corresponding with a face output of approximately 30 t per manshift has already been obtained over a longer period of time from various faces.

The following table (Fig. 18) summarizes some recent data on face output, face output per manshift and shift requirement for stowing of some modern pneumatic stowing faces in the Federal Republic.

Colliery	Seam/thickness (m)	face length (m)	Daily saleable output (t/d)	Daily face advance (m/d)	face performance (t/MS)	Shift expenditure for stowing (MS/1001)	Face equipment
A	3,09	173	3037	5,23	27	0,3	Shield support and mechanised buttock discharge
B	2,70	240	4458	5,64	30	0,54	Chock support and stowing screens (flat Formation)
C	2,30	180	2223	3,74	21,7	0,7	Chock support and stowing screens (inclined Formation)
D	3,30	198	2042	2,55	36	0,78	Shild support and manual buttock discharge
1983	Average monthly results of mechanised stowing faces						Stbv TB 13136

Fig. 18

Considering the earlier-mentioned advantages and the saving potential of pneumatic stowing, modern pneumatic stowing faces are economical already today taking into account the consequential charges of caving. This is confirmed, among others, by the favourable operational cost situation of some hard coal mines in the Ruhr and the Saar districts which use pneumatic stowing on a larger scale.

Some essential experience gained with current pneumatic stowing faces will be described in the following:

- Particularly in thick seams, compact stowing is far better for roof control than caving. This is mainly due to the fact that the face roof finds a good abutment in the goaf. With caving, the immediate roof strata may travel in goaf direction. This may frequently entail roof breaks, particularly in front of the canopy tips of the supports, as a result of the fissures opening in the face roof.
- A comparison of dust measurements showed that the dust content of the air is lower on pneumatic stowing faces than on caving faces, provided the stowing material is sufficiently wetted. This mainly applies to airborne fine dust.
- In individual cases, the effective air temperature on the face could be lowered by 6 - 8° by pneumatic stowing. Such temperature reductions

287

could notably be realized by use of mine cars for the transport of the stowing material close to the face. With this transport system, the stowing material can largely be kept at the low temperature level of preparation.

- In order to obtain a compact early-bearing pack, it is essential that the goaf is stowed in narrow sections and that the roof in the goaf is effectively supported. This support should always reach the previously stowed zone.
- The use of two-layer pipes for the face pipeline, with the inner pipeline consisting of 20 mm thick chilled cast iron bushings, has proved to be very economical and can be used wherever the face pipeline needs not be shifted by hand.
  Related to lifetime, total throughflow volumes of 300,000 - 400,000 m³ of stowing material can be achieved today with such pipelines.
- The use of hydraulic telescopic pipes in the gateroad stowing pipelines has contributed to an essential reduction of the shift requirement for reconstruction of this pipeline.
- The use of face pipelines, which are shiftable in a closed string, with mechanized front-end or lateral discharge systems allows for a considerable reduction of operational interruptions of the stowing process and thus for an essential improvement of the utilization rate of pneumatic stowing machines.

5 EFFICIENT WASTE SUPPLY OF SOLID STOWING FACES

The principal pre-condition, in terms of planning and organization, for capacity and economic viability of pneumatic stowing systems is a sufficient and constant waste supply combined with a low shift requirement. To this end, all automation processes currently available must be applied. According to the current state of transport engineering, the waste supply from the surface to the pneumatic stowing machine can be considered as being satisfactory when the shift requirement for waste transport is below 1 manshift per 100 m³ of stowing material. The Federal Republic's hard coal industry has already reached values around 0.5 manshift per 100 m³. These results have been obtained with an equipment range comprising drop pipes in the open shaft, a belt road controlled by the stowing machine operator for horizontal und inclined waste trans-

port, as well as properly dimensioned underground waste bunkers. Where belt roads are not economical due to the underground mine structure, large bottom-dump cars can alternatively be used. Combined with automated loading points, this transport system will also ensure high transport capacities with low shift requirement.

Apart from the above-mentioned high-capacity systems for waste supply, the waste material is transported in mine cars from the surface to the working points at various German hard coal mines still today. This system is useful whenever the empty waste cars are subsequently charged with coal, i.e. whenever waste and coal haulage are run on the same level. Further advantages of this system can be seen in the large bunker capacity without any stationary bunkers, the conservation of the grain size of the waste material, and the higher cooling effect of the waste material on the face since heating of the air, as in the case of belt conveying, can largely be prevented.
Among the main drawbacks are the relatively high shift requirements and the high susceptibility to failure.

Skip winding is also used for vertical transport of stowing material. With today's skip sizes, its capacity is not as high as that of drop pipes but it entails less grain size reduction.

Gravity pipelines and drop pipes with guide spirals have proved to be particularly efficient for vertical downward waste transport. Due to the still unsatisfactory life of these pipelines, various improvements were made in recent years:

The waste drop pipe with guide spirals having a diameter of 500 mm was further developed by replacing the two-layered spiral design (base plate and wear plates) by a single-layered spiral made of a tough-hardened material (Figures 19 and 20). In this design, the wall elements are toothed in such a way that any vertical cracks can not develop in the chute direction. This results in a substantially smoother waste flow and in less wear.

Gravity pipelines for vertical transport of stowing material used to be more frequent in former times due to the higher share of pneumatic stowing in the Ruhr hard coal industry. Currently, gravity pipelines are only used in the Lorraine hard coal industry. In Germany, a two-layer pipe is being developed, the inner pipe of which is provided with 12 through slots (Fig. 21 and 22). The air displaced by the falling waste plug is

| 1983 | Vertical waste pipe incorporating free standing spiral chute (Top view) | Stbv TB 13119 |

Fig. 19

| 1983 | Gravity waste pipe with grooved cross section (Front view) | Stbv TB 13122 |

Fig. 22

| 1983 | Vertical waste pipe incorporating free standing spiral chute (Bottom view) | Stbv TB 13120 |

Fig. 20

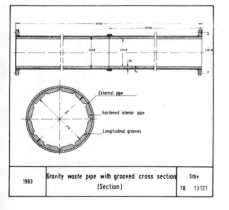

| 1983 | Gravity waste pipe with grooved cross section (Section) | Stbv TB 13121 |

Fig. 21

supposed to flow off upwards through these channels. This should result in a smoother fall and reduced wear of the pipelines. Such a gravity pipeline will be tested shortly.

It is obvious that the stowing process on the face must not be affected by breakdowns of the waste supply; this requires the installation of underground bunkers and buffer rooms. Bunkers for stowing material are installed at the end of the drop pipes and at skip winding plants.

To avoid any interruptions of the waste supply resulting from operational breaks in the pneumatic stowing process or short-time stoppages of stowing work, buffer rooms with capacities for stowing material supplied during 10 - 15 minutes are installed closest possible to the pneumatic stowing machines. Besides, this preventive measure contributes to an essential improvement of the machine running time since the pneumatic stowing machine is hardly affected by short-time interruptions of the waste supply.

6 EARLY-BEARING COMPACT STOWING

Pneumatic stowing in hard coal mines was so far accompanied by an average surface subsidence of 50 % of the seam thickness worked. This subsidence has the following causes:
- Convergence between the roof and the floor resulting from the subsidence pressure before stowing. Roughly 50 % of this convergence is due to the front abutment pressure acting upon the zone in front of the face which

289

cannot be influenced by operational
measures. The remaining 50 % of the
total convergence affect the face area.
- Goaf cavities, i.e., the cavities remaining open after conclusion of the
stowing process. With today's pneumatic stowing systems used in connection
with powered supports, this applies
mainly to cavities at the roof, i.e.
so-called "stowing shadows" caused by
the rather high rear-cantilevered
canopies of the powered supports.
- Compression of the stowed material due
to the super-imposed main roof. The
extent of such subsidence depends on
the pore volume remaining after stowing
Measurements conducted by the Mine Surveying Department of Steinkohlenbergbauverein
in approximately 250 faces of the Federal Republic's hard coal industry showed
that the convergence ranged from 12 -
15 % of the seam thickness and was 13 %
on average. The remaining 37 % of roof
subsidence thus consist of the cavities
in the stowing and in the pore volume
of the waste; this percentage of roof
subsidence is the only one that can be
influenced.

The use of early-bearing compact stowing
is taken into consideration particularly
underneath objects that are sensitive
to mine damage. The most compact possible
stowing pack with the lowest possible
pore volume can only be obtained in the
presence of a grain size distribution
curve of the stowing material that largely coincides with the Fuller curve for
concrete aggregates. It is of particular
importance that the stowing material contains a sufficient amount of fine and
finest grains to allow the formation of
the most body-compressed spherical
packing.

Studies conducted by Bergbau-Forschung,
Essen, on the stowing material used at
various mines have shown that most
stowing materials consisting of washery
tailings have already quite favourable
grain size distribution curves (Fig. 23).

If this is not so, additional fine grain
can be admixed to the stowing material.
Floatation slurries with their particularly high clay content are suited to
this purpose, for example. They are quite
successfully used at various mines already today. Electrostatic filter ash
from power stations with its long-term
hydraulic binding effect can also be used
for admixture.

Further studies of Bergbau-Forschung,
Essen, demonstrated that the clay minerals
(illite and kaolinite) contained in the
washery tailings of hard coal may develop

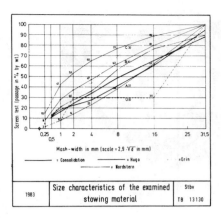

Fig. 23

quite substantial adhesive forces (Van
der Waals forces) which can even be increased by compression, for example by
vibration.

In practice, the substantial adhesive
capacity of argillaceous fine-grain material becomes evident from the fact
that, upon removal of the stowing screen,
the waste slope will remain vertical and
will not slip off even in inclined seams,
provided a waste material with a high
finest grain content is used. In the
laboratory tests conducted, the test
cubes prepared from washery tailings
reached self-stability already after a
vibration period of 1 s. This result
underlines the adhesive forces of clay.

Using washery dirt accruing from hard
coal preparation, any admixture of additional binding agents is not necessary
to obtain economically an early-bearing
compact stowing in flat and moderately
inclined hard coal seams. According to
the current state of the art, optimum compactness and early-bearing capacity can
be obained by the following measures:
- Use of stowing material prepared
from washery tailings having an optimum grain size distribution curve,
and separation-free application of the
stowing material.
- Stowing in narrow strips for proper
compression of the stowing material
by subsequent vibration.
- Filling of the cavities at the roof
of the stowing upon extraction of
the rear-cantilevered goafside canopies
of the powered supports. Packing
equipment attached to the rear-cantilevered canopies, for example, can
be used to this end.

7 FUTURE TRENDS IN THE FIELD OF STOWING

It would be possible to adjust the capacity of pneumatic stowing to the efficiency of current winning machines by the development of pneumatic stowing machines of still higher throughput capacity. At most mines, this would, however, require the installation of new larger compressors and compressed-air pipes with still larger diameters. Consequently, such a development is no longer taken into consideration for cost reasons.

Instead, numerous measures will be taken in the future which will allow a higher rate of utilization of pneumatic stowing machines and improved charging of the pneumatic air stream. Some of these measures have already been mentioned:
- Installation of a waste buffer in front of the pneumatic stowing machine.
- Further development of mechanized discharge systems with the objective of eliminating the need for stopping the stowing material flow when switching to an new discharge point.
- Development of low-cost feed belts with stepwise speed control for the pneumatic stowing machines.
- Development of control systems to regulate the feed flow to the pneumatic stowing machines as a function of the length of the pneumatic stowing pipeline, for example via the back pressure in the pipeline.

Besides, attempts are made to improve the support of the stowing area in such a way that the convergence can be kept as low as possible and that the roof is supported up to the waste by broad rigid rows of bars.

A further main objective of current development efforts is the provision of compact early-bearing stowing packs. This still calls for the development of suitable low-cost vibrators.

At present, the use of current face equipment on pneumatic stowing faces is limited to a seam thickness of approximately 2 m. Recent development work on space-saving stowing equipment is aimed at a reduction of this limit to about 1.5 m.

REFERENCES

Floren, H.: Practical experience on powered support and pneumatic stowing in plough-operated faces of moderate to steep inclinations. Glückauf 106 (1970), pp. 780/92.

Helms, W., und Knissel, W.: Incidence of the binder and water contents on the compressive strength of cement-consolidated washery dirt. Glückauf FH 42 (1981), pp. 250/258.

Leininger, D., and T. Schieder: Utilization of washery and mine dirt. Glückauf 111 (1975), pp. 903/08.

Linde, F.C.: Pneumatic stowing at Hugo colliery. Glückauf 118 (1982), pp. 704/710.

Meyer-Fredrich, A., and W. Hoppstädter: Operational experience on powered support and pneumatic stowing in level coal seams. Glückauf 106 (1970), pp. 984/92.

Mehrhoff and Voss: The future role of solid stowing in west German hard coal mining. Glückauf 112 (1976), pp. 317/322.

Reinshagen and Hahn: Shield support for coal faces with pneumatic stowing. Glückauf 115 (1979), pp. 679 - 688.

Reinshagen and Reinshagen: Ribside technology with extraction of thick coalbeds using pneumatic stowing. Glückauf 115 (1979), pp. 990/994.

Scheidat and K.H. Voss: Innovations in the field of bunkers and vertical flow transport at Bergbau 81. Glückauf 117 (1981), pp. 1216/1217.

Siska, L.: Compressibility of stowage. Glückauf-Forsch.H. 36 (1975), pp. 58/62.

Voss, K.H.: The hard coal industry in Lorraine. Glückauf 108 (1972), pp. 415/20.

Voss, K.H.: Working faces to the rise in conjunction with powered support and pneumatic stowing at Folschviller colliery. Glückauf 108 (1072), pp.123/43.

Voss, K.H.: Working faces to the strike in conjunction with powered support and pneumatic stowing at Wendel-Marienau colliery. Glückauf 109 (1973), p. 1209.

Voss, J.: Improving mine climate by pneumatic stowing. Glückauf 110 (1974), pp. 121/25.

Voss, K.H.: Adaptation of pneumatic stowing to fully mechanized face operations. Glückauf 111 (1975), pp. 780/81.

Voss, K.H.: Powered support and pneumatic stowing in the Saar coalfield. Glückauf 111 (1975), p. 1082.

Voss, K.H.: Shifting stowing ducts with lateral discharges. Pocketbook for mining engineers 1976, pp. 143/46.

Voss, K.H.: Hydraulic transport of material for pneumatic stowing from the surface. Glückauf 113 (1977), No. 1, pp.30/31.

Voss, K.H., and Sielaff: Nordstern colliery: Fully mechanized goal-getting from thick seams using solid stowing. Glückauf 113 (1977), No. 19, pp. 933/937.

Voss, K.H.: Planning of the lay-out of an efficient material supply for pneumatic stowing areas. Glückauf FH 41 (1980), Vol. 3, pp. 82/88.

The design and practice of cut-and-fill method at Fankou Lead-zinc Mine

WU LIANGDUAN
Design & Research Institute of Non-ferrous Metallurgy, Changsha, China

ABSTRACT: The Fankou Lead-zinc Mine is the largest lead-zinc concentrates producer in China. It is a high-grade, high-sulphur content deposit with competent ore and rock, moderate width and depth, and complex spatial positions. Since its construction in 1965, different scales of experiment and design were carried out in mining methods and filling technologies. The present filling technology includes crushed rock hydraulic stowing, concrete filling, cemented tailing filling. Stoping methods are chiefly cut-and-fill with T2G loader and mechanized cut-and-fill with LHD and ramp preparation. During the past 14 years of production, valuable experiences and economical benefit are attained.

1 DESCRIPTION OF MINING METHODS

The Fankou lead-zinc deposit occurs in limestone, with many shape variations and complex attitude due to faulting, etc. The main orebody averages mostly 20 to 50 m in width, with a maximum of over 100 m. The strike length is 200 to 600 m, and the extension downdip 100 to 400 m. The dip angle is from $60°$ to $70°$, but in local places only $30°$ to $40°$, or even smaller. Ore and country rock both are competent but locally they are less competent, as the hanging-wall rock consists of sandy shale or thin marlite layers with much argilla. In the faulted zone where fractures and joints are developed, both ore and rock are broken and quite incompetent.

Mineral composition is simple. Main minerals are galena, sphalerite and pyrite, accompied by a variety of valuable elements, which can be utilized comprehensively.

In consideration of the spatial characteristics of the deposit, the technical conditions of mining, together with other features such as high-grade, high-sulphur ore, well-developed karst with abundant ground water, a higher-recovery fill mining method is adopted in order to reduce ore dilution and prevent ground fall.

Ore blocks are devided into transverse stopes. The rooms have a width of 7 to 10 m, averaging 8 m, and 14m at maximum. The rib pillars are 7 to 8 m wide, crown and floor pillars 6 to 13 m, where orebody width is less than 10 m, stopes are laid out longitudinally. The first step is to mine the room, using cut-and-fill method or shorthole shrinkage with backfill. After the room is completed, it is filled with concrete or cemented tailing to provide an artificial pillar support. Rib pillar mining is done with overhand cut-and-cemented fill or hydraulic rock filling. Recent years have seen the experimentation on VCR method and the wide application of a mechanized cut-and-fill method. The development system is shown in Fig. 1.

2 CALCULATION OF FILL RATE AND SELECTION OF FILL MATERIALS

2.1 Fill rate annual fill rate Q_y:

$$Q_y = K_1 \cdot K_2 \cdot \frac{Q_k}{R_k} \cdot Z$$

Where Q_k = annual ore production by

fill method;

R_k = volumn weight of ore;
z = production to fill ratio, m^3/m^3
K_1 = coefficient of initial volume shrinkage;
K_2 = loss factor.

According to the above calculation, the annual fill rate Q_y will be 326000 m^3 and daily average fill rate Q_d will be 1065 m^3. Room to pillar ratio is 1:1. The room is filled with cemented fill exclusively. During pillar filling, cemented fill is used in the sill pillar and floor surface. The quantity of cemented fill for the pillar amounts to 10.7%. Hence for the whole mine the cemented fill rate and hydraulic sandfilling rate are respectively:

Q_c = 185000 m^3/year
Q_n = Q_y - Q_c = 141000 m^3/year

2.2 Selection of fill materials

In the mine fill design, priority should be given to the possibility of using tailings and waste. Its economics should also be taken into acount. The fill materials available for the Fankou Mine are classified mill tailing sand and development waste underground, the rest supplemented by the rock from the quarry and the river sand purchased.

1. Tailing sand: Mill tailings contain small amounts of lead, zinc and other values according to analysis. And it is difficult to select a tailing pond site around the mine district, so the tailings are classified, with the coarse fraction for use in underground filling. The amount of coarse tailings Qw may be computed as follows:

$$Q_w = Q \cdot \eta_c \cdot \eta_t \cdot P \frac{1}{V_w \cdot K_{st2} \cdot l_q} \qquad (2)$$

Where Q=annual throughout of the mill;
η_c=tailing sand total yield; obtainable from the mill test report;
η_t=classification effifiency of coarse tailing sand, which could be calculated from classification tests;
P =coarse tailings yield in relation to the total tailings, which could be obtained from classification test reports. It may be obtained from the particle size curves of the total tailings.

Vw=weight of the coarse tailing sand per m^3 of fill;
K_{st2} = underground loss factor;
L_{st2} = ground surface loss factor;
From the above calculation, 80000 m^3 coarse tailing sand will be available for mine fill.

2. Underground development waste rock : The wall rock is limestone, a somewhat ideal fill material after crushing or fine grinding. The waste rock is hoisted to the surface and then crushed before use for filling. Its amount available for filling operations may be calculated.

$$Q_f = Q_k \cdot N_b \cdot (1-K_1) \cdot 4 \cdot \eta \cdot K_2 \qquad (3)$$

Where Q_k=annual ore production from the mine
N_b = development to ore produced ratio, development work is calculated by length of 4 m^2 cross section;
K_1 = Ore from development;
K_2 = coefficient of bulk increase of the waste;
η = utilization of waste rock as fill.

From the above equation, the waste available for filling operations will be approximately 100000 m^3 per year.

3. Rock from the quarry: Insufficient supply of tailing sand and waste necessitates applying the crushed finely ground rock from the quarry. The quantity of this fill material supplemented by quarry operations is:

$$Q_c \geq \frac{Q_k}{R_k} - (Q_w + Q_f) \qquad (4)$$

Daily production capacity of the quarry may be calculated from the following formular.

$$Q_T = \frac{Q_c \cdot \gamma_T}{T \cdot K} (1+\varepsilon) \cdot F \cdot L \qquad (5)$$

Where γ_T=specific gravity of rock;
T =annual operating days for the quarry;
K =coefficient of bulk increase of rock
ε =mud percentage, according to actual measurements
F =coefficient of uniformity for production;
L =loss factor.

There is a quarry at the mine, pro-

ducing quartz sandstone and siltstone. The production rate is calculated at 1200 tons/per day.
    4. River sand: Since production,
river sand has been purchased for
filling. Because of long distance
from the sand source,as well as
seasonal influence on the purchase
business,high price and transportation problems,river sand can only
be used on a temporary and supple-
mentary base.

3 DESCRIPTION OF FILLING MODE AND FILLING SYSTEM

The filling mode at the Fankou Mine
consists of cemented fill and hydraulic rockfill. Base on the fill material selected and differences in
transport technology, the former is
devided into cemented tailings fill
and concrete fill,the latter hydraulic rockfill and hydraulic tailings
fill.

3.1 The Jinxingling cemented tailings fill system

This system uses +30μ classified
tailings and cement. Cement to tail-
ratio is 1:5 to 1:30 mainly determined by strength need. Mill tailings
pass through cyclones, +30μ fraction
pumped to the tailings pond at the
Jinxingling fill mixing station for
dewatering and storage. The stored
tailings are delivered to a screw
conveyor by scrapers. The screw conveyor also acts as a metering device
while conveying. Cement is transported by cement tankers and pneumatically fed into a cement silo. Cement
is discharged through a gate. After
passing through a single-tube screw
feeder which services as a metering
device, and a screw conveyor, it is
slurried and fed into the mixer with
the tailings,with the slurry concentration controlled at about 70%. The
slurry flows down a borehole and is
transported by gravity in 3 inch pipelines to stopes at a rate of 60 to
70 m^3 per hour. The schematic diagram
of this system is shown in Fig.2.

3.2 The Shiling concrete filling system

Orebody in the Shiling area are short

and concentrated. The mixing station
is at the center above the orebody
to facilitate concrete transportation. The concrete is prepared on
surface. The rock is crushed and
screened to 0 to 25mm and trucked
to a 700 m^3 crushed rock bin,+30μ
course tailings are pumped to a
tailings pond for dewatering and
storage. Crushed rock and tailings
(or river sand) are metered through
disk feeders and belt conveyored
into a continuous mixer. Cement is
unbagged automatizally and elevated
to the cement bin in the upper portion of the mixing station. Cement
is discharged through a chute gate
and metered by a double tube screw
feeder into the continuous mixers.
Concrete is gravitated through 5
surface chutes and boreholes into
underground filling areas. The 5
fill boreholes cover the filling
area of the main orebodies at Shiling. The 2 continuous mixers at
the mixing station can work simultaneously or independently. Concrete
in transported underground by incline chutes. The schematic diagram
of this system is shown in Fig.3.

3.3 Hydraulic rockfill system

The Jinxingling and Shiling areas
have their own hydraulic rockfill
systems.Crushed rock is trucked and
unloaded into a 10m dia. bin (capacity 700 m^3). The crushed rock is
outlet via discharge gates at the
bottom of the bin in a chamber. It
is then mixed with water in the
mixing ditch declined 15-30°, and
delivered by gravity to underground
stopes. The largest particle of rock
should be less than 1/3 of the pipeline diameter. To suit this requirement, 0 to 60 mm particle size is
adopted for 178 mm ID chromized cast
iron pipe. A lightweight steel pipe
with a wall thickness of 4 mm is
used in stopes for easy movement,
connection and disconnection. The
ratio of the total pipe length to
the length of the vertical pipe section is 1.5 to 2.0, and water to
rock ratio 2.1 to 3.9,actual filling rate in the order of 200 to 250
m^3 per hour. The schematic diagram
of this system is shown in Fig.4.

4 APPLICATIONS OF CEMENTED TAILINGS

FILL AT FANKOU MINE

Because of its easy preparation, transportation and placing, tailings are used at the mine in the early 1970. Operating practices in recent years have shown it basically a success, contributing much to the underground filling work.

4.1 physical and chemical properties of tailings

Requirements for the tailings used for cemented fill are that it must be low in metall content and stable in chemical properties. It must also be less susceptive to become weathered or fluidified when placed. It must be difficult to be oxidized to release large amounts of poisonous gases, or to produce harmful components which may injure cement stability and lower strength. The sulphur content in the tailings must be under strict control. The tests at Fankou have demonstrated that the strength of tailings containing 1% sulphur is 3 times as much as that of tailings containing 9% sulphur when both are mixed with the same amount of cement. In another test, samples were prepared with tailings of sulphur content as high as 12.35% mixed with silicate cement in 7 ratios from 1:2 to 1:10. The compresive strength at curing period of 90 days was 1.4 to 1.78 times that of 28 days, but after 90 days, many samples collapsed. The finely ground pyrite and other sulphides in the tailings are oxidized at the presence of oxigen and water to form sulphate ions (SO_4--). These ions, when reached a certain concentration, cause chemical reactions with aluminates in the cement to produce sulphate crystalls which result in tripling its volume. The enormous internal stress thus produced leads to cemented fill cracking or collapsing. According to some informations from aboard, pyrite in the tailings for cemented fill should not exceed 8%. At Fankou, the fine fraction of classified tailings should have a sulphur content of less than 5% in view of using them as silicon additive for cement production.

4.2 Preparation and transport of tailings

Mill tailings are pumped through pipelines to the classification station where onestage classification by 350 mm dia. rubber-lined and 500 mm dia. diabase-lined cyclones is conducted, with underflow stored in a 2200 m^3 capacity bin (7 days dry sand storage). During filling it is monitored by giants into a 2x2.5 m mixer. The slurry is then pumped to the Jinxingling area by a 5 inch dia. surface pipeline at a designed concentration of 50% and a rate of 40 to 50 ton/hour (dry tailing). To achieve uniformity between the continuous mill operation and intermittent filling operation, the storage pond and bin should have a certain surge capacity, there -fore a 12000 m^3 temporary stockpile on the surface was built, when needed, the tailings sand stored can be reclaimed by giants and pumped to the filling station. The pond is devided into 2 to 4 comartments and drained by decantation.

4.3 Ratio, concentration and strength

The strength of cemented tailings mainly depends on cement to tailings ratio, concentration and segregation. At present no method is in general use for calculation of the cemented fill strength requirements. Normally the strength is determined by experiments and operating experiences, based upon the role to be played by the fill. In back filling, cement to tailings ratio is generally 1:20 to 1:30. In placing a strong floor, the ratio is 1:5 to 1:6, which is equivalent to the compressive strength of R_{28}=40-50 kg/cm^2. The reason for this is a necessity of early strength for large mining equipment and less ore loss and dilution. For a free stand height of 6 m when rib pillar being mined with cut-and-fill the cement to tailings ratio is always selected from 1:8 to 1;10 which is equivalent to a compressive strength from 20 to 30 kg/cm^2 at curing period of 90 days. On these basis, the cement to tailings ratio adopted in recent years ranges from 1:5 to 1:8, a figure which is equivalent to a cement consumption of 160 to 300 kg/m^3 of cemented tailing fill.
Strength test results from the

samples made up of various ratios of cement-tailing mixtures are listed in the Table 1.

According to the mine statistics in the past years, the criterion of cemented fill strength (i.e. R_{28} = 25 kg/cm^2, R_{90}=40 kg/cm^2) has amounted to more than 70%.

5 EXPERIMENTS AND APPLICATIONS OF A MECHANIZED HORIZONTAL CUT-AND-FILL METHOD

The key to raise the mine production lies in increasing the ore-removal capacity in stopes. The experiments on mechanized cut-and-fill method was initiated in the No.4 orebody of the Shiling area in 1977.

The No.4 orebody, an isolated one between the Jinxingling and Shiling areas, occurs at an elevation ranging from -60 to -200m. It is 110 m long, and east-west strikes and northerly dips 70° to 80°. The horizontal width is 20 to 30 m, with 50m at maximum. Confined by faulting, the orebody is complex in shape with moderately competent ore and wall rock, but well-developed joints and fractures. Experiment mining was carried out from -120 to -60 m level, the level interval was 50 to 80m. The whole orebody was designed as a panel, devided into 5 rooms and 4 pillars. The stopes were transversal to strike, with the room 14 m wide and pillars 8 m as shown in Fig. 5.

5.1 Stope preparation

A combined preparation was adopted with openings both in ore and in rock. The preparation in ore consists of a ventilation raise and 2 built-up raises in each room, while fringe preparation includes a footwall ramp of gradient 20%, sublevels access and fringe ore passes.

The stope ramp was mainly for equipment, man and ventilation. The cross-section of ramps is 2.8 x 2.6 m for 1.3m^3 LHD, 3.4 x 3.2 m for 2 m^3 LHD with a grade from 20 to 25% and a turning radius from 8 to 12m, both ramps and sublevels are drifted in the footwall. Man and equipment enterance into stopes from sublevels is via access. Ramps are connected to each sublevel, going

up as a switch-back. The ramp floor is placed with a layer of 0.2m thick concrete. Operating practices have proved that such ramps can work very well.

5.2 Extraction technology

Horizontal holes, 1.8 m in length, are drilled with YT-30 rock drills. Cuts are 4 m high, advancing from footwall to hangingwall. Blasting is done with conventional ANFO, primed by detonators.

Ore removal is done by 2 LHD units, one in operation and the other standby. On average they remove 400 tons per day with 133 ton machine shift. When the stope is cleared for fill, the cemented tailing slurry will be prepared at the Jinxingling mixing station and passed through boreholes and underground pipelines to the upper levels in the No.4 orebody. Then the slurry drops down the boreholes in each stope to the working places. 2 slices is stoped and filled, resulting in a 4 m filling height. Before pouring of slurry, orepasses are built up and fixed with forms, while the access on the lower side is sealed at the same time. Water is decanted through the built-up orepasses.

5.3 Some technical-economic data (experiment data)

Some technical-economic data are as follows:
stope output   133 ton/day
panel output   400 ton/day
drill performance   64.3 ton/drill
productivity (stope) 12.3 ton/man
productivity(underground)6.1ton/man
ore loss   1%
ore dilution   9.7%
filling cost   RMB 30.97/m^3
direct mining cost   RMB 11.24/ton

5.4 Improvements in mechanized cut-and-fill method

Experiments have demonstrated that when ore removal is only by LHDs with no matched equipment, stope output has already reached an average of 133 ton/day, doubling stope output where T2G loaders are used in conventionally filled stopes.

This fact has changed the parlance that cemented fill mining is complex in technology and low in productivity . In continued testing in 1978, the maximum stope output averaged 167 ton/day, demonstrating the superiority of using high-capacity equipment. It also proved that the employment of advanced equipment for ore extraction is an important way to raise stope output. Supposed equipment such as stope drill jumbos, explosive loading trucks and other service vehicles had been well-matched, stope output and productivity would have been increased over more appreciably. Nowadays, 3 panels have been in production at the east and west ends of the -160 m level,plus the north part of the Shiling orebody (at the -200 m level) in order to achieve a production increase. According to the design,each panel will be provided with two 2 m³ LHDs (one in spare), one 2-drill jumbo and service units like explosive-loading vehicles,maintenance vehicles, supply vehicles, roof-scaling

flatforms,etc.. With the combined operations of all these machines, the stope and panel outputs will be raised to 200 ton/day and 600 ton/day respectively.

With this aim in mind and constant improvements, the mechanical cut-and -fill method has become the main mining method in the Fankou Lead-zinc Mine.

REFERENCES (all in Chinese)

1980 Reference for design of fill mining method at metal mines. Beijing. Metallurgical Industry Publishing House
1978. Fill method of mining. Beijing Metallurgical Industry Publishing House
Pan Jian. 1981. Calculation of tailing sand fill system. Changsha. The Society of Metals of Hunan Province.
1978. Experiments on mechanized cut-and-cemented fill mining method. Non-ferrous Metals, No.4.

Table 1   Strength of cemented tailing fill

cement tailings ratio	concentration %	water cement ratio	cement consumption (kg/m³)	compressive strength (kg/cm²) R_{28}	R_{90}
1 : 2	75	0.065	643.7	140.0	——
1 . 3	75	1.ooo	466.6	124.0	——
1 : 4	75	1.330	378.5	85.0	118.9
1 : 5	75	1.66	311.8	50.7	90.0
1 : 6	75	2.00	280.1	44.0	74.0
1 : 8	75	2.66	209.4	29.3	43.0
1 : 10	75	3.33	164.8	23.0	34.8

Cement strength : $R_7 = 233$, $R_{28} = 340$, $R_{90} = 363$ kg/cm².

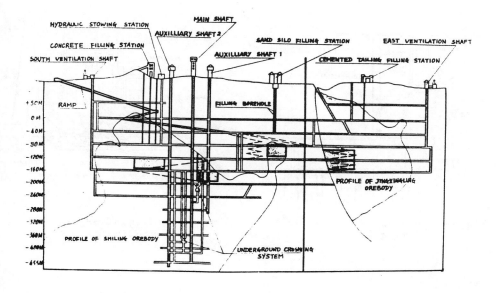

Fig.1   Diagram of development system

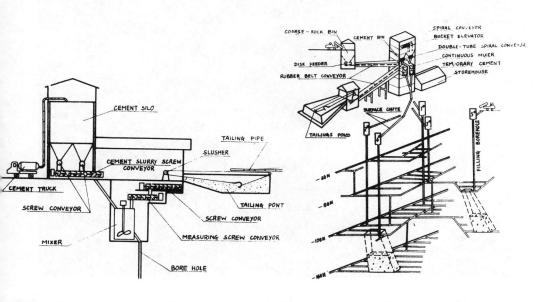

Fig.2   Diagram of cemented
         tailing fill

Fig.3   Diagram of concrete
         filling system

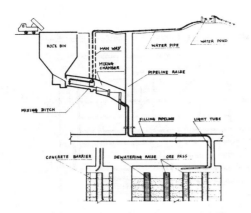

Fig. 4 Diagram of hydraulic stowing

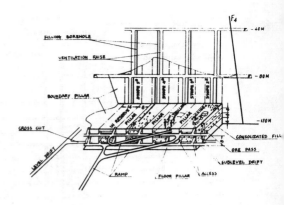

Fig. 5 Mechanized cut-and-fill stoping

Fig.6 LHD in a cut-and-fill stope,ladder and a filling pipe down from ventilation raise

3. Geomechanics

Design and properties of stiff fill for lateral support of pillars

G.E.BLIGHT
University of Witwatersrand, South Africa

L.E.CLARKE
Johannesburg Consolidated Investment Co. Ltd., South Africa

ABSTRACT: A method is described for designing and testing a stiff fill material to provide lateral support to reef pillars in a deep gold mine. A simple theory for assessing the increase in strength of the pillars as a result of lateral support by the fill is developed.

INTRODUCTION

Computer simulations of the mining situation at a deep gold mine have shown the stability of a planned extension to be critically dependent on the lateral support that can be provided to stabilizing rock pillars by surrounding them with a stiff, high modulus fill. This paper will describe the basis for the design of the fill, as well as methods for measuring the fill properties. It will also describe the results and an analysis of tests on model fill-pillar systems to assess the strengthening effect of the fill.

Because of the availability of fill materials it was necessary also to consider the design of a lower modulus fill for use in areas where stresses are expected to be less than critical.

DESIGN OF FILLS

The available fill materials, namely a fine silt-sized tailings and a development waste rock, lend themselves to the design of the two distinct types of fill which will be referred to as "stiff" and "soft". Figure 1 shows the grading analyses of the tailings and the crushed waste rock used in the investigation. The diagram also shows the gradings of the binders (ordinary Portland cement (OPC) and pulverized fuel ash (PFA)) as well as the combined grading of a typical stiff fill.

Stiff fills were designed to consist of a relatively rigid skeleton of waste rock particles having the interstices filled with cemented tailings. To proportion the

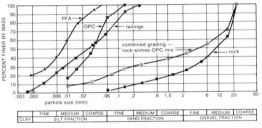

Figure 1: Grading curves for components of a stiff fill.

fill, the void content of the packed waste rock was first determined, and sufficient cemented slimes was provided to fill the voids and leave a small surplus of fines to render the freshly mixed material pumpable.

Mix proportions that appeared to give a good compromise between stiffness and binder cost were:
5 parts by dry mass of waste rock
0,4 part tailings
0,6 part OPC
0,6 part PFA
0,8 part water (12% by dry mass).

An initial indication of the water required for pumpability was obtained by subjecting the mix to the Vebe vibratory compaction test and the standard slump test used in concrete technology[1].

Figure 2 shows the results of a typical set of Vebe and slump tests. Initially, based on experience with pumping concrete, a Vebe time of 5 seconds or a slump of greater than 75 mm was taken to indicate a consistency suitable for pumping. In the light of later experience with the pumping tests, it appears that a Vebe time of 3 seconds would have been more appropriate.

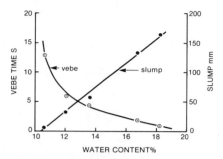

Figure 2: Preliminary determination of water requirement for pumpability of mixes.

Soft fills consisted of cemented tailings. The cement for both stiff and soft fills consisted of ordinary Portland cement (OPC) or a mixture of OPC and PFA.

The mix proportions that appeared to be most effective in terms of binder cost were

10 parts tailings
0,5 parts OPC
1,0 part PFA
3,3 parts water

proportions being by mass of dry material.

TESTING OF FILLS

When the fill is in place in the excavation it will be subjected to essentially one-dimensional vertical compression. To simulate this condition in the laboratory, trial mixes of fill were cast in standard steel compaction moulds to form specimens 150 mm in diameter and 100 mm high. The specimens were sealed against moisture loss by means of a 20 mm thick layer of paraffin wax and were cured at 40°C (the approximate underground temperature) for 7 days before testing. The cylindrical walls of the moulds were instrumented with electric resistance strain gauges spaced at 120° around the circumference of the mould. This enabled both vertical and circumferential strains in the mould cylinder to be measured as the fill was compressed vertically. Vertical strains were used to correct the applied load for friction between the mould and the fill and circumferential strains to assess the lateral stress developed in the fill. As the vertical compression can only occur by expulsion of water from the pores of the fill, compression tests were carried out at a rate slow enough to enable the pore water to escape without building up any water pressure in the fill[2].

Figure 3 summarizes the results of the

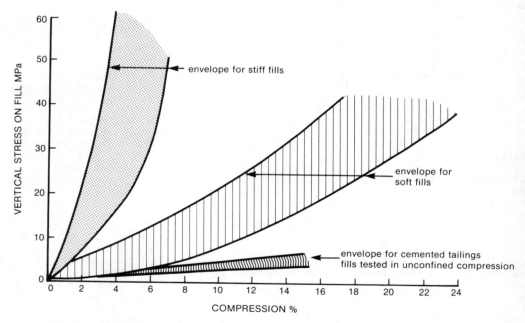

Figure 3: Confined compression tests on stiff and soft fills: stress-compression envelopes.

304

compression tests on soft and stiff fills and shows how both materials progressively stiffen as they approach a limiting strain at which the fill becomes, to all intents and purposes, incompressible. Figure 3 also shows the results of earlier tests[3] in which the compression characteristics of a soft fill were assessed by compressing flat unconfined slabs of material. The comparison clearly illustrates the stiffening effect on the fill of lateral confinement. Figure 4 summarizes the relationship between vertical and lateral stresses in both stiff and soft fills as they are subjected to laterally confined compression. The ratio of lateral to vertical stress in stiff fills varies from about 0,1 to 0,45 whereas in soft fills the ratio varies from about 0,4 to 0,7. These results illustrate the supportive effect that a fill can potentially have on a loaded pillar, although it must be remembered that appreciable vertical strains are necessary to develop the lateral stresses.

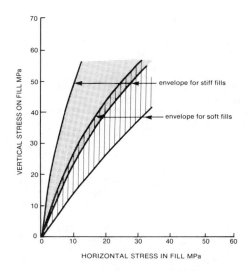

Figure 4: Confined Compression Tests on stiff and soft fills: Vertical stress — Horizontal stress envelopes.

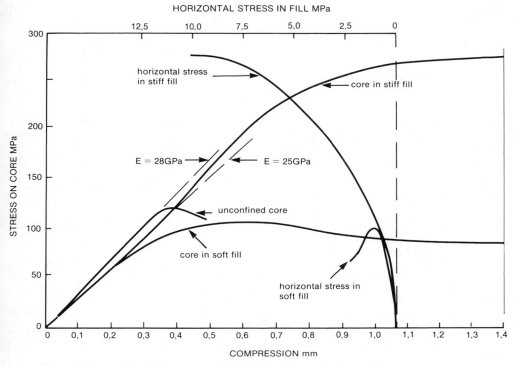

Figure 5: Comparison of stress-compression behaviour of unconfined core with fill-supported cores.

MODELING OF FILL PROPERTIES

For the purposes of the present paper, a pseudo-elastic approach has been adopted, with the stiffness of the fill being represented by a equivalent elastic modulus. If necessary, the complete stress-strain relationship for the material can accurately be represented by the expression:

$$\varepsilon_v = A(1-e^{-\sigma_v/B}) \quad \dots\dots\dots\dots\dots\dots\dots (1)$$

in which ε_v is the vertical compression of the fill, σ_v is the vertical stress and A and B are compression parameters. A represents the limiting strain (referred to above) at which the fill becomes incompressible, while B is a compressibility modulus with units of stress. A can be used as a figure of merit for the fill - the smaller A, the stiffer the fill. For the stiff fills tested in this programme A varied from about 5% to 10% while B varied from 30 MPa to 50 MPa.

For the soft fills A varied from about 20% to 40% while B varied from 15 MPa to 30 MPa.

A given fill subjected to confined compression will have an upper limit to A equal to the porosity (volume of voids ÷ total volume) of the fill when placed.

LATERAL SUPPORT PROVIDED TO PILLARS BY FILL

To assess the extent and effect of lateral support expected to be provided to rock pillars by the fill, a series of tests were performed in which quartzite drill cores representing the pillars were embedded in either soft or stiff fill contained in the moulds that had earlier been used to assess the properties of the fill. The composite core-fill system was loaded via a rigid piston. The stress-compression relationship for the core and the relationship between compression of the core and lateral stress in the fill were recorded.

A typical set of these relationships is shown in Figure 5. The figure shows that the soft fill provided very little lateral support to the core, as the strengths of the unconfined core and the core supported by soft fill were almost identical. Relatively little lateral stress was generated in the soft fill, but there was enough lateral support to maintain a post-failure strength of about 85 per cent of the peak strength.

A considerable lateral stress was genera-

ted in the stiff fill and Figure 5 shows that a peak strength was not reached by the core even at a compression of 1,4 mm or 1,75 per cent. It therefore appears that surrounding a highly stressed pillar with a soft fill will not materially improve the strength of the pillar, but will provide a high level of residual resistance. A stiff fill, on the other hand, can considerably increase the strength of a pillar.

Figure 5 also shows that because the fill must be laterally compressed in order to provide lateral support, surrounding a pillar with fill have little or no effect on the pre-failure compression modulus of the pillar.

MODELING THE LATERAL SUPPORT MECHANISM

Figure 6 represents a cylindrical pillar of initial height S and diameter D which is compressed at approximately constant volume by an amount v. The lateral expansion h of the pillar is then given approximately by:

$$h = v \cdot \frac{D}{2S} \quad \dots\dots\dots\dots\dots\dots\dots\dots\dots\dots\dots (2)$$

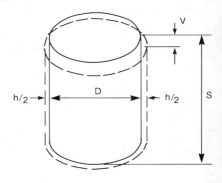

Figure 6: Dimensions used in analysis of pillar-fill system

The pressure required to expand a cylindrical pillar into a surrounding elastic fill is given by

$$\Delta\sigma_h = 2G \cdot \frac{h}{D} = G \cdot \frac{v}{S} \quad \dots\dots\dots\dots\dots\dots\dots (3)$$

(The process of expansion can be likened to the expansion of a cylindrical cavity in an elastic medium[4].)

$\Delta\sigma_h$ is the expanding pressure, i.e. the supporting pressure on the pillar and

G is the shear modulus of the fill.

For an elastic material, the shear modulus is related to the elastic modulus E by

$$G= \frac{E}{2(1+\nu)} \quad \dots\dots\dots\dots\dots\dots (4)$$

where ν is Poisson's ratio.

ν and E for the fill can be evaluated from the results of one-dimensional compression tests such as those described by Figures 3 and 4.

It can be shown that if a material is subjected to vertical one dimensional compression the ratio of horizontal to vertical stresses is

$$\sigma_h/\sigma_v = K_o = \frac{\nu}{1-\nu} \quad \dots\dots\dots\dots\dots (5a)$$

$$\text{or } \nu = \frac{K_o}{1+K_o} \quad \dots\dots\dots\dots\dots (5b)$$

Also, for such a process, the vertical strain

$$\varepsilon_v = v/s = \frac{\sigma_v}{E} \cdot \frac{(1+2K_o)}{(1+K_o)}$$

Hence $G = \frac{\sigma_v S}{2v} \cdot \frac{(1-K_o)(1+K_o)}{(1+2K_o)} \quad \dots\dots\dots (6)$

or, from equation (3)

$$\Delta\sigma_h = \sigma_v \cdot \frac{(1-K_o)(1+K_o)}{2(1+2K_o)} \quad \dots\dots\dots\dots (7)$$

In equation (7) σ_v represents the vertical stress in the fill surrounding the pillar, while $\Delta\sigma_h$ represents the additional horizontal stress generated in the fill when the pillar and the fill have both compressed by an amount v.

The additional vertical strength of the pillar at this stage will be given by

$$\Delta\sigma_v^P = \Delta\sigma_h \frac{1+\sin\phi^P}{1-\sin\phi^P} \quad \dots\dots\dots\dots\dots (8)$$

where ϕ^P is the angle of shearing resistance of the cohesionless fractured rock of which the pillar now consists.

For a very stiff fill at a compression of % (see Figure 3 and 4) $K_o=0,2$ and $\sigma_v=20$MPa. Hence

$$\sigma_h = 7\text{MPa}$$

Reference to Figure 5 will show that in the model tests referred to earlier, $\Delta\sigma_v^P$ was about 100 MPa when $\Delta\sigma_h$ was 7 MPa. i.e.

the ratio $\frac{1 + \sin\phi^P}{1 - \sin\phi^P}$ was 14 which corresponds

to a value for ϕ^P of 60°.

This is a perfectly possible value for a confined fractured rock.

For a good soft fill at a compression of 2%, $K_o=0,45$ and $\sigma_v=5$MPa. Here

$\Delta\sigma_h=1$MPa which is again similar to the measured value in Figure 5.

If the failure strain for the confined pillar is taken to be the same as that for the unconfined core, $\Delta\sigma_v^P$ was about 12 MPa which corresponds to a value for ϕ^P of 58°.

Hence there is a reasonable correspondence between the results of the model tests and the predictions of equations (7) and (8) if suitable values are used for ϕ^P.

CONCLUSION

A basis has been described for designing stiff fills and for evaluating their properties. The additional strength imparted to a rock pillar by the lateral support of a fill can be predicted from the measured parameters for the fill. Model tests show that the predictive procedure described above yields credible results.

REFERENCES

1. Neville, A.M. "Properties of Concrete" Pitman International, London, 3rd Edition, 1981, pp 208 and 214.
2. Blight, G.E. "The Effect of Nonuniform Pore Pressures on Laboratory Measurements of the Shear Strength of Soils" Laboratory Shear Testing of Soils. ASTM Special Technical Publication No. 361, 1963, pp 173-191.
3. Blight, G.E., More O'Ferral, R.C. and Avalle, D.L. "Cemented Tailings Fill for Mining Excavations", Proceedings, 9th International Conference on Soil Mechanics and Foundation Engineering, Tokyo, 1977, Vol I, pp 47-54.
4. Baguelin, F, Jezequel, J.F. and Shields, D.H. "The Pressuremeter and Foundation Engineering", Trans Tech, Switzerland, 1978.

The application of the finite element model of the Näsliden mine
to the prediction of future mining conditions

T.BORG
University of Luleå, Sweden

N.KRAULAND
Boliden Mineral AB, Sweden

ABSTRACT: Finite element models of the Näsliden mine, Sweden, were developed and their agreement with mine behaviour was checked in the Näsliden project. This paper describes the application of an elastic FEM-model to the prediction of mining conditions to the end of the mine's life. The extension strain failure criterion is used for prediction of potential failure zones. The criterion is calibrated against actual mine behaviour in one stope. Stages of failure are predicted for the remaining stopes and the predictions are checked against available evidence. A comparative prediction based on the Coulomb failure criterion is also given.

1 INTRODUCTION

The present work is a continuation of the Näsliden project, which aimed at evaluating the suitability and reliability of finite element models as a rock mechanics planning instrument in mining. In the project, a linearly elastic model and a model, in which joint elements were included, simulating the weak alteration zones adjacent to the orebody, were developed. The degree of agreement with actual mine behaviour was established by comparison with rock mechanics observations and measurements carried out in the Näsliden mine from the start of mining in 1970. A comprehensive presentation of the Näsliden project has been given at the Conference on the Application of Rock Mechanics to Cut and Fill Mining, Luleå, June 1980, Stephansson and Jones, 1981.

For the purpose of the present study the main results of the Näsliden project were
- the development of a linearly elastic model with a known degree of agreement between model and mine behaviour
- a joint element model giving better qualitative information on the rock behaviour in the immediate vicinity of an excavation when the rock mass approaches a state of failure. However, introduction of joint elements increased quantitative uncertainty mainly due to difficulties in determining the necessary mechanical parameters of joints. Complexity and cost of model work increased significantly.

The Näsliden project covered the early stages of mining. Only one stope was approaching the final stage of excavation and experiencing rock fall problems.

The aim of the present study is to develop a method for the prediction of future mining conditions. In cut and fill mining the stresses across the remaining orebody increase as mining proceeds. This process continues until loading is sufficient to cause failure, either in the orebody (starting in the roof of the stope) or in the adjoining sidewall. From the miner's point of view it is important to obtain a prediction of those stages of mining, at which mining has to be stopped or the mining process has to be changed either by changing the mining method or introducing new stabilization measures. The study is therefore concentrated on the prediction of those types of failure that are due to excess loading of the rock mass.

2 CUT AND FILL MINING IN THE NÄSLIDEN MINE

The geology and structures of the Näsliden mine have been described by Stephansson, 1981, Fig. 1. Cut and fill mining is used and the orebody is mined in slices 3.5 m high. The open stope is backfilled with either sand or mill tailings. After backfilling mining of the next slice starts with entries from ramps in the footwall, Fig. 2.

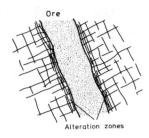

Fig. 1 Vertical section across orebody, alteration zones and sidewall of the Näsliden mine, Stephansson, 1981.

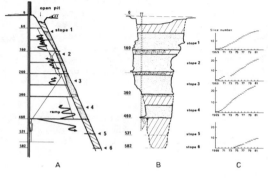

Fig. 2 Vertical sections of the Näsliden mine. A) cross section, B) section along the strike, C) mining sequences.

At present mining is carried out in four stopes, starting at 460 m, 355 m 255 m and 155 m levels. Ramps are being driven for the opening of stopes 5 and 6 in years 1983 to 1984. An open pit was mined to the 40 m level during 1972-1973. The annual production in 1975 was 250 000 tonnes of ore.

3 METHOD

Considering the aim of the present study, namely to develop a prediction method suitable for application to practical mining problems, it was decided to use the concept of potential failure zones described by Hoek 1967. Brittle fracture behaviour of the rock is assumed. This approach comprises the following steps:
    a) Calculate the elastic response of the rock mass to mining in terms of stresses and/or strains.
    b) Determine a failure criterion for the rock mass.

c) Critical stages in the development of failure are defined from observations in the mine:
    - onset of failure (brittle fracturing, often violent)
    - onset of rock fall problems; mining can still be continued
    - ultimate failure; mining has to be stopped or changed
The criteria for these levels are quantified by comparing observations from stope 3 in the mine with calculated values of stresses/strains for the same stage of mining.
    d) Compare elastic stresses/strains with rock strength expressed in terms of a failure criterion. The loci of points where the stresses/strains equal the values given by the failure criterion define the boundary of the potential failure zone.
    e) Prediction of the critical levels for all mine stopes up to the end of mining. Validity of this prediction is checked by comparing predicted critical levels for the stopes with practical experience from the mine.

4 OBSERVATIONS OF FAILURES IN THE MINE

An important part of the follow up program in the Näsliden mine is the observation of failure phenomena. These were presented by Nilsson and Krauland, 1981.

4.1 Types of failure

Fig. 3 shows the common types of failure in cut and fill mining as observed in Näsliden and other cut and fill mines of the Boliden Mineral AB.

4.2 Sequence of events

Considering roof failure due to high, nearly horizontal stresses the following critical levels in the development of failure can be distinguished in the stopes

Stage 1: Brittle fracture of the roof occurs, often violently; nearly horizontal fracture surfaces are created thereby. Stability of the stopes is maintained by fairly sparse rock reinforcement, mainly rock bolting.

Stage 2: Failure of the face consisting of subsidence of large wedges and continue roof failure. Mining can still be continue but frequent scaling and an increased amount of support are necessary.

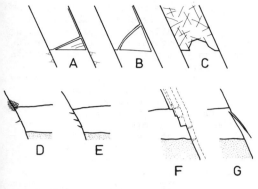

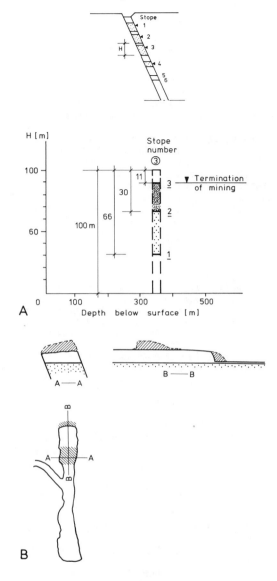

Fig. 3 Types of failure. A) wedge formation in roof, often violent, B) roof failure when hanging wall is very weak, C,
roof failure due to structural features
in the roof, D, E) footwall failures,
F, G) hangingwall failures, Nilsson et al,
1981.

Stage 3: Ultimate failure, continuous roof
failure including crushing, also at a
large distance behind the face. Stable
roof conditions cannot be achieved by
scaling and roof bolting. Termination of
mining or changeover to another mining
method/stabilizing measures.
    A graphical presentation of the field
observations defining stage 1-3 is given
in Fig. 4, where also the failures termi-
nating the mining in stope 3 are illustrated.

4.3 Variations in failure phenomena and
     stability conditions

There is, of course, a fairly wide variation in geological conditions and hence
in mechanical properties of the rock mass,
resulting in variations in rock behaviour.
One significant difference is the composition of the footwall alteration zone.
In the southern part of stope 3 it consists of hard sericitic quartzite. In
this area of the mine violent failure of
the roof occurred, whereas the footwall
remained intact. In the northern part of
the orebody the footwall alteration zone
consists of soft chloritic quartzite with
low strength and a low modulus of elasticity. Here the roof in the ore body remained intact, whereas failure of the
footwall occurred frequently.

Fig. 4 A) Development of failure stages
in stope 3 in Näsliden, B) failure of face
and roof far behind the face at termination of mining in stope 3.

5 FAILURE PREDICTION

The extension strain failure criterion
(ESFC) suggested by Stacey, 1981, was used
in this study for the prediction of failure.
The ESFC is stated by Stacey as follows:
"Fracture of brittle rock will initiate

311

when the total extension strain in the rock exceeds a critical value which is characteristic of that rock type. This may be expressed as follows: Fracture initiates when

$$\varepsilon_3 \geq \varepsilon_c \qquad (1)$$

where ε_c is the critical value of extension strain and ε_3 is the minimum principal strain.

The fractures will form in planes normal to the direction of the extension strain, which corresponds with the direction of the minimum principal strain. For a material which shows ideal linear elastic deformation behaviour, the strain in this direction is related to the three principal stresses by the following equation:

$$\varepsilon_3 = \frac{1}{E}\left[\sigma_3 - \nu\,(\sigma_1 + \sigma_2)\right] \qquad (2)$$

where σ_1, σ_2 and σ_3 are the principal stresses, E is the modulus of elasticity, and ν is Poisson's ratio.

From this it can be seen that if $\nu(\sigma_1 + \sigma_2) > \sigma_3$ then an extension strain will occur. This illustrates that extension fractures can form when all three principal stresses are compressive and thus also in planes across which the net macro stress is compressive. Moreover, after the fracture has formed the net macro stress across the fracture may remain compressive."

The Coulomb failure criterion (CFC) has also been applied. The CFC is given by the equation:

$$\tau = c + \sigma_n \tan \phi \qquad (3)$$

The angle of the failure plane is given by the relation:

$$\beta = \pm(45 - \phi/2) \qquad (4)$$

Equation 3 is a linear approximation to the nonlinear Mohr's envelope which is exhibited by most hard rock types. Roof failure in cut and fill stopes occurs in areas where the confining pressure, σ_3, has very small values. The linear approximation is therefore considered acceptable.

Comparing the ESFC with the CFC the following aspects came forward with regard to the ESFC:
- tensile stressed do not present any difficulties
- the direction of the failure plane is clearly defined whereas in the case of the Coulomb criterion agreement between criterion and observations has been subject to discussion

- only one constant has to be determined There is no experience available with regard to suitable methods of determining the critical extension strain. The influence of size on critical extension strain is also largely unknown. Stacey states, however, that good agreement was obtained when the same value of critical strain (0.0002) was applied to both physical models and to in situ observations
- the ESFC is limited in this application to small confining stresses as indicated by the relationship $\nu(\sigma_1 + \sigma_2) > \sigma_3$. This limitation does not apply to the CFC.

Inspection of failure surfaces in the mine did not give any clear indications that any of these criteria should be excluded or preferred.

On the basis of these considerations it was decided to use both criteria and carry out a comparison of the results.

6 FINITE ELEMENT MODEL OF THE NÄSLIDEN MINE

Failure prediction is based on determination of the elastic response of the rock mass to mining. By comparing this response with a failure criterion, potential failure zones are determined. The criterion is calibrated against practical experience from stope 3 in Näsliden, where mining had to be stopped due to severe stability problems. Predictions are then made for the other stopes for the remaining lifetime of the mine.

The final model of the Näsliden project was used for the present study, Groth and Jonasson, 1981. The composition of the model with regard to rock types and elastic properties is given in Fig. 5, which also shows the properties of the fill and the data for the virgin stress field based on stress measurements, Leijon et al, 1981. Plane strain conditions were assumed for the model. The size of the model was chosen so as to allow complete recovery of the ore to a depth of 582 m.

6.1 Mining sequences

The mining stages were chosen as follows, Fig. 6:
- Three historic mining sequences, 1-3, were selected in which failures important for the calibration of the failure criterion occurred.
- On the basis of long term mining plans three future mining situations, 4-6, were selected, at which significant changes in

mine behaviour were expected from the rock mechanics point of view.

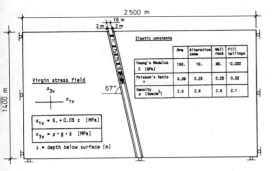

Fig. 5 Virgin stress field, material data, model geometry, and boundary conditions of finite element model.

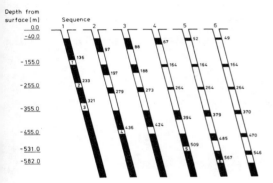

Fig. 6 Mining sequences simulated in the finite element model.

7 APPLICATION OF THE EXTENSION STRAIN FAILURE CRITERIA (ESFC)

With regard to practical application in mining failure prediction should aim primarily at prediction of ultimate failure (stage 3). This stage as well as stage 2 are affected by mining technology. Criteria for these stages have to be determined therefore empirically by calibration against mine observations. Stage 1 (onset of failure) is assumed to be independent of mining technology.

7.1 Quantification of the ESFC

The ESFC has been calibrated against impor-

tant events in stope 3 of the Näsliden mine, defined as the three critical stages discussed in Section 4.2. Observations of these stages are summarized in Fig. 4 A and in Table 1. Figure 4 B shows failure occurrence in stope 3 at ultimate failure.

Table 1. Stages of failure as observed in stope 3 and extension strain levels in model.

Stage of failure	Year	Height of back-filled excava-tion [m]	Height of ore between stope 2 and 3 [m]	Eleva-tion of roof [m]	Exten-sion strain ε_3 $[10^{-6}]$	Cohesion c [MPa]
1	1973	34.	66.	321.	150.	12
2	1979	70.	30.	285.	210.	16
3	1981	89.	11.	266.	300.	20

The position in the model where the maximum extension strain occurs is used for quantification of the failure criterion. The minimum principal strain, ε_3, in the lowest row of finite elements across the roof of the stope is plotted in Fig. 7 A for mining sequences 1-4 as shown in Fig. 6. The maximum values of ε_3 at the early mining stages occur at a distance of about 3 m from the hangingwall. This point is referred to as Q_1. At a late stage in mining, sequence 4, and when the sill pillar condition is reached, the position of the maximum value shifts to the proximity of the footwall, Q_2.

Fig. 7 B shows ε_3 as a function of the height above the roof. Only a narrow zone of the roof is subjected to high extension strain until the sill pillar condition is reached when high ε_3 values occur over a large part of the orebody. Development of ε_3 at point Q_1 for all six stopes as a function of the height of th orebody between adjacent stopes is shown in Fig. 7 C. Determining ε_3 for stope 3 for the three defined stages of Table 1 gives the ε_3 values, shown in the same table.

Stress measurements were performed in the roof of stope 3 at roof elevation 295 m, at a distance of about 5 m from the footwall, Leijon et al 1981. In the ore adjacent to the roof surface, violent failure occurred at the time of the measurements. Core discing was observed during the rock stress measurements and only measurements with door-stoppers were successful. At a depth of 2.25 m from the surface of the roof the first successful measurement with a three-dimensional Leeman cell was obtained, giving strain values of 160 μs. This result is in good agreement with the estimated critical strain value 150 μs, Table 1, earlier estimated from model results.

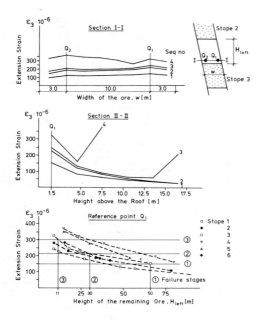

Fig. 7 Development of extension strains in the ore.

7.2 Failure prediction with application of ESFC

The potential failure zones were determined by calculating for each element the ratio $\varepsilon_3/\varepsilon_{3c}$ where ε_{3c} is the critical strain, e.g. for stage 1 $\varepsilon_{3c} = 150 \cdot 10^{-6}$. Elements with $\varepsilon_3/\varepsilon_{3c} > 1$ are within the potential failure zone. The development of critical extensions around the stopes for mining sequences 1–6 is plotted in Fig. 8.

In a similar way as for the ore, the critical extensions were determined for the alteration zones, $\varepsilon_c = 600$ µs, and the sidewall rock, $\varepsilon_c = 400$ µs. The results indicate a narrow critical zone in the roof of the stopes in the early stages of mining. Critical strains also occur at the hangingwall side in the alteration zone above and immediately beneath the roof level and on the footwall side beneath the roof level.

At a late stage in mining, when the sill pillar condition is reached, critical strains also occur in the sidewall at the levels of the pillar and within large part of the pillars. This indicates that failure of the pillars alternatively punching

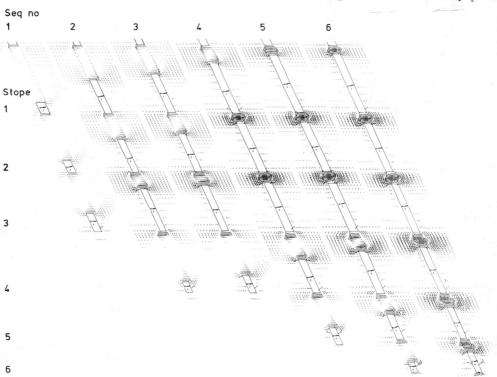

Fig. 8 Development of zones of critical extensions in the FEM model for mining sequences 1–6 in the Näsliden mine.

314

of the pillars into the sidewall are possible modes of failure.

Finally the development of extensions at
the reference point Q_1 in Fig. 7 C is used
to determine the critical stages for stope
1, 2, 4, 5 and 6, as shown in Fig. 9.

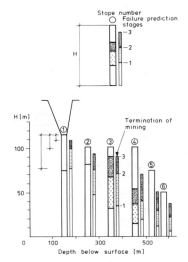

Fig. 9 Comparison of failure observations
in Näsliden up to 1982 with prediction of
roof failure using the ESFC.

7.3 Comparisons of the prediction and
in situ observations

Comparing failure observations (Fig. 4)
with predictions as presented in Fig. 9
shows:

Stope 1: Roof failures have occurred in
few locations. Good agreement.

Stope 2: Fracturing has been observed only
during mining of the last two slices.

Stope 3: Not applicable to this analysis
as the observations are used for the calibration of the model.

Stope 4: Violent failures started somewhat
earlier than predicted. Serious failures
in the hangingwall and in the roof of the
mining face were experienced from a height
of the backfilled excavation of 36 m. An
unfavourable joint orientation affects the
situation.

Footwall drifts along the orebody exist
at 160 m, 260 m and 360 m levels. They are
situated mainly in the alteration zone at
a distance of 0 to 4 m from the orebody.
Extensive damage to these drifts is ob-

served as mining approaches from below and
as the height of the orebody between the
stopes above and below the drifts decreases.
Previous investigations have shown that
onset of failure in the drifts is well
correlated to the maximum principal stress
of approximately 40 MPa at the position of
the drift, Krauland et al, 1981.

8 FAILURE PREDICTION USING THE CFC

8.1 Quantification of the CFC

The CFC was calibrated against mine behaviour in a similar manner as the ESFC. Calibration requires the determination of the
angle of internal friction, ϕ, and the
cohesion intercept, c. On the basis of
triaxial compression tests of brittle rock
under low confining pressure reported by
Hoek and Brown, 1980, an angle of internal
friction of $\phi = 50°$ was assumed. This angle
is in good agreement with the angle calculated from failure plane observations in
the Näsliden mine. The cohesion, c, was
determined from the τ values in Q_1 at the
three stages of failure, Table 1. Figure
10 shows the development during mining of
τ/τ_c in point Q_1 for all the stopes. Stope
3 was used for calibration.

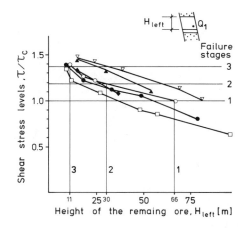

Fig. 10 Development of shear stresses in
the ore for stope 1–6 and determination of
critical shear stresses for the three failure stages in stope 3. Notations, see
Fig. 7.

8.2 Failure prediction with the CFC and
comparison with the ESFC

From Fig. 10 a prediction was obtained for
the onset of the three failure stages in
all the stopes. The results are practically

315

identical to the results obtained with the
ESFC shown in Fig. 9.

The development of failure (stage 1)
around the stopes as predicted by the CFC
was plotted in an analogous way as that
shown for the ESFC in Fig. 8. Figure 11
shows a comparison of failure predictions
based on CFC and ESFC for stage 1 failure
in the roof of stope 3 for three mining
stages. Figure 11 shows that the direction
of the failure plane is predicted reasonably well by both criteria. There is a
difference in the extent of the potential
failure zone. This is shown more clearly
in Fig. 12.

There are too few mine observations to
validate the predictions. Observations
from the final stages of stope 3 indicate
that the sill pillar is in or very near a
state of failure. No major deformations
have occurred, however.

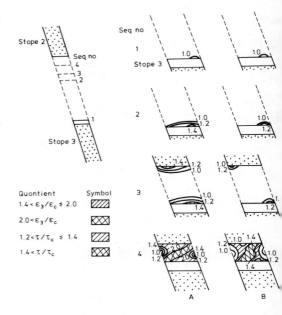

Fig 12 Development of potential failure
zones in stope 3 based on ESFC (left) and
CFC (right).

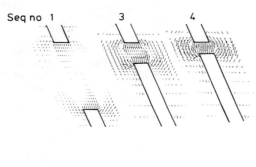

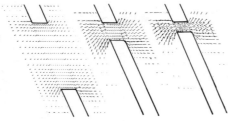

Fig. 11 Comparison of predicted failure
development in stope 3 based on ESFC (top)
and CFC (bottom).

9 DISCUSSION

9.1 Brittle failure of the roof and its modelling

The elastic model that was used in this
study has only been validated for mining
stages up to 1979. At later stages of

mining failure processes become more dominant and deviations from the elastic model
are likely to increase, especially in the
vicinity of the stopes. For these late
stages of mining dilatation will result in
redistribution of stresses and strains.
From triaxial compression tests it is well
known that axial strain often shows linearity up to very near peak strength. The
lateral strain, ε_3, however, starts to
increase beyond linearity already from the
stage of fracture initiation, Jaeger and
Cook, 1981. Beyond strength failure stress
redistribution is enhanced as large scale
failure surfaces form and the volume of
rock involved in failure increases. This
stress redistribution combined with continued loading may lead to progressive
failure and continuous rock falls.

The implications of these processes for
the model simulation are that calculated
strains ε_3, at the boundary of the potential failure zone are expected to be smaller than the strains at strength failure
in the structure. The shape of the potential failure zone is expected to reflect
the real failure zone in the early stages
of failure fairly well, but deviations
increase as failure advances. Progressive
failure is not simulated.

9.2 Alternative types of failure

Besides brittle failure of the roof, discussed in chapter 9.1, failure or rock fall of the sidewalls present serious problems in cut and fill mining. They induce downward movements in the orebody above the roof, which results in additional extension strains in dip direction, high above the roof of the stope, Krauland, 1975. Falls of slices of ore up to 3-5 m in thickness have been observed in several cut and fill mines of the Boliden Mineral AB.

The joint model presented by Groth and Jonasson, 1981, allows to simulate the influence of the sidewalls and provides a better insight into the strain and stress distributions in the vicinity of the stope. However, the tested joint model is a slightly improved approximation of the real conditions, as several other types of failure are feasible. Detailed modelling of all failure modes is a very difficult problem. The determination of relevant rock mass properties in the post failure regime and the simulation of progressive failure are not yet applicable to mine planning.

9.3 Failure criteria

The extension strain criterion was applied successfully. Comparison with the Coulomb criterion showed very similar results with regard to onset of the roof failure stages. The potential failure zones predicted by CFC and ESFC are rather narrow close to the roof until a late stage of mining, when they extend over the whole sill pillar. This is in agreement with observations from stope 3.

The method of calibrating the failure criteria against mine observations has proven to be valuable. It takes care of some difficult problems:
- the failure stages 2 and 3 in Fig. 4 are dependent on the mining method and the support method
- the size dependency of rock strength; factors of influence on rock strength such as jointing, volume of critically loaded rock are included
- approximations and simplifications due to the modelling technique (e.g. dilatation, non-linear material behaviour) are compensated.

The obvious drawback of determining failure criteria from mine observations is that this method is not applicable at the design stage.

10 CONCLUSIONS

1) An elastic finite element model and the extension strain criterion by Stacey, 1981, have successfully been applied to predict brittle roof failures in the Näsliden mine.

2) The model of critical stages in the mine can be used to determine:
- the duration of mining with minor disturbances
- the stage of mining when an increased amount of conventional support is required
- the stage when a change to other support systems or other mining methods is necessary
- the time for close-down of a stope and the start of new production areas.

11 ACKNOWLEDGEMENT

This study forms part of a doctoral thesis by the first author. Financial support by the Swedish Rock Engineering Research Foundation is acknowledged. Thanks are due to the Boliden Mineral AB for permission to use data from the Näsliden mine. The authors would like to thank Dr. Sten G. A. Bergman for his great assistance during the project and Professor Ove Stephansson for proposing improvements in the manuscript.

REFERENCES

Borg, T. 1982. The evaluation and application of a finite element model of the Näsliden mine to the prediction of future mining conditions. Luleå, University of Luleå, Div. of Rock Mechanics (Doctoral thesis) To be published.

Groth, T. & Jonasson, P. 1981. Application of the BEFEM code to the Näsliden mine models. In O. Stephansson & M. Jones (eds), Proc. Application of Rock Mechanics to cut and fill mining, p. 226-232. London, The Institution of Mining and Metallurgy.

Hoek, E. A. 1967. A photoelastic technique for the determination of potential fracture zones in rock structures. In C. Fairhurst (ed), Proc. 8th Rock Mech. Symp., p. 94-112. New York, AIME.

Hoek, E. & Brown, E. T. 1980. Underground excavations in rock. London, The Institution of Mining and Metallurgy.

Jaeger, J. C. & Cook, N. G. W. 1979. Fundamentals of Rock Mechanics. 3rd Edition. London, Chapman and Hall.

Krauland, N. 1975. Deformations around a
  cut and fill stope - experience derived
  from in situ observations: papers presented at Rock Mechanics Meeting, Stockholm, 14 Febr. 1975, p. 202-216. Swedish
  Rock Mechanics Res. Found. Report No. 20
  (In Swedish).
Krauland, N., Nilsson, G. & Jonasson, P.
  1981. Comparison of rock mechanics observations and measurements with FEM calculations. In O. Stephansson & M. Jones
  (eds), Proc. Application of Rock Mechanics to cut and fill mining, p. 250-260.
  London, The Institution of Mining and
  Metallurgy.
Leijon, B., Carlsson, H. & Myrvang, A.
  1981. Stress measurements in Näsliden
  mine. In O. Stephansson & M. Jones (eds),
  Proc. Application of Rock Mechanics to
  cut and fill mining, p. 162-168. London,
  The Institution of Mining and Metallurgy.
Nilsson, G. & Krauland, N. 1981. Rock mech-
  nics observations and measurements in
  Näsliden mine. In O. Stephansson & M
  Jones (eds), Proc. Application of Rock
  Mechanics to cut and fill mining, p. 233-
  249. London, The Institution of Mining
  and Metallurgy.
Stacey, T. R. 1981. A simple extension
  strain criterion for fracture of brittle
  rock. Int. J. Rock Mech. Min. Sci. &
  Geomech. Abstr. (16), 6, 469-474.
Stephansson, O. & Jones, M. 1981. Proc.
  Application of Rock Mechanics to cut and
  fill mining. London, The Institution of
  Mining and Metallurgy.
Stephansson, O. 1981. The Näsliden project
  - rock mass investigations. In O. Stephansson & M. Jones (eds), Proc. Application of Rock Mechanics to cut and fill
  mining, p. 145-161. London, The Institution of Mining and Metallurgy.

Investigation of an experimental sublevel stoping method utilizing a discrete cemented backfill in a vein-shaped orebody

M.CARTA, P.P.MANCA, G.MASSACCI & G.ROSSI
Universita' di Cagliari, Italy

S.PUTZOLU
SAMIM S.p.A., Miniera di Montevecchio, Italy

ABSTRACT: The technological progress achieved in the use of low-binder concretes as mine backfill over the last decade and the growing need to attain high productivities has prompted some mining enterprises to develop mining methods utilizing cemented backfill sills. The paper reports on the results of an in-situ measurement campaign carried out in an experimental stope of the Montevecchio mine where the method was tested. The analysis of collected data highlighted some of the essential requirements for the applicability of the method. Among them, the possibility of adapting the stope morphology to the geometrical characteristics of the orebody and the homogeneity attainable in the cemented backfill preparation and emplacement appear to play a major role.

1 INTRODUCTION

The present work was conceived and carried out in the framework of a research programme aimed at investigating the possibilities of improving stoping methods in orebodies where the morphological characteristics, the rock mechanical conditions of both mineralization and country rock and the low ore grade pose technological and economical problems. Such conditions are encountered in the Montevecchio orebody. The investigations reported in this paper were carried out in the "Casargiu" section. The leanness of the ore contained therein calls for the application of high-productivity, low-cost stoping methods in a situation characterized by a country rock whose lack of stability over wide exposed surfaces does not allow the application of conventional, open-stope mining methods.

In that stope a variant of the downward-moving sublevel stoping method is being tested; this variant entails the installation of a supporting structure composed of a series of cemented backfill sills. Object of this stage of the investigation was the assessment of the possibility of increasing the distance between the cemented backfill sills, in consistency with the stope statics and with a view to obtaining economically sound productive levels.

In the first place the rock-mechanical and structural characteristics of the country rock were identified through the examination of both the exposed surfaces and drill cores. The effects produced by the advancement of stoping operations on the exposed surfaces as well as on the supporting structures, composed of sills, were then analysed by measuring the variations in the distances between pegs located along the perimeter of suitably chosen sections.

2 THE MONTEVECCHIO MINE

2.1 The orebody

The orebody belongs to a family of branched steeply dipping, irregularly shaped veins extending for several kilometres produced by the filling of a system of fractures originating in the post-Gothlandian schists cut by granitic intrusives.

Some zones of these veins, with thicknesses of the order of 10 m have been mined out in the past.

The thickness of the zone presently mined ranges from 2.5 to 4.5 m. The main mineral values are lead and zinc sulphides, often accompanied by minor amounts of chalcopyrite and pyrite. The most common gangue minerals are siderite and quartz, calcite and baryte being very scarce and restricted to small areas. Near the contacts the country rock is frequently altered, tectonized and foliated. The vein filling is generally characterized by alternating bands of irregularly distributed sulphide minerals, up to several centimetres wide, and quartz. As a whole the rock characteristics are related to its heterogeneity and to the occurrence of small cavities.

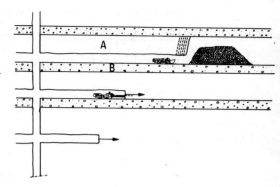

Fig.1 Sublevel mining with sills of cemented backfill. A = ore bench - blasting stage, B = cemented backfill sill

2.2 Previous stoping methods

For several decades the longitudinal overhand stoping method has been employed. With the aim of achieving the desired productivity levels within the limits imposed by the rock mechanics of the orebody and economic constraints, several alternatives of this method were adopted. These concerned the blasthole layout, either vertical or horizontal, or the backfill, hydraulic or made up of dumped gravel.

Changing economic conditions over recent years have led to the present necessity to develop stoping methods characterized by higher productivity potential.

2.3 Experimental method

The above mentioned economic requirements presently impose the utilization, whenever possible, of highly efficient rock-drilling and ore-loading equipment. This entails, however, the excavation of tunnels and stopes with cross-sections of some tens of square metres and consequently the resort to long-hole drilling and to blasting of volumes of rock of up to several hundreds of cubic metres. Due to the small thickness of the veins of the Montevecchio orebody, exposure of the wallrock is inevitable and the statics of the latter become the limiting factor of the productivity of the stoping method.

Amongst the options possible, the down-ward moving stoping method with blasting of ore from sublevels with floors composed of cemented backfill sills was considered appropriate for the Montevecchio mine. A similar layout was discussed by Carta et al. (1980) to which the reader is referred for more details. Similar layouts, though related to different static conditions, have been adopted in some foreign mines (Avril 1979; Durocher 1979; Massacci 1980).

Figure 1 illustrates the method. The cemented backfill sills play a multiplicity of roles:
- preservation of the general stability of the series of cavities left by stoping operations;
- limitation of wall exposure to such heights as to reduce the occurrence and propagation of local slabbing;
- creation of a regular and strong roof to the stope.

The geometric parameters of the layout - and, in particular, the thickness of the sills and their vertical intervals - should be a compromise between the stope statics and their productivity. It should be pointed out that no particular difficulties arise from the variation of the values of these geometrical parameters; the method would therefore seem to match the rock properties of the Montevecchio orebody, which, due to its extension, may differ from one stope to another.

With the aim of defining the geometrical parameters of the stopes and the operating conditions, the mine company started, in 1976, a field test during which the vertical interval between adjacent cemented backfill sills was to be increased with increasing stope depth.

This field test was carried out in a 240 metre long section of the zone called "Casargiu", located at the western edge of the mine, comprised between the "Fais" drift, 160 m a.s.l., and the top of the hydraulic backfill of old mined out stopes, 124 m a.s.l. Access of drilling and ore-loading equipment to the stopes is ensured by short cross-cuts which connect them to a ramp located in the hanging-wall country rock at a distance of 20 m from the vein contact.

The first stage of the stoping test consisted in the excavation of development drifts 3 m high where a 1.5 m cemented backfill layer was laid; this layout, corresponding to a nominal 0.50 : 1 filling ratio, was repeated five times (Fig.2). Subsequently, the sixth development drift was excavated leaving a 3.5 m thick pillar of ore between its roof and the cemented backfill of the fifth drift.

The pillar was to be blasted following the emplacement, in the sixth drift, of a 1.5 m thick layer of cemented backfill (Fig.2). The filling ratio is reduced, in this case, to 0.25 : 1. Figure 3 shows the plan of sublevels 5 and 6, with their respective access crosscuts; most of the stope is developed outwards.

The first five sills were cast laying the cemented backfill by means of pneumatic stowing machines. The aggregate, composed of gravel in the size range 5-30 mm and by a minor proportion of 0.3 mm sand, was fed directly to the stowing machine, the mixing with cement milk being accomplished in the pipeline downstream from the latter.

Prior to placing the backfill, the stope was prepared by inserting ribbed iron rods, 24 mm in diameter, into suitable boreholes drilled in the sublevel walls at 1 m intervals and at a height of 0.5 and 1.2 m alternatively from the floor. The boreholes dipped downwards and iron rods, 2.0-2.5 m long, extended into the sublevel for about 1 m to be subsequently submerged by backfill. A wire net was finally laid on a layer of ore several centimetres thick arranged on the floor. Circular ventilation raises were opened in the backfill at regular intervals. In order to avoid segregation of the mix components in the sixth sill the backfill was laid in place by means of autoloaders, the stowing machine only being employed for transporting the mix to a collection point.

The ore bench was then mined out with horizontal long blastholes: this choice was dictated by space limitations which prevented driving vertical blastholes. However, this latter layout will be adopted in the final design of the stoping method.

The research - whose results are discussed in the present paper - was undertaken on occasion of the preparation of the sixth sublevel.

3. TEST PROCEDURE

The programme concerning the investigation of stope stability consisted in:

1. the characterization of the rock mechanics of the country rock and the evaluation of the homogeneity and strength characteristics of the fifth sill;

2. the measurement of displacements in the fifth sublevel as mining progressed in the sixth.

The programme was implemented by locating nine stations, at about 30 m intervals (Fig.3) in the sixth sublevel. In those stations bearing an odd number - at intervals of about 60 m,- two horizontal core drillings

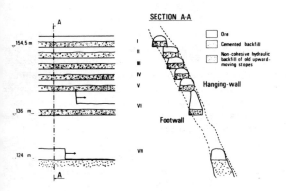

Fig.2. a) Along-strike section and b) cross-section of the experimental stope of Casargiu zone of the Montevecchio mine.

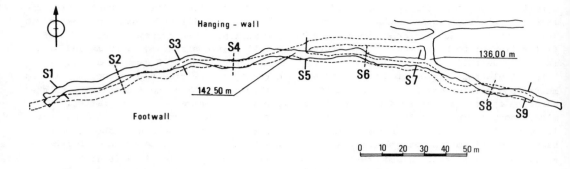

Fig.3. Plan of 5th and 6th sublevels

50 mm external diameter, 10-12 m long, were made, one into the hanging wall rock, the other into the footwall rock, with their collars at a height of about 0.6 m above the floor. An additional core borehole, laying in the plane of the vein and directed upwards, was drilled. These drillholes 5-6 m long, passed through the ore bench and the overlying cemented backfill.

On the perimeter of each of the nine sections seven pegs - three on the floor, one at mid-height on each wall and two on the

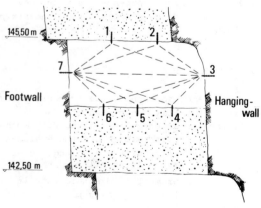

Fig.4. Typical pattern of the positions of pegs.

roof - were grouted for a depth of about 25 cm, according to the typical pattern shown in Fig.4.

The analysis of the drill cores, along with the results of mechanical strength tests carried out on the rock specimens obtained from the former and the visual examination of the exposed rock (as far as fracture properties and the occurrence of water are concerned) enabled the classification of those rocks according to the C.S.I.R procedure (Bieniawski, 1979). In addition, the recoveries of the drill cores taken from the cemented backfill were calculated and mechanical strength tests performed on a specimen cut from them. These measurements and tests were carried out according to I.S.R.M. suggested procedures (Brown 1981).

The distances between the pegs were measured by means of a K15 Interfels convergence measuring device, prior to the passage of the mining face in the underlying stope and were repeated several times subsequent to this passage. In every station the number of measurements was limited to the minimum necessary to allow the reconstruction of the corresponding stope section: it can easily be shown that for n pegs places on the stope perimeter, this number is equal to $(2n - 1)$. Accuracy in the measurement of lengths was one tenth of a millimetre.

322

4 RESULTS OF IN-SITU AND LABORATORY MEASUREMENTS AND OBSERVATIONS

4.1 Results of in-situ observations prior to mining the sixth sublevel

The overlying cavities, corresponding to the first five stopes, were carefully inspected prior to starting mining of the sixth stope. A common feature of the first four cemented backfill sills appears to be the inhomogeneity in the distribution of concrete in the upper parts and the compactness of their lower surfaces, to be attributed to the gravity induced segregation of the finer mix components. This segregation is the cause of the detachment of lower slabs about ten centimetres thick. In addition, in some areas of the upper parts of the sills short fractures were detected, a few millimetres wide, mainly parallel to the vein contacts, and usually located at two thirds of the stope width towards the hanging-wall.

The country rock exhibited, in the zones with low quartz content, marked vertical foliation parallel to the vein with plane and fairly smooth discontinuity surfaces characterized by weak cohesion and frictional resistance.

Since the sublevel drifts are not aligned dipwise, the contact between sill and rock is frequently exposed. In these cases, whenever the nature of the rock was as described above, slabbing of the wall rock comprised between two adjacent sills and, in a few cases, local cavings of roof rock occurred. The continuity between the lateral surfaces of the sills and the wall rock always appeared very good, even when some iron rods were exposed.

In those zones where the contact between the wall rock and the sill remained intact, a discontinuity of up to 5-6 mm wide was frequently observed between the lower surface of the sill and the wall rock. This fact can be ascribed to the loosening of the rock adjacent to the surface.

Water was present in all of the sublevel drifts but was more abundant in the western zone and in the upper parts, whereas in the sixth sublevel, prior to blasting the ore bench, it was confined to those areas with small flow rates.

4.2 Characterization of the rock and of the cemented backfill sills

In Table 1 the values of the RMR indexes, the corresponding predictions in terms of maximum unsupported length and of the respective stability time intervals are shown. The data concerning the hanging-wall and foot-wall rock of the five zones into which the stope was divided by the core-drilling stations are reported separately in the table. In the last column the average support densities achieved during the mining stage are shown.

The average core-drill recovery in the cemented backfill sills was around 40%; the average value of uniaxial compressive strengths of the specimen obtained from drill cores was 16 MPa.

4.3 Results of in-situ observations during and after mining of the sixth sublevel

Mining of the sixth sublevel ore bench started from the entry and proceeded first along the eastern branch and then along the western one (Fig.3) - (In the final layout the ore bench will be mined out in retreat with vertical blastholes).

One necessary prerequisite to forward mining is the accessibility of the whole stope and this prompted the installation of suitable supports, indispensable in that the duration of stoping operations, which lasted about eight months, gave rise to stability problems. The details of the support structures are reported in Table 2. As can be observed, no support was needed along 140 m of the eastern branch of the stope - except for two isolated instances - whereas the remaining 100 m of the western branch has to be supported mostly by means of half sets, with caps resting in hitches in the footwall rock, at intervals of 1.20 m. The support frames were usually installed about ten days after the passage of the stope face.

During mining of the western branch particular care was taken not to expose the contact of the vein with the hanging-wall where the geometrical irregularity of the vein allowed to do so. In fact, the schists forming the hanging-wall contact, when ac-

Table 1. Classification of wallrock according to C.S.I.R.

Distance of stope from western end of sublevel in metres		Hanging-wall rock				Footwall rock				
		RMR	Class	Max. un-supported span	Average stand-up time	RMR	Class	Max. un-supported span	Average stand-up time	Average timbering density
from	to			m				m		frames/m
0	30	36	IV	2.5	16 hours	43	III	3.3	2 days	0.64
30	90	38	IV	2.7	1 day	48	III	4.0	4 days	0.61
90	150	51	III	4.3	1. week	56	III	5.1	2 weeks	0.38
150	210	58	III	5.3	2-3 weeks	65	II	6.6	2 months	0.00
210	240	58	III	5.3	2-3 weeks	63	II	6.3	6 weeks	0.00

Table 2. Supports installed during mining of the sixth sublevel.

Key

Q   = wood frame

QF = iron frame

QZ1 = half-set - hitch in footwall

QZt   half-set - hitch in hanging-wall

Cc, Cl, Ct = wooden, fascine and/or poly-styrene cribs installed to support roof, footwall and hanging wall respectively

Distance of stope from western end of sublevel in metres		Type of support	Support density
from	to		frames/m
0	8	not mined	–
8	47	26 QZl + Ct	0.64
47	61	7 Q + 7 QF + Ct + Cc	0.93
61	77	14 QZl + Cc	0,81
77	90	unsupported	0
90	109	1 Q	1.16
		8 QZt + Cl	
		3 Q + 6 QF + Cl	
		5 Q + Cl + Ct	
109	114	unsupported	0
114	115	2 Q + Cl	1,00
115	131	unsupported	0
131	132	2 QZl	1,00
132	250	unsupported	0

cidentally exposed proved unstable and collapses occurred, which, however, never extended beyond the stope height.

For the purpose of limiting the height of the exposed contacts and of creating an adequate floor for drilling horizontal blast-holes a layer of blasted ore, about 1.50 m thick, was left in place in the western branch. On completion of mining operations this ore was hauled proceeding in retreat.

When mining operations were nearing completion the detachment - already mentioned in section 4.1 - of a cemented backfill slab from the lower part of the fourth sill (which forms the roof of the fifth sublevel) propagated from section 2 westwards over the whole width of the stope and for a length of about 15 m. This phenomenon occurred almost two months after the passage of the face through section 2.

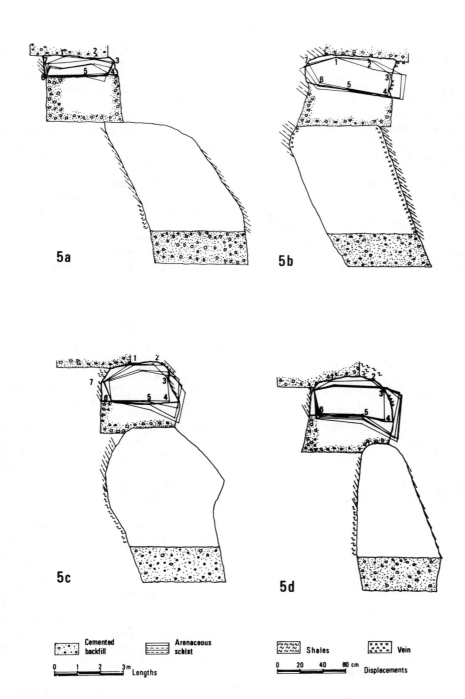

5a

5b

5c

5d

		Cemented backfill			Arenaceous schist

0　1　2　3 m Lengths

	Shales		Vein

0　20　40　60 cm Displacements

Fig.5. Cross-sections of the orebody at stations 2, 5, 6 and 7 respectively.

Furthermore, slabbing of limited dimensions from the hanging-wall and the widening of some pre-existing discontinuities between the sill and country rock were noticed in the fifth stope. It should be emphasized that the lower part of the fifth sill was never affected by detachments throughout the duration of the mining operations in the sixth stope.

4.4 Movements of sills and country rock

Figure 5 shows, for some typical stations, the cross-sections of the orebody and the configurations occupied, as time passed, by the set of pegs under control. The displacements of the latter have been drawn for the sake of clarity to a scale five times that adopted for the undeformed perimeters of the cavities.

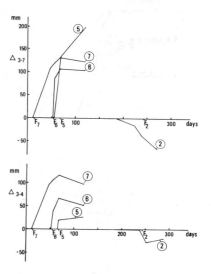

Fig.6. a) Evolution of distance between pegs 3 and 7 vs. time in stations 2,5,6,7. b) Evolution of distance between pegs 3 and 4 vs. time in stations 2,5,6 and 7.
F_x = passage of the face in correspondence with station X.

In all sections the majority of the largest displacements - which usually never exceeded 15 cm - were exhibited by those pegs located in the hanging-wall rock (peg 3) and in the sill in the vicinity of the hanging wall (peg 4). It should be pointed out that the blasting of the ore bench adjacent to the backfill caused the exposure, along most of the stope, of the contact between the former and the hanging-wall rock, as shown by the examples of Fig.5.

Therefore it was considered particularly interesting, in order to show the relative displacements between the cemented backfill sill and the hanging-wall rock, to plot, in Fig.6, the variations vs. time of the distances between peg 3 and pegs 4 and 7 for sections 2,5,6 and 7.

5. COMMENTS AND CONCLUSIONS

On the grounds of the investigations carried out in the stope where mining was in progress and of the results of measurements the following statements can be made.

a) The country rock adjacent to the stopes has variable characteristics: its quality can be defined as poor for the first 90 m starting from the western end of the vein, average between 90 and 150 m and reasonably good in the remaining 90 m corresponding to the eastern branch. These characteristics were determined according to the C.S.I.R. classification, which also permitted predictions concerning the stability of the exposed walls. These predictions appear to be corroborated by the results achieved during production mining. However, the values of the parameters which characterize stope stability (maximum unsupported span and average stand-up time), estimated by means of the calculated RMR values, were conservative for the specific case considered. The application of rock mass classification procedures to mining problems is, in fact, quite recent and presently under development (Laubscher, Taylor 1976; Manca, Massacci 1981); further developments will have to account for the temporary nature of the excavations and the consequent need for less conservative values for corresponding safety factors. In the particular case of the Montevecchio mine the developments of this method appear promising for the definition of the geometry of future stopes.

b) In the cemented backfill sills inhomogeneities were observed, to which the loosening and breaking off of slabs from the lower parts of the sills (which occurred, however, after long time intervals) can be attributed. The fact that gravity is the only important stress field acting on the sills, suggests some modifications of the mix composition and of the methods employed in laying the cemented backfill, with a view to improving the homogeneity of the latter without affecting its strength characteristics.

c) In all the stope sections displacements of the measurement points occurred, almost without exception, at the moment of the passage of the face and diminished considerably afterwards. The most significant displacements in all sections were exhibited by pegs 3 and 4, located in the hangingwall rock and in the sill in the vicinity of the hanging wall respectively. Lesser displacements were recorded for pegs 1 and 2, attributable for the most part to the detachment, mentioned under point b), of the lower slab of the sill where they were grouted. Even smaller and sometimes insignificant displacements were measured for pegs 5,6 and 7, the former two located on the sill floor - the first in a central position and the second close to the footwall rock - and the latter grouted in the hanging-wall rock. The curves of Fig.6 clearly show the downward slipping of the portion of sill close to the hanging wall with respect to the adjacent rock.

The preceding observations lead to assume that the parts of sill and rock exposed in the roof settle under the action of their weights; an equilibrium arrangement is eventually achieved which allows them to act as components of an arch-like structure discharging their weights laterally.

The consequent rotation of the corbel portion of the sill is very likely the cause of the fractures sometimes observed in the extrados of other sills.

The case of section 2 merits particular mention; here the displacements continued to occur with the same order of magnitude even after the passage of the face. In correspondence with that section, the roof of the sixth sublevel exhibited a marked shift towards the hanging wall with respect to the overlying stope and was almost completely located in the country rock (Fig.5). Therefore, the latter slipped downwards with respect to the cemented backfill sill following the passage of the face.

The displacements detected do not therefore appear so serious as to detrimentally affect the stability of the roof in the case where the latter is composed mostly of artificial sill. In this instance, the sill with its movements - produced by the passage of the face - exerts a confining action on the parts of rock with which it is in contact; the latter, due to the geometrical situation and under the effect of its own weight, would appear, on the contrary, potentially unstable. In this way the movements as well as the possible detachments of rock from the hanging-wall appear limited.

The volume of potentially unstable rock increases with increasing exposure of rock in the roof, whereas the corbel portion of sill capable of contributing to the stability correspondingly decreases.

When most of the roof is in rock, the latter starts moving first: its movement, more difficult to prevent than in the former case, continues for longer time intervals and local conditions of instability may ensue.

d) The resort to timbering in some zones was dictated by the necessity to ensure

1. the stability of the rock which had remained exposed for too long on the walls or on the roof due to the protraction of mining times;

2. the accessibility of the stope where the ore bench was mined in advance.

Support was mainly required in those cases where the geometry of the cavities either caused the exposure of large areas of rock in the roof or involved zones of the hanging-wall where the country rock was characterized by poor rock mechanic properties.

In the light of the above considerations and taken for granted the possibility of improving the homogeneity of the cemented backfill sills, it may be concluded that the technical feasibility of this method, also in view of a further increase in the height of the stope, depends on the rock mechanics characteristics of the walls and on the possibility of developing suitable geometrical patterns which ensure the desired cooperation between sills and rock.

In the sixth sublevel the static conditions due to the nature of the contacts appeared satisfactory in the eastern branch - which represents two thirds of the entire length of the stope - whereas they appeared precarious in the western branch.

Classification according to C.S.I.R. seems to be a suitable procedure for predicting the behaviour of the walls. Further developments made possible by its continuing application will allow a better adaptation to the specific case of the Montevecchio mine and consequently less conservative predictions.

The morphology of the stopes, as far as both cross-section and extension are concerned, appears to be a decisive factor for the success of the method. This implies a knowledge of the distribution of the mineralization adequate enough to be used as a basis for the planning of the relative positions of the stopes and the design of the vertical blasthole patterns for production mining rounds.

ACKNOWLEDGEMENTS

The present investigation was financed by the Consiglio Nazionale delle Ricerche, Centro Studi Geominerari e Mineralurgici, Engineering Faculty, University of Cagliari and by contract CCE-CREST No.100-79-1-MPPI.

REFERENCES

Avril, R. 1979. L'uranium français en contexte granitique. Description des gisements exploités par la COGEMA. X Congrès Minier Mondial, Istanbul. Mem.III-21.

Bieniawski, Z.T. 1979. The geomechanics classification in rock engineering applications. Proc. IV Int. Congress on Rock Mechanics, Montreux, 2:41-48.

Brown, E.T. (ed.) 1981. Rock characterization testing and monitoring, p.111-121. Oxford: Pergamon.

Carta, M., R. Cotza, S. Giuliani, P.P. Manc G. Massacci & G. Rossi 1980. Rock mechanics topics in cut and fill mining of veins of varying dip. In O. Stephansson & M.J. Jones (eds.), Application of rock mechanics to cut and fill mining, p.49-54. London: I.M.M.

Durocher, M. 1979. Choix et mise au point d'une méthode d'exploitation descendante sous remblais cimentés pour la mine filonienne de la Société Penarroya à Noailhac-St. Salvy (France). X Congrès Minier Mondial, Istanbul.

Laubscher, D.H. & H.W. Taylor 1976. The importance of geomechanics classification of jointed rock masses in mining operations. In Z.T. Bieniawski (ed.) Proc. Symposium exploration for rock engineering, p.119-128. Rotterdam: Balkema.

Manca, P.P. & G. Massacci 1981. La classificazione geomeccanica delle rocce ed i problemi di sostegno delle opere minerarie sotterranee. Determinazioni sperimentali, analisi critiche e previsioni di comportamento per opere nelle rocce di copertura del giacimento carbonifero del Sulcis. Res.Ass.Min.Sarda. 86(2): 131-157.

Massacci, G. 1980. Note sulla coltivazione con impiego di ripiena cementata discreta della miniera di Saint Salvy, Francia. Atti della Fac. d'Ingegneria, Univ. Cagliari 13(1): 303-320.

Proceedings of the International Symposium on Mining with Backfill / Luleå / 7-9 June 1983

Experience with cemented fill stability at Mount Isa Mines

R.COWLING & G.J.AULD
Mount Isa Mines Ltd., Australia

J.L.MEEK
University of Queensland, Australia

ABSTRACT: Cemented hydraulic fill at Mount Isa Mines was first used in 1969 to aid the recovery of a pillar in the 650 copper orebody. Since 1973 cemented fill has become an integral component in the recovery of the 1100 copper orebody and open stoping areas in numbers 5, 7 and 8 lead orebodies below 13 level. Mining and filling practices have evolved as more information on the properties and behaviour of cemented fill has been obtained by operating experience and research. Almost 20 million tonnes of cemented fill have been placed in 100 stopes and pillars, and 81 wall exposures have been created in 50 stopes/pillars. Factors which resulted in changes to mining and filling practices are identified and an analysis of exposure performance is presented.

INTRODUCTION

Mount Isa Mines Limited currently produces at the rate of 5.5 million tonnes per annum of copper ore and 3.7 million tonnes per annum of lead-zinc-silver ore. The annual metal production is 155 000 tonnes of copper, 180 000 tonnes of lead, 175 000 tonnes of zinc and 465 000 kilograms of silver. (Lead-zinc-silver ore is locally refered to as lead ore and that convention will be used throughout this paper.)

Copper is produced from the 1100 and 1900 orebodies, which lie to the south of a central shaft complex. In financial year 1983/84 lead ore will be mined from 19 orebodies, which are located to the north of the shaft complex.

All copper ore production is by open stoping with pillar recovery and, currently, 65 percent of lead mining is also by open stoping with pillar recovery. In both cases, pillar recovery is usually by a variation of open stoping, typically using a vertical expansion slot into which the remainder of the pillar is massblasted.

Pillar recovery, in both copper and lead open stoping areas, is facilitated by the use of cemented fill in previous stope and pillar voids.

HISTORY OF CEMENTED FILLING

Cement stabilised fill was first used in Mount Isa Mines in 1969 to aid recovery of a floor pillar in the 650 orebody, a now worked-out copper deposit (1). This was a small scale, successful operation with maximum exposure dimensions in the range of 15 to 20 metres. However, because of the limited knowledge of cemented fill properties and behaviour, a high cement content was used - 12 percent by weight.

Following extensive testing (2, 3) and detailed design and feasibility studies (4), it was decided to recover the 1100 copper orebody with large open stopes (up to 200 m high) and pillar recovery. On the initial designs the open stopes accounted for about 35 percent of the ore, and so pillar recovery was dependent on the free standing ability of the cemented fill which was to be placed in the open stopes. The final phase of pillar recovery envisaged that the massblasted ore would be drawn under introduced rockfill. The same rockfill was to be a component of the cemented fill and along with cemented hydraulic fill produce cemented rockfill. Cemented hydraulic filling of the 1100 orebody commenced in January 1973 and rockfill placement started in August 1973. Since then some 70 copper stope and pillar voids have been filled with 8.6 million tonnes of rockfill and 8.8 million tonnes of hydraulic and cemented hydraulic fill. A total of 64 walls in 36 stopes and pillars have been exposed.

The decision to recover lead orebodies numbers 5, 7 and 8 between 13 and 15 levels by open stoping with pillar recovery, resulted in cemented hydraulic fill being first placed in 5 orebody in 1975. For reasons relating to problems with fill material distribution, smaller exposure dimensions and different mining methods (these last two in comparison to the 1100 orebody), rockfill was not required for lead open stopes. To date 35 stope and pillar voids have been filled with 2.1 million tonnes of hydraulic and cemented hydraulic fill. Seventeen exposures of cemented hydraulic fill have been created by pillar recovery.

FILL MATERIALS - PRODUCTION AND PLACEMENT

Production and placement of the constituents of the various types of fill have been described elsewhere (5). Only a brief description of the materials and processes will be presented here.

Hydraulic fill is the name used for the product resulting from the partial dewatering of the combined tailings from the copper and lead concentrators. This material has to have less than 9 percent by weight less than 10 microns, and pulp density is maintained at about 67 to 70 percent solids by weight.

Cemented hydraulic fill is produced by mixing hydraulic fill and cementing agents. The cementing agents are Portland Cement and Copper Reverberatory Furnace Slag; the latter being a waste product from the copper smelting process. It is ground to the same nominal surface area as Portland Cement. Several ratios of cementing agents have been used in the past, but

at present a standard mix of 3 percent Portland Cement, 6 percent Slag and 91 percent hydraulic fill, by weight is used. This mixture has strength properties equivalent to a mix with 6 percent Portland Cement and 94 percent hydraulic fill, and is produced at 60 percent of the cost.

Annual hydraulic fill production is 2.2 million tonnes, and 43 000 tonnes of Portland Cement is scheduled to be used in 1983/84. Hydraulic fill is also used in cut-and-fill operations to recover narrow lead orebodies and to fill pillar voids in the final stages of recovery.

Both of the above fill types are distributed to underground via 300 mm diameter vertical holes from surface. Reticulation underground is by 150 mm diameter flanged pipes and 150 mm diameter vertical boreholes. Two lines of 200 tonnes per hour can be produced simultaneously. There is no storage capacity, fill being placed when the concentrators are operating.

Rockfill is produced from a siltstone quarry to the north of the mine area. The siltstone is crushed to 300 mm and screened at 25 mm. The minus 25 mm fraction is discarded. From the quarry, the rockfill is conveyed 2.5 km on a 1.2 m wide conveyor belt to one of two 2.4 m diameter vertical passes. Each pass is located in the base of a conical pit, giving the rockfill system a combined storage capacity of about 50 000 tonnes. Rockfill is choke fed via these passes to two conveyor systems on 13C sublevel and 15 level. The former services the northern part of the 1100 orebody and the latter services the southern part. The rockfill is carried by the conveyor belts to passes at the top of the stopes to be filled. Where cemented rockfill is required, the rockfill and cemented hydraulic fill are introduced together through the pass. There is no prior mixing. Present production rate from the quarry is 1.7 million tonnes per annum.

COPPER OREBODIES - PRACTICE AND EXPERIENCE

Of the 70 copper stopes and pillars which have been filled since the introduction of cemented rockfill, all but one (which is in the 1900 orebody) are located in the 1100 orebody. The remainder of this section will concentrate on experience gained in this orebody.

The 1100 orebody is a massive copper orebody which at the beginning of production had proven reserves of 100 million tonnes. It has a known strike length of 2.0 km and has typical cross-section dimensions of 300 m high and 300 m wide. A cross-section showing stope and pillar outlines for a northern part of the orebody is shown in Figure 1.

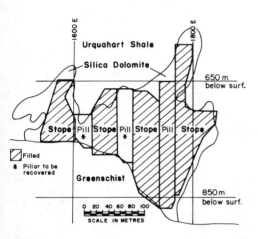

Figure 1. Cross-section of 1100 orebody at 5232N, looking north.

Development of the 1100 orebody commenced in 1967 and filling started in 1973. Extraction and filling started near the northern end of the orebody and proceeded southwards. Just as extraction methods and sequences have been modified with experience (6), so to have practices related to the placement and exposure of cemented fill.

When filling started the ratio of rockfill to cemented hydraulic fill to be placed in the initial open stopes was to be 2 to 1 by weight. Further, to allow for the degradation of rockfill that was envisaged to occur with the length of drop into the stopes, the cemented hydraulic fill strength varied from bottom to top of the stope. The bottom third was filled with cemented hydraulic fill of 8 percent equivalent Portland Cement strength, the middle third with 7 percent equivalent strength and the top third with 6 percent

equivalent strength. It was expected that exposures of the above fill would be stable at dimensions of 15 m wide by 50 m high. Pillar recovery methods were designed to take account of this (4).

During the period 1975 to 1978 many observations and measurements were made of cemented rockfill during placement. In parallel, an extensive research project was undertaken with the CSIRO Division of Applied Geomechanics to determine the important properties of cemented rockfill and establish models which would permit prediction of its behaviour under various operating conditions. Much of the operating and research experience gained during that time is reported by Barrett (7).

Changes which occurred during that time included the use of 6 percent strength cemented hydraulic fill, regardless of length of drop of rockfill, when it was established that fines generation from rockfill degradation was insignificant. Another important change resulted in the rockfill to cemented hydraulic fill ratio being calculated for individual stopes, and even different parts of the same stope. This arose when it was discovered that the filling process resulted in zones within the rockfill core which were only partly cemented, and in the extreme not cemented at all. This condition was aggravated when rockfill was allowed to build up against a wall during filling. To overcome this problem, the rockfill ratio is now calculated to permit the rockfill core to just touch any wall which is to be subsequently exposed.

However, the most important influence on present filling and pillar recovery practices is the performance of the cemented rockfill when it is exposed.

Of the 69 stopes and pillars which have been filled in the 1100 orebody, 5 have been filled with rockfill only. Another 3 were initially filled with rockfill and at a later stage attempts were made to stabilise one or more walls by pouring cemented hydraulic fill into them, at the relevant walls. (This latter technique is known locally as 'Geco' after the mine in Canada which used it as a standard filling technique (8).)

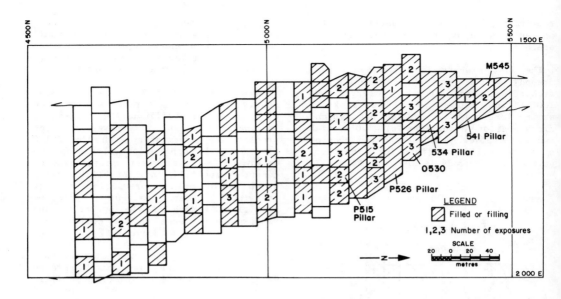

Figure 2. Summary of 1100 orebody exposures, schematic only.

No. of Walls Exposed	No. of Stopes	No. of Exposures
1	16	16
2	12	28 (1)
3	8	26 (2)
TOTALS	36	70

Table I. Summary of stopes/pillars which have been exposed on various walls.

(Notes: 1. One stope had 3 separate exposures on one wall, and two had 2 separate exposures on one wall.

2. Two stopes had 2 separate exposures on one wall.)

Thus 64 stopes have been filled wholly or partially by some type of cemented fill and, of these, 36 have been exposed on 1, 2 or 3 walls. The status of filling in the 1100 orebody along with the location of stopes and pillars with various numbers of exposures is shown schematically in Figure 2. There has been a total of 70 exposures made up as shown in Table I.

As explained in the notes, several stopes have had more than one exposure on a particular wall. This occurs where the ore is less massive, typically in the hangingwall (Figure 3a), or where stope and pillar lines are not at the same co-ordinates (Figure 3b).

Before discussing the performance of fill exposures it is important to differentiate between the nominal, or design, exposure and the visible, or actual, exposure. Typically, the visible exposure is less than the nominal exposure. This arises from several causes. In some pillar firings a thin diaphragm of ore remains in contact against the fill. Usually when this occurs the diaphragm is around the perimeter of the exposure, and only occasionally does it cover a large part of the nominal exposure. The mechanical

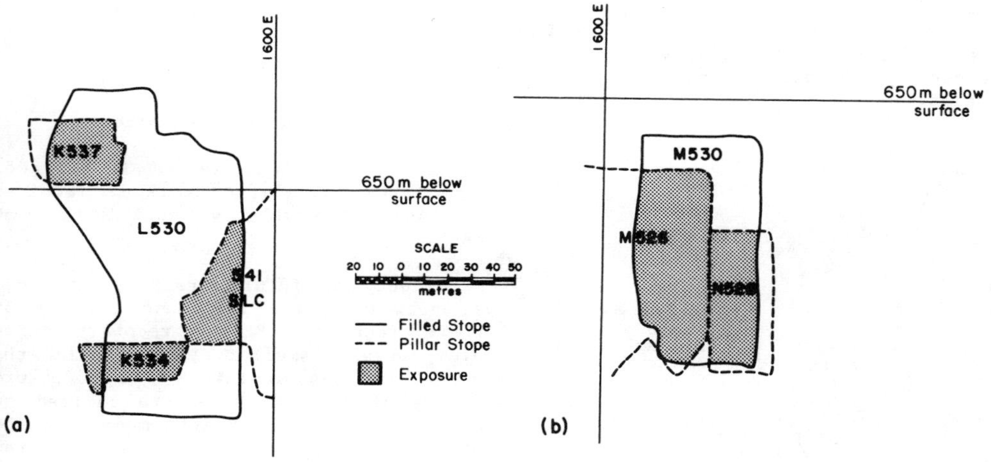
1600 E
650m below surface
K537
L530
K534
SCALE
20 10 0 10 20 30 40 50
metres
Filled Stope
Pillar Stope
Exposure
(a)
1600 E
650m below surface
M530
(b)

Figure 3. Examples of stopes which have had more than one exposure of a wall.

effect of an ore diaphragm on the stability of a cemented fill mass cannot be quantified at this time, but in any case it must be extremely variable and unpredictable. In the early history of pillar recovery, when it had been the plan to recover the broken ore below introduced rockfill, the broken ore and rockfill covered varying amounts of the nominal exposure. A similar situation is encountered on the several occasions where there has been a large collapse of fill onto the broken ore. An extreme example of a nominal exposure covered by a diaphragm and broken ore/rockfill is depicted in Figure 4. For the most part, nominal and visible exposures are in reasonable agreement. In the remainder of the discussion all references to exposures are taken to refer to visible exposure.

Exposure performances are most realistically compared in terms of visible area, although maximum width and height are useful descriptors. In the instances where there have been more than one exposure on a face, their areas are combined and referred to as one exposure. Failure is considered to have occurred when a measureable amount of fill has collapsed from the exposed face. However, failure of the fill does not imply failure of the pillar recovery operation. The largest fill failure occurred when 0530 stope was exposed

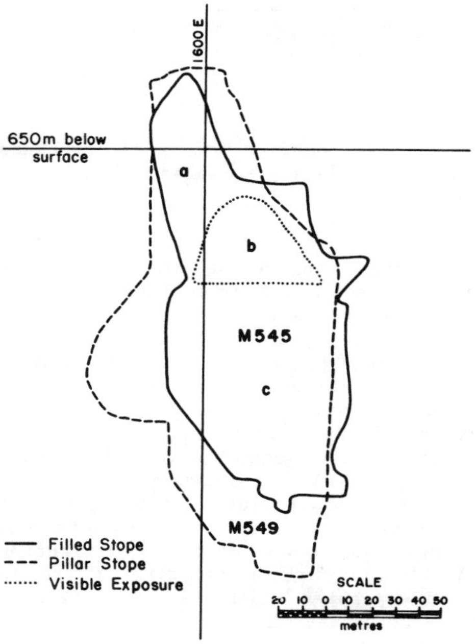
1600 E
650m below surface
a
b
M545
c
M549
Filled Stope
Pillar Stope
Visible Exposure
SCALE
20 10 0 10 20 30 40 50
metres

Figure 4. Exposure partially covered by diaphragm and broken ore/rockfill. (a. diaphragm, b. visible exposure, c. broken ore and rockfill)

during the recovery of P526 pillar. About 200 000 tonnes of fill fell off the south face of 0530 (Figure 5), yet in excess of 99 percent ore/metal was recovered.

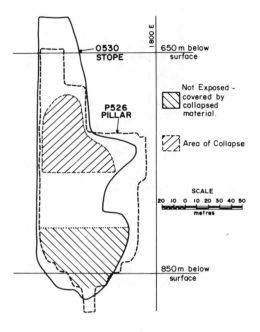

Figure 5. Location of fill exposure when P526 pillar recovered adjacent to 0530 stope.

The total tonnage of ore, with fill dilution, extracted from the stope was 1.2 million tonnes.

The exposures which have been made have been analysed in terms of visible exposure area. Figure 6 presents a summary of the exposures, and failures, for all, first, second, and third wall exposures. An assessment of the exposures and failures leads to the conclusion that area in itself is not the most important factor contributing to failure. Further analysis of the 15 failed exposures indicates the following contributing factors:

- maximum width across exposure;

- total exposed area from first, second and third exposures;

- large increment in exposure area;

- size of massblast in pillar recovery;

- quality of fill;

- drilling and blasting practice.

The above factors are not in order of importance and further work is required to assess the relative contribution of each.

Current and future practice is aimed at decreasing the influence of some of these factors. Pillar recovery units are generally smaller, thus reducing the size of massblasts. Fill exposure widths at any one time are limited to about 40 m. Research into modelling the stability of fill exposures has continued and is reported in a later section.

Overall, the behaviour of cemented rockfill in the 1100 orebody and its role in pillar recovery is considered to be very successful. For short periods the whole of mine copper production has come from pillar recovery, and for prolonged periods the proportion from pillar recovery has been as high as 70 percent. The main factors contributing to fill failure have been identified, and practices and research to minimise and assess their influence are in progress.

LEAD OREBODIES – PRACTICE AND EXPERIENCE

Open stoping history and practice in the lead orebodies have been recorded by Goddard (9). This section concerns the experience with cemented hydraulic fill in the mining of 5, 7 and 8 orebodies between 13 and 15 levels. The relative locations of the orebodies are shown in Figure 7; the stope dimensions cover the range of height and width in this area.

Based on limited operating and geotechnical experience available when filling of 5 orebody commenced in 1975, it was decided to use a 10 percent equivalent strength Portland Cement fill. At most, a block of cemented fill in the lead orebodies would have only two exposures, ie. north and south wall, but it would experience east–west closure resulting from the deformation of the hangingwall and footwall. Although stoping was limited to about 100 m heights and widths of 35 m, it was

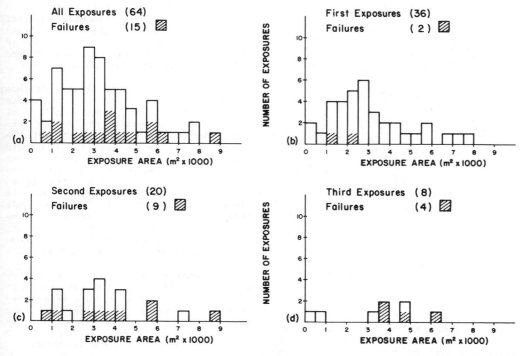

Figure 6. Visible exposure areas and failures for (a) all, (b) first, (c) second and (d) third exposures.

decided to recover the pillars in 2 parts. Subsequent experience and research (10) showed that the pillars could be recovered in one unit, and that cement content could be reduced to 6 percent equivalent strength.

Open stoping and pillar recovery of the 3 orebodies between 13 and 15 levels is nearing completion, Figure 8. Thirty-five stopes and pillars have been filled and 17 exposures created, with no instances of measureable collapse. Like the 1100 orebody, diaphragms have occasionally been formed, but generally the fill has been exposed according to design.

Exposure areas and dimensions are much less for these orebodies, having maximum area, width and height of 4400 m^2, 35 m and 140 m respectively. Pillar massblasts are not usually greater than 150 000 tonnes.

Number 8 orebody below 15 level commenced production in 1981 and plans are for some stopes to be up to 180 m high. Pillar recovery will expose cemented hydraulic fill over these dimensions and, based on experience above 15 level, 6 percent equivalent strength fill will be used.

RESEARCH INTO CEMENTED FILL STABILITY

The majority of the research, undertaken by and for Mount Isa Mines Limited, into the prediction of stability of cemented fill exposures was reported by Barrett (7). Since that time Mount Isa Mines has continued to investigate modelling of cemented fill stability. As a block of cemented fill can have variable geometry, be exposed in complicated sequences and experience varying stress paths, the emphasis has continued to be on finite element models, rather than equilibrium or physical models.

335

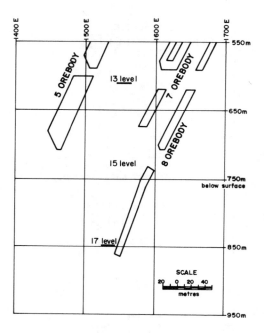

Figure 7. Cross-section, looking north, of 5, 7 and 8 orebodies, showing relative positions between 13 and 15 levels.

The work at Mount Isa has been undertaken in conjunction with the Department of Civil Engineering at the University of Queensland. Based on work reported by Coulthard (11), a three-dimensional finite element program has been used to investigate the development of initial stresses in a filled stope.

Average stable exposures in the 1100 orebody, irrespective of which wall, are about 3100 m^2. This is much larger than the original expectation of about 750 m^2, and can mostly be explained by the fact that depth stresses are not generated during filling. Measurements by Neindorf and Cowling (7) lead to the conclusion that the weight of the fill is partially carried by shear forces generated on the stope walls during filling. Given the evidence of stable exposures, it is clear that depth stresses are not generated during filling. For a 200 m high stope, the vertical stress for depth stress conditions would be 3.5 MPa

for cemented hydraulic fill, and between 3.5 and 4.5 MPa for a cemented rockfill stope. However, the unconfined compressive strength of cemented hydraulic fill and cemented rockfill are 1.1 MPa and 2.3 MPa respectively.

Using a finite element program capable of sequential construction, to simulate the filling process, several models of filling were investigated. These are summarised in Table II. Several of the terms need to be explained. Coulthard has shown that the final results ar significantly influenced by permitting the properties of the fill materials to increase, from zero upwards, as the stope is filled. In these models, only modulus was allowed to change and did so in three steps from 0.35 to 1.00 of the final value. In some models it was kept constant throughout at the final value.

When cemented fills are placed they are in a liquid state and can have high water contents for some time after they have been placed. An attempt to take this into consideration was made by allowing the fill when first placed to be either fully- or half-hydrostatic.

Model No.	No. of Lifts	Modulus	Hydro-Static Option	Fill Type
1	1	Constant	–	CHF
2	8	Constant	–	CHF
3	8	Variable	–	CHF
4	8	Constant	Half	CHF
5	8	Constant	Full	CHF
6	8	Variable	Full	CRF
7	8	Constant	Full	CRF

Table II. Summary of models used to investigate initial stresses in filled stope.

Finally, the fill distribution within the filled stope could either be homogeneous cemented hydraulic fill (CHF) or zoned cemented rockfill (CRF).

The mesh used for the analyses is shown in Figure 9 and the actual and modelled CRF zoning is shown in Figures 10a and 10b. The finite element mesh was composed of 465 8-node elements, requiring 701 nodes. Computer memory requirements were 110 k, 36 bit words on a UNIVAC 1100/82 computer. Central processor time ranged from 11 to 77 minutes. Properties used in the analyses were reported by Dight (12) and are summarised in Table III. Results from the models are summarised in Figure 11, where they can be compared with the measured results (curve 8) and depth stress for both CRF and CHF.

Properties

Density t/m^3	Cohesion MPa	Friction Angle Degrees	Modulus MPa
CHF	1.80	0.30	35	700
CRF	2.36	0.40	40	1150

Table III. Cemented fill properties used in finite element analyses.

The location of the stress cells which measured the development of stresses during filling are indicated in Figure 9. The results in Figure 11 are the stresses computed at that level, for the various models, as subsequent lifts of fill are added.

The choice of height of each lift is perhaps fortuitous for the hydrostatic solutions, but it is apparent that the important factors are:

. number of lifts;
. fill hydrostatic when first placed;
. zoning of CHF and CRF.

As can be seen from Figures 10a and 10b the zoning that was used was a coarse approximation of the real situation. An automatic mesh generation facility with face joining is included in the pre-processor for the finite element program. Using a key diagram as indicated in Figure 10c it is possible to map this geometry into that indicated in Figure 10d. Suitable subdivision of the diagonals permits construction of the mesh shown in Figure 10e. With this it will be possible to achieve a better approximation of the zoning and to investigate different CRF/CHF ratios.

The results indicate that it is possible to obtain a good estimate of the initial stresses within a filled stope. Work is in progress to model the stress redistribution resulting from wall removal, subsequently leading to a better understanding of the influence of exposure dimensions and stress path on exposure stability.

FUTURE DEVELOPMENTS

In August 1982 pre-concentration of the lead ore by heavy medium separation commenced. Reject from the process has a typical particle size of 12 mm and is known locally as aggregate. Previous testwork at Mount Isa has demonstrated that this aggregate can replace up to 35 percent by weight of the solids in hydraulic and cemented hydraulic fill, and still be distributed by the borehole and pipeline system described earlier.

Strength testing of mixtures of cemented hydraulic fill and aggregate, known as cemented aggregate fill, has established that this new fill is intermediate in strength to cemented hydraulic fill and cemented rockfill. At the time of writing, a new mixing plant to produce and distribute cemented aggregate fill had just been commissioned. Preliminary production was of an uncemented product, to gain experience with the plant and pipelines, but eventually cemented aggregate fill will be used to fill lead open stopes and possibly combined with rockfill for copper stopes.

These new materials will require further investigation, particularly with respect to their mass properties and behaviour.

CONCLUSIONS

Experience has been gained with the production and placement of almost 20 million tonnes of cemented fill. One hundred stopes and pillars have been filled in both copper and lead areas. To date 50 have been exposed on 1, 2 or 3 sides. In all, 81 walls have been exposed and maximum stable exposure of about 8 000 m^2 has been achieved. Failure of 15 exposures has been recorded, ranging from several hundred tonnes to about 200 000 tonnes. However, failure of a fill exposure does not imply failure of the stoping operations.

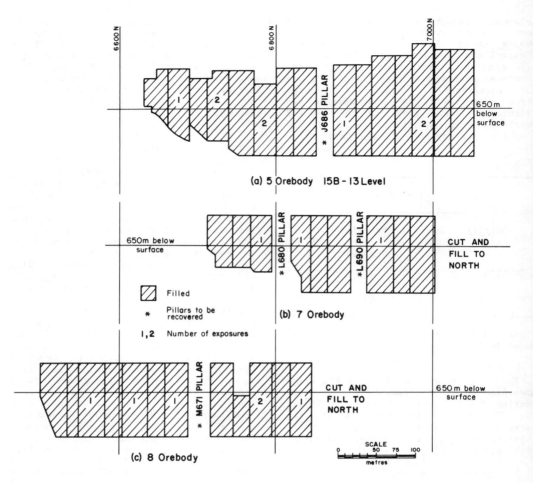

Figure 8. Long sections, looking west, of 5, 7 and 8 orebodies 13 to 15 levels at final stage of pillar recovery.

Factors which appear to influence the failure of fill and subsequent collapse have been identified. Changes to filling and pillar recovery techniques have been implemented and will minimise the effect of the various factors. Mathematical models which can simulate the stability of fill are being tested and will help to achieve the full potential of cemented fill in pillar recovery.

Changes to milling practice have resulted in the introduction of new fill types which will require a significant amount of further research.

The introduction of large scale cemented filling operations at Mount Isa has proved to be very successful, permitting pillar recovery operations to progress at the same pace as the initial stoping operations.

ACKNOWLEDGEMENTS

The permission of the management of Mount Isa Mines to prepare this paper is gratefully acknowledged. The work reported has involved the efforts of many people both at Mount Isa Mines and several research centres.

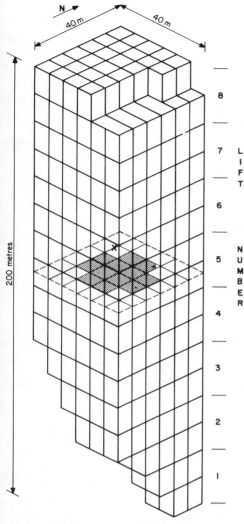

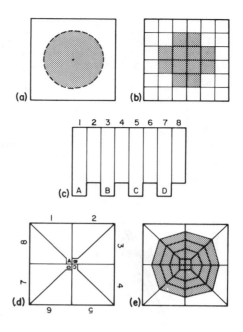

Figure 10. Actual and modelled zoning of cemented hydraulic fill and cemented rockfill within filled stope.

X , Location of Stress Measurement
465, 8-node elements
701, nodes

Figure 9. Finite element mesh for analysis of initial stresses.

REFERENCES

1. Jones, T.O. 'Pillar Recovery in a Section of the 650 Orebody Floor Pillar Using Cemented Fill', Australasian Inst. of Mining and Metallurgy, Regional Conference, June 1970.

2. Thomas, E.G. 'A Review of Cementing Agents for Hydraulic Fill', Jubilee Symposium on Mine Filling, pp 65–75, Australasian Inst. of Mining and Metallurgy, August 1973.

3. Russel, R.E., Barrett, J.R. and Blair, J.R. 'Design and Evaluation of Dry Fill', Jubilee Symp. on Mine Filling, pp 129–137, Australasian Inst. of Mining and Metallurgy, August 1973.

4. Mathews, K.E. and Kaesehagen, F.E. 'The Development and Design of a Cemented Rock Filling System at the Mount Isa Mines, Australia', Jubilee Symposium on Mine Filling, pp 13–23, Australasian Inst. of Mining and Metallurgy, August 1973.

5. Neindorf, L.B. 'Fill Operating Practices at Mount Isa Mines', Proceedings, Symposium 'Mining with Backfill', University of Lulea, 1983.

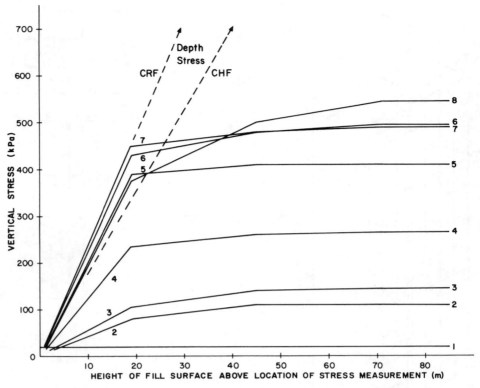

Figure 11. Initial vertical stresses for models (1) to (7) compared with measured stress (8).

6. Alexander, E.G. and Fabjanczyk, M.W. 'Extraction Design Using Open Stoping for Pillar Recovery in the 1100 Orebody at Mount Isa', Proceedings, Symposium 'Caving and Sublevel Open Stoping', Denver, 1981.

7. Barrett, J.R. and Cowling, R. 'Investigations of Cemented Fill Stability in 1100 Orebody, Mount Isa Mines Limited, Queensland, Australia', Transactions, Inst. of Mining and Metallurgy, London, Vol 89, 1980.

8. Schwartz, A. 'Pillar Recoveries Using Consolidated Fill at Noranda Mines Limited, Maintonwadge, Ontario', Proceedings, Symposium 'Mining with Backfill', Canadian Inst. of Mining and Metallurgy, May 1978.

9. Goddard, I.A. 'The Development of Open Stoping in Lead Orebodies at Mount Isa Mines Limited', Proceedings, Symposium 'Caving and Sublevel Open Stoping', Denver, 1981.

10. Meek, J.L.; Beer, G. and Cowling, R. 'Prediction of Stress Levels in Cemented Fill', Proceedings, Symp. 'Implementation of Computer Procedures and Stress Strain Laws in Geotechnical Engineering', Chicago, Illinois, August 1981.

11. Coulthard, M.A. 'Numerical Analysis of Fill Pillar Stability - Three Dimensional Linear Elastic Finite Element Calculations', Technical Report, CSIRO, Division of Applied Geomechanics, No. 97, 1979.

12. Dight, P.M. and Cowling, R. 'Determination of Material Parameters in Cemented Fill', Proceedings, 4th ISRM Congress, Montreaux, 1979.

Evaluation of the stability of back-fill faces

J.P.DIXIT & N.M.RAJU
Central Mining Research Station, Dhanbad, India

ABSTRACT: Cement stabilized fills can provide the required ground control if their strength distribution functions match with the probability of strength of the wall rocks. The critical stress at or above which the backfill faces fail is found out in terms of the probability of the most representative strength distribution in the wall rocks. This probability as derived from the cumulative strength distribution functions of the rocks gives the critical stress at or exceeding which the bearing capacity of the backfill faces will be gradually exhausted. It is, therefore, proposed to produce a reinforced cemented backfill that must have a strength to match with the probability of the wall rock strength which is necessary in order to extract the adjacent ore pillars achieving the greater stability and economics.

1 INTRODUCTION

Placement of cemented classified tailings backfill is in practice in underground metalliferous mining operations and its importance has been increased with the development of large scale bulk mining methods. Stabilization of such backfills is necessary in order to maintain good ground control for removing the ore pillars. A number of theoretical considerations for strength requirement of the cement stabilized fills have been given by different authors in different ways, but neither of the approaches accounts for the problem accurately. The present work reports an analysis of a new reinforced cemented backfill design system assuming that the stability of fill is affected by the wall rock strength and movement and also by the dimension of the fill at the wall-backfill contact.

2 DISCUSSION OF THE PROBLEM

Ground support can be provided by use of cement stabilized fills for enabling the extraction of adjacent ore pillars. Generally the cemented fill is designed as a freestanding vertical face where the unconfined compressive strength required at any depth in the fill is given by

$$c_o \geqslant \gamma z$$

where γ = bulk unit weight of the backfill
z = depth from the fill surface

In this case freestanding wall has a fill of variable strength that increases linearly with depth which results in its heterogeneous and anisotropic behaviour to create the risk of surficial failure and ore dilution due to blasting operations. Therefore, assuming ϕ = 0, a vertical slope of constant strength fill is suggested which requires the well known strength expressed as

$$c_o \geqslant \gamma H/2$$

where H = overall height of the fill

Mitchell et.al. (1982) have given the following equation to estimate the unconfined compressive strength requirement for stability of vertical fill faces whereby reducing the cement requirement as

$$C_o = y \, H/(1+H/L)$$

where L = lateral dimension of the fill

This equation indicates that when $L \rightarrow 0$, no cement is required. Thus, this design approach accounts for such conditions only when rock walls are sufficiently close together to help support the backfill at the wall-backfill contact. Also the equation assumes that the cemented backfill is a frictionless material ($\phi = 0$), no contribution of wall rock shear could be derived.

Since the fill height (H) varies much and may be larger or much smaller than its lateral dimension, whatever the case may be, and as also no account of fill width (W) is taken into, neither of above design approaches can exactly determine the strength requirements for cement stablized fills.

It has been well established from laboratory tests on rock specimens that the specimen size significantly affects the strength. It has been found out that if H/D ratio decreases, the strength of the cylindrical specimen increases, where D is the diameter of the specimen (Dixit, 1981). This indicates that the fill dimension must have a significant effect on the strength. This aspect has been discussed in greater detail elsewhere in the paper.

Observations indicate that the wall rocks as also the backfill faces gradually develope the tensile cracking and surface spalling. Because the stiffness of the backfill is small compared with that of the rock, the strains imposed due to shearing movement at the wall-backfill contact cause rupture in cement bond of the fill. Since the backfill failure occurs progressively by the cracks propagation, the bearing capacity of the fill is exhausted at certain level of fracturing. This level may be expressed as 'Critical stress level' at or above which the fill is to be failed. Therefore, the strength requirement of the fill is here defined by certain probability that such a strength or even lower strength can be found in a representative sample of the wall

rock. In a given population of the specimens, the probability of a certain or lower value of strength is defined by cumulative distribution function.

3 INVESTIGATION PROCEDURE

For investigation into most probable and representative strength distribution in wall rocks, the specimens should preserve all kinds of original and/or secondary defects in such a proportion as to be used as representative samples and should not be disturbed by sampling. Therefore, wall rock samples should be taken in a predetermined spatial scheme irrespective of their structural and textural variations. About 100 specimens will be sufficient for this purpose. Every specimen should be examined geologically with special attention to natural fractures and secondary cracks that constitute the weak zones as these features affect the stability of wall rock as well as of the fill. Consequently, specimens prepared for uniaxial compressive strength test can be classified into two groups, i.e., sound and defective. When natural and/or secondary closed fractures opened during the test, the specimens are put in defective group while rest are sound.

To obtain undisturbed data about strength variation, other possible sources of variations are kept preserved as usual.

Based on such data on uniaxial compressive strengths of two groups of rocks, their cumulative distribution functions and probability are computed (Tables 1 and 2) followed by their histograms (Figures 1 to 6) For the analytical interpretation of the resulting strength distribution, correlation coefficient (r), standard deviations (σ_x and σ_y), variance (σ_x^2 and σ_y^2) and mean of strengths of two groups of rocks are also computed. These data and illustrations presented here are based on strength distribution of the rock of Chikla manganese mines, District Bhandara, Maharastra, India (Dixit, 1981). The data on strength distribution of exposed wall rocks-gondite schist and gneiss as shown in these illustrations were collected for

Table 1. Cumulative distribution function of uniaxial compressive strength. All specimens of gondite.

C_o^* × $10^4 KN/m^2$ (Class)	Frequency (fi)	Cumulative frequency (ci)	Probability P(I)
0-2	0	0	0
2-4	7	7	0.10
4-6	11	18	0.26
6-8	15	33	0.48
8-10	13	46	0.67
10-12	6	52	0.75
12-14	9	61	0.88
14-16	3	64	0.93
16-18	5	69	1.00

*Uniaxial compressive strength.

Table 2. Cumulative distribution function of uniaxial compressive strength. All specimens of schist and gneiss.

C_o × $10^4 KN/m^2$ (Class)	Frequency (fi)	Cumulative frequency (ci)	Probability P(I)
0-1	9	9	0.35
1-2	12	21	0.81
2-3	2	23	0.89
3-4	1	24	0.92
4-5	1	25	0.96
5-6	0	25	0.96
6-7	0	25	0.96
7-8	1	26	1.00

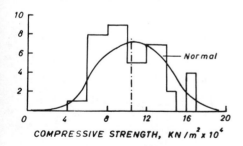

Fig.1. Histogram of uniaxial compressive strength. Only sound specimens of gondite.

different purpose and not for back-fill design, they are presented here to illustrate only as how to design an economical and stable backfill based on such data.

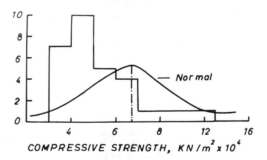

Fig.2. Histogram of uniaxial compressive strength. Only defective specimens of gondite.

3.1 Analysis of the resulting strength distribution in wall rocks

When uniaxial compressive strength distribution of two groups, i.e., sound and defective, are considered separately, they are found to be normally distributed for gondites (Figures 1 and 2) as well as for schist and gneiss (Figures 3 and 4). However, some greater skewness in right hand side of these histograms may be observed in actual distribution of defective specimens (Figures 2 and 4) as they would obviously have lower strength. But when two groups are considered together, they make a good fit to the histogram of all specimens of gondite (Figure 5) and schist and gneiss (Figure 6). The resulting distributions give the composite distribution curves for both types of rocks as seen in the above figures.

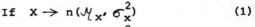

If $\quad X \rightarrow n(\mathcal{M}_x, \sigma_x^2)$ \hfill (1)

$\quad Y \rightarrow n(\mathcal{M}_y, \sigma_y^2)$ \hfill (2)

where $\mathcal{M}_x$ = mean strength of sound specimens

σ_x^2 = variance of strength of sound specimens

$\mathcal{M}_y$ = mean of strength of defective specimens

σ_y^2 = variance of strength of defective specimens

n denotes the normal strength distribution

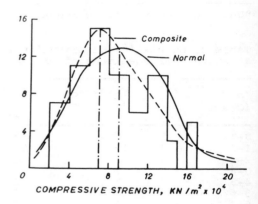

Fig.5. Histogram of uniaxial compressive strength. All specimens of gondite.

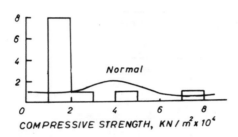

Fig.3. Uniaxial compressive strength histogram of only sound specimens of schist and gneiss.

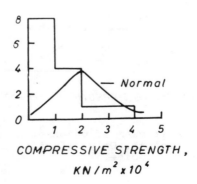

Fig.4. Histogram of uniaxial compressive strength. Only defective specimens of schist and gneiss.

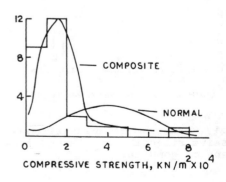

Fig.6. Histogram of uniaxial compressive strength. All specimens of schist and gneiss.

and if, X is the frequency of occurrence of strength of sound specimens and Y is the frequency of occurrence of strength of defective specimens and both are normally distributed, then,

X + Y will also represent a normal distribution with mean $\mathcal{M}_x + \mathcal{M}_y$ and variance $\sigma_x^2 + \sigma_y^2$.

If Y(1) = frequency of occurrence, when the total number of observations is 1, then, the distribution

$$a\,X + b\,Y \rightarrow n(a\mathcal{M}_x + b\mathcal{M}_y,\ a^2\sigma_x^2 + b^2\sigma_y^2) \quad (3)$$

is also normalized, where a and b are the number of specimens in sound and defective groups respectively.

The normal (Gaussian) distribution of uniaxial compressive strengths of sound and defective specimens may be expressed in the form of equations (4) and (5) respectively.

$$X = \frac{1}{\sqrt{2\pi\sigma_x^2}}\ \exp\left[-\tfrac{1}{2}\ \frac{(C_o - \mathcal{M}_x)^2}{\sigma_x^2}\right] \quad (4)$$

and,

$$Y = \frac{1}{\sqrt{2\pi\sigma_y^2}}\ \exp\left[-\tfrac{1}{2}\ \frac{(C_o - \mathcal{M}_y)^2}{\sigma_y^2}\right] \quad (5)$$

where C_o = Uniaxial compressive strength

The distribution found as a total of the two normal distributions fit very well to the histogram of all specimens of gondite (Figure 5) and schist and gneiss (Figure 6) whose resulting total distribution may be expressed as a sum total of the equations (4) and (5) which reads as follows

$$X+Y = Y = \frac{1}{\sqrt{2\pi\sigma_x^2}}\ \exp\left[-\tfrac{1}{2}\ \frac{(C_o - \mathcal{M}_x)^2}{\sigma_x^2}\right]$$
$$+\ \frac{1}{\sqrt{2\pi\sigma_y^2}}\ \exp\left[-\tfrac{1}{2}\ \frac{(C_o - \mathcal{M}_y)^2}{\sigma_y^2}\right] \quad (6)$$

As the conditions given above, i.e., when total number of observations is 1, the equation (6) is normalized, the analytical form of resulting distributions in gondite and schist and gneiss is represented by the

following equations respectively.

$$1 = 1.1813 \times 10^{-5}\exp\ \frac{-(C_o - 10.76 \times 10^4)^2}{21.78 \times 10^8}$$
$$+\ \frac{-(C_o - 6.66 \times 10^4)^2}{21.91 \times 10^8} \quad (7)$$

and,

$$1 = 1.0182 \times 10^{-5}\exp\ \frac{-(C_o - 2.16 \times 10^4)^2}{6.55 \times 10^8}$$
$$+\ 2.6174 \times 10^{-5}\exp\ \frac{-(C_o - 1.17 \times 10^4)^2}{1.3448 \times 10^8} \quad (8)$$

These equations give the normal uniaxial compressive strengths,

$C_o = 8.0740 \times 10^4\,KN/m^2$ for gondite and
$C_o = 1.2432 \times 10^4\,KN/m^2$ for schist and gneiss.

The two normal distributions of two groups, sound and defective show two different populations, which corroborates the separation of these rocks in the above two groups.

Each of the equations (7) and (8) give two different values of strength. One is the real and the other is the imaginary. Hence, only real value (+ve) can be considered and imaginary (-ve) values of strengths are rejected.

4 STRENGTH REQUIREMENT FOR REINFORCED CEMENTED BACKFILL

Assuming the rock strengths of $C_o = 80740\ KN/m^2$ for gondite and $C_o = 12432\ KN/m^2$ for schist and gneiss, as derived from the resulting probability density functions from equations (7) and (8) respectively and relating them to the probability of cumulative distribution function of uniaxial compressive strengths of these rocks, there is a probability of 61% for gondite and a probability of 64% for schist and gneiss (Figure 7). This means that the values of compressive strength of 80740 KN/m^2 and 12432 KN/m^2 or less than these will be observed in these rocks respectively. Similarly the probability of above strength values may also be found in sound and defective specimens separately which would give an idea of approximate strength requirement

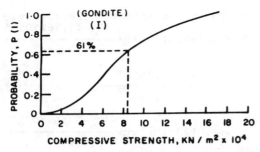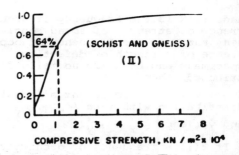

Fig.7. Probability related to rockwall strength in the cumulative strength distribution function.

for the cemented backfill. Strength increases more rapidly when the cement content exceeds about 8% (Weaver and Luka, 1970). Therefore, any such fill strength requirement to match with the probability of unconfined strength distribution function of the wall rocks can be achieved by controlled alternate pouring of cement in two groups, i.e., thick layers of low cement fill and thin layers of high cement fill to produce a reinforced cemented backfill corresponding to the probability of the strengths of two groups of specimens of wall rocks. The match for the strength requirement can be obtained by certain corrections for the size affect of the backfill which has been discussed below.

5 SIZE EFFECT OF BACKFILL ON STRENGTH REQUIREMENT

Skinner (1959) has derived the formula (equation 9) assuming that the strength is distributed according to the statistical distribution of Weibull (1951).

$$\log C_o = K - 1/\beta \log V \qquad (9)$$

where V = Volume of the specimen
K and β = Constants.

In Skinner's formula, the size of the specimen is expressed in terms of volume. However, the effect of volume on the strength cannot be defined unless H/D ratio is taken into consideration (Dixit, 1981). Therefore, the equation (9) may take the following form

$$\log C_o = K - 1/\beta \log N \qquad (10)$$

where N = Height/diameter ratio of the fill assuming the solid cylindrical dimension of the fill.

An approximate value of diameter of the backfill can be calculated taking both length (L) and width (W) of the fill into consideration and assuming the cylindrical diamension of the fill, as no accurate design for strength requirement can be given by considering H/L and H/W ratios independently. Thus, the fill size is defined by H/D ratio, where diameter D arbitrarily represents both length and width of the fill.

The Weibull's distribution does not differ much from the normal (Gaussian) distribution and also the negative strength values as were found in the normal distributions in equations (7) and (8) are impossible in Weibull's distribution. Of course, the normal distribution is symmetrical and unlimited, therefore, certain negative strength values are possible in it.

Hence, the equation (10) was thought to describe the observations on size effect of the backfill in a very satisfactory way. The two groups of specimens may be made to fit according to Weibull's distribution and their sum total may be found as in Gaussian distribution.

The size effect on uniaxial compressive strength of gondite evaluated in the form of equation (10) is expressed as

$$\log_{10} 8.074 o \times 10^4 KN/m^2 = 5.2707 - 0.8926 \log_{10}$$
$$\qquad (11)$$

It is easy to find the values of constants K and β with the help of at least two values of strengths of two groups of rocks and their corresponding Height/Diameter ratios. The size effect, thus, evaluated in schist and gneiss is expressed in the form of equation (12).

$$\log_{10} 1.2432 \times 10^{4} \text{KN/m}^{2} = 4.905 - 1.5725 \log_{10}N$$

(12)

These results are shown graphically in Figure 8. From this linear extrapolation between logarithms of strengths and H/D ratios, the reasonable critical backfill sizes are predicted corresponding to the 'critical stress levels' derived from equations (7) and (8) expressed in terms of wall rock strength probability. A 'critical backfill size' is defined by a certain height/diameter ratio of the fill at or exceeding which, the bearing capacity of the fill is exhausted at the 'critical stress level' derived from the probability density functions.

The critical backfill sizes now can be predicted by substituting the values of C_O, derived from the probability density functions, in equation (10).

The values of C_O obtained from probability density functions, i.e., 8.0740×10^{4} KN/m^{2} of gondite and 1.2432×10^{4} KN/m^{2} of schist and gneiss correspond exactly to N = 2.555 and N = 3.276 respectively on the plots (Figure 8) and one need not substitute the values of C_O in

equation (10) to find out the critical backfill size which can be read directly from the plot also. This observation thus appears quite unique and proves the validity of this concept of backfill strength requirement prediction.

In general the fill height is larger than the fill width or of the same order of magnitude which can, however, exceed its lateral dimension (L) in underground metalligerous mines using blast hole stoping methods. Considering an opening mined to H=4L, as opposed to four stoping blocks with H=L, predicts that the cement strength requirement would increase by more than 50% by having H=4L (instead of H=L). Similarly, the fill width would also significantly affect the cement strength requirement. It is, therefore, inferred that the cement stabilized strength requirement need not be equal to the wall rock strength distribution functions. This requirement can be corrected according to the height, length and width of the fill, using the equation (10). Alternatively, the equation (10) can be used to reduce the cement requirement by modifying the size parameters, depending upon the size of opening, according to the case whatsoever may be existing.

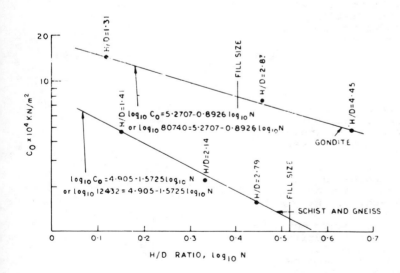

Fig.8. Size effect of backfill on strength requirement.

6 CONCLUSION

A system of reinforced cemented backfill design made up of alternate thick layers of low cement fill and thin layers of high cement fill has been proposed based on controlled cement pouring in two alternate groups so as to correspond the probability of two such most representative strength distributions in wall rocks respectively. The method can be applied for cemented backfill design by considering the normal distribution of two different groups, namely, sound and defective, as representative samples of any type of rock can be separated in two such groups. The cement requirement in two layers may be further reduced after certain corrections for size effect on strength requirement using the equation (10). The proposed design may achieve greater economics in cemented backfill practice. Though the system may require a close integration with mining field conditions, the analysis provides a link between laboratory investigations and actual mining situations.

ACKNOWLEDGEMENTS

The authors gratefully acknowledge the continued support and encouragement by Dr. B.Singh, Director, Central Mining Research Station, Dhanbad and are thankful to him for granting the permission to present this paper.

REFERENCES

Dixit, J.P. 1981. A geomechanical study of Chikla mine rocks,District Bhandara, Maharastra, India. Ph.D.Thesis. Banaras Hindu University, Varanasi. 275 p.

Mitchell, R.J., R.S.Olsen & J.D. Smith 1982. Model studies on cemented tailings used in mine backfill. Can. Geotech. J. 19: 14-28.

Skinner, W.J. 1959. Experiments on compressive strength of anhydrite. Engineer 207:255-259, 288-292.

Weaver, W.S. & R.Luka 1970. Laboratory studies of cement stablized mine tailings. Can. Min. Met. Bull. 63:988-1001.

Weibull, W.A. 1951. Statistical distribution function of wide applicability. Journal of Applied Mechanics. 18(3):293-297.

Proceedings of the International Symposium on Mining with Backfill / Luleå / 7-9 June 1983

Supports of reinforced granular fill

J.A.HAHN & L.DISON
Stratafix (Pty) Ltd., Johannesburg, South Africa

G.E.BLIGHT
University of Witwatersrand, Johannesburg, South Africa

ABSTRACT: A system of support is described which utilizes waste material reinforced with either horizontal steel mesh or spirally-wound high tensile steel wire. Supports can be accurately designed to carry a pre-determined load with a given factor of safety. A design theory for both types of support is developed and the load-compression characteristics are described.

INTRODUCTION

Most of the coal mines in South Africa are relatively shallow (30m to 100m) below surface) and may contain a sequence of superimposed coal seams.

The top or shallowest seam of a sequence can be almost completely extracted by long wall mining or stooping (following bord and pillar mining) if the severe surface subsidence that ensues can be tolerated. Deeper seams, however, cannot be totally extracted unless they are extracted in sequence from the shallowest to the deepest. The coal contained by the different seams is usually of different qualities and types, and commercial demand for coal of specific qualities dictates the sequence in which the seams are extracted. As a result, most multi-seam mines are mined using the bord and pillar method. This practice results in percentages of extraction varying from 75 per cent in very shallow seams to 40 per cent or less in deeper seams.

Although coal provides a cheap form of roof support, coal pillars represent a natural energy resource that has been sterilized and much research is currently in progress to find ways and means of winning this sterilized coal.

This paper describes one of the techniques that is currently under development. The method consists of replacing coal pillars by artificial pillars or walls of horizontally reinforced granular material. In principle, any granular material can be used for this purpose. However, from environmental and cost aspects, it is preferable to use

mining or other waste such as power station bottom ash. Ash is a particularly appropriate choice if the colliery is "captive" to a power station, i.e. if its main purpose is to supply a nearby power station.

STRENGTHENING A GRANULAR MATERIAL BY HORIZONTAL REINFORCING

The effect of horizontal reinforcing on a granular material is to develop a horizontal confining stress when the material is subjected to vertical loading. The horizontally reinforced granular material will tend to go into a state of failure as soon as a vertical stress is applied to it. Actual failure will, however, not occur because of the development of tension in the horizontal reinforcing. As Figure 1 indicates, on a p-q diagram, the stress path oa of the granular

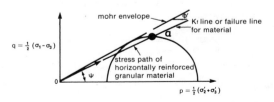

Figure 1: Stress path for reinforced granular material.

material will follow the failure line for the material. On the stress path: a represents the point at which the horizontal reinforcement reaches its yield stress σ_{yR}.

At this stage the vertical stress σ will be related to the horizontal stress σ_h by the relationship

$$\sigma = K_p \sigma_h \quad\ldots\ldots\ldots\ldots\ldots\ldots\ldots \quad (1)$$

where K_p is the passive pressure coefficient of the granular material, i.e. the ratio σ_1/σ_3 at failure in a triaxial compression test on the material.

Equating the tension in the horizontal reinforcing to the compression in the granular material:

$$\sigma_{yR} A_R = \sigma_h (A_m - A_R)$$

where A_R is the cross-sectional area of a reinforcing wire and A_m is the area of reinforced material corresponding to A_R

$$\text{or } \sigma_h = \frac{\sigma_{yR} A_R}{(A_m - A_R)} \quad\ldots\ldots\ldots\ldots\ldots \quad (2a)$$

Hence

$$\sigma = \sigma_{max} = \frac{K_p \sigma_{yR} A_R}{(A_m - A_R)} \quad\ldots\ldots\ldots\ldots \quad (2b)$$

If A_R is small as compared with A_m (A_R is usually a few tenths of a percent of A_m)

$$\sigma = K_p . \sigma_{yR} . \frac{A_R}{A_m} \quad\ldots\ldots\ldots\ldots\ldots \quad (2c)$$

The strength of a horizontally reinforced granular material is thus directly proportional to the reinforcing ratio A_R/A_m and the yield stress of the reinforcing σ_{yR}.

DESIGN OF WALLS OR PILLARS OF HORIZONTALLY REINFORCED GRANULAR MATERIAL

Artificial pillars and walls can be constructed of granular materials reinforced with layers of steel mesh placed at a vertical spacing v. If h is the horizontal spacing of the wires in the mesh and A_R is the cross-sectional area of each wire, it follows that

$$\frac{A_R}{A_m} = \frac{A_R}{vh}$$

and equation (2c) becomes

$$\sigma_{max} = K_p \sigma_{yR} . \frac{A_R}{vh} = F\sigma \quad\ldots\ldots\ldots\ldots \quad (3)$$

where F is the factor of safety on the design stress σ. For a selected steel quality and mesh dimensions, v can be calculated or alternatively, for a selected v, a suitable mesh may be selected.

The properties of horizontally reinforced granular materials have been extensively investigated by means of both model scale and full scale tests. Figure 2 shows a comparison of σ calculated from equation (3) with corresponding measured values.

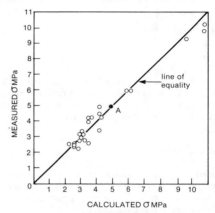

Figure 2: Relationship between calculated and measured strengths of walls built of horizontally reinforced granular material.

It will be seen from Figure 2 that there is an excellent correlation between calculated and measured values over a wide range of stresses.

Figure 3 (test A on Figure 2) represents the results of a typical compression test on a reinforced wall. This diagram illustrates an important feature of walls or pillars of horizontally reinforced granular material. The wall was designed to have a factor of safety of 1,6 on a design vertical stress of 2,39 MPa. The wall was loaded to the design load and was then unloaded and reloaded to failure.

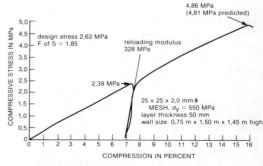

Figure 3: Stress-compression curve for wall of horizontally reinforced ash

The stress-compression curve for the first loading indicates a low compression modulus. However, the reloading curve is considerably steeper, hence the reloading modulus is much higher. Figure 3 illustrates how it is intended to use the walls of reinforced granular material in practice. After building the walls in-situ, they will be precompressed and test-loaded to the design load by jacking off the roof. The space created between the top of the wall and the roof by the precompression will then be filled by horizontally reinforced cemented slabs and wedges which will be bedded against the roof by grout. As the load is subsequently transferred to the wall when the roof tends to sag, the wall will be recompressed along the recompression curve, back to the design load. The support system thus created will be both relatively stiff and pre-tested to the design load. The stiffness is particular-ly important in limiting bending deflexions and therefore bending stresses in the roof.

The recompression modulus of horizontally reinforced walls is related to their strength as shown by Figure 4. The scatter of results illustrated in this figure is caused mainly by variation in the characteristics of the granular materials being used. The upper band of results corresponds to walls built of a highly

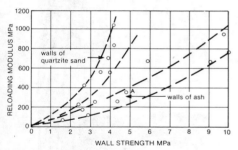

Figure 4: Relationship between wall strength and compression modulus on reloading for walls of horizontally reinforced granular materials

frictional weathered quartzite sand. The lower band of results corresponds to walls built of power station bottom ash which although highly frictional, is considerably more compressible than the sand.

Alternatively, in situations where a yielding support is required, the stiffness and yield stress of the support can be designed to give the required qualities.

BOND OF REINFORCING TO REINFORCED MATERIAL

The tension developed in the reinforcing of a horizontally reinforced granular material depends on stress transfer by bond between the reinforcing and the reinforced material. The mechanism of bond develop-ment is illustrated in Figure 5a.

The average bond stress transmitted to a reinforcing wire over a length x by friction is

$$\frac{\sigma}{2}.\{1 + \frac{1}{K_p}\}\tan\phi\pi D.\frac{x}{2}$$

in which the diameter of the wire is D.

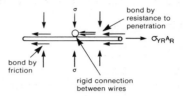

Figure 5a: Mechanism of bond development in reinforced granular material

The most convenient form of horizontal reinforcing for a wall consists of a system of orthogonal wires rigidly bonded to each other at points where they cross. Readily available systems consist of either square or rectangular welded mesh or twisted diamond mesh. The wires that run at an angle to the direction of tension under consideration contribute considerably to the bond because they must penetrate the reinforced material before the reinforcing can slip relative to the material. If the bond wires in an orthogonal system are spaced ℓ apart, the load transferred through these wires in length x will be a minimum of:

$$\frac{x}{\ell}(\sigma.d + 2d\sigma\tan\phi)\frac{h}{K_p}$$

in which d is the diameter of the bond wires that are spaced h apart. Full bond resistance of the wire is therefore developed in a length x given by

$$\frac{\pi D^2}{4}\sigma_{yR} = \frac{\sigma(1 + \frac{1}{K_p})\tan\phi\pi Dx}{4} + \frac{\sigma}{4}\frac{d(4 + 8\tan\phi)h.x}{\ell K_p}$$

i.e. $x = \dfrac{\pi D^2 \ \sigma_{yR}/\sigma}{\pi D(1 + \frac{1}{K_p})\tan\phi + \frac{4dh}{\ell}(1 + 2\tan\phi)\frac{1}{K_p}}$ .... (4)

The value of x is astonishingly small. For example

if $\sigma = 5$MPa $\sigma_{yR} = 600$MPa

$D = d = 2,5$ mm $K_p = 3,5$

$h = \ell = 25$ mm $\tan\phi = 0,7$

then x = 98 mm.

Figure 5b illustrates the results of a series of tests on reinforced walls that was designed to investigate the validity of equation (4). The figure shows the stress at failure in a series of tests on walls in which the number of bond wires at each side of the wall was increased progressively from zero. The spacing of the bond wires was 12,5 mm and the theoretical bond length was 60 mm, hence full bond could theoretically be achieved with 5 bond wires. The results in Figure 5b confirm this.

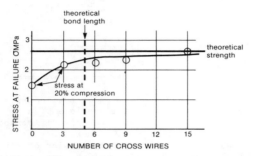

Figure 5b: Investigation of bond length requirements for reinforced granular material

FAILURE MECHANISM OF REINFORCED GRANULAR WALLS

As the design theory indicates, reinforced granular walls fail when the tensile strength of the horizontal reinforcing is reached. When this occurs, the reinforcing wires fracture and the granular material shears along diagonal planes inclined approximately at $(45°-\phi/2)$ to the direction of the major principal or vertical stress.

Figure 6a illustrates the observed positions at which the reinforcing wires fractured in a typical test wall. The multiple fractures of the reinforcing define a system of multiple shear planes inclined at a mean angle of 26° to the direction of the major principal stress. This mean angle corresponds exactly to the theoretical angle $(45°-\phi/2)$ for the fill material for which $\phi = 38°$. Figure 6b is a photograph of one of the mats of mesh taken from the wall analysed in Figure 6a. The multiple fractures of the steel are clearly visible.

Theoretically, it is possible for an infinite number of shear planes to develop in the granular material. In practice, however, the development of shear planes is affected by the frictional restraint exerted on the top and bottom of the wall. The combination of end restraint and the

limited height to width ratio of the walls

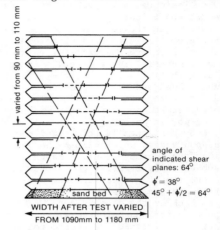

Figure 6a: Positions of tensile failures in reinforcing of horizontally reinforced wall of granular material

prevent shear planes from developing fully except in the diagonal corner-to-corner configuration illustrated in Figure 6a.

Figure 6b: Mat of reinforcing steel taken from wall referred to in Figure 24a showing multiple fractures of transverse wires

SPIRALLY OR HOOP REINFORCED COLUMNS

As an alternative to embedding reinforcing mesh in the fill, a cylindrical column of fill can be reinforced by surrounding it with a steel spiral or a series of steel hoops. A radial confining stress is then developed as shown in Figure 7.

For horizontal equilibrium of the column,

$$2Rp\sigma_h = 2T = \frac{2A_R\sigma_{yR}}{F}$$

i.e. $\sigma_h = \dfrac{A_R\sigma_{yR}}{FRp}$

and $\sigma = \dfrac{K_p A_R \sigma_{yR}}{FRp}$ .... (5)

in which p is the pitch or spacing of the reinforcing hoops or coils.

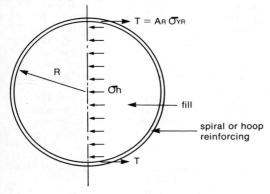

Figure 7: Principle of calculating the strength of a spirally reinforced column.

Figure 8 shows a comparison of measured strengths of hoop and spirally reinforced columns with values predicted from equation (5). The agreement between theory and measurement is excellent and the comparison also shows that equation (5) applies equally well to both spiral and hoop

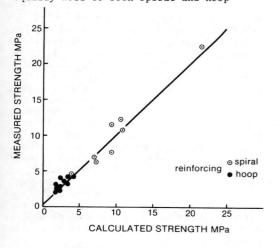

Figure 8: Comparison of calculated and measured strengths of spirally and hoop reinforced columns.

reinforcement.

Bond between the reinforcement and the fill plays no part in a spirally reinforced column. However, to transfer the confining stress into the fill, it is necessary to have a retaining membrane between the fill and the reinforcing. In the laboratory tests carried out so far, a woven polypropylene hessian has been used for this purpose.

The reloading moduli of spirally or hoop reinforced columns are comparable with those of mesh reinforced walls. This is shown in Figure 9 which compares reloading moduli measured on reinforced ash walls with corresponding measurements on spirally reinforced ash columns.

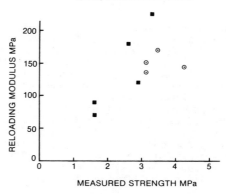

Figure 9: Comparison of reloading moduli for mesh reinforced walls and spirally reinforced columns.

COMPARISON OF WALL AND COLUMN SYSTEMS

According to equation (3) the load carried by unit plan area of horizontally reinforced wall is

$$\sigma = \frac{K_p \sigma_{yR}}{F}\cdot\frac{A_R}{vh} \dots\dots\dots\dots\dots\dots (3a)$$

A_R/vh is then the volume of reinforcement per unit volume of wall.

Similarly, from equation (5) the load carried per unit area of spirally reinforced column is

$$\sigma = \frac{K_p \sigma_{yR}}{F}\cdot\frac{A_R}{Rp} \dots\dots\dots\dots (5a)$$

In this case, $2A_R/Rp$ is the volume of reinforcement per unit volume of column.

The major cost item in a reinforced granular fill support is the reinforcing steel, hence the ratios $^AR/vh$ or $^{2A}R/Rp$ are, to a large extent, the key to the economic viability of the system. For the same quality of fill and reinforcing and the same factor of safety $K_p\sigma_{yR}/F$ will be the same. Hence comparing horizontally reinforced walls with spirally reinforced columns, the two systems will be equally viable economically if

(1)
$$\frac{2A_{R(col)}}{Rp} = \frac{A_{R(wall)}}{vh} \cdots\cdots\cdots (6)$$

and (2) equally strong if

$$\frac{A_{R(col)}}{Rp} = \frac{A_{R(wall)}}{vh} \cdots\cdots\cdots (7)$$

The two requirements clearly cannot be met simultaneously. For equal strength, if $A_{R(col)} = A_{R(wall)}$, then $Rp = vh$. It then follows from equation (6) that a spirally reinforced column will require twice the steel required by the equivalent horizontally reinforced wall.

In practical terms p will be of the same order of magnitude as v (25 to 100 mm). Hence R must of necessity be of the same order of magnitude as h if $A_{R(col)}$ is to be kept within realistic bounds. The situation is best illustrated by an example.

Consider a wall for which

$\sigma = 5$ MPa $K_p = 3,5$ $\sigma_{yR} = 1000$ MPa
$F = 1,75$ $h = 50$ mm and $A_{R(wall)} =$
   20 mm² (equivalent to a wire
   diameter of 5 mm)

Then $v = \dfrac{K_p \cdot \sigma_{yR} A_R}{Fh\sigma} = 160$ mm.

Now consider an equivalent spirally reinforced column having p = 160 mm.

$R = \dfrac{K_p \sigma_{yR} A_R}{\sigma Fp} = 50$ mm.

which is clearly impractical.

If p is reduced to 30 mm, R = 267 mm, a more reasonable value. The example shows that whereas a horizontally reinforced wall may be constructed to any plan dimensions, the radius of a spirally reinforced column is limited to a maximum of about a metre by practical considerations.

Calculating the ratio of volume of reinforcement to volume of fill for the above example, we find that for the wall:

$A_R/vh = 0,25\%$

whereas for the column

$^{2A}R/Rp = 0,5\%$

which agrees with the conclusion arrived at above.

A square or rectangular horizontally reinforced pillar will need to have the same reinforcement normal to each of its pairs of sides. Hence the ratio of volume of reinforcement to volume of fill will be $^{2A}R/vh$ in this case. (Because of the necessity of providing bond steel in a wall, the ratio of reinforcement volume to fill volume is actually slightly higher than A_R/vh).

It is seen, therefore, that for isolated supports both systems require the same amount of reinforcing. The spirally reinforced column does, however, have potential advantages in that

(1) ultra-high tensile steel can be used in spiral form whereas this is not possible in mesh form because welding of the nodes locally reduces the strength.

(2) the cylindrical mass of fill can be placed and compacted either before or after placing the reinforcing; and

(3) the steel is open for inspection.

CONCLUSIONS

It is possible to build very effective artificial supports for shallow mining by using horizontally or spirally reinforced granular fill. The economic viability of the method will depend on local cost and circumstances and the selling price of the product. It is believed that at present the method has most promise for specialist applications. Walls, or rows of columns, could provide artificial barrier pillars for stooping operations in bord-and-pillar mines or for travelways in longwall mines. Spirally- or mesh-reinforced pillars could provide permanent or temporary point supports, or yielding supports. However, when the system is linked to the extraction of a higher-priced product (e.g. in South Africa, export coal) the method may well have wider applicability in primary mining operations.

On several subjects in flat-back cut-and-fill stoping

HUANG WENDIAN & MEI RENZHONG
Jiangxi Institute of Metallurgy, Guanzhou, China

ABSTRACT: The article has expounded several main subjects in flat-back cut-and-fill stepped stoping. The reasonable differentiation for the sizes of interchamber pillars is not
the key to the safety actural mining, but also of the important influence upon comprehensive technical and economical results. On the basis of the datum actually measured, it
puts forward to define the width of room in hyperbola functional relation according to
the exposed area allowed by orebody and orebody thickness. The strength and steadiness
of the filled body were followed with interest long before at home and abroad. Taking
the room pillar stoping for example, it shows that the cemented filled body, used by now
existing mines, are of enough strength and bulk steadiness, and poses a question that
the room pillar stoping can be guaranteed until the strength of filled body is up to
some 5 kg/cm^2 and fillmaterial proportion should be lowered to 1,20 and even to 1,30.
The upright exposed height of filled body is the important mark for entirety steady
degree of filled body. It raises a subject to analyse the fixed state of filled body in
bearing force of breast wall theory and to calculate exposed height. By comparing in
accordance with analysis datum of filled body and overseas datum, it confirms the use of
the low strength filled body the necessity and possibility of cutting down cement consumption. Finally, it analyses the relations between cement mortar density and cement
detachment, and strength of filled body. And also points out that to use cement mortar
of higher density is the important measure of raising filled body strength, improving
fill quality and reducing pollution in gallery, and that cement mortar density must be
fixed above 65%. According to experimental datum, it proposes an empiric formula for the
use of qualitative or semiquantitative analysis in production, design and study.

PREFACE

As a means of roof management and mined-
out area handling, fill has been more and
and more widely taken by mines because of
constant development and reaching perfec-
tion in its technology. It is not only the
powerful measures of the economy with
reason used to mine rare metallic ore and
non-ferrous metal rich ore by operating
mines, but also the effective method to
mine the special ore body. At present, the
flat-back fill is temporarily in the majority of the mining schemes in our filling
mines. It possesses the characteristics of
simpel technology, low filling cost, little
labour intensity and good ventilating condition. Our stoping schemes of backfill in
existence may be roughly divided into the
following:

1. The stoping scheme of flat-back fill
long workin face mechanization (part).

Ore body is stoped for one time along
the strike with flat-back fill, interchamber pillars and side pillars not left.

2. The joint stoping scheme of flat-back
fill multi-working face mechanization (part).

Ore body is divided into certain working
sections and a working section is separated
into several rooms and interchamber pillars,
with first step stoping several rooms and
with second stoping several interchchamber
pillars at the same time, drawing and filling operations are simultaneously performed in turn.

3. The stoping scheme of flat-back interchamber pillar fill.

Permanent pillars are left in the stope
to support the collapsible ore body and
the rooms are stoped succesively or simultaneously.

4. The flat-back fill stepped stoping
scheme.

The ore body is devided into rooms and interchamber pillar, with first step stoping interchamber pillars. In recent years, combined with mines, the scientific research and designing departments have perfected and developed backfill stope technology, got gratifying achievements and accumulated a wealth of experience. But in stepped stoping the ore body with backfill under the conditions of unsteady ore and surrounding rock, there are still some questions, such as the reasonable size division of rooms and interchamber pillars, interchamber pillar stoping and steadiness of filled body, the strength requirement of filled body, the cement consumption, mining damage, ore dilution and so on. There is a great need for us to go further into these to be able to solve them reasonably. This paper specially makes a discussion on several subjects in stepped stoping with backfill.

1 THE SELECTION OF ROOM AND INTERCHAMBER PILLAR SIZES

1.1 A discussion about the division of room and interchamber pillar sizes

Room and interchamber pillar are all the base units of two step stoping. The varing sizes and dimension scale of an unit are both the necessary condition to guarantee safety stoping and of a great influence upon comprehensive technical and economical results of stepped stoping. On the division of room and interchamber pillar, there are two opinions. One is large room and small interchamber pillar. In the steady range or exposed area allowed by roof, enlarge the room area as fully as possible to increase the mining quantity of first step stoping and to raise the mining specific gravity of first step stoping. This division is suitable for the mining condition of unsteady ore body and surrounding rock. Another is small room and large interchamber pillar. The cementing fill quantity in room is reduced, and filling cost is lowered. This dividing method is more suitable for the condition of steady ore body and surrounding rock. As viewed from our stepped filling experience, rooms and interchamber pillars being divided, the dividing methods of both large room and small interchamber pillar have been used. From the point of view of the interchamber pillar bearing and roof management, this kind of division is reasonable. On first step stoping, interchamber pillar becomes

ore prop, on second step stoping, man-made wall pillar (filled body) gets interchamber pillar. The safety degree of two-steps stoping mainly depends on the steadiness of hese two interchamber pillars, except for the relation with the roof steadiness, that is, it has a relation to their strength The strength of interchamber pillar can be expressed with the following formula.

$$\sigma_0 = \frac{4}{\sqrt{3}}K + \frac{K}{\sqrt{3}} \cdot \frac{L}{h_0} \qquad (1)$$

where

σ_0 is the strength of interchamber pillar

K the strength of ore body (or cemented filled body)

L the width of interchamber pillar and

h_0 the height of interchamber pillar

The strength of interchamber pillar has something to do with the dimension scale and its material strength. The interchamber pillars (ore pillar) with the same dimension scale are higher than the man-made wall pillars in strength. In the interchamber pillars made of same material, the greater the ratio of $\frac{L}{h_0}$, the higher the strength. Therefore, the degree of safety is higher on primary stoping on interchamber pillar stoping. The man-made wall pillar and ore pillar should be of equal strength so as to have two steps stoping be basically same in the safety degree, while the width of man-made wall pillar

should be $L_2 = \frac{L_1 K_1 + 4h_0(K_1 - K_2)}{K_2}$,

K_1 and K_2 are the strength of ore and filled body respectively, L_1 is the width of pillar, and h_0 is the height of ore pillar.

$\frac{L_2}{L_1} = \frac{K_1}{K_2} + \frac{4h_0(K_1 - K_2)}{L_1 K_2} > 1$, thus it can be

seen that the width of man-made wall pillar (that is, the width of room) is greater than that of ore pillar. Obviously, from the point view of mechanics, the dividing methods of large room and small interchamber pillar are profitable to either first step stoping or second step stoping, and using the large room stoping is possessed of evident superiority in undertaking the comprehensive extraction.

1.2 The selection of room sizes

As is well known, the room sizes are restricted by the factors of mechanical properties of ore and surrounding rock, geological structure, allowable exposed area, etc. Allowable exposed area is the base of selecting room sizes. The question on how to evaluate the roof exposed area has not been thoroughly solved now in theory and practice. When definiting room sizes, they will be selected generally by analogue, referring to the practical experience of similar operating mines. The basis selected in essence is still based on spot observation. In accordance with the observational data from spot, it is known that the stoping area of room has a close bearing on the exposed area of ore body. In a certain ore section, fixed value is desirable to the stoping area of room, but can not be over the number of allowable exposed area. And the hyperbola functional relation appears between the width of room and the thickness of ore body.

For calculation this relationship is followed.

$$\frac{1}{Y} = a + b\frac{1}{X} \qquad (2)$$

where

- Y – the width of room
- X – the horizontal thickness of ore body
- a, b – the speciality coefficient relative to the mechanical properties of ore body (a and b possess different values to separate ore bodies).

During the testing period of one-step stoping, for example, mine No 5 has made an on-the-spot investigation and practical test. Having been summed up and sorted out, the data measured practically are placed in table 1.

Table 1.

The horizontal thickness of ore body (m)	The width of room (m)	The exposed area of roof (m^2)
6	30	180
7	25	175
8	22	176
10	20	200
12	16	192
14	13	182
16	12	192
18	10	130

Through the regressive analysis of the data measured practically, it is made clearer that the relation of hyperbola function is assumed between the width of room and the horizontal thickness of ore body. This can be indicated in formula 2, in which a is equal to 0.1165 and b is equal to -547. The relationship between these two may be seen, as illustated in Fig. 1.

Thus it can be known that as far as the intermediate ore body is concerned, the width of room may be selected by formula 2 for calculation, and that it should be changed according to the different changes of the thickness of ore body, and it is not suitable for following same pattern and being invariable. The thickness of ore body is 6-8 m, while the width of room is 20-25 m; analogously 10-12 m, 14-16 m, 12-16m, 10-12 m, 16-18 m, 8-10 m. The practice of production has proved that cave-in will not happen to the back of a stope and safety production can be guaranteed when the width of room is not in outstriping the values above and anchor poles are used for partial roof supports.

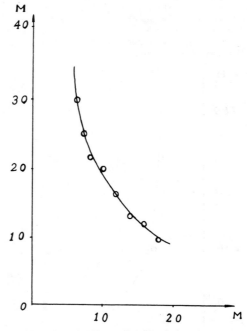

Fig.1. The relationship between the width of room and the horizontal thickness of ore body at the -70 m level in mine No.5.
X – the horixontal thickness of ore body
Y – the width of room

357

At 5 fill stopes in mine No. 8, the data observed on the spot also reflect the regular pattern stated above. The 5 stopes are arranged along the strike. Their total length is 110-180 m, the thickness of ore body 3–30 m, and their total area 1000-2256 m². On the subsistence condition of ore body and the essential factors of stope formation, see table 2.

Table 2.

| Place | The name of stope | The subsistence condition of ore body | | | The essential factors of stope formation | | | |
		Slope angle (degree)	The mean thickness (m)	The max thickness (m)	Length (m)	Width (m)	The mean exposed area (m²)	The max exposed area (m²)
-107	93010	60-70	8	18	130	3-13	1000	1500
-167	103001	60-70	14	22	130	3-22	1856	2256
-227	113001	60-70	10	20	180	5-20	1882	2048
-227	110103	70-75	15	30	110	8-30	1686	1740
-287	123001	60-70	8	16	180	8-16	1400	1600

The data above show that there is also a certain relation between the length of a stope and the thickness of ore body. Through the regressive analysis, it is known that both of them present hyperbola functional relation, too. This can be shown by formula 2, in which a is equal to 0.0126 and b is equal to -0.1115. The relationship between these two may be made out, as illustrated in Figure 2.

As another example, when observing the steadiness of roof exposed face at worked-out section, Nijes-Kum copper-nicked mine in the Kora peninsula also found that there was a certain relation between the length of worked-out section along strike ant its slant size. Both of them present hyperbola functional relation, too. This can be indicated by formula 2, in which a is equal to 0.0013 and b -0.8698. The relationship between these two is shown in Figure 3.

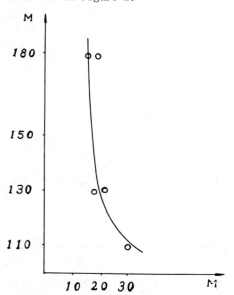

X - the thickness of ore body
Y - the length of stope
Fig. 2. The relationship between the length of stope and the thickness of ore body.

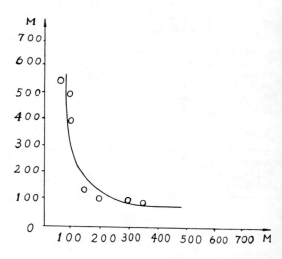

Fig. 3. The relationship between the length of worked-out section along strike and its slant size.
X - the slant size
Y - the size along strike

The practical instances stated above make it clear that because of the immense variety of the mechanical properties of ore and surrounding rock and their construction, the coefficient a differs greatly from b and the shapes of curve are not completely identical, but the relation between the width of room and the thickness of ore body can be summarized into the hyperbola functional relation.

In a word, the width of room should be reasonably selected in accordance with the changes of the thickness of ore body on the basis of the allowable exposed area of ore and surrounding rock.

1.3 The selection of interhamber pillar sizes

The interchamber pillar stoping is operated among the filled bodies of room. Being influenced by the factors of whether fill material contact to the top of worked-out section, fill denseness, strength of filled body, etc. the roofs of room on both sides of interchamber pillar will bring about deformation and displacement in varying degrees, and will shift the load of upper part on to interchamber pillar. The interchamber pillar stoping is operated in the ore body of stress concentration. During the operation of this stoping there is still the pressure action on the working face. Therefore, on selecting the sizes of interchamber pillar, its enough strength must be guaranteed. It is seen from formula 1 that the reasonable value of L and h_0 must be selected for the purpose of $\sigma_0 > \sigma_1 + \sigma_2 + \sigma_3$, in which σ_1 is the original pressure, σ_2 the additional pressure put on interchamber pillar by the roof of room, and σ_3 the stope pressure in interchamber

pillar stoping. When dividing the interchamber pillars and rooms, as stated above, the principles of division of large room and small interchamber pillar should be followed, but the width of interchamber pillar should not be too small, otherwise it is not of enough strength, that is, $\frac{L_1}{L_2} < 1$ is appropriate, and its concrete

value should be considered as a shole on the authority of the steady condition of ore and surrounding rock, and the comprehensive technical requirements. In mine No. 5, for example, the width of room is over 10m while the width of interchamber pillar is appropriate to 6-8 m. Practice proved that the interchamber pillar with

a certain strength can ensure the dependable stoping of second step.

2 THE STEADINESS ANALYSIS OF FILLED BODY AND THE DEFINITION OF EXPOSED HEIGHT

In back-fill stoping rooms, the tailings cementing fill materials are mostly used in our fill mines. For calculation of the strength of filled body, the Moore-Kullen law is followed.

$$S_0 = C + \sigma_n \, tg\varphi \qquad (3)$$

where
S_0 is the shearing strength of filled body
σ_n the positive normal stress on the arbitrary face in filled body and
φ the angle of friction in filled body.

The shearing strength of filled body is somposed of inner friction force and cohesive force. The inner friction force refers to the surface friction force and the gripping force among the particles of fill material, while the cohesive force includes the original cohesive force, the cement-solidifying cohesive force and the capillary cohesive force of the particles. The cohesive force of cement filled body is chiefly the cement-solidifying cohesive force. The increasing of cohesive and inner friction force will raise the shearing strength of filled body without doubt. The action of cement or other cementing agents is to increase the cohesive force of filled body. For example, the testing results of the shearing of tailings-cementing filled body in the ration 1:10 and 1:20, as in Fig. 4 make the following clear.

1) The cohesive force of filled body increases as cement content goes up, such as to the filled body in the ration 1:10 the cohesive force of curing for 7 days is 0.98 kg/sq.cm, and 1:20, is 0.23 kg/sq.cm while 1:10, that of curing for 28 days is 1.04 kg/sq.cm, and 1:20, is 0.45 kg/sq.cm.

2) There is greater cohesive force in the filled body of curing for 28 days than that for 7 days.

3) The value of friction angle in filled body is not remarkably influenced by the curing time and cement content.

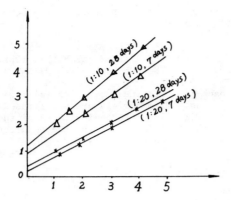

Fig. 4. The relationship between the shearing strength of filled body and its positive normal stress on the arbitrary face.

The allowable exposed height of filled body should be definited according the following.

The destruction of filled body is made by shearing force under the action of external force. In interchamber pillar stoping, the bearing state of filled body will be changed. Under the action of upright and horizontal stress it will produce deformation and displacement in the direction of interchamber pillar-stoping space. This deformation increases with the exposed height of filled body. When the exposed height is up to a limiting value, filled body will cause sticking-out and scale-off. Therfore, exposed height may be considered to be the important mark of filled-body steadiness. Filled body is the loose medium with a certain stickability. This filled body will bring about horizontal deformation under the action of external force and put horizontal side pressure on interchamber pillar. The bearing state can be analysed with the theory of breast wall being pressed. Arbitrarily taking an unit in Z deeps within filled body, the upright and horizontal stresses acting on the unit are respectively σ_2 and σ_x (Fig.5).

In a state of ultimate balance, the upright and horizontal stresses are respectively maximum and minimum principal stress. The values of σ_2 and σ_x are as follows.

$$\sigma_2 = r \cdot z + P_0$$

$$\sigma_x = (r \cdot z + P_0)\frac{1-Sin\varphi}{1+Sin\varphi} - \frac{2 \cdot C \cdot Cos\varphi}{1+Sin\varphi} \qquad (4)$$

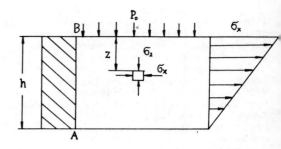

Fig. 5. The analysis of the bearing state in filled body.

where

  P – the bearing pressure on the filled body
  r – the volume weight of filled body
  φ – the angle of internal friction in filled body
  z – the depth of an unit in filled body
  C – the cohesive force of filled body

It is known from the formula above that the horizontal stress is negative value when the cohesive force is great enough and keeps $C > (r \cdot z + P_0)\frac{1-Sin\varphi}{2Cos\varphi}$, and then filled body will not change its shape or displace to the outside, that is, filled body will not produce the horixontal side-pressure upon interchamber pillar until the lowest point z = h and the horizontal stress is zero. In this range filled body does not bring about horizon action on upright wall AB (interchamber pillar), that is to say, filled body can support itself without any breast wall. For this reason, for calculation the maximum allowable exposed height h of filled body may be followed.

$$h = \frac{1}{r} \left(\frac{2 \cdot C \cdot Cos\varphi}{1-Sin\varphi} - P_0\right) \qquad (5)$$

In mine No. 5, filled body is in the ratio 1:10 and its exposed height is 2.6 m, while in mine No. 3, filled body is in the ratio 1:7 and its maximum exposed height is 7.7 m and its steadiness is good, too. From the steadiness-study data of cement filled body in ore body No. 1100, Mount-Isa Mine Limited Company, Queensland Australia, it thus can be known that the cement content in the cement filled body is 3.5%, the slag of reflection is 7%, therest is graded tailings the density of tailings transported is70%, the cohesive force of filled body is

0.51 kg/sq.cm, the compressive strength of
single axis is 0.8 - 1 kg/sq.cm, and the
exposed height of filled body is 40-90 m.
Therefore, in stoping interchamber pillar
with the flat-back cut-and-fill method,
filled body may be completely exposed and
it can guarantee safety stoping of interchamber pillar.

3 THE POSSIBILITY OF USING THE FILLED BODY WITH LOW STRENGTH AND CUTTING DOWN CEMENT CONSUMPTION

The main factors relative to the strength
of filled body are the cement content, the
density of tailings mortar and the grading
of tailings. When the latter two are rated
values, the relationship between the
strength of filled body and the cement content can be represented by exponential fun-
tion.

$$Y = Ae^{B \cdot X} \qquad (6)$$

where
  Y - the strength of filled body
  X - the distribution ratio of cement to
      tailings
  A, B - coefficient

Now our fill mines have no practical
experience of interchamber pillar stoping
and are lacking in the knowledge of the
steadiness of filled body. When working out
the distribution ratio of fill material,
for the sake of insurance, they blindly
concentrate on the filled body with greater
strength so that the selection is generally
on the higher side of cement content. It
si known that in our fill mines the lowest
sidtribution ratio of cement to tailings
is 1:10 (refers to the fill material of
room), and the highest is 1:6-1:8. So the
strength is higher. In mine No. 5 when fill
material of room is in the ratio 1:10, the
strength Y for 18 months is 10 -15 kr/sq.cm.
In mine No.3 when that of man-made pillar
is 1:7, strength Y for 27 months is 11-42
kg/sq.cm. In accordance with the observed
data to the steadiness of room filled-body
in interchamber pillar stoping, in mine No.
5 the maximum exposed height of room filled-
body is 2.6, and that of man-made pillar is
up to 7.7 m. Both of them are possessed of
enough steadiness for the safety stoping of
interhamber pillar. We hold that now in our
fill mines the distribution ratio of cement
to tailings is too high, and it is great
necessity for them to reduce fill cost. At
present, cement of 150-200 kg, even more,
should be added into fill material per
cubic metre. The fill cost is up to

30 RMBY/m cube, which is a great waste.
The reasonableness in distribution ratio
of cement to tailings reflects meeting
with the strength demand of man-made pillar,
and then does the lowest cement consumption.
Now taking the spot of mine No. 5 for an
example, we want to talk about the possibility to use the filled body with low
strength. Proof sample must be taken out
from filled body for actually testing so
as to determine the strength in both sides
of filled body and to analyse the influence
factors upon it in interchamber pillar
stoping. The designing distribution-ratio
of filled body was 1:10, the actual density
of tailings mortar was 60-65% and filled
body was formed one year and a half ago.
From the data tested actually it can be
seen that the detachment of cement is grave
and inhomogeneous coefficients of cement
distribution are 0.1-0.54. The distribution
ratio of cement to tailings in the proof
sample changed a great deal and there was
a great disparity in strength. When the
lowest distribution ratio of cement to
tailings was 1:21.6, the strength of filled
body for 18 months was 3.81-20.40 kg/sq.cm.
After making a regressive analysis treatment on the data tested actually, it thus
could be found that there is a exponential
functional relation between the strength of
filled body and the cement content. The
relation may be shown in formular (6), in
which A equals 2.3596 and B equals 14.494.
In the calculation of formula (6), when
the distribution ratio of cement to tailings
is 1:15, the last strength for 18 months is
6.14 kg/sq.cm, while 1:20, 4.80 kg/sq.cm.
It thus can be recognized that the filled
body formed under the condition that the
distribution ratio of fill material is
1:15-1:20 and the density of tailings mortar is over 65% will be of enough strength
and be able to guarantee the safety stoping
of interchamber pillar. This has been proved
by the interchamber pillar stoping-practice
of mine No. 5.
  In a word, the strength 5 kg/sq.cm of
filled body and the distribution ratio
1:15-1:20 of cement to tailings may be
approximately taken according to the spot
experience.

4 RAISING THE DENSITY OF FILL-TAILINGS MORTAR AND IMPROVING FILL QUALITY

The strength and the whole steadiness of
filled body are in close relationship with
fill quality. The contents to check fill
quality include the detachability of
cement, the homogenization of particle
distribution of fill material, the mud

content dewartered and so on.

There are two ways to improve the fill quality, that is, to raise the density of fill-tailings mortar and to add in an amount of polyacrylamide. The action of polyacrylamide is notable only in the tailings mortar with low density. And therefore, the economical and practical way is to increase the density of tailings mortar. Now the influence of tailings-mortar's density upon fill quality are respectively stated as follows.

4.1 The relationship between the density of tailings mortar and the detachability of cement

An experiment of stope model was made, using the fill material of the cement tailings (1:10) in mine No. 5. In the result, it was pointed out that the detachability of cement is closely relative to the density of tailings mortar and there is an exponential functional relation between these two. This relation can be represented by the following formula.

$$Y = A \cdot e^{BX} \tag{7}$$

in which

Y is the inhomogeneous coefficient of cement distribution
X the density of tailings mortar
A and B the coefficients (respectively 393.86 and −12.68)

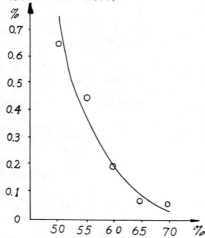

X − the density of tailings mortar
Y − the inhomogeneous coefficient of cement distribution
Fig. 6. The relationship between the density of tailings and the inhomogeneous coefficient of cement distribution.

The relationship between these two is shown in Fig. 6. When the density of tailings mortar is 65%, cement only produces a light detachment, while is over 65-70%, the detachment is basically eliminated.

4.2 The relationship between the density of tailings mortar and the homogenization of particle distribution

The tailings mortar with high density is possessed of greater viscosity and resistance to the subsiding of rough particles, and the subsiding speed is lowered and the subsiding difference is shortened owing to the reducing of water in tailings mortar, thus resulting in a rise in the homogeneity of particle distribution. The model experiment pointed it out that there is a power functional relation between the density of tailings mortar and the homogeneous coefficient of particle distribution. For calculation the formula (8) is followed

$$Y = A \cdot X^{B} \tag{8}$$

where

Y − the inhomogeneous coefficient of particle distribution
X − the density of tailings mortar
A − coefficient
B − coefficient

Fig. 7. shows the relationship between the dendity of tailings mortar and the inhomogeneous coefficient of particle distribution.

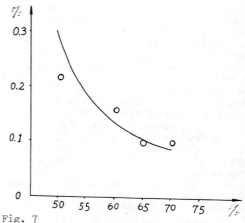

Fig. 7
X − the density of tailings mortar
Y − the inhomogeneous coefficient of particle distribution

4.3 The relationship between the density of tailings mortar and the mud content in water

When the cementing fill is made, the mud in water seperated from the fill material is mainly fine cement particles of -360 mesh, which is often washed away by the water. The results of model experiment have made it clear that there is a logrithmic functional relation between the density of tailings mortar and the mud content. This can be marked by the following formula.

$$Y = A + B \cdot \log^{X} \qquad (9)$$

in which

Y – the mud content in water (%)
X – the density of tailings mortar (%)
A – coefficient
B – coefficient

With respect to the cementing fill of mine No 5, A is equal to -o.026 and B is equal to -0.148, and the relationship between both of them is illustrated in Fig. 8.

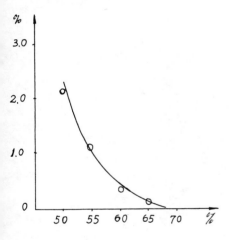

X – the density of tailings mortar
Y – mud content

Fig. 8 The relationship between the density of tailings mortar and the mud content

4.4 The relationship between the density of tailings mortar and the strength of filled body

After heightening the density of tailings mortar cement detachment is reduced, which gets granula be well-distributed and lowers the quantity of the running off of cement. For that reason, the strength of filled body is increased. The result of regressive analysis showed an exponential function relation presented between these two. It can be illustrated with the following formula.

$$Y = ae^{bX} \qquad (10)$$

where
Y is the compressive strength
X the density of tailings mortar and
a, b the coefficients

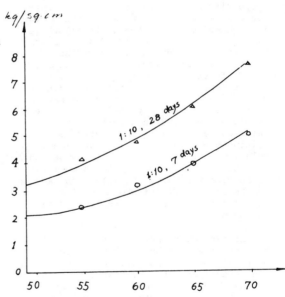

X – the density of tailings mortar
Y – compressive strength

Fig. 9. The relationship between the density of tailings and the compressive strength.

In connection with the compressive strength for 7 days curing to the cementing filled-body of mine No. 5, a is equal to 0.212 and b is equal to 4.512, while for 28 days, a = 0.408, b = 4.172 their relationship is shown in Fig. 9.

363

Thus it can be seen from Fig. 9 that about the fill material in the ratio 1:10, the strength for 7 days is enlarged from 2 kg/sq.cm to 5 kg/sq.cm when the density is raised from 50% to 70% and similarly, the strength for 28 days is increased from 3.34 kg/ sq.cm to 7.68 kg/sq.cm. This proves that to increase the density of tailings mortar is the effective measure to raise the strength of filled body. The tailings mortar with high density can improve the fill quality and reduce the running off of water and workings pollution. So the advanced fill system should be used to supply it with the fill tailings mortar of high density.

CONCLUSION

1. In flat-back cut-and-fill stepped stoping the reasonable selection of chamber sizes should be on the basis of the mechanical properties of ore body (mainly refers to the allowable exposed area). Because of the inhomogeneity of ore body, its allowable exposed area is possessed of stochastic property. And therefore, the widths of chambers are not completely identical at the different ore sections and mining depth in the same mine. Under the circumstances without evident changes in the mechanical properties of ore body, the width of chamber should be selected in hyperbola functional relation according to the mode of ore origin (ore thickness). The functional relation between the width of chamber and the thickness of ore body has been established on the basis of field testing data. So the final definition of the width should take the field practical testeing data as the standard.

2. The exposed height and area of filled body are the mark by which the steadiness of filled body is judged. The exposed height can be analysed and calculated with the stress theory of breast wall. The cement content is the important factor influential upon the strength and steadiness of filled body. According to the observational data upon the steadiness of filled body in the stoping of interchamber pillar, the cement contents are all higher in the fill material used by our fill mines and the ultimate strength of filled bodies are all over 10 kg/sq.cm and they are possessed of enough steadiness. In practise, it can meet the demands of two-steps stoping to use the fill material in the cement distribution ratio 1:20. For this reason, we make a suggestion to use the fill material in the cement distribution ratio 1:20 for the purpose of reducing cement consumption.

3. It is the economical and practical measure of increasing the strength of filled body to make filling with the tailings mortar of higher density. When the density of tailings mortar goes up to 70% from 50%, the strength of filled body can increases by 1.5 times. Therefore, the existen fill system should be reasonably remade or the advanced fill system should be newly set up so as to offer the filling tailings mortar with higher density.

Assessment of support performance of consolidated backfills in different mining and geotechnical conditions

JUN-YAN CHEN & DI-WEN CHEN
Central-South Institute of Mining & Metallurgy, Changsha, China

STEFAN H.BOSHKOV
Columbia University, New York, USA

INTRODUCTION

The development of cemented tailing fill in Canada during the late 1950's has made a dramatic impact on underground mining methods and ground control techniques. The use of cemented tailings has permitted the complete extraction of high-grade orebodies with a minimum of ore dilution and has led to the development of highly mechanized mining techniques. These techniques are demonstrated by the mechanized cut-and-fill method, the block cut-and-fill method, which features a large block or panel consisting of several small cut and fill stopes, and the application of the vertical crater retreat method for recovery of the pillars between backfilled stopes.

In the early 1970's, an important and continuing development in consolidated backfill, after the development of cemented tailing fill, was to mix the cemented tailings with dry rockfill, later called cemented rockfill. This type of consolidated backfill is usually used in the delayed backfilling operation and is developed for specific applications related to pillar recovery. These applications are associated with open stoping methods, where pillars are recovered alongside cemented rockfill walls.

It is generally recognized that the ground support function is the most important role of consolidated backfill. It is also recognized that the use of high-strength backfill (containing a higher cement ratio) would modify the stress pattern around a stope. However, very little reliable information on the magnitude of the reduction of the stress concentration exists, even though it is a most important consideration in practical mine design.

In the area of fill mechanics, many attempts to understand the role and the behavior of backfill in an underground environment have been made since the 1960's. Numerous papers and several symposia [1,2,3] attest to the prolonged interest of the mining community in backfill. With regard to theoretical studies, numerical methods were usually selected to investigate the problems. Pariseau et. al. [4,5] perhaps were the first to apply the finite element method to analyze the support function of hydraulic fill in cut-and-fill mining. It was found that the support capability of fill increases dramatically with the increase of the stiffness ratio, i.e. the ratio of fill modulus to rock modulus.

Mine planners, however, need to know the exact inter-relationship of design and performance of consolidated backfills against the physical properties of the wallrock and the nature of the pre-mining stress field. A definition of such inter-relationships may ensure optimal design of the consolidated backfill composition in order to reduce cement cost and match support capability to the requirements of ground control.

This paper summarizes the results of non-linear finite element analyses of the support performance of consolidated backfills in different mining and geotechnical conditions. The study employed three kinds of mining models: room-and-pillar mining, blasthole open stoping

TABLE 1
MINING GEOMETRIES IN FINITE ELEMENT MODELS

Model	Case	Mining Method	Stope or Room Width (m)	Pillar Width (m)	Stope Height or Ore Seam Thickness (m)
1	1-9	Room-and-pillar mining with de-layed backfilling	18.24	13.68	6.84
2	10-25	Room-and-pillar mining with de-layed backfilling	18.24	13.68	6.84
	26	Room-and-pillar mining with de-layed backfilling	13.68	13.68	6.84
	27	Room-and-pillar mining with de-layed backfilling	13.68	18.24	6.84
3	28-29	Blasthole open stoping with de-layed backfilling	12	12	45
	30-31	Blasthole open stoping with de-layed backfilling	20	20	45
4	32-33	Blasthole open stoping with de-layed backfilling	12	12	45
5	34	Benching cut-and-fill stoping	20	--	22.5

TABLE 2
PHYSICAL CONSTANTS OF ROCK PROPERTIES

Model	Properties		Ore	Wallrock
1	Elastic Modulus	MPa	11,721.5	23,443
		(lb/in^2)	(1.7×10^6)	(3.4×10^6)
	Poisson's Ratio		0.22	0.18
	Density	Kg/m^3	2,491	2,408
		(lb/in^3)	(0.09)	(0.087)
	Cohesion	KPa	4,137	5,861
		(lb/in^2)	(600)	(850)
	Friction Angle		43°	45°
2	Elastic Modulus	MPa	33,096	23,443
		(lb/in^2)	(4.8×10^6)	(3.4×10^6)
	Poisson's Ratio		0.26	0.18
	Density	Kg/m^3	26.30	2,408
		(lb/in^3)	(0.095)	(0.087)
	Cohesion	KPa	9,653	5,861
		(lb/in^2)	(1,400)	(850)
	Friction Angle		45°	45°
3	Elastic Modulus	MPa	13,376	8,963
		(lb/in^2)	(1.94×10^6)	(1.3×10^6)
	Poisson's Ratio		0.07	0.06
	Density	Kg/m^3	2,463	2,630
		(lb/in^3)	(0.089)	(0.095)
	Cohesion	KPa	4,826	4,826
		(lb/in^2)	(700)	(700)
	Friction Angle		38°	34°

and sublevel cut-and-fill stoping. De-layed backfilling and pillar recovery were practiced in all of these models. The behavior of backfills of different moduli, and their interactions with the host wallrocks were studied under various postulates of the pre-mining stress field and physical properties of the wallrock.

The authors believe that this kind of study could give the mining design engineers a better theoretical approach for the selection of consolidated back-fills in underground mining.

MODELING TECHNIQUES AND INPUT DATA

In this study, three kinds of mining methods which involve delayed backfill-ing operations using consolidated fill were simulated: room-and-pillar mining, blasthole open stoping and sublevel cut-and-fill stoping. The representative example of room-and-pillar mining with delayed concrete filling is the Keretti mine in Finland. Other applications can be found in Mogul (Ireland), Xikuang Shan (China) and some Russian non-ferrous metal mines.

Blasthole open stoping with delayed backfilling is widely used in North America to mine massive, nonferrous metal orebodies, such as Geco, Fox, Kidd Creek, some INCO and Falconbridge opera-tions (Canada), and Carr Fork (U.S.).

Sublevel cut-and-fill stoping is a rel-atively new mining method. It was devel-oped for the Avoca mine, and later ap-plied to the Navan mine, both in Ireland.

A total of 34 cases using five models, have been employed to simulate these three delayed backfilling mining methods. The mining geometries for each model are tabulated in Table 1.

Models 1 and 2 simulate room-and-pillar mining with delayed backfilling. Model 1 features a "soft" ore and "hard" roof, and model 2 depicts the case of "harder" ore and "hard" roof. Models 3 and 4, both consider jointed and fractured ore and wallrock conditions in blasthole open stoping with delayed backfilling. The difference between the two models is the mining sequence (see Fig. 1). Model 5 is the sublevel cut-and-fill

stoping case. It also considers jointed and fractured ore and wallrock.

The input data of rock properties for the computer runs are shown in Table 2.

It is a well-known fact that the influence of fill on ground support would not be seen unless the ground clo-sure had continued after the stope back-filling was completed. In other words, backfill will display its support role only under circumstances in which mine workings continously advance or the neighbouring stopes or pillars are re-moved subsequently. For this reason, all these models are designed for sequen-tial mining with multiple passes, i.e. primary mining, backfilling and pillar recovery, as shown in Fig. 1.

The results of stress variation and the behavior of fill of each model are compared while employing different con-solidated backfills, applied loading conditions and the stiffness between ore and wall rock. The theoretical rela-tionship of support performance of consolidated backfill against the nature of the pre-mining stress field and the physical properties of ore and the host rock can thus be obtained.

Three types of fill materials have been considered in this study: cemented tailings, cemented rockfill and concrete. Each type fill is then subdivided into several kinds, based on its elastic modulus, ranging from the weakest for 1:30 cement/tailing fill with a modulus of 41.4 MPa (0.006 x 10^6 psi) to the strongest concrete fill with a modulus of 8205 MPa (1.19x 10^6 psi). Physical properties of backfills, which were mostly obtained from laboratory and field measurements,[6] are shown in Table 3.

As one of the purposes of fill is to permit the mining out of ore pillars, the consolidated backfill itself then acts as an artificial pillar. There-fore, its stiffness, or elastic modulus, is of some significance. The stiffness ratio of the consolidated fill to the wallrock will dominate the support ef-fect of this kind of fill pillar and affect the stress distributions in the fill pillar as well as those in the walls and abutments. The influence of

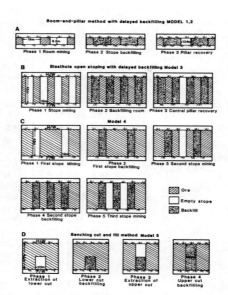

Fig. 1 Mining sequences of five mining models.

TABLE 3
PHYSICAL CONSTANTS OF FILL PROPERTIES

Fill		Elastic Modulus		Poisson's Ratio	Density		Cohesion		Friction Angle
		MPa	(lb/in^2)		Kg/m^3	(lb/in^3)	KPa	(lb/in^2)	
TYPE 1									
Cemented tailing I	(1:30)	41.4	(0.006 x 10^6)	0.15	1,503	(0.0543)	41.37	(6)	30°
Cemented tailing II	(1:6)	317.2	(0.046 x 10^6)	0.25	2,001	(0.0723)	117.2	(17)	35°
Cemented tailing III	(1:2)	586.0	(0.085 x 10^6)	0.28	2,103	(0.0760)	586.0	(85)	35°
TYPE 2									
Cemented rockfill I		862.0	(0.125 x 10^6)	0.28	2,159	(0.0780)	620.0	(90)	35°
Cemented rockfill II		979.0	(0.142 x 10^6)	0.30	2,214	(0.08)	655.0	(95)	35°
Cemented rockfill III		1,517.0	(0.22 x 10^6)	0.33	2,214	(0.08)	931.0	(135)	35°
Cemented rockfill IV		1,958.0	(0.284 x 10^6)	0.33	2,214	(0.08)	1,310.0	(190)	38°
Cemented rockfill V		2,344.0	(0.34 x 10^6)	0.30	2,270	(0.082)	1,448.0	(210)	38°
TYPE 3									
Concrete fill I		2,813.2	(0.408 x 10^6)	0.30	2,270	(0.082)	2,620.0	(380)	40°
Concrete fill II		4,688.6	(0.68 x 10^6)	0.30	2,270	(0.082)	2,620.0	(380)	42°
Concrete fill III		5,860.7	(0.85 x 10^6)	0.26	2,270	(0.082)	2,620.0	(380)	42°
Concrete fill IV		8,205.0	(1.19 x 10^6)	0.26	2,297	(0.083)	3,171.7	(460)	43°

changes in the modulus of the consolidated backfill were investigated in detail in this study.

Several kinds of the pre-mining stress states were considered in the finite element models. For shallow mining, where the mining depth was less than 500 m below the surface, a gravitational stress field with 1/3 ratio of horizontal stress to vertical stress was assumed. For medium-deep mining, where the mining depth was between 500 m and 1000 m, a gravitational stress field with pre-mining stress ratios of 1/2 and 1/3 and a hydrostatic stress field were investigated. For the deep mining model, a gravitational stress field with a stress ratio of 1/2 was assumed at a mining depth of 1500 m. The shallow and medium deep mining conditions were both considered in models 1 and 2. The details of the stress fields, which are also the external loading conditions for the finite element models, are shown in Table 4.

Two other cases, 26 and 27, are specifically designed to investigate the influence of the geometries of room and pillar mining on the support performance of backfill and on the stress distributions in the roof and abutments.

A final consideration in the numerical modeling technique which should be pointed out is that all models were designed in symmetrical shape in order to save computer units by using only a half or a quarter of the model. An additional advantage of a symmetrically shaped model is that the real three-dimensional mine structure problem can be approximated by a two-dimensional model.

ANALYSIS METHODOLOGY

The two-dimensional, non-linear elastic-plastic finite element method was selected to investigate the support performance of consolidated backfill under various mining and geotechnical conditions.

Because the backfill is usually surrounded by different conditions and the actual problem is three-dimensional, the plane-strain assumption makes this approximation not quite realistic. However, the total number of cases simulated would have been limited severely if performed by three dimensional finite element analyses, because of its inherent extremely large need for computer units to effect such analyses. On the other hand, theoretical studies have shown that stress concentrations around various symmetrically shaped openings in two-dimensional cases are always slightly larger than those prevailing in three-dimensional cases. Therefore, support performance data of consolidated backfill in symmetrically shaped openings of three-dimensional character can be safely related to the equivalent two-dimensional cases.

The stress-strain curves for mine backfill are non-linear and strains are not fully recoverable. This means that the actual stress-strain relationship is either time-dependent (viscoelastic) or a portion of the strain is irrecoverable plastic flow. In an effective stress-strain analysis (fully drained conditions), the stress-strain properties of backfill can be considered as time-independent for all practical purposes. Therefore, one can assume that the non-linear stress-strain characteristics of the backfill are the results of continuous elastic-plastic deformation up to the yield point.

The elastic-plastic finite element program EPFE which was developed by CANMET for mine backfilling problems has been modified by the authors for this study. Some other programs, such as ADINA and NONSAP were tried and were found to be unsuitable for use in backfilling and sequential mining problems.

The major assumptions in these analyses are:
(1) The analyses in this study are limited to static problems in plane-strain.
(2) Both the rock and backfill behave as elasto-plastic solids, with the yield surface specified by the Drucker-Prager criterion.
(3) The rock and backfill are both perfectly plastic. At the yield point, no loss of strength and no volume changes accompany plastic flow. The residual strength equals the peak strength of the reference material.
(4) The rock and backfill are homogeneous and isotropic. Plastic flow does not affect isothropy.
(5) The deformation is time-independent.

A procedure including primary mining, backfilling and pillar recovery was

369

TABLE 4
EXTERNAL LOADING CONDITIONS AND FILL TYPES USED FOR FINITE ELEMENT MODELS

Model	Case	Fill Type	Gravitational loading MPa (lb/in^2)	Horizontal loading, MPa (lb/in^2)	Ratio of applied loadings (m)
1	1	Cemented tailing I			
	2	Cemented tailing II	10.34	3.45	1/3
	3	Cemented rockfill II	(1500)	(500)	
	4	Cemented rockfill III			
	5	Cemented tailing II			
	6	Cemented rockfill II	20.68	6.89	1/3
	7	Concrete fill I	(3000)	(1000)	
	8	Concrete fill III			
	9	Concrete fill IV			
2	10	Cemented tailing I			
	11	Cemented tailing II			
	12	Cemented tailing III			
	13	Cemented rockfill I	10.34	3.45	1/3
	14	Cemented rockfill II	(1500)	(500)	
	15	Cemented rockfill III			
	16	Cemented rockfill V			
	17	Concrete fill II			
	18	Cemented rockfill II			
	19	Cemented rockfill IV			
	20	Concrete fill I	20.68	6.89	1/3
	21	Concrete fill II	(3000)	(1000)	
	22	Concrete fill III			
	23	Concrete fill IV			
	24	Cemented rockfill II	20.68	10.34	1/2
	25	Cemented rockfill II	(3000)	(1500)	
				20.68	1
				(3000)	
	26	Cemented rockfill II	20.68	6.89	1/3
	27	Cemented rockfill II	(3000)	(1000)	
3	28	Cemented tailing I			
	29	Cemented rockfill III	31.00	15.50	1/2
	30	Cemented tailing I	(4500)	(2250)	
	31	Cemented rockfill III			
4	32	Cemented rockfill III	31.00	15.50	1/2
	33	Cemented tailing I	(4500)	(2250)	
5	34	Cemented rockfill III	31.00	15.50	1/2
			(45.00)	(2250)	

adopted for most cases. At each mining stage, in which a disturbance was introduced by extraction or backfilling, sufficient time steps or approximations were pre-set for calculation cycles to converge to equilibrium. For most cases, fifteen approximations were sufficient, but, for some cases, in which the span of the mine workings was too large or the modulus of the fill was too low compared to the modulus of the rock material, serious ground and backfill failures occurred. In such cases the series of approximations reached 20 or more.

In each mining phase, the model adjusted to the new conditions and calculations continued until the unbalanced load approached zero.

Discussion of Results

Model 1

Model 1 simulates a room-pillar mining in a "soft" ore bed surrounded with a "hard" roof and floor. The typical stress distributions for three mining phases in model 1 are illustrated in Figs. 2 to 5. In all cases, the contour values are expressed as ratios of the maximum and minimum principal stresses, σ_{max} and σ_{min} respectively, to the pre-mining horizontal stress S_h and vertical stress S_v. The plots show that the stress patterns are almost the same from phase 1 - room mining to phase 2 - room backfilling, and independent of the type consolidated backfill used. This means that delayed backfilling provides very little immediate support effects to the surrounding wallrock where the pre-mining stress state has been changed due

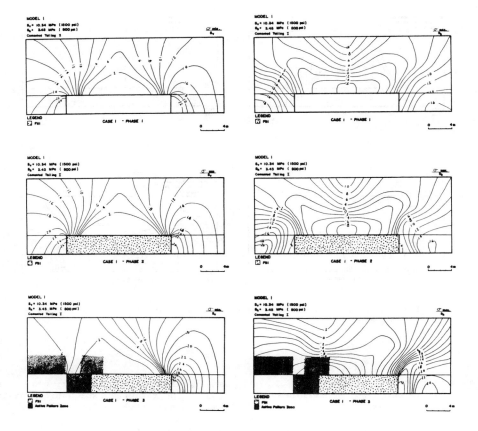

Fig. 2   Principal stress distributions in case 1, room-and-pillar mining model 1. Only a quarter of the model is shown.

to room mining. In other words, the reduction in wallrock closure, after backfilling the rooms, is negligible even
with use of high-modulus, high-strength
cemented rockfill or concrete fill, as
shown in Figs. 3 and 5.

As anticipated, the ground support
function of the backfill is mobilized
under circumstances when the mine workings are enlarged due to the mining of
other stopes or pillars are recovered.
In phase 3 of all cases in model 1,
stress distributions and roof failures
were significantly related to the type
of backfill used. Comparing the stress
patterns of phase 3 cases of case 1 to
those of phase 3 of case 4, the following observations can be made (see
Table 5).

As the modulus ratio percentage of consolidated backfill to wallrock increases
from 0.17% to 6.47%, respectively for
1:30 cemented tailing and a very strong
cemented rockfill, the load-bearing
capacity of the fill increases rapidly
from 0.02 to 0.82, expressed as the
ratio of vertical stress concentration;
the minimum principal stress concentration in the roof of the pillar excavation decreases rapidly from 0.68 to 0.02
and the maximum principal stress concentration increases from -1.29 to -1.71
in tension. The vertical stress concentration in the abutments also decreases from 3.45 to 3.05. As a result,
the active shear failure zone in the
roof of the pillar excavation is dramatically reduced with the increase of the
fill load-bearing capacity, and eventually roof failures are eliminated as

371

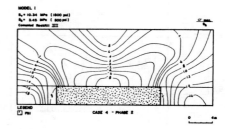
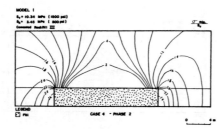

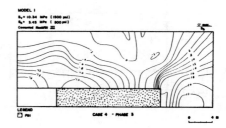
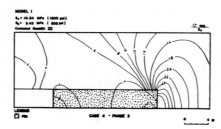

Fig. 3 Principal stress distributions in case 4, room-and-pillar mining model 1. Only a quarter of the model is shown.

depicted in case 4 in which cemented rockfill III is used and the stiffness ratio is 6.5% (Fig.3).

The results of case 1 to case 4 were obtained under shallow mining depth conditions where pre-mining vertical stress S_v and horizontal stress S_h were assumed to be equal to 10.34 MPa (1500 psi) and 3.45 MPa (500 psi), respectively.

Similar variations of boundary stresses, abutment stresses, load-bearing capacities of backfill and roof failure can be observed in results in cases 5 to 9, as shown in Figs. 4 and 5 and Table 6, where the modulus ratio percentage changes from 1.35% to 35% under conditions of a greater mining depth, with S_v of 20.68 MPa (3000 psi) and S_h of 6.89 MPa (1000 psi). Under deep mining conditions, the results of room-and-pillar model 1 indicate that roof failure during pillar extraction occurred extensively until the fill modulus reached 35% of the wallrock modulus. In case 9, although the ratio of the load-bearing capacity of the back-fill exceeded 1.0 and reached 1.84, a

small failure zone still existed in the roof. The rational explanation for this fact is that rock failure normally occurs in deep mining because of heavy ground pressures, and rock failure would be reduced much less with increasing fill modulus and strength because it is still essentially weak compared to the high values of the ground stresses. This important fact indicates that back-filling in deep mining would not be as effective as in shallow mining for control of roof failure.

Comparing the stress distributions in Fig. 2 with those in Fig. 4, one finds that the pillar roof vertical stress concentration in phase 3 is 0.68 and 0.91, respectively for case 1 and case 5. They are almost the same in phase 2 for both cases. The abutment stress concentration in case 1 is 3.45 and 3.09 in phase 3 of case 5. This means the vertical load is still increasing on the roof during the stage of pillar recovery in case 5 which depicts a deep mining condition, and results in higher roof stress concentrations and more extensive roof failure.

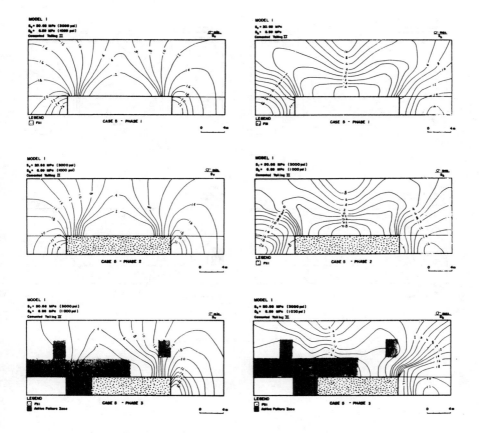

Fig. 4 Principal stress distributions in case 5, room-and-pillar mining model 1. Only a quarter of the model is shown.

Model 2

This model features a "harder" ore and a "hard" rock condition in room-and-pillar mining.

Similar relationships, as described in model 1, between support performance of consolidated backfills, applied loading conditions, and rock properties, are obtained and listed in Tables 5 and 6. Because of the harder and stiffer ore, more vertical load is transferred to the ore abutments and higher abutment stress concentration result, when compared to the cases of model 1. This explains why roof failure is more rapidly eliminated as the fill modulus increases. As shown in Fig. 6, the critical stiffness ratio in model 2 was found to be 4.2% at shallow depth and 35% at great depth for the elimination

of roof failure. Both ratios are lower than that in model 1.

Figs. 7 and 8 show the yielding and active failure zones in model 2, for two mining depth conditions.

The general results drawn from room-and-pillar models 1 and 2 can be stated as follows:

For shallow mining,
(1) The load-bearing capacity of consolidated backfills is linearly proportional to the increase of the modulus ratio of fill to rock until the ratio percentage reaches about 4.2%. Beyond this limit, it becomes nonlinearly proportional to the increase of the modulus ratio (Fig. 9).

373

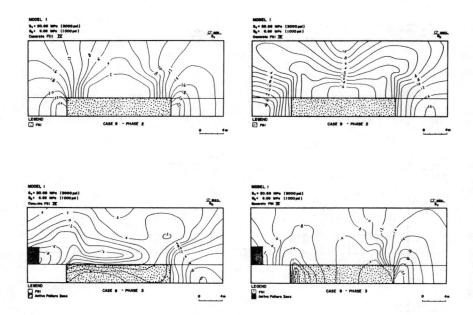

Fig. 5 Principal stress distributions in case 9, room-and-pillar mining model 1. Only a quarter of the model is shown.

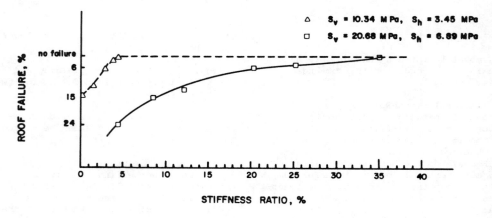

Fig. 6 Influence of fill modulus upon roof failure, room-and-pillar mining model 2

TABLE 5
SUPPORT PERFORMANCE DATA OF CONSOLIDATED BACKFILLS IN ROOM-AND-PILLAR
MINING MODELS AT SHALLOW MINING DEPTH

Model	Case No.	Fill modulus (lb/in x 10^6)	Modulus Ratio (%)	Fill load-bearing capacity ratio	Maximum abutment stress ratio	Pillar roof stress ratio σ_{max}/S_h	σ_{min}/S_v	Pillar roof closure (mm)	Room roof closure (mm)	Roof Failure (%)
1	1	0.006	0.17	0.02	3.45	-1.29	0.68	21.13	10.46	22.7
	2	0.046	1.35	0.22	3.78	-1.94	0.12	22.70	9.47	13.6
	3	0.142	4.17	0.59	3.31	-1.91	0.04	24.96	9.19	4.5
	4	0.220	6.47	0.82	3.05	-1.71	0.02	22.29	6.90	0.0
2	10	0.006	0.17	0.0025	4.58	-1.62	----	30.65	14.43	22.7
	11	0.046	1.35	0.19	4.26	-1.61	0.03	28.27	12.28	13.6
	12	0.085	2.50	0.346	4.05	-1.60	0.04	26.46	10.68	9.0
	13	0.125	3.67	0.465	3.85	-1.59	0.01	24.86	9.34	4.5
	14	0.142	4.17	0.52	3.76	-1.56	0.004	24.20	8.76	0.0
	15	0.220	6.47	0.72	3.42	-1.33	0.01	21.74	6.63	0.0
	16	0.340	10.0	0.91	3.23	-1.19	0.02	18.10	5.18	0.0
	17	0.680	20.0	1.32	2.88	-0.91	0.03	14.98	2.82	0.0

TABLE 6
SUPPORT PERFORMANCE DATA OF CONSOLIDATED BACKFILLS IN ROOM-AND-PILLAR
MINING MODELS AT GREAT DEPTH

Model	Case No.	Fill Modulus (lb/in x 10^6)	Modulus Ratio (%)	Fill load-bearing capacity ratio	Maximum abutment stress ratio	Pillar roof stress ratio σ_{max}/S_h	σ_{min}/S_v	Pillar roof closure (mm)	Room roof closure (mm)	Roof Failure (%)
1	5	0.046	1.35	0.12	3.09	-0.48	0.91	30.48	14.20	45.5
	6	0.142	4.17	0.41	2.98	-0.73	0.66	32.51	12.24	31.8
	7	0.408	12.0	1.13	2.81	-0.99	0.10	38.86	9.17	27.3
	8	0.850	25.0	1.59	2.59	-0.98	0.05	33.27	4.90	13.6
	9	1.190	35.0	1.84	2.50	-0.96	0.09	30.73	3.35	4.5
2	18	0.142	4.17	0.38	3.72	-0.80	0.64	33.83	31.14	36.4
	19	0.284	8.35	0.74	3.50	-0.87	0.42	34.54	28.47	22.7
	20	0.408	12.0	0.95	3.40	-0.98	0.14	41.20	27.13	18.2
	21	0.680	20.0	1.40	3.23	-0.96	0.09	36.73	25.30	4.5
	22	0.850	25.0	1.55	3.18	-0.96	0.08	33.70	24.20	4.5
	23	1.190	35.0	1.73	3.06	-0.87	0.05	32.69	22.73	0.0

(2) When the modulus ratio increases to 4.2% and the fill load-bearing capacity ratio reaches 0.52 to 0.59, roof failure in pillar recovery is eliminated. Thus, 4.2% is the upper bound of the critical modulus ratio in room-and-pillar mining for the hard ore case and about 6% for soft ore case (Fig. 10).

(3) The minimum principal stress in the back of the pillar excavation decreases rapidly with the increase of the fill modulus in model 1, and shows no obvious changes in model 2. The maximum principal stress is always tensile in both models. It decreases with the increase of the fill modulus in model 2; in model 1 it increases rapidly until the critical modulus ratio is approached and then decreases, as shown in Fig. 11.

(4) The variations of roof closure of the pillar excavation show the same tendency with the variations of pillar roof stress, as shown in Fig. 12. In the curve of pillar roof closure vs. stiffness ratio of model 2, the extreme point as seen in model 1 does not exist. It approaches an exponential curve. The curve of room roof closure with the fill modulus can be seen in Fig. 13.

For deep mining,

(1) The load-bearing capacity of consolidated backfill also increases as the fill modulus increases, however, it becomes nonlinear after the ratio percentage of 12% is reached where the load-bearing capacity of the fill approaches or exceeds 1.0 (Fig. 14). Also, the load-bearing capacity of fill in "harder" ore cases (model 2) is lower than in "soft" ore cases (model 1).

(2) Roof failure occurs naturally and extensively until the modulus ratio percentage approaches 35% or more (Figs. 6 and 8). For the geometries analyzed,

375

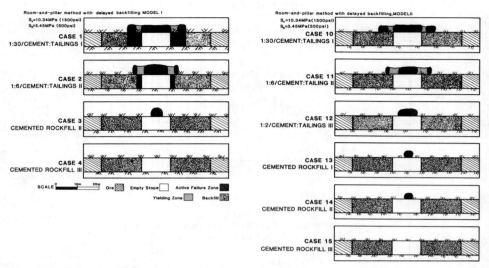

Fig. 7 Yielding and active failure in roof (pre-mining stress field: $S_v=10.34$ MPa, $S_h=3.45$ MPa)

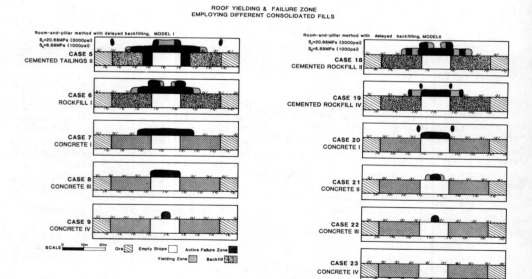

Fig. 8 Yielding and active failure in roof (pre-mining stress field: $S_v=20.68$ MPa, $S_h=6.89$ MPa)

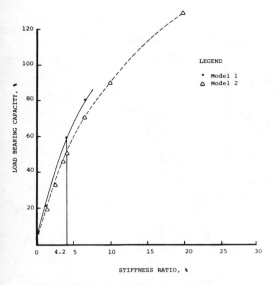

Fig. 9 Load-bearing capacity ratio of backfill versus modulus ratio (pre-mining stress field: S_v=10.34 MPa, S_h=3.45 MPa)

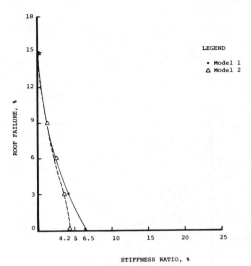

Fig. 10 **Roof failure versus modulus ratio** (pre-mining stress field: S_v=10.34 MPa, S_h=3.45 MPa)

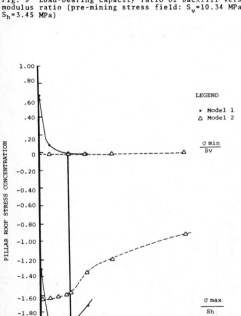

Fig. 11 Relationship between roof stress concentration of the pillar excavation and the fill modulus (pre-mining stress field: S_v=10.34 MPa, S_h= 3.45 MPa)

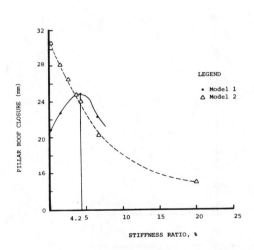

Fig. 12 **Roof closure of the pillar excavation versus modulus ratio** (pre-mining stress field: S_v=10.34 MPa, S_h=3.45 MPa)

377

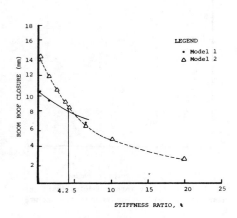

Fig. 13 Room roof closure versus modulus ratio (pre-mining stress field: S_v=10.34 MPa, S_h=3.45 MPa)

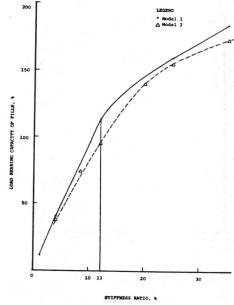

Fig. 14 Load-bearing capacity of backfill versus modulus ratio (pre-mining stress field: S_v=20.68 MPa, S_h=6.89 MPa)

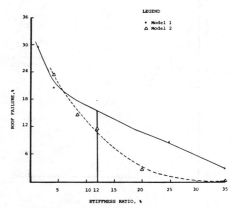

Fig. 15 Roof failure versus modulus ratio (pre-mining stress field: S_v=20.68 MPa, S_h=6.89 MPa)

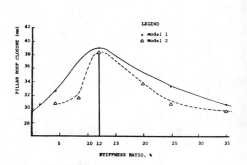

Fig. 16 Roof closure of the pillar excavation versus modulus ratio (pre-mining stress field: S_v=20.68 MPa, S_h=6.89 MPa)

378

roof failure is relatively difficult to be controlled in deep mining conditions by using backfilling. Hence, the critical fill modulus for the purpose of eliminating roof failure would be very high under deep mining conditions and may not be practical.

(3) In general, roof convergence can be controlled as the fill modulus increases, as shown in Figs. 16 and 17. However, it is found in Fig. 16, that the roof closure of the pillar excavation first increases with the increase of fill modulus, with a modulus ratio percentage of 12% as an extreme point. The roof closure curves in Figs. 16 and 12 demonstrate that the actual situations of wallrock convergence would be complicated under the conditions of sequential mining with different applied loadings and stiffnesses of rock and fill. The ratio of 0.12 is estimated to be the lower bound of the critical modulus ratio and can be used for practical room-and-pillar mining in deep mines. It can also be seen from Fig. 15, that at the point of a 12% modulus ratio, roof failure decreases to 11% for model 1 and 18% for model 2.

(4) As the fill modulus increases, the minimum principal stress in the roof of the pillar excavation decreases and the maximum principal stress, which is tensile in both models, increases (Fig. 18).

INFLUENCE OF THE NATURE OF THE PRE-MINING STRESS FIELD

Cases 18, 24 and 25 were planned to investigate the influence of the premining stress field on the support effect of consolidated backfill. The ratio of the pre-mining horizontal stress to vertical stress are 1/3, 1/2, and 1, respectively for cases 18, 24 and 25. The analysis results are shown in Table 7.

As the pre-mining horizontal stress increases, the load-bearing capacity of the backfill increases as much as 34 to 50%; the abutment stress level will also increase slightly, about 8 to 9%. During the stage of pillar recovery, the maximum principal stress concentration in the roof center will change dramatically from -0.8 (for M=1/3) to -0.62 (for M=1/2), and eventually drop to zero for a hydrostatic stress field. Because of this change, the minimum factor of safety in the roof will increase from 0.64 to more than 1.0. Consequently,, roof failure is reduced with increases

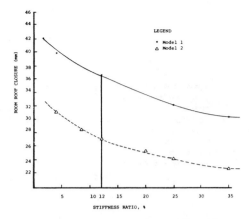

Fig. 17 Room roof closure versus modulus ratio (pre-mining stress field: S_v=20.68 MPa, S_h=6.89 MPa)

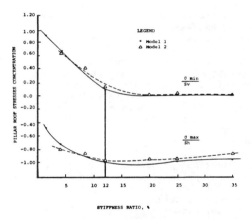

Fig. 18 Relationship between roof stress concentration of the pillar excavation and the fill modulus (pre-mining stress field: S_v=20.68 MPa, S_h=6.89 MPa)

of the horizontal applied stress, and roof failures will not occur in the hydrostatic stress field, illustrated by case 25.

INFLUENCE OF MINING GEOMETRIES

The influence of mining geometries on the effect of backfilling was also investigated in the studies. A room-and-pillar mining model under two types of stress field conditions were assumed for the five cases presented in Table 8.

TABLE 7

SUPPORT PERFORMANCE DATA OF CONSOLIDATED BACKFILL
IN DIFFERENT STRESS FIELDS

	Case 18 (M=1/3)	Case 24 (M=1/2)	Case 25 (M=1)
Fill load-bearing capacity ratio	0.38	0.51	0.58
Maximum abutment stress ratio	3.72	4.01	4.07
Pillar roof stress ratio			
$\sigma max/S_h$	-0.80	-0.62	0.01
$\sigma min/S_v$	0.64	0.23	0.16
Pillar roof closure (mm)	33.8	44.5	35.3
Room roof closure (mm)	31.1	34.6	48.9
Roof failure (%)	36.4	9.0	0.0
Minimum roof safety factor	0.64	0.87	>1.0

In any mining method employing delayed backfilling, different arrangements of the width of stope and pillar will result in different roof and pillar stability conditions, ore losses, dilution, and different backfilling costs, while using the same fill material under the same stress field condition. From the analysis, it was found that reducing the mining width of the primary mining opening will result in improvement of overall roof stability and the saving of cement cost. This is much more obvious and effective in a high stress field than in a lower stress field.

Models 3 and 4

As described previously, one of the primary applications of consolidated backfills has been in delayed backfilling operations in blasthole open stoping, not only for the general purpose of ground stability and subsidence control, but for pillar recovery as well. The stoping methods and ore recovery of pillars are tied to the nature of the fill that is used. If the adjacent fill pillar can stand vertically, a mass firing method or the VCR mining method can be employed to mine vertical pillars and little dilution occurs. If the fill pillar collapses, high dilution and ore losses will occur, requiring the use of an undercut and fill method.

In this study, the support effect of consolidated backfill in blasthole open stoping was investigated in two models, models 3 and 4. These simulated four stopes mined simultaneously and a multiple-passes form of the blasthole open stoping method with delayed backfilling, respectively.

The analysis results from model 3 and 4 can be summarized as follow:

(1) As shown in Fig. 19, the load-bearing capacity of a high-modulus cemented rockfill with a modulus ratio of 17% can reach 7.7 MPa (1,110 psi), corresponding to 25% of the vertical applied load; and the commonly used 1:30 cemented tailing fill has almost no load-bearing ability and cannot perform well as an artificial fill pillar.

(2) In these narrower and taller rib pillars of blasthole open stoping, the loss of lateral confining stress in the middle of the sidewall of the rib pillar, resulting in spalling and collapse starting in these specific localities, will be the main cause of the failure of vertical pillars[8]. In the six cases of these two models, it can be seen that the minimum horizontal stress concentration ratio in the sidewalls of rib pillars usually dropped down to less than 0.01, resulting in a tensile stress at worst (-0.16 in case 30). However, the use of backfills will provide lateral support to the rib pillar to improve its stability, particularly if high-modulus consolidated backfills are used. The minimum horizontal stress ratio in case 29 is 0.147, where high-strength cemented rockfill is used, as opposed to 0.02 for case 28 in which 1:30 cemented tailing is used (See Tables 9 and 10).

(3) With high-modulus cemented rockfill the failure zone in the stope back and rib pillar will be reduced 15%, when compared to use of low-modulus cemented tailings. When compared to room-and-pillar models, the ground support function of high-modulus backfills in blasthole open stoping is not as obvious. However, as shown in phases 2,3 and 4

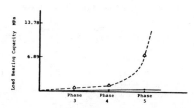

Fig. 19 Variation of load-bearing capacity of backfill in different mining stages

TABLE 8
SUPPORT PERFORMANCE DATA OF CONSOLIDATED BACKFILLS
WITH MINING GEOMETRIES IN ROOM-AND-PILLAR MODELS

Stress Field	S_v = 10.34 MPa S_h = 3.45 MPa		Stress Field	S_v = 20.68 MPa S_h = 6.89 MPa		
Fill	Cemented rockfill II		Fill	Cemented rockfill II		
Model	2		Model	2		
	Case 14	Case 14-A	Case 18	Case 26	Case 27
Ratio of W_s to W_p	1.33	0.75	1.33	1	0.75
Total width of mining workings (m)	50.16	45.6	50.16	41	45.6
Fill load-bearing capacity ratio	0.515	0.467	0.381	0.336	0.347
Maximum abutment stress ratio	---	---	3.72	3.43	3.50
Maximum tensile stress concentration in roof	---	---	-0.908	-0.983	-0.856
Maximum shear stress concentration in roof (%)	100	98.3	100	77	83.2
Roof closure (%)	100	93.0	100	91	94.7
Roof failure (%)	0	0	36.4	18.2	27.3
Safety factor of roof (%)	100	101.8	100	105.7	104
Minimum safety factor of roof	---	---	0.64	0.74	0.66

TABLE 9

SUPPORT PERFORMANCE DATA OF CONSOLIDATED BACKFILLS
IN BLASTHOLE OPEN STOPING MODELS AT GREAT DEPTH

		Case 28 (3) 1:30 cemented tailing	Case 29 (3) cemented rockfill III	Case 30 (4) 1:30 cemented tailing	Case 31 (4) cemented rockfill III
Phase 1 - Stope mining	Failure zone (%)	37.5	37.5	38.8	38.8
	Maximum vertical stress ratio in abutment	1.59	1.59	1.54	1.54
	Minimum horizontal stress ratio in rib pillar	0.013	0.013	0.21	0.21
Phase 2 - Backfilling stopes	Failure zone (%)	33.3	30.5	----	----
	Maximum vertical stress ratio in abutment	2.17	1.75	----	----
	Minimum horizontal stress ratio in rib pillar	0.002	0.186	----	----
	Fill load-bearing capacity ratio	0.003	0.137	----	----
Phase 3 - Central pillar recovery	Failure zone (%)	54.7	39.0	59.3	45.3
	Maximum vertical stress ratio in abutment	2.49	2.17	2.41	2.43
	Minimum horizontal stress ratio in rib pillar	0.02	0.147	-0.16	0.06
	Fill load-bearing capcity ratio	0.007	0.25	0.012	0.26

(1) Mining Depth = 1500 m, (2) S_v = 31 MPa, S_h = 15.5 MPa, (3) W_s = W_p = 12 m, H = 45 m,
(4) W_s = W_p = 20, H= 45 m

TABLE 10

SUPPORT PERFORMANCE DATA OF CONSOLIDATED BACKFILLS
IN MULTI-PASSES BLASTHOLE OPEN STOPING MODEL

		Case 32 (Cemented rockfill III)	Case 33 (1:30 cemented tailing)
Phase 3 - Central stope extraction	Failure zone (%)	38.4	46.2
	Maximum vertical stress ratio in abutment	1.56	1.57
	Minimum horizontal stress ratio in rib pillar	0.03	0.005
	Fill load-bearing capacity ratio	0.023	0.001
Phase 4 - Backfilling central stope	Failure zone (%)	34.6	50.0
	Maximum vertical stress ratio in abutment	1.65	1.72
	Minimum horizontal stress ratio in rib pillar	0.01	-0.04
	Fill load-bearing capacity ratio	0.04	0.004
Phase 5 - Pillar recovery	Failure zone (%)	55.5	60.0
	Maximum vertical stress ratio in abutment	2.14	1.93
	Minimum horizontal stress ratio in rib pillar	0.14	0.03
	Fill load-bearing capacity ratio	0.24	0.006

(1) Mining depth = 1500 m, (2) S_v = 31 MPa, S_h = 15.5 MPa, (3) W_s = W_p = 12 m, H = 45 m

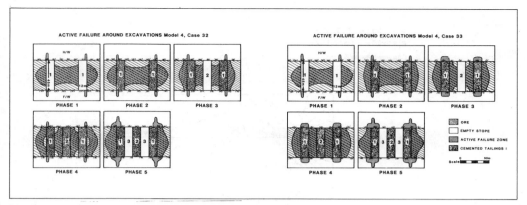

Fig. 20   Failure pattern--cases 32 and 33.

in Fig. 20, the basic role of high-modulus backfill in blasthole open stoping is to retard the development of failure in the rib pillar, not in the roof. The failure zone in rib pillars in case 32 is significantly less than in case 33.

In deep mines, the first and foremost function of delayed backfilling in blasthole open stoping is to improve the mining conditions for pillar recovery because of its freestanding ability and the provision of lateral support to pillars. The second function is to provide overall ground support and to limit surface subsidence.

Model 5

In order to improve the stress states around the stope and to maximize the utilization of the support potential of consolidated backfill in bad ground conditions, a variation of sublevel cut-and-fill stoping, called the benching cut-and-fill method, was developed[8]. Case 34 in model 5 was specifically designed to simulate this method (Fig. 1c). The open stope was excavated in two steps, by mining the lower half with immediate backfilling and, then mining the upper half.

The results of a finite element analysis for this method indicate that the maximum load-bearing capacity ratio of the backfill (high-modulus cemented rockfill) reached 0.27, as shown in Fig. 22.

Active shear failure did not occur in the sidewalls of the stope, which is attributed to the effect of reducing stope height and the use of high-modulus cemented rockfill (Fig. 21). The total failure zone around the stope was found to be reduced to 9% as opposed to 55% for a 45 m-high stope mined by blasthole open stoping (Figs. 22 and 23). Boundary stress states were also improved. The maximum vertical stress concentration in the corners of the stope back decreased from 1.97 to 1.65. Particularly, the lateral confining stress concentration on the sidewall increased to 0.57 as opposed 0.02 for a 45 m-high stope without backfilling. This increase of lateral stress constrain greatly improves the stability of the stope and rib pillar. Therefore, it can be concluded that in bad ground conditions, the stope and pillar in sublevel cut-and-fill stoping would be more stable than in conventional blasthole open stoping with or without delayed backfilling.

CONCLUSIONS

Numerical modelling methods can be used to predict the stress within the backfill and the stress redistribution in the wallrock in the sequential stope and pillar mining with delayed backfilling. Non-linear finite element solutions may be powerful tools in the representation of the behavior of backfills in underground mining environments.

The computer analyses presented herein show that the support performance of

383

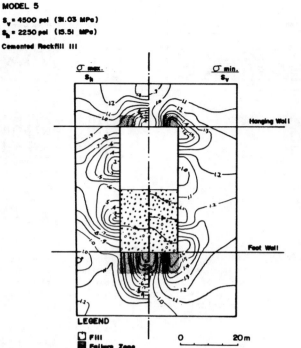

MODEL 5

S_v = 4500 psi (31.03 MPa)

S_h = 2250 psi (15.51 MPa)

Cemented Rockfill III

σ max.
S_h

σ min.
S_v

Hanging Wall

Foot Wall

LEGEND

☐ Fill
■ Failure Zone

0   20 m

BENCHING CUT AND FILL METHOD (Case 34 - Phase 3)

Fig. 21 Principal stress distributions in phase 3 of case 34, benching cut and fill mining model

consolidated backfills depends on its modulus and strength, and also depends on the conditions that are encountered, such as the mining depth, the nature of stress field and the geometry of the stope. In general, the support property of consolidated backfills is higher and more effective under shallow mining conditions, than in deep mining cases. Furthermore, the ground support function of backfill is particularly effective in a hydrostatic and a high horizontal stress fields, compared with the cases of gravitational stress field conditions. As a structural component, a consolidated fill pillar in a horizontal bedded orebody will possess a higher load-bearing capacity and will be more stable than in a steeply dipping vein mine or a massive orebody because of the influence of the geometry and the shape of the stope. The most important role of consolidated backfilling in blasthole open stoping is to create better mining conditions for pillar recovery and to provide lateral confining stress to vertical pillars, which results in improvement of the stability of ore pillars. Another important conclusion drawn from the study is that reducing the mining span or width of primary stoping will result in an improvement of overall ground stability and savings in cement cost in the sequential mining methods using delayed backfilling.

In order to improve the effectiveness of delayed backfilling and the load-bearing capacity of the artificial fill pillar in a massive orebody at great depth, the sublevel cut-and-fill method will be a more effective and safer alternative.

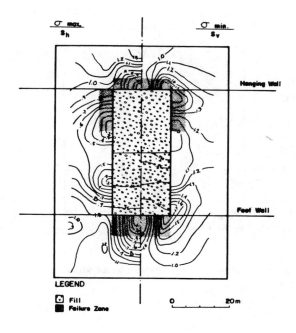

MODEL 5
S_v = 4500 psi (31.03 MPa)
S_h = 2250 psi (15.51 MPa)
Cemented Rockfill III

σ max.
S_h

σ min.
S_v

Hanging Wall

Foot Wall

LEGEND

Fill

Failure Zone

0   20 m

BENCHING CUT AND FILL METHOD (Case 34 - Phase 4)

Fig. 22 Principal stress distributions in phase 4 of case 34,
benching cut and fill mining model

Some estimates of lower bounds or upper bounds on the critical modulus ratio or consolidated backfill to wallrock in different mining and geotechnical conditions presented in this paper are useful for practical mine design. The methodology used in the study can be used for specification of an optimal consolidated backfill under defined mining conditions. Because of computer budget constraints, a hydrostatic stress condition for shallow mining and high horizontal stress conditions were not analysed in the study.

Further research work in this area is required. The imput parameters in the finite element model are critical for obtaining the correct solution. It is important that the pre-mining stress field and the rock properties be determined. Satisfactory modeling for fill behavior requires further development. Greater knowledge is required of changes in stiffness and strength of the backfill with time. An appropriate failure criterion and sophisticated computer program for non-linear mine fill properties should be developed. In order to evaluate the results from the finite element method, monitoring of displacements and failure should be done at the mine so that the finite element method can be modified to better predict future mining problems.

ACKNOWLEDGEMENTS

The work described in this paper was sponsored by the U. S. Bureau of Mines,

385

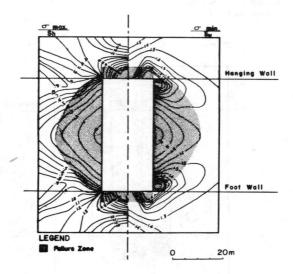

$S_v = 31.00$ MPa (4500 psi)

$S_h = 15.50$ MPa (2250 psi)

σ max.
Sh

σ min.
Sv

Hanging Wall

Foot Wall

LEGEND
■ Failure Zone

0   20m

BLASTHOLE OPEN STOPING - SINGLE STOPE

Fig. 23 Principal stress distributions around
a sublevel open stope

Kennecott Minerals Company and Columbia
University. The financial sponsorship
and technical cooperation received from
the above is gratefully acknowledged.
The authors wish to thank Dr. J. Daemen,
Dr. R. D. Call and Mr. D. E. Nicholas
for their review and helpful discussions
in preparing this paper.

REFERENCES

1."Jubilee Symposium on Mine Filling",
   Mt. Isa. August, 1973.

2."Mining with Backfill", 12th Canadian
   Rock Mechanics Symposium, Sudbury,
   Ontario, May, 1978.

3."Application of Rock Mechanics to Cut
   and Fill Mining", the University of
   Lulea, Sweden, 1980. Proceedings
   published by the Institution of
   Mining and Metallurgy, 1981.

4.Pariseau, W. G. and Kealy, C. D.,
   "Support Potential of Hydraulic
   Backfill", New Horizons in Rock
   Mechanics, 14th U. S. Symposium on
   Rock Mechanics, 1973, pp. 501-504.

5.Pariseau, W. G. et. al., "A Support-
   Performance Prediction Method for
   Hydraulic Backfill", USBM RI 8161,
   1976.

6.Stout, K. et. al., "Early High-Strength
   Hydraulic Backfill", Engineering and
   Mining Journal, July, 1980, pp. 93-103.

7.Meek, J. L. et. al., "Finite Element
   Analysis of Cemented Fill Exposures",
   Numerical Methods in Geomechanics,
   1978, pp. 901-903.

8.Chen, Di-Wen et. al., "Stability
   Analysis of Sublevel Open Stopes at
   Great Depth", 24th U. S. Symposium
   on Rock Mechanics, Texas A & M
   University, College Station, Texas,
   June 20, 1983.

Computer optimization and evaluation of cut-and-fill method

PAN JIAN
Research & Design Institute of Non-ferrous Metallurgy, Changsha, China

ABSTRACT: The underground mining is a problem of " large-scale system ". It is difficult to optimize with such technic as linear programing. We have compiled a program on the basis of "decision theory" to optimize filling system and evaluate stoping methods. In our program such methods are used as "finite element" for verification of stope size,"step regression" for statistical description of technological results, "expectation effect" for the uncertainty of cost and return,five decision-making criteria to draw objective conclusion for evaluation of mining method alternatives. This program has been proven to be effective from the feasibility study to detailed engineering in mining method design.

1 BACKGROUND AND UNDERSTANDING

It is rarely reported on a complete and comprehensive program to optimize underground mining. Because of the conditions encountered and problems to be solved,underground mining belongs to the "large-scale system". As correctly pointed out in the reference (1), mining method alone is also a problem of large-scale system. As an example set in that paper,the stoping method of a certain sublevel caving may have 144x50x60=432000 alternatives,only considering the options of three blastings, 2 hole patterns, 4 drills,2 explosives and 3 types of mucking machines. Suppose an optimization such as linear programing is used, it takes time in solving multielement polynomial equations and restricts its usage on small computers, not to speak of a satisfactory result if the generalized mathematical model uncomformable with mine conditions. In selecting an optimization method for mining design, the following pointsshould be paid attention.

1.1 The same method applied to different mines will yield variations. Take the cut-and-fill method with tailing filling for instance,according to the width of orebody and sta-bility of ore and rock,a block may be arranged parallel or perpendicular to the strike, it may be stoped overhand or underhand. It is meaningless to optimize mining methods ignoring the existing deposit conditions.

1.2 Most problems in mining methods are characterized by scattered pointwise variables of specific value,but not continuous functions. E.g. the surface locations of filling facilities at uneven terrain, the diameter of blast holes, or the bucket volume of LHDs, there are only several choises. If an intermediate value is obtained through any optimization,no matter how precise it may be and important for future research,but at present,one has to select a value ajacent to the optimum so as to use the available technique. Thus, some optimization methods,although perfect in theory but complex in calculation, seem to be unpracticable for mining design.

1.3 Because of the differences in deposit, management and regional economical circumstances, the same mining equipment may yield outstanding differences in rate and cost. Therefore, experience is preferred to complex calculation by some mining engineers in determining technical figures for a certain mine. How to meet this demand in the program is also a factor to gain satisfactory results.

1.4 In the period of design, the reserves of some metal mines may not be so definite, the stoping method selected or optimized in the design should be tested in the experiment run. Uncertainty exists in the production of a mine from its beginning. As to the method of investment research underground mining is much more worth than any other enterprices to apply the theory of uncertainty.

The general method of decision theory is shown in Fig.1. From its conceptual principle, it is possible to simplify the large-scale system of mining method into several independent alternatives. That is to say, we may propose several alternatives of mining methods which are safe in production and feasible in technology, then we carry out technical and economical comparison so as to select the optimum mining method and filling system.

From the preceding understanding, we have compiled a program of 3 subprograms and data input in series. Hydraulic filling system is optimized and selected by the possible largest slurry concentration and then regulated to the ordinary concentration to calchlate material flow in slurry. Waste filling is selected by the least underground transport work. The environmental problems caused by surface caving and drawing ore under the capping (important for selective mining of ajacent veins) are not considered in the program. The level's output of caving method is approximately calculated by the block output times block utility factor (the ratio of blocks stoped simultaneously to the total blocks of the level). Caving method in the program is not optimized and only as an alternative to compare with filled stope. The methodical flowsheet of the program is shown in Fig.2.

2 FINITE ELEMENT ANALYSIS TO VERIFY STOP SIZE

The stoping method selected should be safe. In the period of design, it is sometimes not available to collect all data influencing the safety of stoping face. Besides the stope size and ore strength, the safety of a cut-and-fill stope is governed by such mechanical defects as fractures, joints and small faults which can hardly be expressed by simple mathemetical means. In the conventional design, the choise of stope size is more from experience of a similar mine than from calculation. The so-called optimum stope size in this meaning is under the prerequisite of safety a broader area and higher slice so as to increase the rate and output. The largest size is always avoided because one can not make sure ofall the defects. The stope size may be chosen large or small according to the designer's intention without proper identification. The finite element analysis gives stress and strain distributions around the stope Of course, in this method the country rock,ore and fill are considered homogeneous elastical bodies. The loads exerted on may not be so exact as the stope condition. But our finite element program can compute many cases in one run. Through repeated calculation and revision of loads and restraints of boundary nodes, qaulitative or even quantitative conclusion of proximation can be drawn. To compare the outputs of various stope sizes in the same orebody will contribute to determine the optimum size. For instance, we studied a cut-and-fill method of continuous stoping with following mechanical properties, consolidated fill: tensile strength σ_T= 2.42 kg/cm^2, elasticity modulus E=2.34x10^2 kg/cm^2, poissons ratio μ=0.292, specific weight γ=1.778 T/m^3 ore: σ_T=30, E=5.25x10^5, μ=0.2, γ=2.92. The stope is 400 m below surface. The overlapping strata are chiefly sandy shale of γ=2.7. The finite element analysis is applied to the stope sizes of 6M free stand fill times width of 8 M, 10 M and 12M respectively with the assumption that the ore is loaded by the gravity of all its overlapping strata, while the fill its gravity load of one leve The maximum roof deformation induced by greater tensile stress is 13.8, 16.7 and 19.6 mm respectively. The spot is in the right upper corner near the fill. Its tensile strength is only 44%, 36% and 31% of the induced stress there. That spot is evidently dangerous not only from theory but also from experience.The elements in which the main stress surpasses the working stress reach 44%, 49% and 49%. As a conclusion, the stope width of 8 M or even smaller is advisable.

3 STEP REGRESSION FOR STATISTICAL DESCRIPTION OF TECHNOLOGICAL RESULTS

The technological results of a mining method such as productivity,cost, material consumption,production rate and so on, are governed by many factors among which some are principal and some are subordinate. The conventional method to describe quantatively a technological result is to derive an equation according to its technological sequence with measurable arguments and coefficients standing for unmeasurable arguments of principal factors. For example, the stope output of cut-and-fill method was studied in the reference (3) and an equation was given by technological sequence with 19 variables. But the filled stope has many deviations, differing from thickness and dip angle, stoping equipment and direction, fill material and technology. If we insert an equation for each variation according its technological sequence, the program will be redundent. Analyzing and comparing many formulae of stope output, the principal variables are s (stope area), h(slice height), w (specific weight of ore) and f (a coefficient to denote the soundness of intact ore). In our program, we represent the numerical relation between a technological result and principal variables with a polynomial in three elements and second order. The stope output Os (tonnes per day) is stated as follows.

$$Os= a_0 + a_1 \cdot f + a_2 \cdot s + a_3 \cdot h + a_4 \cdot f \cdot s + $$
$$+ a_5 \cdot f \cdot h + a_6 \cdot s \cdot h + a_7 \cdot f^2 + a_8 \cdot s^2$$
$$+ a_9 \cdot h^2 \tag{1}$$

The array a(0:9) in Eq(1) is solved by "step regression" (4) from measured data of each variation.The advantage of step regression is to reject (if necessary) step by step

the negligible variables so as to achieve optimum numerical simulation. Owing to the statistical description the following merits are attained in the program.

3.1 Simplification of program

Almost all of the technological results is calculated by equations like Eq.(1) which is a "procedure" symbolized SUS in the program. The stope output of various methods is calculated by the procedure SUS (f, s,h,a). For a cut-and-fill stope of horizontal short hole blasting and T4G peumatic loader mucking, the array A is -35.69; 28.91; 0.158; -20.49; 0.0072; -0.376; -0.026;-2.04; 0.00003; 5.91. It should be pointed out that since the stope output is an important factor in the optimization of mining methods, its calculation is also arranged by the whole suite of equations in the reference(5).

3.2 Flexibility of data preparation

The array A in Eq.(2) can be collected and calculated from time to time, filled into a document,revised and renewed to meet the changing demand of technology,cost equipment and management.

3.3 To build up confidence by capability of substituting key figures of the user's own

To compute four mining alternatives by our main program will take less than 30 minutes. To calculate the small procedure like SUS may be finished in an instant, but its result involves capital investment of millions of yuans. A designer or his authority may "doubt" the result. As to the key figures like stope output, production cost etc.,he may prefer values of his own investigation to the computed. It is very simple in this case for our program. E.g. if he insists on the stope output should be 250 T/day, then the array A may be simply input as 250; 0;0;0;0;0;0;0;0; 0. Besides, the program can compute many problems in one run. At first, data from step regression are input and computed, and then figures from the designer's own, compare the results, revise some key figures and compute once more. In this way satisfactory result can be eventually reached.

4 "EXPERTS' COMMENTARY" FOR TECHNICAL COMPARISON OF ALTERNATIVES

The coefficient of systhetic evaluation CSE is used for technical comparison among alternatives. Invite several experts (including the designer) and ask them to give the weight factor for each parameter. The experts are certainly different in authority technically and administratively. In order to represent the different degree of authority in the portion of synthetic evaluation, a second weight factor is sugested. If the mine condition is simple, there is no need for a exports' commentary, the designer is qualified to prepare the evaluation, however, this is a special case of exports' commentary in which there is only one export, the designer himself, 12 parameters for synthetic evaluation are show in table 2. The relative deviation of each parameter D_{ji} is as follows

$$D_{ji} = \frac{\left| C_i^{opt} - C_{ji} \right|}{C_i^{max} + C_i^{min}} \qquad (2)$$

The coefficient of synthetic evaluation of jth alternative CSE_j is stated:

$$CSE_j = \sum_{1=1}^{NK} \frac{W_i}{P} \sqrt{\sum_{i=1}^{M} \left(\frac{D_{ji} \cdot Pl_i}{Pl} \right)^2} \qquad (3)$$

where C_i^{opt}, C_i^{max}, C_i^{min} = optimum, maximum and minimum of parameter i among alternatives:

C_{ji} = ith parameter in jth alternative;

Wl = weight factor of export l;

P = sum of weight factors of NK experts;

Pl_i = weight factor of ith parameter given by expert l;

Pl = mean value of 12 weight factor given by expert l.

Since the relative deviation is compared with the optimum, the least among CSE_1, CSE_2, ... CSE_j corresponds to the best alternative. In principle, the CSE derived from 12 parameters by simplification of the

problem had better be served for reference only. The decisive factor should be the economical benefit.

5 "EXPECTATION EFFECT" FOR UNCERTAINTY OF ECONOMICAL BENEFIT

The economical benefit of a underground mine is marked by uncertainty. It may be less, equal or more than the expected income due to the unexpected alternation of ore grade, ground water, geothermal escalation, fluctuation of price, renewal of equipment and management. Expectation effect(5) is a method to study the uncertainty of capital investment. The cash flow is handled as a stochastic variable which is denoted by the mean value and variance. The expectation or expectation effect of the variable is directly proportional to the mean value and inversely proportional to the standard deviation. Under some uncertain factors the economical benefit fluctuates and is represented by the fluctuation coefficients A and C. The A stands for the best favourable, while the C the least favourable. Apply a method like PERT:

$$\mu_{jt} = \frac{Y_{jt}}{6} (A + 4 + C) \qquad (4)$$

$$V_{jt} = \frac{Y_{jt} (C-A)}{36} \qquad (5)$$

where μ_{jt}, Y_{jt}, V_{jt} = mean value, mode (or the value designed), variance of the net cash flow of jth alternative in year t.

If the indifferent curve is represented by an exponential effect function of second power, the expectation effect E(Y) of cash flow Y is stated as follows

$$E(Y) = 1 - e^{-0.5(2B \cdot \mu - B^2 \cdot V)}$$

where B=risk coeffecient;
μ, V =mean value, variance of cash flow Y.

The value on the expectation effect curve where variance V=0 corresponds to its equivalent value EV_{jt}

$$EV_{jt} = \frac{LN\left[1 - E(Y_{jt})\right]}{-B} \qquad (7)$$

If the rate of interest (or discounted rate) is R, the life of the mine is n, for the jth alternative, the sum of the present equivalent value $SPEV_j$, the expectation effect EE_j, the rate of return RR_j(7) and the interest ratio IR_j are mathematically stated:

$$SPEV_j = \sum_{t=0}^{n-1} \frac{EV_{jt}}{(1+R)^t} \qquad (8)$$

$$EE_j = 1 - e^{-0.5(2B\mu_s - B^2 V_s)} \qquad (9)$$

$$\mu_s = \sum_{t=0}^{n-1} \mu_{jt} \qquad (10)$$

$$V_s = \sum_{t=o}^{n-1} \frac{V_{jt} \cdot_{jt}}{(1+R)^{2t}} \qquad (11)$$

$$IR_j = \frac{YA_j}{CI_j} \qquad (12)$$

where YA_j =annual average of net cash flow within the production years; CI_j = total capital investment
If the mine condition is simple,or its life is short,the economical benefit calbe considered as certainty, then A=C=1. That is to say,certainty is a specialcase of the uncertainty analysis in which the fluctuation coefficient equal one.

6 FIVE DECISION-MAKING CRITERIA TO DRAW OBJECTIVE CONCLUSION

A return table (decision table) is shown as table 3 in which SPEV from Eq(8) is filled for 4 alternatives under 3 uncertainties. How to select the best one is up to the view point of decision maker. In general there are five decision rules as follows.

6.1 Maximum expectation

If P_i is the probability of the ith uncertainty,the expectation of the alternative j is the sum products of return $SPEV_j$ times P_i. As shown in table 3,the expectation of hydraulic stowing is 1.02x0.166+1.061x0.5+ +1.14x0.334=1.081, the same way for the others. This rule is to select the alternative with maximum expectation, as the sublevel stoping in table 3. If the return $SPEV_j$ stands for the element of the ith row and

jth column in the return table, the judgement criterion of this rule is
$$\underset{j}{Max} \quad SPEV_{ij} \cdot P_i$$

6.2 Max Min

To use this rule has the minimum return i in mind and select the maximum out of them. The minimum returns of 4 alternatives are shown in the column "Min" in table 3. Select the maximum among them (sublevel stoping). In doing so,no matter what uncertainty happens,the best outof the worst will be obtained. This rule is said to be conservative but safe. Its judgment criterion is
$$\underset{j}{Max} \quad \underset{i}{Min} \quad SPEV_{ij}$$

6.3 Max Max

This is the most ambitious rule. It has the maximum return in mind, and select the maximum among them. As shown in table 3 the maximum of figures in column "Max" is selected (sublevel stoping). If the returns are markedly different, this rule is said to be risky and try one's luck. Its judgment criterion is
$$\underset{j}{Max} \quad \underset{i}{Max} \quad SPEV_{ij}$$

6.4 Mean index

This rule is a compromise value of the former two. Suggest a constant k ($0 \leq k \leq 1$) and X_j is the index of k,then:

$$X_j = k\underset{i}{Min}\ SPEV_{ij} + (1-k)\underset{i}{Max}\ SPEV_{ij} \qquad (14)$$

With a specific value of k,select the alternative of maximum X_j. Its judgment criterion is

$$\underset{j}{Max}\left\{ k\underset{i}{Min} SPEV_{ij} + (1-k)\underset{i}{Max} SPEV_{ij} \right\}$$

If k=1, this is the rule of Max Min: if K=0,that of Max Max. If a worse result is expected,choose a larger k($0.5 \leq k \leq 1$) and a pessimestic index is produced. If a better result seems dominant, choose a smaller K ($0 \leq k \leq 0.5$) and an optimestic index is obtained. If K=0.5 that is the mean index and filled in table 3. Select the largest (sublevel stoping).

The rule of mean index is said to be steady.

6.5 Min Max

Faults in work is inevitable. Suppose an alternative is selected and proved to be regret (unsatisfactory). Then we determine the maximum degree of regret and take the least one. Among them. Take an example of hydraulic stowing in table 3, if production price escalates, the return will be SPEV=1.14. If there is no escalation, but normal condition, $SPEV_{12}= 1.061$, then the regret is evaluated by $1.14-1.061=0.079$ or the regret degree $G_{12}=0.079$. If there is neither escalation nor normal condition, but only mismanagement, the return $SPEV_{11}=1.02$, or the regret degree $G_{11}=1.14-1.02$, obviously $G_{13}=0$. Therefore, the maximum regret degree of hydraulic stowing is $G_{11}=0.12$ as shown in table 3. Select the minimum among the maximum regret degree (waste fill). This rule will lead to the least gain and loss. Its economical benefit can be surely realized, but may lose obtainable interest. So it is said to be conservative. Its judgment criterion is

$$\underset{j}{Min} \ \underset{i}{Max} \ G_{ij}$$

From the results of five decision rules, the precedence numbers of 4 alternatives are shown in table 3. In reference of the current economical situations such as market, finance and labour, an objective conclusion for decision-making can be drawn.

7 CASE EXAMPLE

Suppose a large lead-zinc deposit. The length of orebody is about 1000m along strike, average thickness 30m, average dip angle 60, laying 270-510m below surface. The ore and country rock are moderately competent. The proved reserves are 35.18 MT, with ore grade 4-6% Pb, 14-16% Zn, 22-24% S. Grades of country rock are 0-1% Pb, 1% Zn, 5-12% S. Products price:1810 yuans per ton of Pb in Pb concentrate, 995 yuans per ton of Zn in Zn concentrate, 120 yuans per ton of S in S concentrate. 4 mining method alternatives are considered: they are cut-and-fill with hydraulic

stowing, sublevel stoping with back fill, cut-and-fill with waste fill and block caving. The filling facility may locate on 3 surface sites, 2 in the foot wall and 1 in the hanging wall. The construction lasts 4 years. Full production is scheduled after 2 or 3 years. Repayment begins at the third year after production, and pay off within 5 years. Annual interest is 8%. There are 2 uncertainties in its economical benefit. The first is the possibility of increase income, chiefly because of the expected rise in metal price and the construction of a rare metall recovery shop which has been planned but not eventually decided. The second uncertainty involves mismanagement due to ground water, oxidation and ignition of ore. Fluctuation coefficients are suggested for the construction period, the first half and the rest of the production period .for each alternative under different uncertainties. The main economical figures are in table 1 and CSEs in table 2. Three experts are invited, their second weight factors are 1,3 and 2. The least CSE belongs to the block caving. The precedence number of each alternative is respectively 0;4;1;0; as shown in table 3. It shows that the sublevel stoping is the optimum. From this example we can see that to select an alternative by our program with table 3 is much more convincible than by any conventional method with table 1. According to the metilosity and reliability of data input, the program can be used in the mine design from feasibility study to detailed engineering.

REFERENCES

1. Sung Xiaotian, 1980 Optimization of underground mining. Symposium of CMSMD. (Chinese).
2. Chikafuji, S. 1981 Management engineering. CSSP. (Chinese).
3. Li Yihoung, and ent al. 1981, The effect and improvement of different mucking machines applied to cut-and-fill stope in steep deposit. Mining and Metallurgical engineering, No.2. p.44-49. (Chinese)
4. Computer centre of China Academy of Sciences. 1979. Probability and statistical calculations. CSP. P118-145 (Chinese).

5. Chang Zhongbo, 1981. Expectation effect for alternative selection. Optimization of capital construction. No.6. P22-26 (Chinese)
6. Burman,G.M. and ent al. Determination of annual output of metall mines awd with cemented filling. Mining Magazine. No. 11 (Russian).

7. Pleider,E.P. and ent al. 1969. Effect of different financing methods on the profitability of mining investment. A decade of digital computing in the mineral industry. AIME. New York.P255-274.

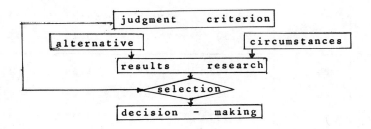

Fig. 1, general method of decision-making

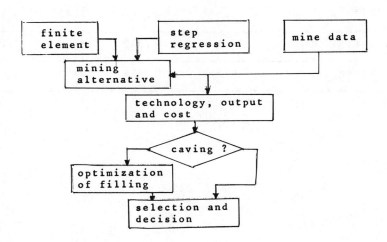

Fig. 2, methodical flowsheet of program

393

Table 1   main economical figures (normal)

index \ alternative		hydrailic filling	sublevel stoping	waste fill	block caving
ore mined (M T)		36.96	34.16	38.61	31.68
grade of crude ore (%)	Pb	4.65	4.45	4.45	4.04
	Zn	13.24	12.54	12.54	11.13
	S	21.53	20.70	20.70	19.04
metalls produced (M T)	Pb	1.562	1.39	1.565	1.173
	Zn	4.009	3.516	3.969	2.876
	S	3.987	3.517	3.992	2.994
life of mine (year)		39	29	46	30
investment (MY)		139.83	136.88	130.27	121.86
mean interest (MY / year)		158.64	212.01	138.49	185.68
total interest (BY)		5.552	5.300	5.817	4.828
interest ratio		1.134	1.550	1.063	1.524
rate of return		0.356	0.488	0.337	0.483

Table 2.   comparison of alternatives

alternative parameter	Hydraulic stowing	Sublevel stoping	Waste fill	Block caving
block output (t/d)	66.7	158.1	63.1	300
block preparation (shifts)	75	156	78	150
annual tonnage (M T)	1.155	1.485	0.99	1.32
preparation ratio (M / K T)	23.3	32.8	23.9	34
recovery (%)	0.95	0.85	0.95	0.7
dilution (%)	0.1	0.15	0.15	0.25
productivity	5	15	3	15
cost of mining	44.9	30.63	35.6	12.5
unit investment	121.1	92.2	131.6	92.3
mine investment	5983	5888	5327	3886
fill investment*	1098	1117	432	20
fill cost**	32.18	19.69	15.65	0.5
CSE in arthm. mean	1.85	1.7	1.43	0.89
CSE in weight mean	1.99	1.71	1.33	0.85

* investment from caving for caving method,
** cost from caving for caving method.

Table 3,   return table (SPEV, billions of yuans**)

item altern.	uncertainties			judgment criteria					prece- dence
	mismana gement	normal	rise of income	expec- tation	Min	Max	mean index	regret	
hydraulic stowing	1.02	1.061	1.14	1.081	1.02	1.14	1.08	0.12	0
sublevel stoping	1.414	1.445	1.615	1.497*	1.414*	1.515*	1.515*	0.201	4
waste fill	0.911	0.942	0.991	0.953	0.911	0.991	0.951	0.080*	1
block caving	1.251	1.283	1.393	1.314	1.251	1.393	1.322	0.142	0
probability	0.166	0.5	0.334						

* Selected
** 1 yuan = 0.503 U.S. dollar

Some aspects of recovery of pillars adjacent to uncemented sandfilled stopes

Y.V.A.RAO & V.S.VUTUKURI
University of New South Wales, Broken Hill Division, Australia

ABSTRACT: A substantial amount of ore reserves in the form of pillars are located adjacent to the primary stopes filled with uncemented sandfill in many underground metal mines. At present these pillars are recovered by relatively expensive methods such as undercut-and-fill.

The possibility of using mass blasting to recover pillars is explored by considering grouting a certain thickness of fillmass, so that the fill will not fail when the grouted side is exposed.

Ester-hardened sodium silicate grouts were selected in this investigation and laboratory tests were conducted to determine the strengths of grouted mine fill specimens.

Numerical methods have been adopted to determine the stability of the grouted fillmass when one side has been exposed. Initial stresses in the fillmass are considered to be very important for subsequent stability determinations. Results from two-dimensional finite element analysis showed initial stress distribution similar to that reported in the past for cemented sandfill, namely arching. Results obtained from three-dimensional finite difference method showed a more gradual arching pattern in the stress-distribution.

A 5 m thick grouted zone was introduced after the sequential construction of the filled stope and the face adjacent to the grouted zone exposed for 50 m from the top out of the total height of 100 m. Results from both two- and three-dimensional analyses showed failure.

1 INTRODUCTION

A substantial amount of pillar ore reserves located adjacent to the primary stopes filled with uncemented sandfill in many underground metal mines are presently being extracted by expensive undercut-and-fill (UCAF) method of mining. As this is a relatively expensive method coupled with a disadvantage of lower rates of production it would be advantageous to employ cheaper mass-mining methods.

Obviously, mass-mining methods work only if the exposed sandfill face remains stable as in the case of methods employing cemented sandfill. As most primary stopes already are filled with uncemented sandfill, the strengthening of sandfill has to be done in-situ.

Various techniques of in-situ stabilisation of soil exist, viz., sheet-piling, slurry-trenching to build a retaining wall and, chemical grouting. All these methods are expensive from mining point of view though it is thought that chemical grouting using sodium-silicate based grouts may offer some advantages as regards to overall benefits.

Even if only a certain thickness of uncemented fill mass adjacent to the secondary pillar has to be exposed, it will still be an ambitious exercise.

However, a theoretical exercise as to if the above-mentioned approach is feasible is nevertheless interesting. Closed form solutions of stresses developed in the in-situ fill mass which control its behaviour after subsequent exposure are invalid owing to the complexities of the problem, viz., complex stope geometry, lack of accurate information regarding the physical properties of sandfill and its stress-

strain behaviour. Therefore numerical
methods of analysis have to be adopted.

This paper presents the results of a
hypothetical problem of a primary-stope
filled with uncemented sandfill when the
adjacent secondary pillar is extracted
using mass-mining methods. Two-dimensional
finite element and three-dimensional finite
difference methods of analyses employing
an elasto-plastic model are presented and
their capability to model initial stresses
in the fill mass and predict stability or
otherwise of the exposed strengthened fill-
face are discussed.

2 THE PROBLEM

The problem under study is a rectangular
stope filled with uncemented sandfill over
a period of time in a number of lifts. In
order to retain the uncemented sandfill
from diluting the extracted pillar ore, a
grouted zone is introduced adjacent to the
future exposure. The aim therefore is to
assess if any such exposure adjacent to
the grouted zone can be modelled at all
using numerical methods in order to find
out if such a method can render the grouted
face stable after exposure.

The problem is schematically represented
in Figure 1.

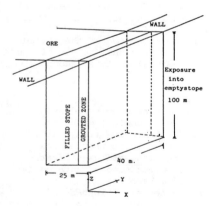

Figure 1. The Problem

3 THEORETICAL CONSIDERATIONS

The stresses present in a filled stope
after it is filled in a series of pours
are distributed in a way that maximum
vertical stress existing at the bottom of
the stope is much less than that due

directly to the overburden pressure. This
phenomenon is known as "arching".

As a consequence of stope geometry, the
time taken for the stope to be filled, and
the increase in strength of fill with time,
especially cemented sandfill, arching
develops in minefill. This phenomenon as
occurring was reported by Mitchell et al
(1974) and Barrett and Cowling (1980), who
measured stresses in cemented sandfill.
Theoretical treatment of arching is
presented by Terzaghi (1943).

It will be logical to assume that arching
will be much less significant in the case
of uncemented sandfill than in cemented
sandfill owing to little or no cohesion
and no significant increase in strength
with time. But if a zone of certain
thickness is strengthened in-situ arching
may be significant. This problem is
amenable to closed form of solution only if
the total stress acting on the strengthened
zone causing it to fail internally or by
overturning is known. Numerical modelling
of arching effect is discussed in next
section.

Another problem which necessitates to be
modelled is the stress-strain behaviour of
minefill. In this study an elastic-
perfectly plastic stress-strain behaviour
is assumed.

4 MODELLING PROCEDURE

Any modelling procedure adopted for this
problem should be able to "satisfactorily"
model arching in minefill. Most of the
work done to date in this direction was on
cemented sandfill (Askew et al, 1978;
Barrett et al (1978), Fitzgerald, 1979;
Coulthard and Dight, 1979; Barrett and
Cowling, 1980). The same modelling
strategy is used for uncemented sandfill
modelling in this study. It is described
in brief below.

4.1 Construction of the fill pillar

When fill is introduced into the stope
afresh it is in the form of fluid with
solids contents of about 70% by weight.
Within 24 hours or so after it may be
assumed to have been completely drained.
Thereafter it gradually gains strength.
The strength increase in the case of
uncemented sandfill is very low (7 kPa is
used in analysis).

The fill is assumed to have been placed
in not less than 10 lifts as this is
suggested as an acceptable figure by
Barrett et al (1978), for cemented-sand-

filled stopes.

As the wall-rock on the sides of the stope is much stiffer than the fill mass, it can be modelled as a completely rigid boundary and the fill is assumed to be in complete contact with it. Therefore, only fill-pillar is included in the finite-element mesh. The newly placed aqueous fill is modelled using vertical roller boundaries at the side and cured fill with fixed boundaries.

Freshly placed fill is assumed to have an elastic modulus of one-tenth (1/10) of that of cured fill, and a Poisson ratio 0.45; an intermediate stage of curing where the modulus is one-half (1/2) and Poisson ratio 0.35; and fully cured fill with full value of modulus and Poisson ratio 0.25. That part of the stope which at any stage is empty or unfilled is represented by very low modulus "air" material.

The first lift of fill placed is modelled in a number of layers of elements of material type 1, say, for which density and the stress-strain characteristics represent the fill. The rest of the elements representing the empty stope are "air" elements with near zero stiffness. To ensure that these elements have zero displacement at the start of analysis of next stage, they are totally fixed and a gravity turn-on analysis is performed on the first lift. The stresses, strains, and displacements are stored as input initial stresses for the next stage.

In the second stage type 1 materials of first stage are now defined, say, as type 3, with stress-strain characteristics, but with zero density; and with initial stresses, etc. from first stage. More layers of fresh fill placed over this are again type 1 elements. Now, the displacements and stresses in the second lift due to its own weight are calculated, and the incremental stresses and displacements in the first lift due to the imposition of the second lift are calculated. This analysis is repeated for subsequent stages until the whole stope is filled.

4.2 Introduction of a grouted zone

Grouted zone of a certain thickness adjacent to the future exposure is modelled simply by changing material properties for the required elements. This only ensures that the grouted zone has higher strength and stiffness but its effect on the overall modelling of the exposure is not known.

4.3 Exposure

After the introduction of strengthened zone, one side of the filled stope is exposed. This means that the virgin-rock support is in effect removed on that side. This is modelled by freeing the nodes along that side and applying the negative of the effective supporting forces which were previously acting against that side.

This modelling procedure as discussed above is due to Coulthard and Warburton (1978) which discusses the procedure in detail.

5 PROGRAMS USED

Program TNJTEP (1SA) developed by Coulthard and Warburton (1978) from the original code by Chang and Nair (1972) was used to model the problem in two-dimensions. This is a finite-element program employing elasto-plastic model using Drucker-Prager's generalisation of Mohr-Coulomb yield criterion.

Program RMS (Fitzgerald, 1980) was used to model the problem in three-dimensions. This is a finite-difference code using elasto-plastic yield criterion of Mohr-Coulomb along with non-associated flow rule. This program was obtained for use with kind permission from the Zinc Corporation/New Broken Hill Consolidated Ltd., Broken Hill.

6 ELASTO-PLASTIC MODEL

For each load increment an elastic analysis is performed and the resulting stresses in each element are checked against the elasto-plastic and limited-tension yield criteria. If one or both of these stresses are exceeded, plastic or tensile yield is assumed to have occurred, and excessive stresses are re-distributed to neighbouring elements. This re-distribution is continued iteratively until specified convergence criteria are satisfied. The degree of convergence is arbitrary and is user-specified. Tension convergence criterion of 5 kPa and elasto-plastic convergence criterion of 0.05 are chosen for the analyses. It means that:

1. all excess stresses must be below 5.0 kPa; and

2. $-(\alpha J_1 + \sqrt{J_2} - k)/(\alpha J_1 - k) < 0.05$

where

$$\alpha = \frac{\tan \theta}{(9 + 12 \tan^2 \theta)^{\frac{1}{2}}}$$

$$k = \frac{3c}{(9 + 12 \tan^2\theta)^{\frac{1}{2}}}$$

$$J_1 = \sigma_1 + \sigma_2 + \sigma_3$$

$$J_2 = \frac{1}{6}[(\sigma_1 - \sigma_2)^2 + (\sigma_2 - \sigma_3)^2 +$$

$$(\sigma_3 - \sigma_1)^2]$$

where θ = friction angle,
       c = cohesion, and
       σ_i = principal stresses

Material properties used in the analyses are presented in Table 1.

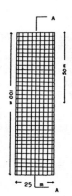

Figure 2. Mesh used in 2-D analysis

Table 1. Material properties used in the analyses.

Material Type	Cohesion (Pa)	Young's Modulus (Pa)	Poisson ratio ν	Bulk * Modulus (Pa)	Shear * Modulus (Pa)	Friction Angle (degrees)
Uncemented sandfill						
1. freshly poured	1×10^3	5.0×10^6	0.45	67.0×10^6	4.0×10^6	
2. intermediate	"	25.0×10^6	0.35	"	20.0×10^6	34^0
3. cured	"	50.0×10^6	0.25	"	40.0×10^6	
Grouted sandfill						
1. cured	370.0×10^3	100.0×10^6	0.15	100.0×10^6	45.0×10^6	
2. intermediate	185.0×10^3	200.0×10^6	0.15	100.0×10^6	90.0×10^6	34^0

* Shear and Bulk modulii are used in 3-D analysis and Young's modulus and Poisson ratio in 2-D analysis.
Also a dialation angle of 0^0 is used in 3-D analysis. Density of 2390 n/m^3 and 2500 n/m^3 are used for uncemented and grouted fill respectively.

7 RESULTS AND DISCUSSION

The finite element mesh used for the analysis in two-dimensions is shown in Figure 2. In three-dimensional case the grid had 16 zones in the third-dimension representing a total depth (wall to wall) of 40 m.

Fill was assumed to have been placed in 13 lifts from the point of view of modelling.

Figure 3 shows vertical stresses developed in the fill mass against the depth. Both, the results of two- and three-dimensional cases are shown. It may be seen that arching develops in uncemented minefill at a depth of about 30 m from top of the stope in two-dimensional analysis. In three-dimensional analysis the vertical stress increases gradually at a relatively lower level though the maximum vertical

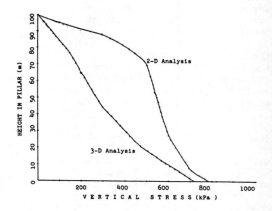

Figure 3. 2-D vs. 3-D analysis. Vertical stresses.

400

stress is more or less the same in both the cases at between 700 and 800 kPa. In Figure 3 the slope of the three-dimensional vertical stress curve is much less than that in two-dimensional case. This is because of the fixed boundary condition in the top lift of the three-dimensional case.

The results of two- and three- dimensional analyses cannot directly be compared owing to different numerical techniques employed for each case. Development of arching is modelled in both cases, though in different fashion. The lower values of vertical stress in the case of three-dimensional analysis is due to the fact that there is a finite third dimension, which is not the case with two-dimensional analysis.

Figure 4 shows the shear stresses at the side of the filled stope. The values seem rather high which suggests that the degree of convergence obtained is not satisfactory. It may be pointed out that adequate convergence in uncemented fill is very difficult to achieve owing to its little or no strength.

and increase below the exposure. Elements which are in yield (unconverged after 50 iterations) are shown in Figure 7. Results show that failure is imminent by plastic and tensile yielding. No particular shape of failure is associated with the results.

Horizontal and shear stresses after exposure in three-dimensional case (Figures 8 and 9) are similar to those in two-dimensional case (Figures 5 and 6). Zones of plastic yield in three-dimensional analysis are shown in Figure 10 which shows a considerable yield suggesting failure.

The thickness of the strengthened zone was chosen as 5 m arbitrarily. It could be well possible that a thicker zone might render the fill mass stable after exposure. The cost considerations were not looked into as the interest is in the application of the numerical methods of analysis.

The limitation of use of elasto-plastic model for uncemented sandfill is only a reasonable approximation of its actual stress-strain behaviour which is complex. Also there is a difference in the failure

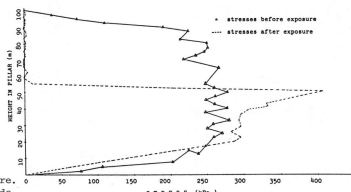

Figure 4. 2-D Analysis. Exposure. Shear stresses at side.

Stresses after a strengthened zone 5 m thick was introduced, differed slightly (±25 kPa) from those before it was introduced (not shown). This discrepancy is again due to insufficient convergence. The strengthened zone was introduced in two stages (as per Table 1).

Vertical and horizontal stresses after sequential construction and after exposure are shown in Figure 5 for two-dimensional case. The exposure was 50 m from top. It may be seen that the horizontal stress (Figure 5) and shear stress (Figure 6) reduce to zero or below on the exposure

surfaces of Drucker-Prager, Mohr-Coulomb in uniaxial, plane-stress, and plane-strain which gives rise to different failure stresses.

Finally, the grouted zone/sandfill inter-face is not modelled in this analysis. Displacements on this interface may give rise to different results. This remains the main drawback of these analyses.

8 CONCLUSIONS

1. The cement fill modelling technique

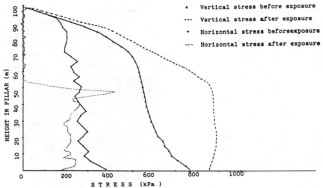

Legend:
- ▲ Vertical stress before exposure
- --- Vertical stress after exposure
- • Horizontal stress beforeexposure
- --- Horizontal stress after exposure

Figure 5. 2-D Analysis. Exposure. Vertical and Horizontal Stresses after Exposure.

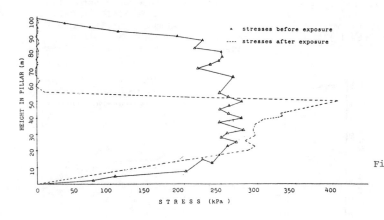

Legend:
- ▲ stresses before exposure
- --- stresses after exposure

Figure 6. 2-D Analysis. Problem 2. Exposure Shear stresses at side.

which was adopted could well be adequate for uncemented sandfill, but many more iterations are required to attain convergence in filling stages of the analysis.

2. Arching is modelled by both two- and three-dimensional analyses, but in different fashion. Three-dimensional analysis predicts lower stress levels in general but the maximum vertical stress at the bottom is not very different in both the analyses.

3. 5 m thick grouted zone as modelled is not adequate to render stability to the fill mass.

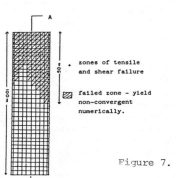

Legend:
- • zones of tensile and shear failure
- ▨ failed zone - yield non-convergent numerically.

Figure 7. 2-D Analysis. Exposure. Failure zones

402

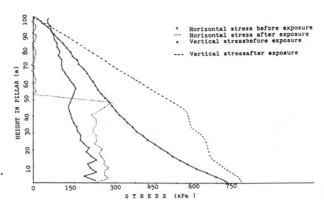

Figure 8. 3-D Analysis.   Exposure.
Vertical and Horizontal
stresses after exposure.

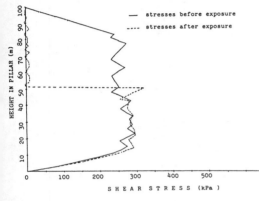

Figure 9. 3-D Analysis.   Exposure.
Shear stresses at side.

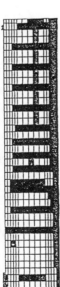

Figure 10. 3-D Analysis.
Failure zones after
exposure.
(plastic yield).

9   ACKNOWLEDGEMENTS

The authors wish to thank the Zinc
Corporation/New Broken Hill Conßolidated
Ltd., Broken Hill for their help during
the project. Financial assistance provided
by the University of New South Wales in
the form of Postgraduate Scholarship to one
of the authors is gratefully acknowledged.
Thanks are also due to Dr. M.A. Coulthard
of C.S.I.R.O., Division of Applied Geomechanics, Victoria for his comments.

REFERENCES

Askew, J.E., McCarthy, P.L. & Fitzgerald,
  D.J. (1978). Backfill research for
  pillar extraction at ZC/NBHC Ltd.,
  Broken Hill. Proceedings, 12th Canadian
  Rock Mechanics Symposium on Mining with
  Backfill; Sudbury, Ontario, 1978: 100-110.
Barrett, J.R., Coulthard, M.A. & Dight, P.M.
  (1978). Determination of fill pillar
  stability. ibid.: 85-91.
Barrett, J.R., Cowling, R.L. (1980).
  Investigations of cemented sandfill
  stability in 1100 orebody, Mt. Isa Mines.
  Transactions, IMM., July 1980: A118-128.
Chang, C.Y. & Nair, K. (1972). A theoretical
  method for evaluating stability of
  openings in rock. Final Report to USBM,
  Contract No. H0210046.
Coulthard, M.A. & Warburton, P.M. (1978).
  Two-dimensional program TNJTEP - hints
  for users and modifications for some
  applications in mining. Unpublished
  Report, CSIRO, Division of Applied
  Geomechanics, Syndal, Victoria.
Coulthard, M.A. & Dight, P.M. (1979).
  Numerical analysis of failed cemented fill
  at ZC/NBHC Ltd., Broken Hill. Proceedings,
  3rd Aust.-N.Z. Conference on Rock
  Mechanics, N.Z., Vol. 2.: 145-151.

Fitzgerald, D.J. (1979). 3rd Sandfill
  stability report, ZC/NBHC Ltd.,
  Broken Hill.
Fitzgerald, D.J. (1980). Program RMS -
  Information for users. Courtesy -
  D.J. Fitzgerald & Associates, Victoria,
  and ZC/NBHC Ltd., Broken Hill.
Mitchell, R.J. et al (1974). Performance
  of a cemented hydraulic backfill.
  Proceedings, 27th Canadian Geotechnical
  Conference, Edmonton, Canada.
Terzaghi, K. (1943). Theoretical soil
  mechanics, John Wiley and Sons, Inc.,
  N.Y.

A simple and convenient method for design of strength of cemented hydraulic fill

SIJING CAI
Jiang-Xi Institute of Metallurgy, China

ABSTRACT: A simple and convenient design method for determination of the strength required of cemented hydraulic fill is discussed, including a summary of the main mechanical effects of cemented hydraulic fill in emptied stopes, based upon more than 10 years of mining practice and numerous experiments and studies at the Fan Kou lead-zinc mine and the Xi Kuang Shan antimony mine, China. Depending on the ore deposit, the country rock conditions and the particular mining operations, mines using stoping methods involving fill may employ this simple and convenient method to determine the strength required of the cemented hydraulic fill, for particular applications.

Potential fill support mechanisms are listed and discussed and disparities between theoretical predictions and actual observations in practice are quantitatively analysed. Formulae are developed to allow prediction of actual fill performance, under different mining conditions.

INTRODUCTION

Use of mining methods employing fill has developed very rapidly in metalliferous mines in China, as described by the Hunan Institute of Metallurgical Research (1980). Filling materials include dry-placed fill, concrete fill and uncemented hydraulic fill as well as cemented hydraulic fill. The main advantages of the cemented hydraulic fill, compared with the other fill materials, are considered to be as follows.
 1. strength is greater,
 2. preparation and transport of fill material are easier, and
 3. at least part of the tailing can be immediately disposed of underground.
 The cost of cemented hydraulic fill is closely related to the content of cement employed. According to Xi Kuang Shan antimony mine (XKSAM) data, the cost of cement makes up 80 per cent of cemented fill direct cost. The content of cement also affects directly the strength of the fill body. Experimentation at Fan Kou lead zinc mine (FKLNM) demonstrated that with other conditions unchanged, when the cement-tailing ratio was increased from 1:8 to 1:4, the average uniaxial compressive strength of fill samples cured to 90 days increased 30 per cent. Therefore, even though cement addition assures

strength, it is a significant problem to decrease the cost of the fill as much as possible, as indicated by Thomas (1973).
 In summary, the strength required of cemented fill is a theoretical problem which has been studied but not yet adequately resolved in China.

THE MAIN MECHANICAL EFFECTS OF THE CEMENTED HYDRAULIC FILL

In 1968, use of cemented hydraulic fill stoping was started at FKLNM and a study begun into such use. At present, cemented fill makes up 80-90 per cent of the total, as reported by the Fan Kou lead zinc mine and the Changsha Institute of Mining Research (1981). XKSAM started an experiment with pillars filled with "concrete" and stopes filled with uncemented hydraulic fill in 1966 and by 1980 90 per cent of production was by such methods, as reported by Xiang (1980). ("Concrete" is actually rough aggregate (0-50 mm) mixed with cemented hydraulic fill). Details of these two mines are shown in Table 1.
 At FKLNM, primary stopes are mined using cut and fill, between pillars. These pillars are then mined, again using cut and fill. At XKSAM, primary stopes are delay

Mine	Mining method	Stope production	Ore extraction	Ore dilution	Total cement fill volume	Filling direct cost	Mining cost
		tpd	%	%	$1000m^3$	$yuan/m^3$	yuan/tonne
FKLNM	Cemented hydraulic fill stoping	63-68	95.6	12.7	460 (from 1968-80)	30.97	37.44
XKSAM	Primary stopes filled with "concrete" and secondary stopes with uncemented hydraulic fill	50-70	86.8	9.9	251 (from 1968-79)	16.43	17.60

Note: 1 yuan = US$0.51

filled with "concrete". If orebody height exceeds 12 m, mining and filling are conducted in two lifts. Pillars are then mined and delay filled with uncemented hydraulic fill.

Some authors in China, as reported by Gao (1979) and the Changsha Institute of Mining Research (1978), hold that the mechanical effects of a mass of cemented fill in an empty stope are the same as those of a primary ore pillar, that is, the fill body bears the whole weight of the overburden above the stope. According to that point view, the strengths of the cemented fills produced at XKSAM and at FKLNM are shown in Table 2.

Table 2 shows that designed strengths are in fact two to three times as large as actual measured strengths, indicating a lack of comparability between design method and practice.

The question of how cemented fills can successfully support the country rock and the overburden, while cemented fill strength is only 1/50 to 1/100 that of the primary pillars and the coefficient of resilience of the fill is only 1/100 to 1/1000 that of the ore, has received much attention, for example, Aitchison, Kurzeme and Willoughby (1973) and Yan and Kan (1980). Mathematics, physics and mechanical models have been used in study-ing the mechanical effects of cemented fills in emptied stopes in China. XKSAM and the FKLNM have been studying this problem and have collected a wealth of data at mine sites as reported by Fan Kou lead zinc mine and Changsha Institute of Mining Research (1981), Yan and Kan (1980) and Xi Kuang Shan antimony mine, Changsha Institute of Mining Research and Wuhan Institute of Metallurgical Safety Research (1980). According to these data, the author considers that the mechanical effects of cemented fills are not the same as those of the pillars. The fill mass cannot rigidly support the full weight of the overburden layers and can only play the role of subsidiary supporting. The main subsidiary mechanical effects are

1. provision of a lateral pressure to separated rock body,
2. bearing the weight of failed rock above the stope, and
3. performing as a self-supporting retaining wall, under XKSAM mining conditions. In the general case, the first two effects coexist, but sometimes one may be considerably more significant than the other, depending on geological conditions. Each of the above functions is now considered in turn.

Table 2

| Mine | Orebody conditions | | | | | Country rock conditions | | | | | | Strength designed of the cemented fill $R_{90\ days}$ (MPa) | Formula for calculation of desired strength | Actual strength measured by sampling at site $R_{90\ days}$ (MPa) |
| | | | | | | hanging wall | | | foot wall | | | | | |
	Ore	Dip	Width (m)	Compressive strength (MPa)	Stability	Rock	Compressive strength (MPa)	Stability	Rock	Compressive strength (MPa)	Stability			
XKSAM	Crude antimony	10°-20°	1.2-20.0 average 4-6	120-160	Stable, broken, brittleness	Shale, limestone	30-60, 80-100	Unstable, stable	Limestone	80-120	Stable	7.3-7.7	$R_{90}=\dfrac{\gamma Hb}{L}$ γ-unit weight of overburden rock H-depth of overburden b-width of stope L-width of pillar	2-4
FKLNM	Block-shaped lead-zinc ore	30°-40°, 50°-70°	15-40	110-130	Stable	Limestone	80	Mid-stable	Limestone	80	Mid-stable	4	Undesigned, select $R_{90}=4$ MPa	$R_{90}<4$ makes up 73.79 per cent of the total, $R_{90}<2$ makes up 39.29 per cent of the total

Note: Safety requirements, both in the mining operations and in the support, can be satisfied with the actual strengths measured.

STRENGTH REQUIRED OF CEMENTED FILL PROVIDING A LATERAL PRESSURE

As shown in Fig. 1, a steeply inclined orebody is mined using overhand cut and fill stoping.

When the emptied stcpe is formed, the pressure in the ore pillars, the hanging wall and the foot wall would increase. To simplify the analysis below, both geological structural stresses and frictional forces on the fractured rock mass above the orebody are neglected. If the pressure in the country rock walls is regarded as unchanged, the increment of stress in the ore pillar should equal the weight of the oberburden layers, if arching effects are not considered.

$$\Delta p = \frac{\gamma Ha}{L} \qquad .. \quad .. \quad .. \quad (a)$$

where, Δp is the stress increment in a ore pillar (MPa)
γ   the average unit weight of overburden layers (kg/m³)
H   the height of overburden layers (m)
L   the width of a pillar (m), and
a   the span of an emptied stope (m).

1   emptied stope filled
2   ore pillar
3   access raise
4   uncemented fill
5   ore pass
6   main drift
7   unmined ore block
8   roof pillar
9   access for hydraulic filling
10   main drift

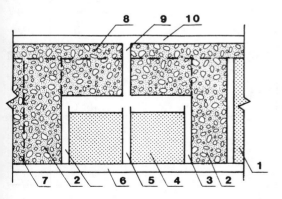

Fig. 1   Uncemented hydraulic fill stoping

Uncemented fill in an emptied stope can only support a fraction of the total weight of the rock mass above the orebody because of its non-cohesive nature and compressibility. The support mechanism of uncemented fill lies simply in increasing the strength of ore pillars by providing lateral confining pressure. Coates (1970) held that this support was comparable with the lateral pressure provided by water on a dam.

That is $\sigma_o = \gamma_o h$ .. .. .. (b)

where, σ_o is the lateral pressure of fill on the pillar (MPa)

γ_o the unit weight of the fill (kg/m^3)

h the height of fill (m).

As is well known, with increased confining pressure, the strength of a rock increases according to the formula

$\sigma_c = \sigma_1 + K\sigma_o^1$ .. .. .. (c)

where, σ_c is the strength of a rock supported by a lateral pressure (MPa)

σ_1 the strength of a rock without a lateral pressure (MPa)

σ_o^1 the lateral pressure (MPa), and

K a coefficient varying with rock type.

$K = \dfrac{1 + \sin \phi}{1 - \sin \phi}$, where ϕ is the internal friction angle of the rock. In the case of a stope filled with uncemented fill, the rock at the top corners readily slides down because in this region the value of σ_o^1 is nearly zero.

In contrast with uncemented fill, cemented fill has unconfined strength due to cohesion, resulting in self-supporting characteristics. As it provides lateral pressure to a pillar, the magnitude of the lateral pressure is nearly equal to the strength of the fill. Therefore, the lateral pressure on the full height of a pillar is evenly distributed, provided fill quality is uniform.

The strength required of the cemented fill can be analysed as follows. The difference between the strength of a pillar provided with a lateral pressure and that of a pillar without a lateral pressure is $\sigma_c - \sigma_1 = K\sigma_o^1$, should equal the stress increment Δp of a pillar. That is

$\sigma_c - \sigma_1 = \Delta p$ .. .. .. (d)

so $K\sigma_o^1 = \dfrac{\gamma Ha}{L}$

and $\sigma_o^1 = \dfrac{\gamma Ha}{KL}$ .. .. .. (e)

Considering the likely segregation of the fill materials during transport and placement and the discontinuous nature of the filling operation, the formula is multiplied by a correction coefficient n, where n=1.2 - 1.5 depending on the nature of the mining operation.

So, $\sigma_o^1 = n\dfrac{\gamma Ha}{KL}$ .. .. .. (f)

The significance of formula (f) is quite different from that of the calculation formula shown in Table 2, though their forms are approximately the same. According to formula (f), the strengths required of the cemented fills at FKLNM and at XKSAM are shown in Table 3.

The results calculated in Table 3 correlate quite well with the actual measured strength of the fill used in FKLNM. The result calculated for XKSAM is smaller than the actual strength used, the reason being that the first mining blocks are in fact the ore pillars.

Table 3

Mine	Unit weight of over-burden γ kg/m^3	Depth of overburden H m	Width of the first ore block mined a m	Width of the second block mined L m	Angle of internal friction ϕ -	K -	N -	σ_o^1 MPa
FKLMN	2,700	200 (-80level)	7-10 (stope)	6-8 (pillar)	40°	4.60	1.4	1.88 to 2.02
XKSAM	2,650	180	5-8 (stope)	8-10 (pillar)	36°	3.94	1.5	1.11 to 1.42

STRENGTH REQUIRED OF THE CEMENTED FILL
SUPPORTING THE WEIGHT OF THE FRACTURED
ROCK MASS ABOVE THE OREBODY

Where the ore deposit is gently inclined,
or the width of the orebody is very large,
the area and the height of the fractured
rock mass above the orebody are large.
Therefore, the weight of such a rock
mass bearing on the cemented fill might
be large enough to destroy the fill. The
strength required of the cemented fill
to support the weight of such a rock mass
may be in excess of the result calculated
by formula (f).

Determination of the height of the
fractured rock mass above the orebody is
a extremely complicated problem. It
depends closely on the roof area exposed,
geological structure, the physical and
mechanical characteris of the roof strata,
ground water, balsting operations, and so
on. According to Gao (1979), this height
may be as much as two to six times the
height of the stopes in the case of open
stopes. Roof-separation survey instrumentation was set in the roof rocks of
some stopes located on three different
levels at XKSAM and ovservations were
made over a period of several years
(1975-1979). The observed results (Xiang
(1980)) showed that the height of the
fractured rock mass above the orebodies
is normally in the range 4-10 m, with a
maximum value of less than 20 m, while
the height of an orebody, that is, the
height of a stope is about 10 m. Therefore, the height of the fractured rock
mass above the orebody in cut and fill
stopes is much lower than with open
stopes, being limited to about twice the
height of the stope, though this cannot
be proved in theory.

When a mining region has only few stoping areas, where the roof strata are
gentle inclined and where the mining
operations are as at XKSAM, the strength
required of the cemented fill can be
calculated as follows:

$$\sigma_o'' = n \left(\frac{2h\gamma}{m} + \gamma_1 h \right) \quad .. \quad .. \quad .. \quad (g)$$

where σ_o'' is the strength required of the
cemented fill (MPa)
  h the height of a stope (m)
  2h the height of the roof fracture
     zone (m)
  γ_1 the unit weight of the cemented
     fill (kg/m^3)
  γ and n as above, and m the ratio
of the cemented fill to the total fill.
When stopes and pillars are cement filled
(as at FKLNM), m=1; when stopes are
cement filled while pillars are uncement
fill (as at XKSAM),

$$m = \frac{B_1}{B_1 + B_2} \quad \text{where } B_1 \text{ is the width of}$$

a stope, B_2 the width of a pillar.

According to formula (g), the strength
required of cemented fill at XKSAM is
shown in Table 4.

The result in Table 4 compares closely
with the actual measured strength of the
fill used at site. In fact, a roof
pressure arch formed when an emptied
stope was formed. The rock body below
the pressure arch is the roof fractured
zone, and this is the only weight supported by the cemented fill. Monitoring
over an extended period of time has
shown that the span of a pressure arch
might cover up to seven ore pillars at
XKSAM, as reported by the Xi Kuang Shan
antimony mine, Changsha Institute of
Mining Research and Wuhan Institute of

Table 4

Mine	Unit weight of roof rock γ	Unit weight of cemented fill γ_1	Height of stope h	Ratio of cemented fill to total fill m	n	σ_o''
	kg/m^3	kg/m^3	m	-	-	MPa
XKSAM	2,650	1,500	12 (the height of the ore layers)	$\frac{6}{6+10}$=0.375	1.5	2.77

409

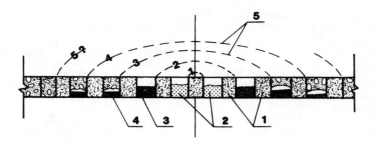

1 cemented fill pillars
2 filling stopes
3 mined stopes
4 mining stope
5 pressure arches

Fig. 2 Development of pressure arches at XKSAM

Metallurgical Safety Research (1980). If the mining region is greater than the length of seven pillars, a single pressure arch cannot be formed, as shown in Fig. 2.

According to Fig. 2, the pressure borne by the cemented fill placed beneath the centre of arch is the largest. The strength required of the cemented fill is

$$\sigma_o'' = n \left(\frac{f_o \cdot \gamma}{m} + \gamma_1 h \right)$$

$$= n \left(\frac{\gamma}{m} \cdot \frac{B_o}{3} + \gamma_1 h \right)$$

$$= n \left(\frac{\gamma}{m} \cdot \frac{7L + 6a}{3} + \gamma_1 h \right) \quad .. \quad .. \quad (h)$$

where f_o is the height of the roof pressure arch, $f_o = \frac{1}{3} B_o$ (m)

B_o the span of the arch, $B_o = 7L + 6a$ (m), and

σ_o'', n, m, L, a, γ, and h are as above.

The strengths required of the cemented fill calculated by formula (h) are shown in Table 5 and the results in Table 5 also agree closely with strengths observed in practice.

STRENGTH REQUIRED OF CEMENTED FILL PERFORMING AS A SELF-SUPPORTING RETAINING WALL

When man-made pillars are formed and the ore blocks mined by the second mining step are extracted by cut and fill stoping, the cemented pillars are exposed on either one or both sides. If the area exposed is large enough, shearing damage to the cemented fill may occur because of a thrust from the uncemented fill or other soft rocks into the cemented fill, as shown in Fig. 3.

Consider the worst possible case, where the height of the last sublevel is highest, contact between the cemented fill and the roof pillar is not achieved

Table 5

Mine	Width of pillar L m	Width of stope a m	Span of roof pressure arch B_o m	Height of roof pressure arch f_o m	Height of stope h m	Unit weight of roof rock γ kg/m³	Unit weight of cemented fill γ_1 kg/m³	m –	n –	σ_o'' MPa
FKLNM	7	10	109	36.3	36	2,700	1,900	1	1.4	2.28
XKSAM	6	10	102	34	12	2,650	1,500	0.375	1.5	3.80

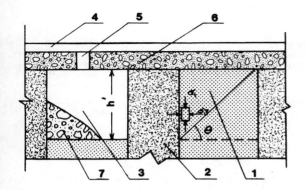

4 5 6

1 sliding body
2 cemented fill pillar
3 mined stope
4 main drift
5 access for hydraulic filling
6 roof pillar
7 ore

7 3 2 1

Fig. 3 Cemented fill performing the role of self-supporting retaining wall

and one side of the fill is exposed while the other side of it is thrust side ways (Fig. 3). If the height of the exposed side of the cemented fill is h' (m), the active pressure on the retaining wall is:

$$\sigma_3 = \gamma_o' \, h' \, tg^2 \left(\frac{90-\phi'}{2}\right) \quad .. \quad .. \quad (i)$$

where σ_3 is the active pressure of the cemented fill on the sliding body (MPa),

γ_o' the unit weight of the sliding body, and

ϕ' the internal friction angle of the sliding body.

Meanwhile, the cemented fill is also loaded by a shearing force

$\sigma_3' = -\sigma_3$. Therefore, the cemented fill should have an antishearing strength σ_3. Therefore, the strength required of the cemented fill is

$$\sigma_o''' = nk\sigma_3$$
$$= nk\sigma_o' h' tg^2 \left(\frac{90-\phi}{2}\right) \quad .. \quad .. \quad (j)$$

where σ_o''' is the strength required of the cemented fill (MPa),

σ_3 the anti-shearing strength of the cemented fill (MPa),

k the ratio of the cemented fill compressive strength to its anti-shearing strength, and

n is as above.

According to formula (j), the strength required of the cemented fill at XKSAM is shown in Table 6.

If the strength required of the cemented fill performing as a self-supporting retaining wall is calculated from formula (b),

$$\sigma_o''' = nk\sigma_o' h' \quad .. \quad .. \quad (k)$$

The result calculated by formula (k) for XKSAM is $\sigma_o''' = 2.47$ MPa.

For one emptied stope at FKLNM, the area of one cemented fill exposure was as large as 128 m^2 (height 4.9 m x width 26.2 m). The emptied stope was filled in mid-1981. The strength of the cemented fill measured by sampling at site was 1.93 Mpa. The self-supporting performance of the cemented fill was quite adequate, though its strength was not high.

Synthesizing the data in Tables 3 to 6 and considering that blasting is operated against cemented fill pours, it has been concluded that at XKSAM the required fill strength is R_{90} 4 MPa and at FKLNM is R_{90} 3 MPa and the design of the cemented fill mixes should be based upon these strength requirements.

CONCLUSIONS

1. The simple and convenient designing method described in the paper is easy to apply and has a high dependability. It allows ready calculation of required cemented fill strength for a variety of mining conditions.
2. In this simple and convenient designing method, the influence of geological structure has not been considered. When such structure is highly significant in a particular mining region, its influence must be considered in determining required cemented fill strength.
3. As operations at XKSAM and FKLNM proceed to greater depths, continuing and extended research will be necessary to establish cemented fill strength characteristics.

411

Table 6

Mine	Unit weight of fill γ_o'	Height of a sublevel h'	ϕ'	k	n	σ_o''
	kg/m^3	m	$^\circ$	–	–	MPa
XKSAM	1,500	12	30	10	1.5	0.883

ACKNOWLEDGEMENTS

The author wishes to acknowledge the support of the Fan Kou lead zinc mine and the Xi Kuang Shan antimony mine and their provision of data which has made this paper possible. Acknowledgement is also accorded Mr. Li Boxian, advanced engineer in the Jiangxi Institute of Non-ferrous Metallurgy Research, for his guidance during writing the paper. Also, Dr. E.G. Thomas, Senior Lecturer in Mining Engineering, University of New South Wales, Australia, has corrected the grammatical presentation of the paper and suggested its inclusion in the Lulea Symposium proceedings.

REFERENCES

Aitchison, G.D., Kurzeme, M. and Willoughby, D.R. 1973, The geomechanics factors to be considered for making the most reasonable use of a fill body in mines. Papers presented at the jubilee symposium on Mine Filling, Mount Isa, Australia.
Changsha Institute of Mining Research 1978, "Mining with Backfills".
Coates, D.F. 1970, "Rock Mechanics Principles", Canada.
Fan Kou lead zinc mine and Changsha Institute of Mining Research 1981. A report on operations and material tests of cemented hydraulic fill. Unpublished.
Gao, L. 1979, "Rock Mechanics in Mining Engineering". Sian Institute of Metallurgy and Architecture.
Hunan Institute of Metallurgical Research 1980, Usage and development of mining with backfills in metal mines of Hunan Province. Unpublished.
Thomas, E.G. 1973, Cemented hydraulic fill: an approach to its use in mining engineering, Mining Magazine No. 128.
Yan, G., Kan, Y. 1980, A primary approach to the mechanical effects of fills at XKSAM. Unpublished.
Xi Kuang Shan antimony mine, Changsha Institute of Mining Research and Wuhan Institute of Metallurgical Safety Research 1980. A study for mining the river safety pillar and mining strata pressure. Unpublished.
Xi Kuang Shan antimony mine, Changsha Institute of Mining Research and Wuhan Institute of Metallurgical Safety Research 1980. A study of mining strata pressure and its control at XKSAM. Unpublished.
Xiang, J. 1980, Cemented hydraulic fill: an approach to its usage possibility at XKSAM. Unpublished.

Large scale model tests to determine backfill strength requirements for pillar recovery at the Black Mountain Mine

J.D.SMITH
John D.Smith Engineering Associates Ltd., Kingston, Canada

C.L.DE JONGH
Gold Fields of South Africa, Johannesburg

R.J.MITCHELL
Queen's University, Kingston, Canada

ABSTRACT: Walls of cemented tailings plus desert sand up to 70 m high and 30 m wide are exposed during pillar recovery. Fill strength requirement for the initial stoping blocks had been designed using a free-standing vertical wall concept. A design strength of 700 kPa was calculated. In 1981, another basis for design was developed, from a series of small scale model tests conducted in Canada, that resulted in a significant cement saving. A large scale model was designed to satisfy mine conditions and a series of nine tests were conducted at the mine in March 1982. Analysis of test data established a design strength at 400 kPa in unconfined compression. A saving of R1,000,000 is anticipated. The first exposure of lower cement content fill is scheduled for April 1983, and results should be available for this meeting.

1. INTRODUCTION

The Black Mountain Mineral Development Company began production in 1980, utilizing a blasthole stoping method to extract the steeper section of the orebody. The orebody is mined in a sequence designed to maintain ground stability by replacing extracted ore. A block is typically 70 m high and contains about 50,000 m³ of fill. The backfill material consists of a 50:50 mix of de-slimed tailings and wind-blown desert sand with Portland cement added to provide the required strength. The stability of the fill is critical for the efficient recovery of the adjacent remaining ore pillars.

Because this mine has high grade ore and no previous pillar extraction experience, a conservative approach to fill strength design was used. The free-standing vertical wall concept was orginally used to provide an unconfined design strength of 700 kPa. This strength was obtained from a dry cement content of 7.14% or a 1:13 cement to sand/tailings mix. In 1981, an alternative design equation for fill strength requirement was developed from a series of small scale model tests conducted in Canada. (Mitchell, Olsen, & Smith, 1982) This design equation considers the three dimensional geometry of the stope and arching effects in the fill to produce a significantly lower strength requirement to

satisfy local mine conditions. A series of nine tests were conducted in a 1:14 scale stope model. Analysis of the data indicated that a fill strength of 400 kPa in unconfined compression would be sufficient (an approximate 1:20 mix).

This paper presents the theoretical considerations and modelling theory used to develop the Black Mountain model and the testing procedures. The model test results are included, along with the development of a modified design equation to suit mine conditions. The use of the model test results in deciding on backfill strength requirements of future stope pours is discussed.

In March 1982, the backfill plant settings were modified to produce a 30 day unconfined strength of 400 kPa. A substantial cement savings has been realized since that time.

2. THEORETICAL CONSIDERATIONS

At shallow depth, when elastic deformation of the hanging wall is minimal, the largest shear stresses within any fill material will be due to self weight. Two design approaches are common for determining strength requirements under such conditions. These are the free standing wall and vertical slope concepts. The free standing wall approach gives a uniaxial compressive strength (U.C.S.) requirement at any depth as: $[\sigma_1] \geq H\gamma$

where $[\sigma_1[$ = U.C.S. of cemented fill

γ = Mass Density × g.

H = Exposed fill height

This approach gives a variable strength requirement, depending on H. The vertical slope approach gives a cement bond shear strength requirement as:

$$c_b = \frac{[\sigma_1[}{2} \text{ for a frictional material}$$

therefore $c_b = \frac{\gamma H}{2}$ where

c_b = cement bond shear strength

The shear strength is given as (U.C.S.)/2 for a frictionless material.

Unconfined test results are normally used to determine the cement requirements. Since the unconfined strength of a given backfill varies approximately linearly with cement content over the range of 3% to 8% cement by dry weight, both approaches give about the same overall cement usage.

However, neither of the above designs account for the true three dimensional geometry of the cemented fill block.

Part of the downward force acting on a sliding block will be resisted by shear along the rough interface between the rock wall and the fill as shown on Fig.1.

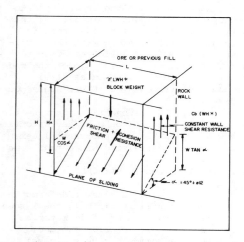

Figure 1   Confined block failure

This resistance can be approximated to the cement bond strength activated over the area of the failure plane concerned. Assuming drained conditions, this wall shear effect can be used in the determination of the U.C.S. requirements for given backfill material and stope geometry as shown below:

Factor of Safety (F)

= $\dfrac{\text{Force resisting failure}}{\text{Force activating failure}}$

= $\dfrac{\text{Frictional resistance \& cohesion on shear surface}}{\text{Net weight of block acting along shear plane}}$

Net weight of the block (Wn) = WH*(Lγ- 2c_b) as derived from the forces shown on Fig. 1.

$$F = \frac{Wn\ Cos\alpha\ tan\ \emptyset + \frac{CLW}{Cos\alpha}}{Wn\ Sin\alpha}$$

where $H* = \dfrac{H - W\ tan\alpha}{2}$

C = apparent cohesion on failure plane

$\alpha = 45° + \dfrac{\emptyset}{2}$

If it is assumed that c_b = C, i.e. the cement bond shear strength is equal to the apparent cohesion on the failure plane, then the required cement bond shear strength for any value of $\emptyset$ when the height (H) is much greater than the width (W) may be given as:

$$(2c_b) = \frac{H*}{\frac{H*}{L} + \frac{1}{1 - \frac{tan\ \emptyset}{tan\alpha}\ Sin\ 2\alpha}}$$

For a frictional material where $\emptyset$ = 0, the U.C.S. requirement is given by:

$$[\sigma_1] = (2c_b) = \frac{\gamma H}{1 + \frac{H}{L}}$$

where L = Length of orebody as shown on Fig. 1.

When H = L the U.C.S. is the same as determined by the vertical slope approach. When H becomes larger than L, a lower U.C. S. value is indicated. As L tends towards zero, so does the required U.C.S. and therefore the cement content, because of the increased effect of arching. For a typical backfill where $\emptyset$ = 35° and H = 3L, the net block weight would be decreased by 20% if both c and $\emptyset$ were mobilized in wall shear. Equation (1) will therefore yield a conservative design for the required U.C.S. of the fill.

The above analysis was verified by Mitchell et al (1982), by means of scale model testing completed in 1981. Although this model simulated a vertical orebody and different backfill materials were used, the confidence in the design principles was sufficient to conduct similar larger scale model studies at Black Mountain.

3. MODEL DESIGN

Physical modelling is well known in engineering research. For this study, the important dimensionless parameters in a stope backfill situation are $\frac{c}{\gamma H}$, $\tan \emptyset$, $\frac{W}{H}$ and $\frac{L}{H}$ where c and $\emptyset$ are strength parameters and W, L, H are width, length and height respectively as shown on Fig. 1. All of these parameters can be scaled, maintaining theoretically perfect similitude between model and prototype, using the same materials. This is possible because the prototype backfill is designed on a 28 day strength, while the model could be tested at any time after pouring when the value of $\frac{c}{\gamma H}$ is equal to the prototype design value. The value of γ being the same in model and prototype, the decrease in c is directly proportional to the linear model scale factor. For example, consider a 50 m high stope designed to have a cementation strength (shear) strength in unconfined compression) of 250 kPa, giving, for γ = 20 KN/m^3, $\frac{c}{\gamma H}$ = 0,25. If the model is exposed after 24 hours of curing and has a strength, at that time, of 12.5 kPa, the model height would then be 2.5 m.

Blasthole stopes at Black Mountain are on average 70 m high (vertical), 28 m on strike and vary between 20 and 30 metres in the hanging wall-footwall direction (true width). The average dip is 55°. It was calculated that a scaling factor of 14 would provide sufficient height in the model to eliminate capillary tension effects. Accordingly, the model was made 5 m high, 2 m on strike and 2,1 m in the H.W. to F.W. direction (See Fig. 2). This geometry was used for tests 1 - 7 inclusive, after which the H.W. was blocked in to produce a H.W. to F.W. distance of 1,4 m (to correspond to a 20 m orebody thickness).

4. BLACK MOUNTAIN TESTING METHODS AND PROCEDURES

The model was constructed in the form of a box supported by an open frame work, structural material being timber with steel reinforcing as shown on Fig. 2. All the pieces were manufactured in the mine workshop on surface. The model was assembled in a special excavation on 3 Level(an enlarged abandoned crosscut). The reasons for underground tests were:
(a) To maintain even temperatures, because curing depends on temperature
(b) Fill could be piped very easily from the underground pipe line on 3 Level.

After preliminary laboratory strength tests had been completed on a range of tailings: cement ratios, it was decided to use 30:1 for all tests. The backfill plant was always informed 30 minutes before start-up of a model pour so that the plant settings could be modified and steady state conditions established.

An adjacent stope was used as a sump for the material produced during this mix stabilizing period and during topping-up periods. When plant instrumentation showed that a constant mix was being produced, the backfill was directed to the model through a flow reduction valve. An effort was made to model the pour rate but this was difficult to do because of sanding in the bleed-off pipe.

When it was decided that the fill had cured to the required strength (estimated to be that necessary to cause failure when the facing height of fill had been exposed), the facing timbers were removed from the top, towards the bottom, one at a time. As the fill was exposed, it was sampled near the H.W. and F.W. by driving a sampling tube into it. Shear box and occasional uniaxial compressive strength tests were conducted immediately after sampling. Decisions were then made, on the basis of the test results, to defer further fill exposure or to proceed with the stripping. Due to the variable curing time, and possible other factors, strength generally increased with depth, making such decisions more difficult. Several shallow failures occurred near the top of the test pours, but by the time the full face was exposed, the strength at the bottom was sufficient to prevent failure. On the other hand, if stripping proceeded after a shorter curing time, small surficial failures occurred as the boards were removed. When full exposure had been reached in all tests, the face was observed carefully for signs of insipient failure. When no further failures had occurred over a reasonable time period, samples were taken from the central portion of the face and from behind any failure surfaces.

Photographs were taken at various stages of face exposure for each test, along with notes on test results and pertinent features. When all data had been recorded, the model stope was turned over to the

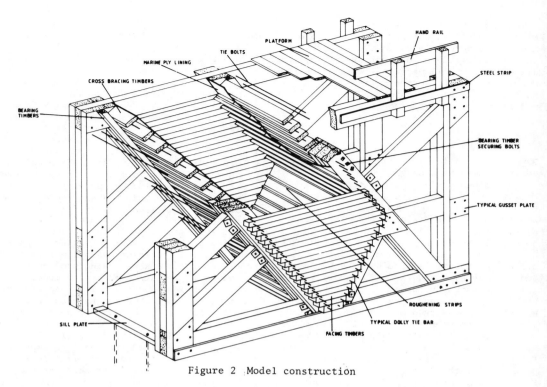

Figure 2 Model construction

clean-up crew. The backfill still remaining in the box was removed by hand shovelling, the inside of the box was cleaned using a water hose, the facing timbers were reinstalled, and filter materials were replaced. This was very arduous work, but the model stope was always prepared so that one test could proceed every 24 hours, including the down-time due to blasting.

5. STRENGTH OF MODEL FILLS

The first and second model pours were too strong to fail and the strength data from test No. 6 was scattered because of segregation due to pouring at a low pulp density. Strength data from the remaining test pours are plotted on Figure 3. The strength data on Figure 3 was obtained by doing direct shear tests on samples taken from the stable fill remaining in the test box after each failure had taken place. Uniaxial compression tests were carried out on some samples and the uniaxial compressive strength was found to be equal to twice the shear strength (as expected in a cohesive material) in all comparisons where uniaxial samples failed without excessive bulging. The theoretical relation of $[\sigma_1]=2c$ (unconfined compressive strength = 2 times shear strength) was also confirmed in the pre-

preliminary test program.

From the data on Figure 3 the average strength in the wider models (tests 3 through 7) was significantly higher than that in the narrower models (tests 8 and 9). If a variation of about 25% of the average strengths (as shown in Figure 3) is assumed and applied to the strength lines on Figure 3, 9 data points fall outside these limits (darker points on Figure 3). In order that no single test result is unduly influenced by erroneous data, the shaded data points are omitted from further data analyses.

It is further observed from Figure 3 that there was a pronounced strength increase with depth, and this is considered to result because of the reduction in curing time from bottom to top of the pour. It generally took about 1 hour to pour a model fill and about 1 hour to strip the facing timbers so that the curing times could be as high as two hours different between the top and the bottom of a pour.

6. MODEL TEST RESULTS

Preliminary testing was carried out to determine the probable curing time before exposure in order to obtain a failure condition in the model tests. Nine model tests

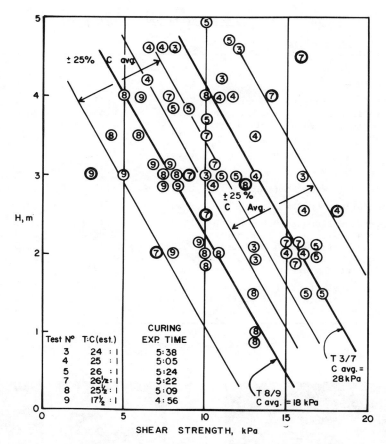

Figure 3   Shear strength in model tests

were carried out and the data are summariz-ed on Table 1.

Photographs of two failures are shown on Figure 4, Test 1 did not fail but Test 2 was considered very close to failure. Model failures were obtained in Tests 3 through 9. Multiple failures were obtained in several tests (particularly with the nar-rower geometry).

Test No. 1 - cured for over 18 hours and was quite stable when exposed. $H=3.8m$, $c \leq 20$ kPa, $\gamma = 21.7$ KN/m³ giving $(c/\gamma H) \geq 0.25$ and $(H/L) = 1.48$

Test No. 2 - cured for 9:35 hours and re-mained stable when exposed, although some bulging was noted. $H = 4.2m$, $c = 13,3$ kPa, $\gamma = 21.9$ Kn/m³ giving $(c/\gamma H) = 0.145$ and $(H/L) = 1.64$

Test No. 3 - cured for 5:38 hours and ex-hibited three distinct failures during exposure. $H = 4m$, $c = 11.8$ kPa, $\gamma = 21,7$ kN/m³. 1st Failure: $c = 10.3$ kPa, $H_f = 1.2m$, probable seepage
$c = 11.4$ kPa, $H_f = 1.5m$, probable seepage
$c = 13$ kPa, $H_f = 2m$, probable seepage

Test No. 4 - cured for 5:5 hours and failed at an exposure height of 3.1m with an aver-age $\gamma = 21.8$ kN/m³, $c = 11.2$ kPa then $(c/\gamma H) = 0.166$ and $(H/L) = 1.21$.

Test No. 5 - cured for 5:24 hours and fail-ed at an exposure height of 1.8m, remained stable for full exposure below this depth. Strength, $c = 9.8$ kPa to $H = 1.8m$. Average strength, $c = 12.6$ kPa and $\gamma = 21.3$ kN/m. For failure $(c/\gamma H) = 0.256$ and $(H/L) = 0.70$. For overall exposure $(c/\gamma H) = 0.13$ and $(H/L) = 1.76$ (stable).

417

TABLE 1 – SUMMARY OF MODEL TEST FAILURES

Test No.	Description of Failure	γ (kN/m³)	H_f/L	$[\sigma_1]$ kPa	$\dfrac{[\sigma_1]}{\gamma H_f}$
1	No failure	21.7			
2	No failure	21.9			
3	Surficial failure to 1.2 m depth	21.7	0.47	20.6	0.79
	Surficial failure to 1.5 m depth	21.7	0.59	22.8	0.70
	2 m Lower failure possibly due to seepage	2.17	0.78	26.0	0.60
4	Failure to 3.1 m depth	21.8	1.21	22.4	0.33
5	Failure to 1.8 m depth	21.3	0.70	19.6	0.51
6	Failure to 3 m depth (seepage)	20.6	1.17	17.0	0.28
7	Major failure to 3 m depth	21.6	1.17	19.4	0.30
8	Surficial failure to 0.8 m depth	21.8	0.47	10.0	0.57
	Surficial failure to 1.4 m depth	21.8	0.82	10.0	0.32
	Surficial failure to 1.8 m depth	21.8	1.05	12.8	0.33
	Surficial failure to 2.7 m depth	21.8	1.58	12.8	0.22
	Full height failure to 3.8 m depth	21.8	2.22	16.8	0.20
9	Very surficial failure to 0.7,	21.7	0.41	10.0	0.66
	Surficial failure to 1.4 depth	21.7	0.82	12.0	0.39
	Deeper failure to depth of 2.8 m	21.7	1.64	14.8	0.24

Figure 4(a) Typical partial failure

Figure 4(b) Typical full height failure

TABLE 2 - SUMMARY OF OVERALL TEST STRENGTH RATIOS

Test No.	$[\sigma_1]$ Avg.	Description of Test	γ kN/m³	H/L	$\dfrac{[\sigma_1]}{\gamma H}$ test avg.	$\dfrac{[\sigma_1]}{\gamma H}$ Fig 2 (avg.)
1	40+	Stable	21.7	1.48	0.48	
2	27	Stable	21.9	1.64	0.29	0.30
3	24	Composite failures	21.7	1.56	0.28	0.32
4	24	Deep failure	21.8	1.76	0.24	0.29
5	25	Mid-depth failure	21.3	1.76	0.26	0.29
6	17	Deep failure (seepage erosion)	20.6	1.60	0.20	0.33
7	24	Deep failure (hanging wall slip)	21.6	1.76	0.25	0.29
8	22	Continuous failure to full depth	21.8	2.51	0.23	0.19
9	14	Continuous failure to cone depth	21.7	2.40	0.16	0.20

Test No. 6 - Cured for 6:53 hours and failed to a depth of 3m with C (avg.) = 8.5kPa, γ = 20.6 kN/m³ giving (c/γH) = 0.138 and (H/L) = 1.17. The strength was quite variable (from 4 to 14 kPa) and seepage erosion due to segregation was evident.

Test No. 7 - cured for 5:22 hours and exhibited three failures: one very surficial failure, a second failure along the footwall and a major failure to 3 m depth. Ignoring the five test results that are outside of the limits defined on Figure 2, c = 9.7 kPa over the failure depth giving (c/γH) = 0.149 and H/L = 1.17 for the major failure(using γ = 21.6 kN/m³).

The footwall failure may be due to a weakness in the fill in this area but no footwall shear tests were carried out since the material in this area was disturbed by the failure.

Test No. 8 - cured for 5:09 hours this test exhibited five distinct failures as the face was exposed. The first four failures were rather surficial while the last failure was full depth to 3.8 m with c = 8.4 kPa (ignoring the two data points lying outside the limits defined on Figure 2) and γ = 21.8 kN/m³. Then, for the major failure (c/γH) = 0.10, (H/L) = 2.2. For the surficial failures the average parameters are listed on Table 1.

Test No. 9 - cured for 4:56 hours this test exhibited three failures progressively deeper into the fill as the face was exposed. With γ = 21.7 kN/m³ (and ignoring the one strength result outside the limits of Figure 2) the average depths and strengths of the failure are 0.7m, 1.4m and 2.8m for c = 5 kPa, 6 kPa and 7.4kPa respectively. The average values characterizing these failures are tabulated on Table 1.

In Table 1, the value of H_f is the height of the observed failures and the failure strength $[\sigma_1]$ is the average of measured strengths over the failure height (with adjustments as noted above and doubling the direct shear result to obtain the equivalent unconfined strength (σ). Although some minor internal erosion was noted in some of the model tests its effect is mainly restricted to shallow surficial failures.

Table 2, summarizes the overall evaluation of the model stability using the full height of the model fill (top of pour level and the bottom of the fill) and two average strength values: the test average strength as measured from direct shear tests on samples and the average from Figure 3 (the latter being 18 kPa for L = 1.71 and 28 kPa for L = 2.56).

7. DISCUSSION OF RESULTS

Previous upper bound limit equilibrium analysis of cement stabilized backfill in vertical walled stopes (1) gave the

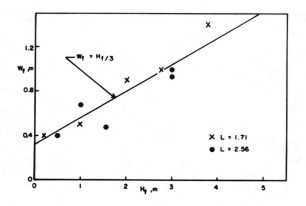

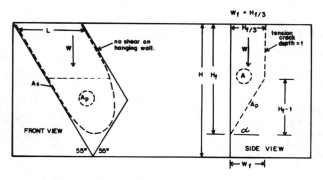

Figure 5 Analysis of observed failures

required value of unconfined compressive strength as:

$$[\sigma_1] = \gamma H / (1 + H/L)$$

where $[\sigma_1]$ = unconfined compressive strength

  H = exposed fill height

  L = strike length of exposed fill.

From test observations, the depth of failure that developed with the 55° sloping stope walls were found to be about $H_f/3$ where H_f is the height of failure (see plot on Figure 5). Tension cracks were also noted to develop and these were assumed to extend vertically up from the failure plane over a depth of t meters. Figure 5 provides a model for an upper bound analysis of this situation as outline below. Defining failure to occur by sliding on planes As and Ap but assuming zero hanging wall shear because of the likelihood of tension failure on this plane, the equations are developed for a material with $\emptyset$ = 0 (shear strength = c_b):

$$A = \frac{H_f}{3}\ \frac{(t + H_f - t)}{2} = \frac{H_f}{6}\ (H_f + t)$$

$$Ap = L\sqrt{(H_f - t)^2 + \left(\frac{H_f}{3}\right)^2}$$

$$As = (A/\sin 55°) = 1.22A \ , \ W = A L \gamma$$

The factor of safety $F = \dfrac{\text{resisting forces}}{\text{driving forces}}$

$$\frac{c_b(As + Ap)}{W \sin 55°}$$

Then for F = 1 (a failure condition) the unconfined strength is given by

$$[\sigma_1] = 2c_b = \frac{2(0.819)L\gamma}{1.22 + Ap/A}$$

and $$\frac{[\sigma_1]}{\gamma H_f} = \frac{1}{X + 0.75\ \dfrac{(H_f)}{L}}$$

where $X = \dfrac{\sqrt{(H_f - t)^2 + (H_f/3)^2}}{0.27\ (H_f + t)}$ is a geometric constant.

Model test data, as tabulated on Table 1,

420

is compared with equations 1 and 2 on Figure 6. Most of the shallow failures producing $(H_f/L) < 1$ were considered to have been unduly influenced by seepage of water impounded on top of the model pours at the time that the facing boards were removed. Such failures are not considered to be a potential risk in the stope pours for two reasons: the stope pours will be better drained before exposure and the curing time variation between the top and bottom of the pour will be proportionally less resulting in a more uniform backfill strength. Many of the less-than-full-height failures that occurred in the models would not be expected in the prototype where the strength variation with depth would be less pronounced. It is clear that all potential prototype failures $(H_f \geq L)$ plotted on Figure 6 are well within the criterion of equation 1 and are well approximated by equation 2 with A = 2.21. This value of A corresponds to a tension crack extending to a depth of $0.3 H_f$ and this depth of tension crack is supported by observations on the model failures as well as by previous experience with vertical slopes in brittle cohesive materials.

The correlation is extended on Figure 7 in which data points from Table 2 are plotted. While some of the open points (representing average strengths from Figure 3) lie above the curve of equation 2 with A = 2.21, all of the closed points where failures developed are enclosed by this curve. Test No. 2 which remained stable provides additional support to the conclusion that the sloped geometry allows a strength reduction below that given by equation 1. The data on Figures 6 and 7 lead to the general conclusion that the strength requirement for the Black Mountain 55° sloped stope fills can be conservatively designed using equation 1 or can be designed with a minimal risk of failure using equation 2 with A = 2.21.

8. BACKFILL STRENGTH REQUIREMENT AT BLACK MOUNTAIN

The choice of a suitable design safety factor will depend on the choice of design criterion. It was considered that equation 2 should be used in design with A = 2.21 and a safety factor of 1.2. A design factor of safety of 1.2 is considered sufficient for the static case because there is no risk of loss of life and only a minor risk of ore loss in the event of a failure in the backfill.

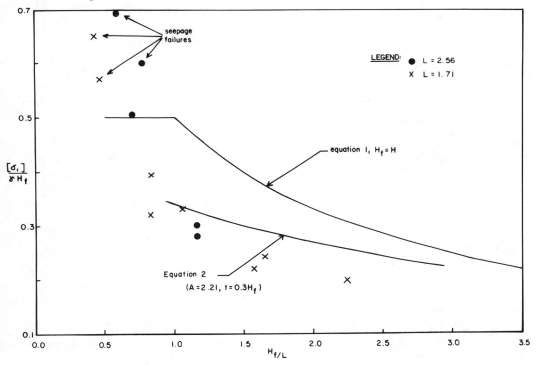

Figure 6   Model test data

421

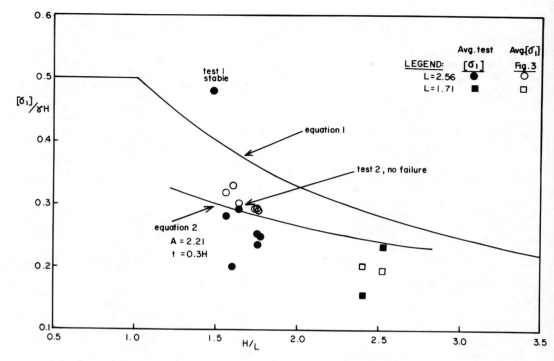

Figure 7 Analysis of model test stability

Experience at other mines has indicated that controlled blasting has little effect on the backfill stability (i.e. the back-fill is capable of absorbing blast energy without cracking to the extent that major sloughing of backfill would be initiated). Laboratory testing and field experience indicate that the "insitu" backfill strength will be marginally higher than that attained in laboratory samples at the same tailings: cement ratio providing the pour pulp density is maintained at a high value and surface drainage of the pour is maintained , by natural percolation or by decant towers, and is adequate such that cement segregation is avoided (cemented fill should never be discharged or allowed to run directly into surface ponded areas or severe segregation will develop).

Design is usually based on 28 day strength values. With good transport and placement techniques, it is estimated that the average "insitu" strength will be about 20% greater than that of laboratory control specimens.

Drained unit weights of prototype back-fills at Black Mountain mine are not expected to exceed 18 kN/m^3. For a 70 m stoping height the minimum strength requirements from equation 2 with A = 2.21 are:

Orebody Thickness	Unconfined Compressive Strength
L = 30m (H/L = 1.91)	415 kPa
L = 20m (H/L = 2.87)	347 kPa

Using the model test data directly, the average strengths under which failures developed (about 18 kPa for L = 1.71 and 28 kPa for L = 2.56) can be multiplied by the scaling factor (14) and the safety factor (1.2) to give:

$[\sigma_1]$ design for L = 30m as 470 kPa

$[\sigma_1]$ design for L = 20m as 300 kPa

It was considered that the 470 kPa was rather high for the L = 30m stopes, be-cause some of the model failures appeared to be promoted by seepage pressures. An appropriate strength for the 30m wide fills would be 400 kPa, and for the 20m wide fills, 350 kPa should be sufficient.

While the above analysis indicated that the narrow stope (L = 20m) required less cement than the wider stope (L = 30m), the strength difference is less than 25% and Test No. 8 (L = 1.71m) failed at almost the same average strength as Tests 3, 4, 5 and 7 (L = 2.56m). In the absence of more data on the L = 1.71m stope models (only two tests) and for backfill plant operation and control, it was considered prudent to use the same tailings: cement ratio for both stope geometries.

It was decided to use an average unconfined compressive strength of (σ_1) = 400 kPa in all cemented backfills at the Black Mountain Mine.

The tailings to cement ratio required to achieve a strength of 400 kPa was another design problem. Form published data, this strength level should be attained with a cement content of about 5% (20:1 T:C) but the actual requirement would depend on the grain size distribution, mineral composition and other characteristics of the sands. From the limited Black Mountain backfill strength data available in this strength range at the time of the model testing, it appeared that 20:1 T:C ratio (5% cement by dry weight) would be suitable for backfilling at Black Mountain. Good quality control with regard to cement content, stope drainage and delivery pulp density are requisite, of course, in producing a uniform quality backfill. Additional laboratory specimen testing and testing of pour samples were done in order to confirm the backfill strengths.

9. INITIAL RESULTS

9.1 The first pillar comprising 127 000 tonnes of ore was removed successfully. Dilution was minor and mainly caused by the additional blasting required due to the poor shape of the adjacent primary stope. This pillar was supported on either side by a fill containing a 13:1 mix.

9.2 By November 1982, 50% of the second pillar had been recovered. The fill on both sides is 13:1. Cracks in the fill parallel to the side walls were noticeable. These are thought to be the result of brittle failure caused by blasting vibrations.

9.3 Laboratory results indicated that a strength of 400 kPa is easily attained with 5% cement content.

9.4 Expected savings on cement costs should be about R1,000,000 as compared to the total study cost of R70,000.

9.5 By the end of 1982, 200 000 tonnes of the 20:1 mix had been poured. This fill will be exposed with pillar recovery in April 1983, and the first full scale

assessment of the new fill stability will then be possible. Mine Management, however, is confident that the 20:1 mix will be as stable in the underground situation as it was in the large scale model. It is hoped that the results of this fill exposure can be presented verbally at this meeting.

REFERENCES

Mitchell, R.J. Olsen, R.S. Smith, J.D. 1982. Model studies on Cemented Tailings Used in Mine Backfill. Canadian Geotechnical Journal. Vol. 19, Number 1. 14 - 28.

Design of post-pillar cut-and-fill mining system

Rock mechanics investigations at Surda Mine, Singhbhum

S.B.SRIVASTAVA & A.K.GHOSE
Indian School of Mines, Dhanbad

ABSTRACT: Until 1974, the principal method of mining in Surda Mine was breast stoping using timber as support. Need for improved productivity and conservation led to the introduction of the post-pillar method of mining. The method, essentially a modified cut-and-fill, offered distinct benefits vis-a-vis ground control and more productive LHD equipment combination. The rock mechanics programme currently underway to validate the design parameters, is outlined.

1 INTRODUCTION

Surda Copper Mine of Hindustan Copper Ltd. is situated in the Singhbhum Thrust Belt, which has been the most important source of copper metal in India since ancient times dating back to 2000 B.C. The narrow rich veins of copper ore were heretofore selectively mined out to maintain a high run-of-mine grade. Mining of such an orebody was being carried out using largely the breast method of stoping with timber supports. The method was simple and capable of following the changing contours of the narrow copper veins. The need to change the mining method was felt when wider orebodies were established at lower paylimit and the breast method of stoping became unsuitable under the changed circumstances. New mining methods with more effective systems of support were evaluated on the basis of characteristics of the orebody, nature and strength of wall rocks etc. The choice of these methods due to the peculiar geological set up of orebody was limited to a few only keeping in view the high rate of output and a higher recovery percentage of mineral with low cost of production and ensuring safer working conditions. Post-pillar method of mining, was first evolved at Strathcona

mine of Falconbridge Nickel Mines Ltd., Canada in 1970. This method offered a viable solution to ground control and permitted of the use of highly productive LHD combinations for mining of the flat dipping wide orebodies. The paper addresses the design aspects and the geomechanical investigations carried out at Surda to examine and optimise the parameters of post-pillar design.

2 THE GEOLOGICAL ENVIRONMENT

Surda Mine is located in the 160 km long Singhbhum Copper Belt (86°28', 22°30' to 86°22', 22°38') along the strike of which five copper mines of Hindustan Copper Ltd. occupy a 25 km stretch with large intervening areas, presently believed to be barren or lean. The belt is primarily a zone of overthrust and shearing. The formations within the area are regionally metamorphosed sediments and metavolcanics, both of Precambrian age (Thomas & Khan, 1981). The rocks within the overthrust zone are highly sheared and comprise mylonitised equivalents of siliceous chlorite-biotite-quartz-schists and quartzites. Copper-bearing hydrothermal solutions possibly derived from the granite magma, are believed to have permeated the openings within the sheared granite to form the lodes. Copper mineralisation has been controlled by the fractures formed during shearing. Chalcopyrite is the predominant sulphide mineral present, along with pyrite, pyrrhotite, pentlandite, marcasite and the oxide minerals.

Two principal areas of mineralisation have been established at Surda within the shear zone along the hanging wall and footwall contacts over a strike length of 1,800 m. The mineralised zones have been exposed by underground development to a maximum strike length of 1,700 m on 4th and 5th levels. Of the two zones of mineralisation, the hangingwall ore zone is better developed along the entire strike length and most of the wide lode zones upto 30 m are confined to this lode. The lodes dip at around 35° towards NE. The mine has been developed upto 13th level which is 487.5 m below the surface. The footwall formation consists of quartz chlorite schist and the hangingwall consists of quartzite, felspathic schist, quartz chlorite schist and chloritised quartzite, micaceous sheeted quartzite etc.

3 THE MINE AND THE MINING METHODS

In Singhbhum Copper belt, room and pillar method of stoping has been adopted for orebodies upto a width of 6 m. With raises at intervals of 13 m along the hanging wall contact, in a working room 1.5 m to 2.5 m slices along the hanging wall contact are advanced along the strike. The exposed hanging wall is bolted on a 1.2m x 1.2m pattern using grouted bolts. The ore left in the footwall is subsequently stripped out. Mechanical scraping is employed for mucking. The maximum permissible span along the strike is at present 10 m leaving 3 m wide rib pillar between the stopes. The excavation is backfilled before the start of the next stope.

Horizontal cut and fill method of mining is practised in ore bodies of widths of about 6 m where the ground conditions warrant adequate support of the hanging wall and exposure of large areas of hanging wall as in room and pillar

method is precluded. The method permits the use of cavo autoloaders with markedly improved productivity. The exposed hanging wall and the back are supported by rockbolts.

Post-pillar method has been used for ore widths exceeding 6 m and the method in essence represents a mix of room and pillar and the horizontal cut and fill methods. The method allows of the utilisation of the benefits of the cut and fill layout in that large capacity LHD machines can be deployed in the stope and wider orebodies can be extracted in a relatively inexpensive manner.

The method, originally conceived at Strathcona mines of Felconbridge Nickel Mines Ltd., (Clealand & Singh, 1973) was later transplanted at Surda Mine where the conditions warranted the extraction of rich narrow veins occuring parallel to each other. By lowering the pay-limit, these narrow veins could be mined combinedly as a single wide orebody thereby increasing the net extraction. The post-pillar method was found to be the most suitable candidate method for the conditions at Surda (Singh, 1977).

In the post-pillar method of stoping, the post failure strength of the pillars is utilised for supporting the workings. As such the post-pillars are designed for a low factor of safety of 0.8 to 1.0. In the layout adopted at Surda, 4m square pillars are spaced at 13m along and 9m across the strike. A stope of such a geometry has been selected for site investigation for deformation measurements of the pillars, hanging wall and stope back in 10th level 25 South post-pillar stope.

For extraction of the mineral, the stope blocks are developed by excavating a stope drive level and connecting this with the haulage level by raises which later serve as orepasses. A sill pillar measuring 8m along the dip is left to support the level. Each stope block is about 120m-150m long and is divided into 2 to 3 panels for ease of operation. Each panel is connected to the upper level and has an ore pass. The sill floor level is widened to the full width of the orebody by stripping, leaving post-pillars 4m x 4m size at a spacing

of 13m along and 9m across strike.
The back is thereafter stripped to
the full permissible height of 4.7m
and the exposed hanging wall and
back are bolted by grouted rockbolts
on 1.5m x 1.5m pattern. The use of
cable bolts for the support of
stope back has been successfully
applied at Surda Mine. The use of
friction rock stabilisers in place
of the grouted type of rock bolts
is presently in the experimental
stage. After removing the broken
material, the orepasses and manways are extended to 2.5m from the
back and classified mill tailings
emplaced upto this height leaving
enough clearance for the movement
of men and machines. The normal
cycle of operations consists of
drilling, blasting and fume clearance, scaling and supporting, mucking and finally preparations for
filling and filling.

4 DESIGN PARAMETERS OF POST-PILLAR STOPES

A post-pillar has been defined as
a rock column open on four sides,
designed to fail in such a manner
that failure would occur below the
fill line. The evaluation of the
load acting on the pillar is based
upon the "tributary area theory"
which assumes that pillars uniformly supported the load of the rock
cover overlying both the pillars
and the mined openings. The post-
failure strength of the pillars
plays a decisive role in the design.
For economic extraction of the mineral using this technique, the post
pillars should be loaded to failure
in such a way that they gradually
fail below the fill level without
causing a sudden collapse. The
factor of safety is near unity. The
post-pillars to-date have been
designed with a safety factor of
0.8 to 1 and depend largely on the
constraint of the fill surrounding
them for stability.

For estimation of pillar strength,
a large number of cored specimens
were tested in the laboratory for
their uniaxial strength. No specimens were rejected, even those
containing fracture, slips, joints
and other visible defects. The minimum value of the uniaxial compressive strength was used for calculation of the pillar strength by the
following formula:

$$CP = C_1 \left(0.778 + 0.222 \frac{WP}{HP}\right), \text{ MPa} \quad ..(1)$$

Where CP = pillar strength in MPa

C_1 = Uniaxial compressive strength
of the specimen, taken at the
minimum value in MPa for
$h/d = 1$ (where h = height of
core specimen and d = diameter
of core specimen)

WP = width of the pillar in metres
HP = height of the pillar in metres

Any underestimation of the pillar
strength is compensated partly by
the following three factors:
1. Development of triaxial stress
conditions in the central core of
the pillar.
2. Reduction of pillar load due
to redistribution of stresses to
the abutments.
3. Confining effect of the fill.
The long axis of the pillars is
vertical and for the stability of
the pillars of increasing height,
it is important that additional
lateral support be provided by fill
to inhibit surface spalling and to
improve pillar strength by inducing
triaxial stress conditions. The
post-pillars are designed to fracture and yield and to support only
the immediate roof strata with the
extra load being transferred to the
surrounding solid rock. This concept
of the post-failure behaviour of
rock which emerged from laboratory
investigations is useful for practical application in the design of
post-pillar stopes. The parameters
of post-pillar method at Surda
have been obtained from simple calculations taking the pillar load
as that due to the superincumbent
strata and computing the pillar
stress from the extraction ratio
and the pillar strength from the
results of the strength characteristics as determined in the laboratory
(Singh, 1977).
The designed dimension of pillars
was considered to be convenient
from the standpoint of formation.
For a system of rooms and pillars
with 4m x 4m pillars, the percentage
extraction was worked out. The
pillar spacing was determined approximately at 13m along and 9m
across strike. For purposes of

HINDUSTAN COPPER LIMITED
Indian Copper Complex
SURDA MINE

EX1,EX2,EX3,EX4,EX5 & EX6:- Points where single point extensometers were installed

S1,S2,S3 & S4:- Points where vibrating wire stressmeters have been installed

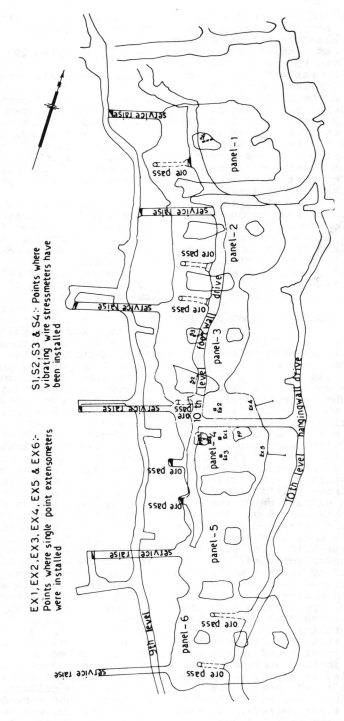

Figure 1. 10th Level 25 South Post-Pillar Stope Plan

428

original estimation, a pillar strength of 37.5 MPa, and an overburden stress of 2.75 MPa was assumed with gravity loading only. The parameters of the stope design however, were not optimised and the present study seeks to address the problem of optimisation of these parameters.

The optimum size of a post-pillar is probably that which will not disintegrate during mining (Hedley & Wilson, 1975). This depends upon the structural geology, accuracy of drilling and overbreak due to blasting which also depend on engineering judgement.

The function of the barrier pillars is to isolate the various sections of the mine in case of necessity and to facilitate the working of the stope. The barrier pillars must be competent to support the weight of the undermined overburden, even without the assistance of the pillars inside the panels. The panel pillars are required merely to maintain the integrity of the roof between the barriers. The rib pillars between the sub-panels in the stope facilitate the filling operation in addition to partially supporting the stope back. The most efficient use is made of the panel pillars if they are designed to exert their maximum supporting action on the back.

5 STOPE BACK STABILITY

The problem of stability of the stope back is germane to the overall optimisation problem. Instability of the back can in general be ascribed to the tensile strength of the rock along the planes of schistocity. The roof beam deflects under its own weight and the failure of the rockmass which is already fractured due to blast shock waves, occurs. The factors which affect stope back stability include inter alia the following (Barret & Chester, 1980):

1. Weak stratigraphic features including bedding planes and fault planes.
2. Differential movement between adjacent post-pillars of different heights.
3. Blast effects from firing within or adjacent to stopes.
4. Improperly placed supports.

The support density has to be increased as mining progresses upwards. The increase in the size of the room also increases the support requirement. For improved ground control, the following measures have been adopted (De Jongh, 1980):

1. Reinforcement by rock bolting.
2. Pre-support of ground by cable bolting.
3. Appropriate stope layout.
4. Ground stabilisation by backfill.

Though rock bolting has been quite successful in post-pillar stopes, the development of tensile fractures, (parallel to stope back beyond the zone of influence of the rockbolts) has in some cases led to rock falls, even after the installation of the rockbolt. In cable bolting, the method of pre-support utilises tension members to secure the stope back before blasting. The time lag between blasting and the installation of the supports is eliminated, this in turn reduces the stresses induced on conventional supports. The use of cable bolts allows long length of reinforcement members (upto 22m) to be inserted and to provide a continuous zone of support. In this case, one additional row of cable bolts is inserted after each lift from the footwall while the previously placed cable bolts still provide a network of support. Experience to date indicates that the use of cable bolts in conjunction with conventional rock bolts results in a stable stope back acting as a competent rock beam which is suspended by a network of tension members.

6 FIELD STUDIES

A suitable location in the deepest post-pillar stope of the Surda Mine was initially chosen for instrumentation. The central portion of the 10th level 25 South post-pillar stope gave a suitable site for the installation of six single-point rod extensometers. Two extensometers were installed in the hanging wall of the two sub-panels of panel No.4 (Fig.1). Likewise, one extensometer each was installed in the centre of the stope back of the two sub-panels. Another extensometer was installed in the stope back between the two post-pillars of the panel. The last extensometer was installed in the footwall side post-pillar of the panel. The extensometers were monitored periodically over a 10 month period using inexpedient measurement systems such as dial gauge, vernier calliper and micrometer screw gauge. The measurements revealed that there was negligible deformation of the strata. Figures 2 & 3 show representative plots of the field data.

DEFORMATION CHARACTERISTIC OF
STOPE BACK: STATION EX2 (4A-S/B)

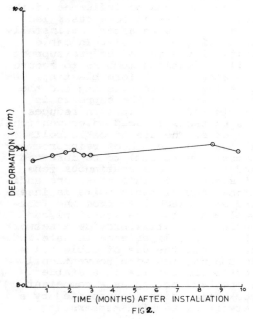

FIG 2.

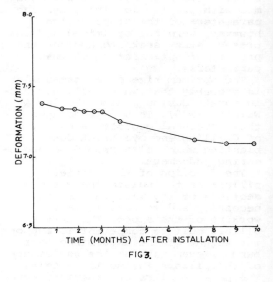

DEFORMATION CHARACTERISTIC OF
POST PILLAR: STATION: Ex6 (PP)

FIG 3.

The rock quality designation (RQD) and core recovery of ore and hangingwall rock of panel 4 of the instrumented stope have been summarised in annexure 1. Vibrating wire stressmeters were installed in the post-pillar of 1st, 3rd and 4th panels of the 10L 25 South stope and also in the rib pillar between panels 3 and 4. Figure 1 shows the location of the stressmeters insta--llations too. The stressmeters are being monitored since August 1982. Some more stressmeters as well as sonic probe extensometers are proposed to be installed in other post-pillar stopes.

Measurements with single point extensometers confirm that there is hardly any deformation of the stope back and of the post-pillar. It may thus be feasible to optimise the post-pillar stoping geometry with altered dimensions of working spans with resultant benefits of augmented recovery.

The trend of the observations from the vibrating wire stressmeters does not show that the pillars are under any distress and are perhaps not "post-pillars" according to definition. These however are preliminary conclusions which can confirmed after observations have been taken for at least 3 cuts.

7 CONCLUDING REMARKS

The post-pillar method applied to the mining of wide, flat-dipping orebodies holds considerable promise as a high production method of improved safety. Optimisation of design parameters is, therefore, an important objective for maximising recovery of the mineral with safety. Efforts are under way to optimise the design parameters at Surda Mine.

430

8 ACKNOWLEDGEMENT

The authors are thankful to the management of Hindustan Copper Ltd. for providing facilities to conduct the field investigations. The authors are also thankful to Director, Indian School of Mines, Dhanbad for permission to present this paper. The study forms a part of a S & T project funded by the Department of Mines, Government of India.

REFERENCES

Cleland, R.S. & H.H. Singh 1973. Development of Post-pillar Mining at Falconbridge Nickel Mines Ltd. C.I.M. Bulletin. 66:57-64.

Thomas, D.E. & M.A. Khan 1981. Introduction of Mechanisation at the Mosaboni Group of Mines, Bihar, India. Asian Mining'81, p.125-137. London:The Institution of Mining and Metallurgy.

Singh, A.K. 1977. Evolution of post-pillar method for extraction of wide orebody at Surda. Paper presented at MGMI Meeting held in April 1977.

Hedley, D.G.F. & J.C. Wilson 1975. Rock Mechanics Application in Canadian Underground Mines. C.I.M. Bulletin. 68:

Srivastava, S.B. & R.D. Singh 1979. On Post-pillar Technique. Minerals and Metals Review, June issue.

De Jongh, C.L. 1980. Design Parameters used and backfill material selected for a new base metal mine in the Republic of South Africa. Conference on Application of Rock Mechanics to cut-and-fill mining. Lulea:The Institution of Mining & Metallurgy.

Barret, J.R. & G. Chester 1980. Post-pillar Cut-and-fill Mining, A Comparision of Theory and Practice. Conference on Application of Rock Mechanics to cut and fill mining. Lulea:The Institution of Mining & Metallurgy.

Annexure I. Rock Quality Designation and Core Recovery of Ore in Stope Back and Hanging Wall Rock of Panel 4 of 10L 25 South Post Pillar Stope.

Location	Hole No.	Core Recovery (percent)	R.Q.D. (Percent)	Description of rock quality
4A-H/W	NX7	93.79	63.65	Fair
Stope back (between stations EX1 & EX2)	NX1	98.54	89.68	Good
Stope back (between station EX2 & North Rib)	NX2	100.00	88.38	Good
4B-H/W	NX8	96.25	56.79	Fair
Stope back (between EX3 & South Rib)	NX4	97.73	71.09	Fair
Stope back (between stations EX1 & EX3)	NX3	99.28	85.78	Good
Stope back (H/W side of PP)	NX5	100	79.14	Good
Stope back (F/W side of PP)	NX6	100	92.05	Excellent

Assessment on support ability of the fill mass

TONG GUANG-XU
Beijing University of Iron & Steel Technology, China

HAN MAO-YUAN
Beijing Mining & Metallurgical Research Institute, China

ABSTRACT: During the development of mining method with backfill in China, the support ability of fill mass has been further investigated to improve the fill technology as well as fill mining method. According to types of fill materials, the fill mass can be devided into loose mass and cemented mass. The different kinds of fill materials would affect compactness and strength of fill mass to produce different kinds of secondary stress fields as well as stress values in the wall rock. Owing to the fill mass itself can not produce an active pressure on the wall rock of stope, it also has very lower strength and elastic modulus as compared with ore body, therefore the fill mass can not cause the wall rock to regain the primary stress field. But it can gradually stabilizes the secondary stress field of wall rock after filling.

Owing to the strength and elastic modulus of cemented fill mass are much lower then ore body, the same amount of strain change will have quite different support abilities. Therefore, either the application of fill mass along or leaving small amount of discontinuous pillars only can not prevent caving of wall rock and subsidence of surface ground. The combination of them would give a better effort to support wall rock, as it applies lateral pressure on pillars to improve stress state formed on pillars and obviously increases support ability on pillar as well. Sometimes, the cemented fill mass contains more cement to increase its strength, for some technological reasons about controlling cement/water ratio, mixing fill constituents, transporting fill materials and presenting cement segregation during filling, the compressive strength value of 28 curing dates R_{28} is still very low. But the underground environment is more favour to the low-grade concrete for curing, it can gradually increase the strength of cemented fill mass up to 2-3 times of initial value at the later period. Meantime, the wall rock also applies a comparatively small load on cemented fill mass at the early stage, but gradually increases its value as the time passed. Therefore, it is allowed to have a comparatively lower strength of fill mass at the early period to reduce the amount of cement used during filling operation. For the practical use, the simple formulae are given to evaluate the strength of "artifical pillars". Meantime, the failure mechanics of cemented fill mass is also discussed.

As to the strength of hydraulic tailings or sand fill mass, it is generally to be considered as in a plastic state and called a rigidplastic mass. Its breakage is due to shear stress. The Mohr's Circle Theory can be used to show the relation between major and minor principal stresses. The formulae for determining required strength of tailings or sand fill mass are given.

The mining method with backfill had been adopted for a long history in China. Hence, in the middle of fifties, a large amount of ore was still produced by this method with dry fill from underground metal mines, about 54.8% in iron mines and 42.3% in nonferrous mines. Since then, its application has been obviously restricted, owing to the mining method with dry fill was used with strenuously manual labour, low output, low output, low worker's productivity and high cost. But, in the middle of sixties, the adoption of this kind of mining method was gradually increased every year, due to the improvement on technologies of hydraulic filling, cemented tailings filling and cemented rock filling, the increase of mechanization and automation for fill preparation and fill transportation, as well as the adoption of loading and hauling machineries in stope, all of these can obviously raise the stope's production and worker's productivity. Although this method nowadays has not been reused yet in the iron mines, its application is increased in non-ferrous metal mines as illustrated in Table 1.

Table 1. Percentage in weight of various types of mining method with backfill in underground production on non-ferrous mines.

Types of fill method	1955 %	1959 %	1961 %	1971 %	1982 %
Dry fill	35.2	11.6	2.3	1.2	1.2
Square set with fill	7.1	5.2	4.8	0.5	0.5
Hydraulic tailings (sand) fill	-	-	-	2.9	4.4
Cemented fill	-	-	-		4.0
Total	42.3	21.8	7.1	4.6	10.1

In China, the mining method with backfill is mainly used for extracting high-grade and high-value ore at the present moment. On the purpose for suitability of extracting technological and hydrological conditions of complicated ore deposits, reduction of ore loss and dilution, protection of ground surface and upper portion of low-grade or oxide ore from damage, avoidance or diminution of ground pressure, etc. the mining methods with dry fill, post-pillar cut-and-fill, overhand hydraulic fill, overhand cemented fill (including tailings fill and cemented rock fill), shrinkage of delayed cemented fill, underhand fill ans so forth are all used nowadays. In most cases, the large ore deposits are divided into room and pillar by using two stages of extraction, which sometimes require the different types of the mining method with backfill, such as the cemented fill stoping is adopted for extraction of room and hydraulic fill stoping for pillar. This kind of combination is considered to be cheaper and more effective. As to the appraisal of these types of mining method with backfill, the dry fill stoping is simple and easily

applies, it is also a best way for disposal of underground waste rock, that is why the reusing mining stoping is popular for extracting very narrow veins, therefore this type of dry fill stoping is still used. But in the large ore body mined by two stages of extraction, the pillar is very difficult to be extracted, which leads to a large amount of ore loss. Meanwhile, the fill material is in a loose state and can not be filled up to the roof of stone, which makes the weak roof easily collapsible and subsiding. Although a trial to install a thickness of 300-500 mm concrete wall between room and pillar, this wall is not easily to be constructed and the quality is also poor, which may lead to an occurrence of local collapse and create a difficulty of pillar recovery, increasing ore loss about 5 times higher than extraction of room and also giving high waste dilution. Besides, the stope daily production is very low, only about 30 tonnes. Therefore the two stages of extraction in dry fill stoping with many difficulties in technology is unable to be conquered at the present moment. About the post-pillar cut-and fill stoping, although the stope daily production can reach 250 tonnes, but the planned ore loss in pillar is about 12-15%, with additional loss of crown and sill pillars totalized about 30%, therefore it does not suit for extracting high-grade and high-value ore deposit. As to the hydraulic fill stoping, the introduction of pipeline transportation and mechanization of filling technology give a high stope daily production about 80-250 tonnes, but the output of unit stope area is still low, the method of desliming and dewatering have to be further improved, the tailings particles of less than 0.037 mm is still hardly to be utilized, which requires to replenish some other fill materials, and the tailings somehow can not be allowed to use as fill materials for a few mines at present times. When the hydraulic fill stoping has to be used in both two stages for extracting room and pillar, the same difficulty is still existed as in the mining method with dry fill. Because of the above reasons, the adoption of hydraulic fill method has not been spread. Nowadays, the cemented fill method is more popularized, it is generally classified as three types, such as overhand cemented fill, shrinkage of delayed cemented fill and underhand cemented fill. According to the fill materials, it can also be devided into cemented tailings fill (including sand fill) and cemented rock fill, but the former is more popular. If according to the stages of extraction, it has one stage and two stages, the latter is more widely used. Owing to trackless machines used in drilling, loading and hauling operation, pipeline transportation for fill materials, and different types of fill stoping chosen to suit all kinds of weak wall rocks and ore bodies as well as complicated geological and hydrological conditions, all of these make stope average daily production reach 90-200 tonnes, ore loss about 1.5-2.0% and waste dilution 7.8-13.8%. Although the mining cost is still comparatively high, it can be reduced by further improvement on technology and management. It has also been proved on research work and production experience in China, that the application of cemented fill stoping in high-grade, high-value and polymetallic ore deposits is very effective.[2] The best relation of extraction stages for securing a better result on economy is to use cemented fill for room and hydraulic fill for pillar.[3]

From the above description, a general scheme on the development of mining method with backfill used in China is given. The reason of rapid spread in recent years and more, except the improvement of production technology and mechanization. Rhe research work on the support performance of fill mass achieved an effective result is also important. In this paper, the assessment on support ability of the fill mass is only to be discussed.

1 SUPPORT MECHANICS OF THE FILL MASS

According to the different types of fill materials, the fill masses can be devided into loose fill and cemented fill. The mass of dry fill and hydraulic fill is belonged to loose type, but the cemented tailings fill and low-grade cemented rock fill are cemented type. Both types of fills have different kinds of compactness, and different strength as well, which creates different shapes and values of secondary stress fields in the wall rock. In Figure 1, it shows the changes of vertical components of stresses above fill masses of different materials.[4] The different types of fill masses have different compressibility, among which the dry fill is the highest and its subsicence can reach 25%;[1] Hydraulic fill is next, not beyond 15%;[1] the cemented fill is the smallest, generally about 2%.[2] The compressibility can influence the fill material to be packed closely up to the roof of stope, which surely reduces support effect of fill mass. Meanwhile, the degree of compactness of fill mass. Meanwhile, the degree of compact-

ness of fill mass increases, it can raise the support strength of neighbouring pillars. In other words, the stability of neighbouring pillars is raised as the compactness of fill mass increases.

Owing to the different types of fill masses, the support mechanics is also different. The loose fill mass to the wall rock can alleviate and protect the deformation and failure continuously developed. If a close packing up to the roof of stope is achieved, the fill mass can stand a portion of pressure coming from the roof.[3] It also gives the pillars inside a stope with a lateral confinement, which can protect the scaling and collapse of pillar. If the lateral confinement is partially or completely taken off, the damage of pillars will be occurred, which causes a great quantity of ore loss and waste dilution inside the stope. About the cemented fill mass, a certain quantity of cemented agent makes the fill mass to be consolidated as a whole body to have sufficient strength. In this case, if the lateral comfinement is partially or wholly dismantled, the cemented fill mass is unable to collapse, it still has a certain support ability. Although someone has considered that the action of cemented fill mass is only to reduce the ore loss and waste dilution, it has very little possibility to resist ground pressure and restrict deformation of wall rock. But the production experience has proved that the cemented fill mass can be used as "artificial pillar".[3] Therefore, the side wall of cemented fill mass after taken off lateral confinement can still stand and does not collapse, the explosure surface of sid wall sometimes can reach about more then 2 000 m[2]. This gives the two stages of extraction of ore body to adopt the cemented fill stoping for room and leaves a good chance for the pillar to select a very suitable kind of recovery method. This is an advantage that the loose fill mass can not be compared and is more useful than support ability to wall rock.

As previously mentioned, the action of fill mass to pillars inside stope is a very important problem. Although the fill mass has low strength and rigidity, it does not easily restrict the movement of wall rock by itself only. Meanwhile, a small quantity of continuous pillar also can not resist the sloughing of wall rock and subsidence of ground surface. Hence, both of them used together can give a comparatively good support effect. Actually the fill mass is unable to support the pressure from wall rock together with pillars, owing to have different elastic

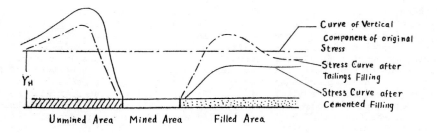

Fig. 1. Changes of vertical components of stress.

moduli. Meantime, the support ability of
fill mass is also tiny. But they are used
together the fill mass aplies a lateral
pressure against pillar, making pillar in
a triaxial stress state and having a much
high resistance to compressive stress as
compared with the biaxial or uniaxial
stress state, their relation is as follows:

$$\sigma_c = \sigma_0 + k\sigma_a \qquad (1)$$

Where: σ_c = Trixial compressive stress of
pillar, kg/cm^2

σ_0 = Uniaxial compressive stress of
pillar, kg/cm^2

σ_a = Lateral pressure from fill
mass, kg/cm^2

k = A constant, determined by the
variety and characteristics of
ore body.

At the Korman's triaxial test condition,
the value of k for a siliceous limestone
of a stibnite mine in China is equal to 4.
This formula proves that the pillar surrounded by the fill mass is able to increase its strength. Meantime, the fill
material also strengthens the support
ability of fractured surface of pillar.
The depth of fractured surface is usually
about 0.4-1.0 m, which decreases strength
of pillar, and is formed by blasting action
from stoping operation. If the lateral
pressure from fill material presses against
the layer of fractured surface, which certainly raises support ability and relevently increases the strength in central portion of pillar. Specially, when the cemented fill material is used, the cement
slurry percolates into the fissures, which
surely consolidates and strengthens the
surface portion of pillar. Besides, the

subsiding, compacting and strength-increa-
sing of fill mass can also raise the lateral pressure on pillars, which relevently
increases strength of pillar as well.[4]
During the process of extraction, the
just mined out opening in the stope can
not be filled immediately with fill material, because the working opening of a
certain height must be kept between stope
roof and fill mass. Usually, the secondary
stress field in wall rock with displacement and deformation around opening has
already taken place before the time of
filling. Owing to the fill mass itself can
not produce an active pressure against
wall rock, its strength and elastic modulus
are also much low than original ore body,
so that the fill mass is unable to recover
the primary stress field of wall rock. But
the fill mass can stabilize the secondary
stress field not further developed after
filling operation. Thus the requirement
for extracting and filling operations of
a stope is as quick as possible to restrict
the development of deformation and reduce
the failure of wall rock around stone
opening. Besides, the extracting and filling
operations during process of production
directly disturbs the stress fields in
fill mass and wall rock, which relevently
creates the change of stone rock pressure
all the time to have different regularities
at the different stages of operations. For
explaining the support mechanics of fill
mass, a slightly dipping stibnite ore body
in China is given as an example. This ore
body has a dip angle α 15-30° and thickness
of 2-30 m, and is mined by a room and
pillar mining method with delayed fill.
The room is of 8 m wide with cemented fill
and pillar mining method with delayed fill.
The room is of 8 m wide with cemented fill
and pillar of 10 m wide with hydraulic fill
separately. The rock pressure change in
this mine can tell the alteration of

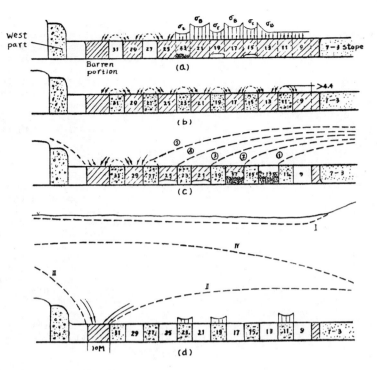

Fig. 2. Rock pressure change during process of extraction.
   a--Room Extraction, b--Cemented Fill, c--Pillar Recovery,
   d--Hydraulic Fill
   I--Subsidence of Ground Surface, II--Boundary of Ground Movement,
   III--Boundary of Ground Movement at West Portion
   IV--Boundary of combined ground movement

secondary stress field between fill mass and wall rock as illustrated in Figure 2. At the first stage of extraction of room, the rock layers on roof of room gradually came off and broke up to a height about 4.4 m before cemented filling. Meantime, the stress concentration was measured in pillars on both sides of room carried the load from overlying strata and became a kind of pressure abutments (Fig.2a). Put, during the operation of cemented filling, the increase of stress concentration in the neighbouring pillars became slowly and gradually stopped. This meant that the extraction of room to create stress concentration on pillars had finished (Fig.2b). At the second stage of pillar recovery, the redistribution of stress inside wall rock was measured. The roof of opening in pillar was also collapsed ut to a certain extent. At this time, the stress measurement in cemented mass proved that the load was very small, specially in the cemented

mass neighbouring to recovering pillar, but the stress in unmined pillar on other side of fill mass had rapidly increased (Fig.2c). Hence, during the second stage of pillar recovery, the unmined portion of ore body bore the rock burden from overlying strata, but the cemented fill mass only undertook the load from broken zone of roof. After the pillar recovered and at the stage of hydraulic filling, the stress in cemented fill mass and wall rock gradually stabilized again, which proved that the process of stress concentration at the second stage of extraction had ended (Fig. 2d). The above-mentioned regularity of rock pressure change had explained that the cemented fill mass onlybore the load from the broken zone of roof, but the unmined portion of ore body or the bed rock at flank undertook the major burden of overlying strata (i.e. pressure abutment), that was why the measurement of load in cemented fill mass was not great and the rock pressure at pillar recovery

437

was also not obvious and safe. Consequently
the pillar recovery would be carried out
safely at the second stage of extraction,
even the low-grade of cemented fill mass
had been used. As the mined out area extended, the roof movement was also gradually
enlarged. When the roof movement had reached
a certain height, the rock strata near the
surface began to deflect and finally subsided. Owing to the cemented fill mass had
a certain support ability to overlying stata,
thus it made the ground movement diminish
and develop slowly. After a formation of
complete breakage of cemented fill mass and
great compactness of hydraulic fill mass,
then the process of ground movement stopped.[7]

2 STRENGTH OF THE CEMENTED FILL MASS

The strength standard of cemented fill mass
inChina is the uniaxial compressive stress
value of 28 curing dates, i.e. R_{28}. At present, the cemented fill materials used are
belonged to the low-grade, the strength is
between 2-10 kg/cm^2 to 30-60 kg/cm^2. As
mentioned above, the strength of ore body
is about 10 times of it or more. Meanwhile,
the elastic modulus E of cemented fill mass
is generally only 10^5 kg/cm^2 and the ore
body is about 10^7 kg/cm^2, the difference
between them is almost about 100 times. If
they have the same quantity of deformation,
the support ability between them will have
100 times difference. Although the strength
of cemented fill mass deponds on the content of cement, for som technological reasons about controlling cement/water ratio,
mixing fill constitutents, transporting fill
materials and appearing cement segregation
during filling, the compressive strength
value of 28 curing dates R_{28} is still low.
But the underground environment of constant
temperature and humidity, as well as consolidation due to compression from dead
weight of fill mass are more favourable for
increase of strength at later period up to
2-3 times of initial value. Meantime, the
wall rock only applies a small load on
cemented fill mass at the early period, and
gradually increases its value as the time
passed. The above two phenomena show that
the low-strength fill mass can be adopted
to reduce the quantity of cement used during
the filling operation. In Chinese mines of
ratio between cement and tailings or sand
is from 1:10 to 1:6 at the present times.
It has been considered that the quantity is
too high, which increases the cost of filling operation. From the viewpoint of
strength of cemented mass, it is also not
necessary, therefore the reduction of cement
is proposed. According to the experience,

the quantity is suggested not less than
1:20 to 1:30.[2,8] In addition, the increase
of fill pulp density (by weight solid) can
also raise the strength of cemented fill
mass. If the cemented fill mass is built
up by fill pulp of low density, the formation of many layers of low cement content
would greatly weaken the shear strength and
stability of cemented mass. The difference
between high and low densities of fill pulp
accords with the criticle density of flow.
If the fill pulp density between the criticle density of flow, the characteristics
of fill pulp would be changed from the
heterogeneous two-phase flow to approximately hemogeneous of structural flow. Therefore, it is required that the fill pulp
density reaches 65% or higher. The production experience also proved that the increase of fill pulp density reduced the
cement segregation, made the particle-dis-
tribution more homogeneous and diminished
the loose of cement, hence the strength of
cemented fill mass would be increased. When
the fill pulp density had been raised from
50% to 70%, the strength increased about
1.0-1.5 times of original value. [8]

About a theoretical calculation of strength
of cemented fill mass, it is very difficult
to derive a general formula, owing to the
fact that because of the various characters
of ore bodies as well as the different
types of extracting methods. For the verification of figures obtained from the practical experience, a calculated result from
an appropriate formula for reference is
still useful. A Chinese stibnite mine of a
slightly dipping ore body with a width of
medium or more is used as an example to explain the calculation of strength of cemented fill mass, while only the load of overlying strata is considered. As stated above,
this mine has adpoted the cemented fill
stoping for room extraction and hydraulic
fill stoping for pillar recovery. After the
room was filled with cemented tailings, it
actually became a kind of "artificial pillar" is equal to the partial or total
weight of the above rock column between
central lines of pillars at both sides.
Hence, the compressive stress σ_c (unit:
kg/m^2) can be derived from the following
formula:

$$\sigma_c = \frac{2 \ arH_0}{B} \qquad (2)$$

where: 2a = width of block, e.i. distance
between central lines of pillars
at both sides of"artificial
pillar", m.
r = specific volume weight of overlying strata, kg/m^3

H_0 = height of fracture zone at bottom of overlying strata, m

B = width of room, e.i. width of "artificial pillar", m.

As to the height of fractured zone at bottom of overlying strata, it may be the total height or only a part of it, depending on the actual condition of mine. As above-mentioned stibnite mine, the height of fractured zone at bottom of overlying strata is 60-107 m, calculated from average failure compressive stress σ_y of the roof rock about 63-98 kg/cm². But som statistical materials pointed out, that the load applied on artificial pillar only about 20-70% of total weight of overlying strata, i.e. about 0.2-0.7 thickness of overlying strata. Therefore a value of 0.4 is suggested to be used in a general circumstance. It is also considered that the total weight of overlying strata will apply on "artificial pillar" about 2.5 times of original load after a certain quantity of blocks (i.e. room and pillar together) have been worked out. But the time factor can raise the strength of cemented fill mass, the cemented fill mass in a triaxial condition also increases its own strength. This total quantity of increment will guarantee the strength requirement for cemented fill mass.

The another case, when a large ore body has an inclined or steeply inclined angle with displacement of overlying strata after mining, a theory of hanging wall sliding prism can be used to check the strength of planned cemented fill mass. The acting load P (unit: tonnes/m) of an unit length along strike by hanging wall sliding prism against cemented fill mass is calculated as follows:

$$P = \frac{Q \sin(\beta - \varphi)}{\cos \varphi} \qquad (3)$$

Where: Q = weight of hanging wall sliding prism of an unit length along strike, tonnes/m

β = breaking angle of hanging wall, degree

φ = internal friction angle of wall rock, degree

The compressive stress (unit: tonnes/m²) due to hanging wall sliding prism applies against the cemented fill mass (i.e. artificial pillar) is as follows:

$$\sigma_c = \frac{P}{S} \qquad (4)$$

Where: S = loading area on cemented fill mass, m², equals to the width b of cemented fill mass timing the perpendicular distance h

between the upper and lower faces of sliding prism, all of which can be measured directly from the surveying.

The calculating results of compressive stress σ_c on cemented fill mass as shown from above two methods can be compared with the allowable compressive stress [σ] of cemented fill mass, the requirement is:

$$\sigma_c = [\sigma] \qquad (5)$$

An example of a nickel mine in China used for the comparison of calculating results from the sliding prism theory with the allowable compressive stress of cemented fill mass is illustrated in Figure 3 and Table 2.

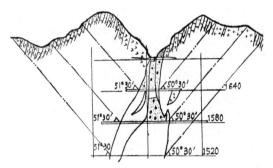

Fig. 3. A cross-section illustrating lines of break angles along No 10 prospecting line of a nickel mine in China.

The strength of cemented fill mass has a certain relation for determining the dimension of room and pillar, when the two stages of extraction are used. As stated above the strength of cemented fill mass is generally lower than ore body. Thus, when the ore strength is high the size of pillar is able to be small; because the strength of cemented fill mass is low, the room size must relevently be large; therefore the Chinese mines of mining method with backfill generally adopted the large room and small pillar. But according to the stability of ore body and wall rock the large room and small pillar are used in the unstable ore body and wall rock, when a certain explosure area of roof is allowable, this way can obviously increase the extracted ore at the first stage. As to the stable ore body and wall rock, the small room and large pillar are favourable, which can reduce the quantity of cemented fill materials to cut down the filling cost.8

Table 2. Calculating results from artificial rib pillars.

Level interval	No. of prospecting line	β	φ	Q (tonnes/m)	$P=\dfrac{Q\sin(\beta-\varphi)}{\cos\varphi}$ (tonnes/m)	Loading area S b x h (m²)	σ_c (tonnes /m²)	$[\sigma]$ (tonnes /m²)	Comparison of the results
1640	8	60°	30°	15876	9166.7	350	261.9	500	
1640	10	51°30'	30°	17018	7123	300	237.4	500	
1640	12	54°30'	30°	14900	7054	325	217.0	500	
1580	10	51°30'	30°	35910	14915.5	350	426.1	500	
1580	12	46°30'	30°	48195	13495	350	385.6	500	
1520	10	51°30'	30°	26887	11255	300	375.2	500	
1520	14	40°	30°	60480	15581	350	445.2	500	
1520	16	50°	30°	33696	13230	300	441.0	500	

As to the failure of cemented fill mass, the low-grade is the reason that the side surface of cemented fill mass is easily fractured and broken, when it is exposed by the extraction of pillar. This feature is subsequently developed from outside surface to core of cemented fill mass. Figure 4 gives the characteristic manner of cemented fill mass after compressive. It shows along the width of cemented fill mass that the great value of stress is on the surface, but gradually decreases towards the center as shown with dotted line. Because the low uniaxial compressive stress of cemented fill mass causes its surface to be broken very rapidly, ti makes the peak stress move to the inside portion as shown with solid line. The non-elastic zone is outside the peak value of stress, but elastic area is inside. From the surface to the central portion of the cemented fill mass, the zones are fractured, plastic, yielded and elastic. During the process of failure in cemented fill mass, the fractured zone on the surface of cemented fill mass would gradually expand to the center, which gives a final result at central portion of the whole width also in a state of plasticity. At this time, the cemented fill mass finally becomes a "yielding abutment".4

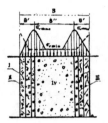

Fig. 4. Stress distribution and characteristic manner of cemented fill mass after compression.
I = fractured zone
II= plastic zone
III= yield zone
IV = elastic zone

3 STRENGTH OF HYDRAULIC TAILINGS OR SAND FILL MASS

As to the strength of hydraulic tailings or sand fill mass, the aggregate of solid particles is presented in a loose state, within which a plenty of pores filled with water and air. But, under the action of gravity of fill mass and overlying load, the solid particles have to dewatering and make new arrangement, which causes the fill mass to subside and becomes more compact, this result will give fill mass a certain strength actually the cohesion among solid particles is much smaller than the strength of particle itself, the failure due to applied load is only occurred along a certain plane inside the fill mass by shear displacement, that has to conquer the strength of fractional force and very small amount of cohesion among the solid particles. Besides, the tailings fill mass has both elastic and plastic deformations. The elas- deformation is due to the compression of air or liquid inside the closure pores and the molecular water on the surface of solid particles. These two can recover their original shapes after abolishing the load, which shows the elastic property of tailings or sand fill mass. Considering the plastic deformation, the solid particles are forced to compress together and diminish the porosity to make the fill mass becoming more compact. This kind of deformation, after removal of applied load, can not be recovered. So the feature of plastic deformation is the decrease of porosity. But the elastic and plastic deformations of tailings fill mass are quite different from the solid body, its elasticity and plasticity may be occurred at the same time, even sometimes occurred alternately. Owing to the displacement of water to create the elastic deformation is very small, if compare with the change of porosity, it can be eliminated. Therefore, the tailings or sand fill mass is generally considered to be in a plastic state, having no deformation before yield state. If once the yield limit has reached, the plastic deformation is immediately occurred, and finally becomes a kind of plastic flow. Therefore, the model of this material is considered as a rigid-plastic mass, its feature is illustrated in fig.5.

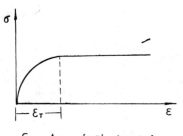

ε_T —Area of Plastic Deformation

Fig. 5. Curve of tailing or sand fill mass.

The load on tailings or sand fill mass is due to itself gravity and pressure from wall rock. Because the shear stress can not be occurred along surface of tailings or sand fill mass (filled surface), and the pressure from overlying strata is larger than gravity of fill mass. Therefore, the plane of minor principal stress is along the surface of fill mass, the major principal stress plane is perpendicular to it. Hence, the action of force on tailings or sand fill mass is in a passive state. Fig.6 illustrates an ideal loose mass (cohesion is almost 0.2-0.3 kg/cm^2 or less) at limit equilibrium condition, the relation between major principal stress and minor principal stress is as follows:

$$\sigma_1 = \sigma_3 \tan^2(45+\varphi/2) \qquad (6)$$

$$\sigma_3 = \sigma_1 \tan^2(45-\varphi/2) \qquad (7)$$

Where: σ_1 = major principal stress, kg/cm^2

σ_3 = minor principal stress kg/cm^2

φ = internal angle of friction, degree

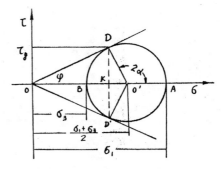

Fig. 6. Mohr's circle of theoretical loose material.

The above two formulae are important for the tailings or sand fill mass. When the wall rock is competent and the gravity of fill mass is greater than rock pressure. Then the tailings fill mass is in an active stress state, the action of fill mass to wall rock (support force) can be calculated by formula (7). If the wall rock is incompetent, even in a broken state, the rock pressure from wall rock is larger then the gravity of fill mass. Then the fill mass is in a passive state, it requires the fill mass with enough strength for maintenance of its stability. This strength can be calculated by formula (6).

During the extraction, after the room has been filled with tailings or sand, if the fill mass has not enough strength and also not separated by concrete wall with neighbouring pillar. The tailings or sand fill mass will be collapsed as the pillar is extracted. At this time, the cement agent must be added into the tailings or sand aggregate for raising the strength of fill mass.

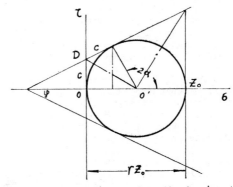

Fig. 7. Stress circle when the horizontal principal stress equals to zero.

As shown in Figure 7, when the stress circle contacts with the vertical axis and rupture line, and horizontal stress is also equal to zero, the weight of tailings or sand fill mass at the depth H m. is:

$$rH = 2c\tan(45+\varphi/2) \qquad (10)$$

Where: r = unit volume weight of tailings or sand fill mass, tonnes/m^3

c = cohesion, tonnes/m^2

From the above formula, the required strength of cemented fill mass, when the extracting pillar has a height H m., can be calculated as follows:

$$c = \tfrac{1}{2}H\tan(45-\varphi/2) \qquad (11)$$

4. CONCLUSION

It has been found out that the support ability of fill masses has the same function for different types of fill materials, although there is a certain difference among varieties of fill masses. At the recent ten years and more, the research on support ability of fill mass whether in China or in other countries has done a considerable amount of works, and the achievement is also obvious. This result has played an important role in the reduction of fill cost and development of mining method with backfill. But most works are on the cemented fill mass, it is now necessary to pay more attention to the hydraulic fill mass in the future.

The materials in this paper are collected from the practical experience in Chinese metal mines, among them some are belonged to the viewpoint of authers. Any comments on it are welcome.

REFERENCE

1. Zhangsha Research Institute of Mining, et al., "Mining Method with Backfill", Metallurgical Publication Company, Beijing, China, 1978.

2. Liu De-Mao, "The Adoption and Development of Cemented Fill Stoping Used in the Steeply Dipning Ore Bodies of Chinese Non-Ferrous Metal Mines", Paper for Symposium of Fill Mining Stoping, Chinese Society of Metals, China, 1982.

3. Yang Guo-Zhang and Kang Yi-Xing, "The Preliminery Discussion of Ability about Fill Mass in Xi Kuang Shan", The Collected papers in Fill Mining Stoping, Hunan Society of Metals, 1981.

4. Yang Zhong-Gong and Chen Zhong-Jing, "Rock Pressure and Ground Control for Underground Metal Mines", Metallurgical Design and Exploration, No.3-4, 1979, Kunming Design Institute of Metallurgy.

5. Wu Tong-Shun, Zhu Yu-Xin, et al., "The Mechanics of Cemented Fill Mass in the Underground Fill Mining Stoping with Large Drift Extraction in Long Shou Mine", Quarterly of Zhangsha Research Institute of Mining, No. 1, 1982.

6. Yu Run-Chang, "Some Problems about Technological Design on Cemented Fill Mass", Paper for Symposium on Fill Mining Stoping, Chinese Society of Metals, China, 1982.

7. Sang Yu-Fa, Bao Dong-Xu and Guan Shen-Lan, "The Ground Pressure about Cemented Fill Mining Stoping of a Slightly Dipping Ore Body", Quarterly of Zhangsha Research Institute of Mining, No.1, 1981.

8. Beijing Research Institute of Mining and Metallurgy, "The Discussion on Some Problems about Overhand Cut- and Fill Mining Stoping", Paper for Symposium on Fill Mining Stoping, Chinese Society of Metals, China, 1982.

9. Ma En-Rong, "The Physical Model of tailings fill mass", Quarterly Shan-Dong Engineering College of Metallurgy, China, No.1, 1982.

10. Ma En-Rong, "The Strength of Tailings Fill Mass", Quarterly of Shan-Dong Engineering College of Metallurgy, China, No.1, 1981.

A consideration on the effect of backfill for the ground stability

U.YAMAGUCHI
University of Tokyo, Japan

J.YAMATOMI
Akita University, Japan

ABSTRACT: To consider the supporting mechanism of backfill, a short discussion on the deformation of underground opening and the state of backfill stowed in the opening. Finite element calculation was performed to analyze stresses and deformations in the rock sourrounding a spherical opening made in rock mass and fill material. The results of calculation were analyzed and discussed. Finally, as a conclusion, it is proposed that the effect of backfill is to limit the deformation of rock, to restrict the progression of failure to the farther extent, and consequently, to generate a slow and moderate "dilatant" failure in the rock mass around the opening.

1 INTRODUCTION

Backfilling for underground mine openings has been one of important techniques and now employed in many mines as a routine work. However, there has been no established theory clearly explaining the supporting mechanism of backfill.

It is generally considered that backfill supports the rock stress and prevents the caving of rock into the opening, and as the results, stabilizes the sourrounding rock mass and minimizes the surface subsidence. However, this simple explanation is not completely understandable, because of the unfavourable mechanical properties of fill materials in comparing with the rock sourrounding the opening.

The packing density of fill is also much smaller than the rock previously existing at the place before excavation. Neverthless, the backfilling, even a partial filling method, is widely adopted for a ground stabilization.

2 DEFORMATION OF UNDERGROUND OPENING AND THE STATE OF BACKFILL

To study the effect of backfill for the ground stability, it is necessarily required to analize the deformation characteristics of opening made in rock mass which have been filled and not filled with fill material. At one time, in these analysis, the follow-ing simple model was generally employed, which was:

An opening was made in rock mass without any loads and then filled or not filled with fill material and finally, the rock mass was stressed by external forces. When the rock and fill material are elastic and the total deformation is small, the above model is enough for the analysis. But, if not so, it will be inadequate.

In actual fact, openings are made in loaded rock mass. Therefore, the opening should be deformed and stressed by the load at the moment of excavation. And then, fill material is stowed into the opening under without any loads. As far as the rock is elastic and the state of stress sourrounding the opening is not changed, the above situation will be unchanged. Therefore, the filled material stowed in the opening will be not deformed and not stressed, at least at the moment.

When the rock is non-elastic and stressed, the opening with and/or without fill will be gradually deformed and fill material stowed in the opening will be also compacted and stressed with the lapse of time. And also, if the state of stress around the opening changes by a mining operation done in the vicinity of the opening and the other reasons, the opening will be additionally stressed and deformed from the very nature of things.

These deformation models are shown in Fig.1.

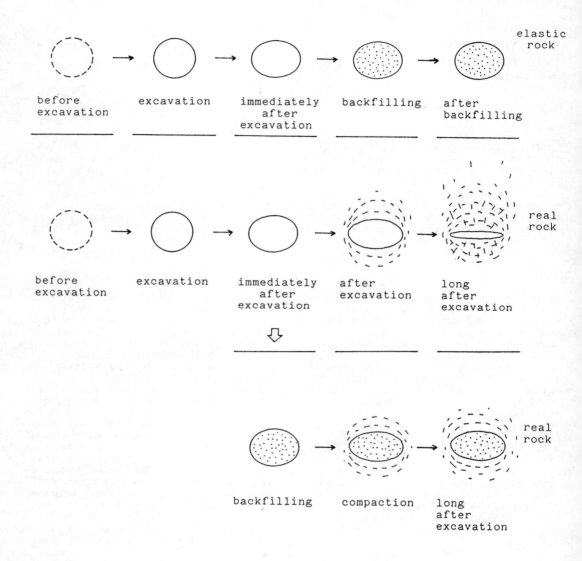

Fig. 1. Deformation model of underground opening for elastic and real rocks.

A spherical cavity was excavated in rock
mass of the model. A stress state shown in
Fig.2 was assumed and stress and deformation
analyses were carried out by the aid of
finite element method. The model was axi-
symmetrical and the section area calculated
was 80 x 80 meters. P in the model were the
external forces acting on the top surface of
the calculation volume of which the numerical values were given and explained in later
sentenses. Body force, namely gravitational
force, acting on each element was considered.
   Calculations were performed on the following three cases:
   Case 1   An opening was excavated under a
   stress. Then, the cavity was not filled
   and kept to be open,
   Case 2   After excavation, the cavity was
   filled with fill material,
and
   Case 3   The opening was excavated and
   filled with fill material without any
   loads. After the excavation and filling,
   external loads were added.
   Two kinds of rock, basalt for a hard rock
and clayey ore for a soft rock, were selected as the sourrounding rock of the opening.
As the fill material, 3 % cement mortar was
employed. This 3 % cement mortar is consisted of mill tailing sand, volcanic ash and
3 % of portland cement and is commonly used
in our "Kuroko ore" mines. Mechanical
properties of these rocks and fill material
are tabulated in Table 1 and shown in Fig.3.
   As explained in the section 2, in an actual excavation, an opening excavated in non-
elastic material will deform with the lapse
of time and the state of stress around the
opening also will change. But in this analysis, it was difficult to realize the time
dependent deformation and stress changes.
Therefore, instead of the time dependency,

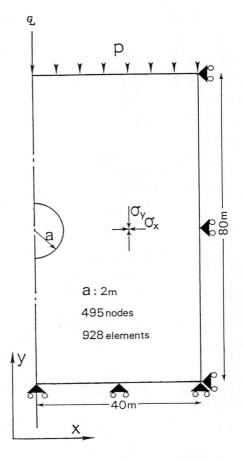

Fig. 2.   Assumed stress state for the
stress and deformation analyses.

Table 1.   Mechanical properties of rocks and fill materials.

Item	Unit	Basalt	Cleyey ore	Fill material
Specific gravity		2.22	3.05	1.91
Young's modulus	MPa	1.10×10^4	6.10×10^3	1.72×10
Poisson's ratio		0.154	0.112	0.313
Compressive strength	MPa	39.3	3.95	0.17
Tensile strength	MPa	4.41	0.35	0.0077

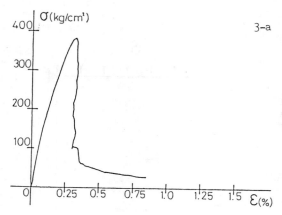

3-a

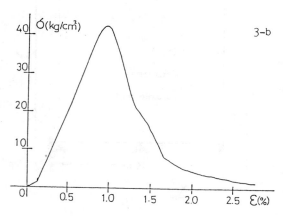

3-b

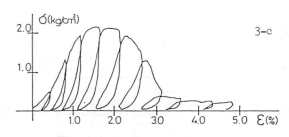

3-c

Fig. 3.   Stress-strain curves of uniaxial compression test.
— a   Basalt
— b   Clayey ore
— c   Fill material

a step-wise stress increase procedure was applied for the calculation. Which was the method that the external forces around the calculation area were step-wisely increased along an assuming program. In this calculations, as the P which are the vertical forces applied on the upper surface of the calculation area, 3 stages for basalt and 5 for clayey ore were selected. These forces produce vertical stress of 39.3 MPa at the centre of the spherical cavity in basalt which is correspond to the compressive strength of the basalt, and also 110 % and 120 % of the strength, and 3.95 MPa in clayey ore which is the same to the compressive strength of the ore and 125 %, 150 %, 175 % and 200 % of the strength.

Results of the calculation are shown in the following figures. Fig.4 is the finite element diagram showing the state of deformation of rock sourrounding the spherical cavity and fill material. The degree of plastic deformation of each element is indicated by hatched lines and the density. Not yielded but elastically deformed elements are not illustrated in these figures as blank parts. In Fig.5, linear displacements around the opening are plotted. In diagrams, CV_1 and CV_2 are showing the horizontal and vertical convergences on the diameter of the sphare which are corresponding to $-2u_1$ and $u_3 - u_4$. EX shows the linear elongation of

Fig. 4.   Finite element diagrams.

— I Stress distribution around the spherical cavity in clayey ore. Excavation was performed under stress generated by P of the Fig.2 (by P, vertical stress at the center of sphere became to be the same as the uniaxial compression strength of clayey ore, 3.95 MPa). After the excavation, backfilling was performed and then the external force P was increased to 2P.

— II Deformation diagram of Fig.— I.

— III Deformation diagram around the spherical cavity in basalt. Direct after the excavation under stress generated by P.

— IV In basalt. Excavation under P. And then P was increased to 1.2 P.

— V In basalt. Excavation under P. Backfilling. And then P was increased to 1.2 P.

— VI In basalt. Excavation without P. Backfilling. And then 1.2 P was loaded.

446

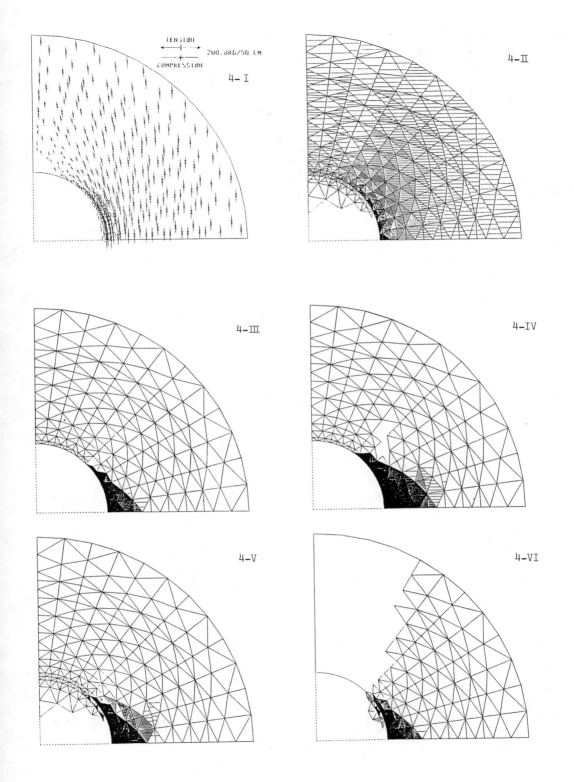

4-I

4-II

4-III

4-IV

4-V

4-VI

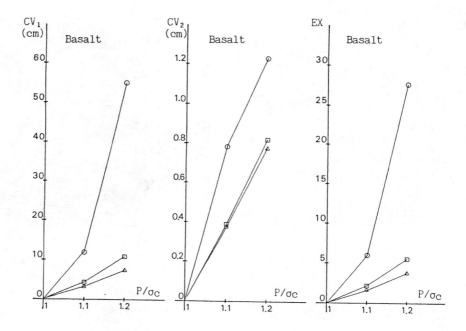

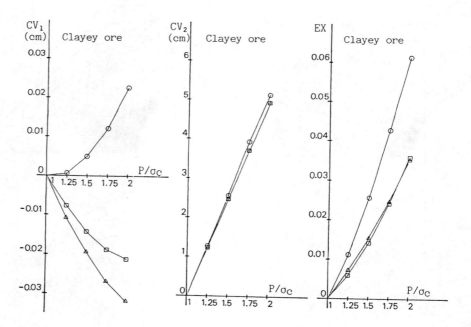

Fig. 5. Linear displacements around the opening. (Basalt and Clayey ore)
CV_1 : Horizontal convergence on the diameter of the spherical cavity ($-2u_1$).
CV_2 : Vertical convergence ($u_3 - u_4$).
EX : Linear elongation of rock along the four meters from the surface ($u_2 - u_1$).
Abscissa shows the external force P of the Fig.2.

rock along the four meters from the surface, that is $u_2 - u_1$. u_1, u_2, u_3 and u_4 are the displacements of each points shown in Fig.6. and the circle is corresponding to the spherical opening of Fig.2. Fig.7 is also showing the stresses, σ_x, σ_y and σ_θ, acting along the horizontal axis from the surface of the cavity to 40 meters deep in basalt.

From these figures, followings are conclusively derived.

1) Significant difference in the degree of deformation is found between cavities with fill and without fill.

2) In the case of with-fill in the spherical cavity, the deformation of cavity is considerably restricted in comparison with the case of without-fill. In any cases, the restriction is more significant in basalt compared with in clayey ore.

3) Two cases, a case in which excavation and backfilling are performed under no load and then the external forces are step-wisely added and the other case in which excavation is performed under a load and fill material is stowed in the spherical cavity and then the rock mass is loaded, are compared in the stress and deformation. By the comparison, it is found

that there is not so large difference between these two cases within the stress range of the calculation. However, if the additional load increases much more, the difference will become larger. Also in this comparison, more significant difference is observed in basalt in comparison with clayey ore.

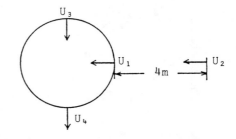

Fig. 6. Explanation of the displacements of each point sourrounding the spherical cavity.

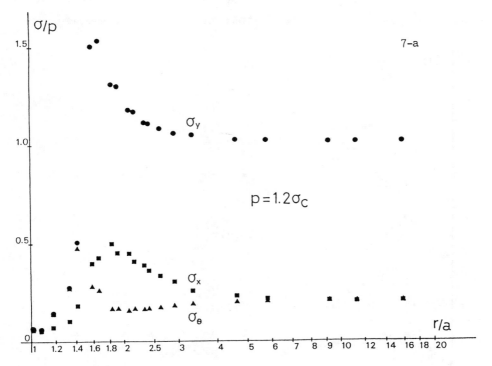

Fig. 7. Stresses, σx, σy, and σ₆, acting along the horizontal axis from the surface of the cavity to 40 meters deep in basalt.
7-a Excavation under P. Backfilling. And then P was increased to 1.2 P.

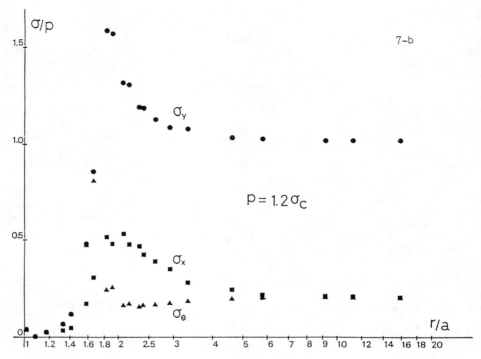

7-b   Excavation under P.   No backfilling.   And then P was increased to 1.2 P.

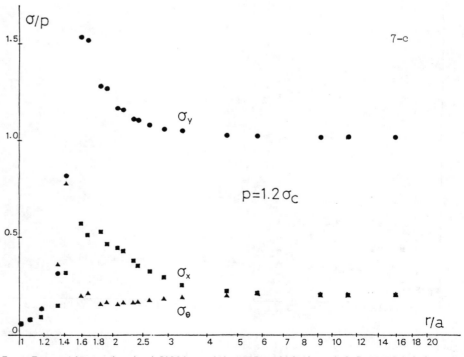

7-c   Excavation under backfilling without P.   And then 1.2 P was loaded.

4 DISCUSSION

Followings may be given as a conclusion of the consideration and calculation done in the sections 2 and 3.

1) Supporting effect of backfill has gradually come into existence after the sourrounding rock deformed sufficiently and squeeze into the opening. At least, at the moment when the fill material is filled into the opening, no stress should be naturally found in the fill material except a light stress due to the gravity.

2) Even if the filled material is compressed, it is rather act to restrict the movement of rock and limit the deformation of sourrounding rock mass than to support the rock stress directly.

3) Limiting and restricting the movement and deformation of rock would yield a stress relaxation of rock in the farther extent and lead the rock to be a softer failure.

Particularly, in connection with the section 3, Jack Parker, an experienced mining engineer of the United States, mentioned an interested version in his useful lecture as follows[2]:

If underground caving progressed in small increments, the surface subsided a moderate amount, with few, if any, cracks in the soil. But if the cave "hung up" for several rows, then there would be another violent plug-like fall, with wide, deep cracks at the surface.

Most of mining engineers have experienced these facts. Violent failure of rock like a rock burst produces a distinct surface subsidence and conversely, a gradual and controlled failure like a roof caving of long wall coal getting makes a moderate subsidence on the surface. As is commonly known, a sudden weakening of pillar by locally existing faults, a deterioration of rock by water penetration, and so on are connected to an unexpected disaster.

In these instances, one of the effects of backfill for ground stability is suggested. It is considered that deformation of rock sourrounding the opening is limited by backfill and progression of failure to the farther extent is restricted, and consequently, a slow and moderate "dilatant" failure generates in the sourrounding rock mass as against Mr. Parker's quick and violent "plug-like" failure. During the slow and moderate "dilatant" failure, rock will increase the volume by cracking or plastic deformation and "Trompeter" zone will be built in the sourrounding rock mass around the opening. Stresses around the opening shall be redistributed and a stress release zone will be built in the newly developed arched rock mass.

In this paper, a consideration on the effect of backfill for ground stability was discussed. The discussion is brief and insufficient. But, if any suggestion to explain the supporting mechanism of backfill and to stimulate more discussions on this problem is found in this discussion, it will be an unexpected success. It is very much appreciated to be given some comments on this problem.

REFERENCES

1) Yamatomi, J., Shimotani, T. and Yamaguchi, U., 1981, A study on the underground opening of Kuroko mines, Proc. of the Internat. Symp. on Weak rock/ Tokyo/21-24 Sept. 1981, 831-836.
2) Parker, J., 1973, Practical rock mechanics for mines, Part II, EMJ, July 1973, 70-73.

4. New development

Backfilling with ice

HENNING FANGEL
Fangel & Co. A/S, Eiksmarka, Norway

ABSTRACT: Just like glaciers in their natural enviroment fill voids and cavities in the landscape, an artificial glacier placed above a mine void move into and fill the void completely. The lateral walls of the void are then supported by the hydrostatic pressures exerted by the glacial ice and thus prevented from caving. The ice in an empty mine room provides the necessary safe support that enable the mine personnel to concentrate on the basic objective of all mining: Economical and safe production of ores.

Ores mined according to the method of backfilling with ice are extracted by the well known and elaborated sub-level-caving systems and equipment, however the drifts are spaced ca.25 m both vertically and laterally.

The required volumes of ice are those needed to fill the mine voids plus those needed to replace the ice losses caused by melting. The aggregated demand is about twice the fix-volume of the extracted ores - less close to the day surface and slowly increasing when the mine is worked deeper. Any required volume of ice may be manufactured through systematic plagiarism of natures own processes. Costs of ice are $1 - 3$ NOK/m^3. The costs of the ores are less than 60 NOK/m^3, i.e. ca.20 NOK/tonn - depending upon the specific gravity of the ores. These costs include the manufacturing of all the ice for both backfilling and replacement of melting losses.

Backfilling with ice restricts mining to the orebody itself. All ores are recovered. Encroachments upon nature are drastically reduced and meet with the demands of enviromentalists. The value of the orebody increases by 30% compared to the use of conventional underground mining methods. All mining operations are carried out in drifts, never in stopes, ensuring drifting safety conditions.

INTRODUCTION

To prevent roof or lateral walls of mines from collapsing after the ores have been extracted, they must be supported. This is achieved by leaving pillars of the ore itself, by erecting concrete columns or make settings of steel or timber. Alternatively, the empty rooms can be backfilled with rocks, sands, tailings etc. These are necessary but slow and costly processes that demand skill and craftmanship. They are unavoidable safety meassures, but interrupt contineous ore production.

The sub-level-caving method has been developed and perfected, to simplify and industrialize the extraction process.

In the sub-level-caving method the lateral walls contineously cave into the outmined rooms in a carefully planned and controlled manner, consistent to the ore extraction process. Ore mining and extraction become more economical than by other traditional mining methods.

Sub-level-caving may be studied at the Kiruna and Malmberget mines of LKAB.

Opening drifts must be driven at short intervals both vertically and laterally, to controle the caving and the blending of wasterock into the ores. LKAB found that spacing of 11 m give an economic optimum.

Ores delivered from drifting amount to ca.1/6 of the total ore production. In spite of the elaborate cave control, only ca.80% of the ores are recovered. The rest remain trapped in the caved rocks in the

JOSTEDALSBREEN, Norway, August 1982.
The photo "freezes" the ice-flow from the mountain plateau, ca.1850 m a.S.L. It may be observed how the top of the glacier cracks and opens in the upper area where acceleration takes place, and again in convex flow areas in the middle of the picture. Where movement decelerates or passes through a concave area, the glacier combine to a compact mass of ice.

1983-02-01

finished rooms. In addition, extracted ores are diluted by 20 - 30% wasterock.

ROCK SUPPORT BY GLACIAL ICE

Nature is using and has been using glacial ice to support rock walls and mountain faces during millions of years. The action may be studied in U-walleys. The rocks that were supported by the glacial ice, after its retreat have slid, avalanched and piled in screes at the walley floor.

Glacial ice conform to the shape and size of the void containing it, fill it out completely and exert hydrostatic pressure against its walls

SYNOPSIS

The backfilling method described in this paper is a practical combination of wall-rock support by glacial ice and ore extraction by sub-level-mining systems.

The method is feasible when sufficient volumes of ice to fill the emptied mine-rooms are supplied at an acceptable price.

PROPERTIES OF ICE

Glaciology is the knowledge og glaciers' behaviour and of the physical laws that explain such behaviour.

Chunks and lumps of ice brought together at freezing temperatures, combine to a compact mass without joints. This may be observed in an ice-cube-bucket or on any glacier. The upper convex part of glaciers are intersected by cracks and flaws that close and disappear completely as soon as the glacier move into a concave area.

Ice under pressures lower than ca.2 bars has similarities to solids: it is brittle and will break when subjected to shocks.

At higher pressures ice is plastic and have viscous properties. Under strain, ice will deform plastically and adapt to the load situation.

Ice changing from solid to viscous state when pressure increases, can be seen in glacier cracks. Cracks open to a depth of ca.20 m. Deeper cracks are results of glacial movement. Explanations to transformation from solid to viscous state may be found in glaciological literature.

Where the annual mean temperature is above 0°C. glaciers invariably are at the melting point of ice i.e. 0°C. Melting or freezing automatically regulates this.

MOVEMENTS OF GLACIAL ICE

Elements of glacier movements are:

1. Viscous flow
2. Sliding over the base rocks
3. Preesure melting and creep
4. Ice recrystalisation

1. Viscous flow of glaciers have been studied and recorded in all parts of the world. In order to express the flow in mathematical form, dr.L.W.Glen conducted laboratory tests to supplement field observations, and proposed the following:

$$\dot{\sigma} = B \cdot e^{-\frac{Q}{RT}} \cdot p^m$$

where:
B = constant = $7 \ 10^{24}$
e = base ln = 2.71
Q = gasconstant = 32 Kcal/gmol
T = temperature in deg.K
p = pressure in bar
R = constant = $8.314 \ 10^7$ Erg/gmol
n = Glen's exponent = 3.2
$\dot{\sigma}$ = ice strain

At the temperature of "temperate" glaciers i.e. 0°C. the formula is simplified to:

$$\dot{\sigma} = \text{Constant} \cdot p^{3.2}$$

The formula has been evaluated by other glaciologists. Some of them have found slightly differing numerical value for the Glen-exponent. All of them confirm the character of the formulae.

The interesting and usefull part of dr.Glen's formulae is that when the pressure doubles, the strain will increase by a factor of roughly ten. Necessary or wanted increases in the flow, or rather: of the potential flow, of the backfill ice, can easily be met by increasing the ice height.

Glaciologists believe an ice height of 40 m in a mine void will produce a potential ice movement of ca. 40 m/yr. I.e. needed increase of ice movement to 50 m/yr, correspond to an icrease in ice height to 43 m.

Glacial flow is directly related to the forces acting in the direction of the flow.

2.Glacial sliding over the base rocks is governed by:

A) the forces acting in the direction of the sliding, and
B) friction resulting from the weight from the ice against the rocks.

As ice is partly solid, partly viscous

and transformation between the two phases is depending upon weight, pressure and strain, friction is not easily calculated.

This situation gets even more complicated because water acts as a lubricant. The fric-friction between ice and rock generates heat that melts heat to water. In this way friction produces its own antidote.

The lubricating effect of water present between a glacier and the base rock is clearly demonstrated in the movement of natural glaciers. The movement seem to follow some sort of 80/20-rule: ca.80% of the annual movemnts takes place during the ca.20% of the year when glaciated areas have summer.

Moisture and meltwater will allways be present in mines using the ice backfill method. The ice-fill will "summer-move" yearround, unrelated to the temperature variations at the day-surface. The actual ice movements can of course never be higher than the rate of ore-extraction and making of emptied voids.

Ice movement in itself also generates movement. The energy level of a descending glacier is reduced. The difference is hardly transformed into velocity. Instead heat and meltwater for lubrication is generated.

This phenomenon is the probable explanation of surging of glaciers.

Friction is further influenced by rocks and sand present in the ice/rock interface. Rock and sand may form local pads, so that friction will be rock against rock, obviously having a different coefficient of frict-tion from that of ice against rock.

3. Pressure melting and ice-creep are other peculiarities of ice. This phenomenon causes ice to pass obstacles and hinderances.

The melting point of ice gets lower under pressure. Ice therefore tend to melt at the windward side of obstacles. The meltwater is forced to the leeward side by the pressure, and refreezes. The melting heat is recuperated and go through the obstacle to the windward side and melt more ice. In this way ice pass obstacles without apparent deformation and without consumption of energy.

Because of this, the roughness and varying contours of the mine voids will not hinder the movements of the ice fill, nor cause the ice to bridge across the mine opening or become suspended at the lateral walls.

4. Ice recrystalisation and forming of long, needlelike crystals that have their axis roughly paralell to the direction of the resulting forces, are going on slowly but contineously. Crystal sliding is much easier along the crystal axis than trans-versal to them. Ice crystals recrystalize to an orientation that give less resistance against movement and lesser use of energy in performing movement and crystal sliding.

Ice recrystalisation as phenomenon may be observed in lake-ice during spring thawin

Calculation of movements of glacial ice and predicting potential rate of movement for a glacier is very complex. The complexity is increased inside a mine void where a third dimension is added. The direction of movement is basically in the direction of the force of gravity and friction is given by hydrostatic pressures. These directions of forces are inverted relative to natural glaciers that are accessible for studies.

MINE BACKFILLING WITH ICE

Glaciology is important because it tells that glacial ice has

POTENTIAL TO MOVE

into the mine cavity at a speed exceeding the mine's oreextraction rate, i.e. the rate of advance of the stoping front.

Efficient mining rates today are up to 30 m/yr sinking of the average extraction level. Modern techniques and organisations may double this figure. The mine-backfill-glacier must thus have the potential to move more than 60 m/yr. The potential movement assures that the mine void is filled with ice, the lateral walls kept from caving and waste rock prevented from mixing into the muck.

Ice is an ideal material for support of the lateral walls. Not only does ice prevent rocks from subsiding but just like glaciers do not mix into the rock screes, moraines or eskers they push, ice backfill stay on top of the pile of blasted ore.

If ice start to penetrate into the openings of a pile of ore, pressure would decrease, strain drop and the ice change from the viscous to the solid state and the penetration stop. The ice will at the most act as a mortar in a crust at the very top of the ore pile.

Drilling, blasting and loading add heat to the ore. This heat is additional to the inherent temperature of the ores. The heat is dissipated in the ore pile.

Blasted ores, by consequence are located below the ice where pressures are highest and the ice melting point lowest. Each of these factors produce preferred ice melting at the ore pile and make ice penetration into the pile impossible and against order of nature.

A terminated open pit or a mine's day opening is a good place to manufacture an ice-backfill glacier that poessessing the required power of locomotion to keep the mine voids filled and the workings stable and safe. The open pit provides a reservoir of ice for the warm seasons and for fluctuating demands.

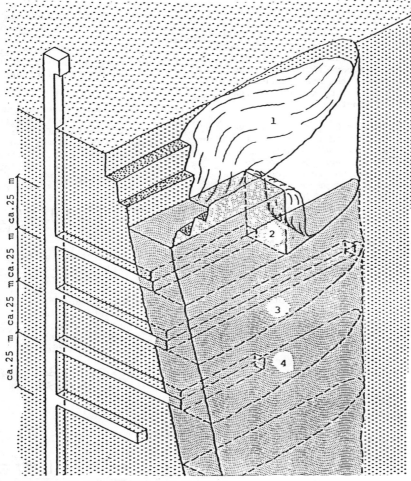

SYSTEMS OF MINING WITH ICE BACKFILL

The drawing shows a vertical section across the orebody and along the hanging wall. Figures indicate:

1. The finished open pit, partly filled with manufactured glacial ice.
2. Ore production level ca.25 m below the floor of the open pit. The ice is shown moving into and filling the emptied stope.
3. Perforation level where systematic stope blasthole drilling is carried out without interference from other jobs.
4. Preparation level where drifting is carried out without any interference on or from ore production.

PRODUCTION AND EXTRACTION OF ORES

Backfilling the mine voids with ice permit oreextraction to be made in the safe, well developed and mechanized way that is stand- ard in sub-level-caving mines. Efficient operations of that type deliver ores at the surface at a cost of ca.15 NOK/tonn. Backfilling with ice permit the abilities and economy of modern long-hole jumbos, drilling tools, mechanical loaders etc. to be fully exploited by spacing the production levels at 20 - 30 meters, perhaps even more if sidewalls are regular and smooth. Hori- sontal distance between drifts may be up to 50 m. In this manner the ores produced by expensive drifting may be reduced to few percent of the production.

More important and of bigger economic value is that ice-filling permit recovery of all ores in an orebody without dilution and admixture from waste rock. Compared to sub-level-caving and to conventional room and pillar methods, the value of an orebody is thereby increased by ca.20%. Compared to sub-level-caving the value of each and every tonn of ore from the mine is increased by another ca.20%.
    Both these advantages are obtained at the cost of ca.15 NOK/tonn of ore.

This cost is still ca.5 kr/t higher than open pit mining costs. The top part of the orebody therefore ought to be mined by open pit or day stope mining.
    Opening of the deeper parts of the ore and preparation for underground mining may be advanced during this phase. This approach enables the mining company to start produc- tion within the shortest time and lowest investments possible.

ICE NEEDS AND CONSUMPTION

Backfilling with ice naturally hinges on the possibilities of supplying the required volumes of ice at an acceptable price.

The required volumes are for:

1. Filling the voids left after extracting the ores.
2. Replacing the losses by melting from heat contained in and delivered from the lateral walls.
3. Replacing ice melted on the day surface during the warm seasons.
4. Replacing the ice melted by energy lost from the mining equipment and from the mine operations.
5. Replacing ice melted by heat lost from ventilating air.

The emptied voids are directly related to the rate of production and to the specific gravity of the ores. It can easily be calcu- lated at each mine. This element of ice- consumption is ca.0.3 m^3/tonn of ore.

2. In calculating the heat delivered from the lateral rocks the thermal gradient of the earth may be omitted. The average is given as 362 Kcal/m^2yr. and will melt 5 mm ice on the surfaces in contact with the walls.
    The heat delivered by the geogradient is even governed by the heat conductivity of the wall rocks and thus included in the calculations below.

The ice/lateral wall interface has the temperature $0°C$. The heat flow from the rocks towards this interface is the result of the temperature gradient and of the heat conductivity of the rock. The temperature gradient, or the temperature flow through rocks may be expressed:

$$T = T_i + (T_r - T_i) \cdot erf \frac{x}{\sqrt{4\alpha t}}$$

where. T = Temperature in $°C$.
T_i = Icetemperature
T_r = inherent rocktemperature
x = distance in m from ice/rock face
K = heat conductivity factor
s = specific gravity
c_p = specific heat
t = time in hours
erf = Errorfunction

$$T = T_r \cdot erf \frac{x}{\sqrt{4\alpha t}}, \text{ as } f(x,t)$$

when $T_i = 0°C$.

Heatflow towards the ice is:

$$\dot{q} = K \frac{dT}{dx} = K \frac{d\left(T_r erf \frac{x}{\sqrt{4\alpha t}}\right)}{dx}$$

$$\left\{ \frac{d \, erf \, p}{dp} = \frac{2}{\sqrt{\pi}} exp\left(-p^2\right) \right\}$$

$$\dot{q} = K T_r \cdot \frac{2}{\sqrt{\pi}} \frac{1}{\sqrt{4\alpha t}} exp\left(-\frac{x^2}{4\alpha t}\right)$$

at interface $x = 0$, thus

$$\boxed{\dot{q} = \frac{K}{\sqrt{\pi \alpha t}} \cdot T_r \qquad \text{Heatflow}}$$

$$\int_0^t dq = \text{Const.} \cdot \int_0^t t^{-\frac{1}{2}} dt$$

$$q = \frac{2K}{\sqrt{\pi \alpha}} \cdot T_r \sqrt{t} \qquad \text{Accumul.} \atop \text{heat supply}$$

The corresponding ice melting:

$$a = \frac{q}{g_i \cdot \Delta H_i} \quad \text{in } m/hr$$

$$A = \frac{q}{g_i \cdot \Delta H_i} = \frac{2K \cdot T_r \cdot \sqrt{t}}{\sqrt{\pi \alpha} \cdot g_i \cdot \Delta H_i}$$

Example for granite:

K = 3 Kcal/m hr $^{\circ}$C.

$$\alpha = \frac{K}{g \cdot C_r} = \frac{3}{2700 \cdot 0.19} = 0.0058 \ m^2/hr$$

ΔH_i = 80 Kcal/kg = 72000 Kcal/m^3

$T_r = 3^{\circ}$C.

$$q = \frac{2 \cdot 3 \cdot 3 \sqrt{24 \cdot 365}}{\sqrt{\pi \cdot 0.0058}} = 12500 \ kcal/m^2$$

a = 0.17 m/ 1st.year.

What time does it take to melt 3 m of ice?

$$3 = \frac{2 \cdot 3 \cdot 3 \sqrt{t_y \cdot 24 \cdot 365}}{\sqrt{\pi \cdot 0.0058} \cdot 72000}$$

t_y = 300 years.

What is the melting rate at depth of 600 m?

$$T = 3 + \frac{1.8 \cdot 600}{100} = 13.8 \ ^{\circ}C.$$

$$q = \frac{2 \cdot 3 \cdot 13.8 \sqrt{1 \cdot 24 \cdot 365}}{\sqrt{\pi \cdot 0.0058}} = 57405 \ kcal/m^2$$

a = 0.80 m/1st.yr.

To melt 3 m of ice at 600 m level, takes

t_y = 14.2 years.

By help of these formulaes ice melting can be computed for every level of the mine as a function of inherent rock temperature and of time.

The following general physical constants and values,(or other pertaining to the specific mine in question), may be used in the calculations:

Melting energy of ice ΔH_i =80 kcal/kg
    =72000 kcal/m^3
Sp.gr. of ice g_i =900 kg/m^3
Sp.gr. of rock g_r =2.8 - 3.0t/m^3

Sp.heat of rock C_r=0.2-0.25 Kcal/kg
    =540-700 Kcal/m^3
Heat conductivity of rock
    K = 3 - 4 Kcal/m hr$^{\circ}$C.
Temperature gradient:
Global average 3.0 $^{\circ}$C/100 m
Scandinavia 1.8 $^{\circ}$C/100 m

3.Ice melting on the day-surface will vary from year to year and depend upon a number of local factors. Melting will be less than in open country because the ice is located in the depression of the open pit and is shielded from the winds and protected by a blanket of cool air resting on top of it.

In Scandinavia glaciers melt less than two meters in course of the warm seasons. For estimating purposes is suggested to use a melting rate of one m/yr on the open air surface of the mine ice-fill-glacier.

4.Ice melting due to energy lost from the mining operations also will vary from mine to mine. It will even to a large extent depend upon the types of equipment and forms of energy in use.

Informations submitted by a number of mines give energy consumptions of:

5 - 15 kWh/tonn of ores extracted

Energy efficiencies in mining are low. To assume that all energy used in a mine is lost and contribute to ice melting is a minor error and one that provide a margin. The ice-fill method is effective and permit use of low 5 kWh/tonn = 15 - 20 kWh/m^3 ore-depending on the specific gravity:

1 kWh = 860 Kcal will melt lo.75 kg ice
5 kWh/t melt 0.06 m^3/tonn
    = 0.18 - 0.24 m^3/m^3 ore.

5.Ventilating air requirements depend to a very large extend upon if electric or diesel machinery are in use. Presence of radioactive isotopes in the mine atmosphere certainly must be considered. In the end applicable safety regulations will probably determine the ventilating air volumes.

The melting effect is given by the difference in temperature of the air entering and leaving the mine. Heat lost by the air:

Q_{air}=$\Delta T \cdot V \cdot 0.31$ per operating hour.

where:
Q_{air} is the heat lost by the air,
V is air volume in m^3/hr,
ΔT is temperature difference in $^{\circ}$C.,
0.31 is specific heat of air in Kcal/m^3

The corresponding volumes of ice are:

$$V_{ice} = \frac{Q}{80 \ 0.9 \ 1000} = \frac{Q}{72000} \ m^3/hr.$$

461

ICE-MELTING by thermal flow through wallrock.

Heatflow:
$$q = \frac{2K}{\sqrt{\pi\alpha}} T_R \cdot \sqrt{t} = \frac{2.3\sqrt{24 \cdot 365}}{\sqrt{\pi \cdot 5.8 \cdot 10^{-3}}} \cdot T_R \sqrt{t_y}$$

Melting:
$$a = \frac{4160.2}{72000} \cdot T_R \cdot \sqrt{t_y} = 0.0578 \cdot T_R \cdot \sqrt{t_y}$$

t = time in hours
t_y = time in years
T_R = initial rock temperature
K = thermal conductivity

Assumptions: Annual mean surface temperature = 3°C
Geothermal gradient = 1.8°C/100m = 0.45°C/25m level
Mining rate = 25m vertical = 1 level/yr.

ICE-MELTING "a" by level in m ice/m² wallrock.

83-02-02 HR.

	Level	T_R°C	1.yr	2.yr	3.yr	4.yr	5.yr	6.yr	7.yr	8.yr	9.yr	10.yr	11.yr
Cum.	0/25	3.45		.28	.35	.40	.45	.49	.53	.56	.60	.63	.66
per yr			.20	.08	.06	.05	.05	.04	.04	.04	.03	.03	.03
Cum.	25/50	3.90			.32	.39	.45	.50	.55	.60	.64	.68	.71
per. yr				.23	.09	.07	.06	.05	.05	.04	.04	.04	.04
Cum.	50/75	4.35				.36	.44	.50	.56	.62	.66	.71	.75
per. yr.					.25	.10	.08	.07	.06	.05	.05	.05	.04
Cum.	75/100	4.80					.39	.48	.55	.62	.68	.73	.78
per yr.						.28	.12	.09	.07	.07	.06	.06	.05
Cum	100/125	5.25						.43	.53	.61	.68	.74	.80
per yr.							.30	.13	.10	.08	.07	.06	.06
Cum	125/150	5.70							.47	.57	.66	.74	.81
per yr.								.33	.14	.11	.09	.08	.07
Cum.	150/175	6.15								.50	.62	.71	.79
per yr.									.36	.15	.11	.10	.08
Cum.	175/200	6.60									.54	.66	.76
per yr.										.38	.16	.12	.10
Cum	200/225	7.05										.58	.71
per yr											.41	.17	.13
Cum.	225/250	7.50											.61
per yr												.43	.18
Cum.	250/275	7.95											
per yr													.46
Cum.	275/300	8.40											
per yr.													
Sum. per yr			.20	.31	.41	.51	.61	.71	.81	.91	1.02	1.13	1.25
Σ m³ ice/yr			4.98	7.68	10.2	12.7	15.1	17.6	20.2	22.8	25.5	28.3	31.2

ICE CONSUMPTION IN SUM:

One cubicmeter of extracted ore requires ca. two cubicmeters of ice backfill, - less in the beginning of the mine's life and increasing slowly as the mine gets deeper.

Accurate computations are made for each location and for the actual operating conditions.

SUPPLYING ICE FOR BACKFILL

Adoption of the ice-fill-method depend upon the ability to supply the large volumes of ice that are required at an acceptable cost.

Ice manufactured by ice-making machines is expencive for this purpose. Electric energy alone amounts to ca.15 kr/m³ ice. Capital costs, wages, maintainance etc. increase the costs of ice to more than 25 kr/m³ and enhance the costs of the extracted ores by at least 50 kr/m³. Artificially manufactured ice is therefore feasible only for certain purposes. The THB method may be such a purpose.

Where the natural conditions are favourable, we may fare better and less expecive by cutting winter ice on a suitable nearby lake, haul the ice blocks to the mine and dump them into the void.

A cheap, safe, convenient and reliable way is to manufacture the required ice from water pumped out of the mine and from a nearby lake. Ice can then be made in the cold winter air over the open pit and/or on the walls of the pit.

Snow making machines in regular use at winter resorts and in ski-slopes, are built in sizes that use up to 1000 m^3/hr of water to make around 3000 m^3/hr of snow. Natural glaciers are made from snow. So can even glacial ice for mine backfill be made. The method has the advantage of being able to work at temperatures of up to $+2^0$C.

ICE MAKING BY WATER MONITORS

During the past winters tests have been made and experience gained in the use of water monitors for ice manufacture. The aim was to make ice islands for petroleum prospecting drilling in arctic waters.

Resulting from these tests twin-water-monitors have been built and delivered, each having a capacity of 3600 m^3/hr of water, i.e. 7200 m^3/hr for the unit, to manufacture ca.8000 m^3/hr of ice within a radius of ca.220 m. At the diameter of 440 m the platform will grow ca.0.05 m/hr or 1.26 m/24hrs.

The user of these monitors intends to manufacture ice down to ca.20^0C below zero.

When such water monitors are employed to make ice in cold places like Kiruna, Kirkenes or Shefferville they can make ca.15 million m^3/yr of ice. This may enable

WATER MONITOR INSTALLATION (example).

Monitors having throw radius of spray up to 220 m have been built. Available capacities range from less than 100 m^3/hr to 4000 m^3/hr. ice.

extraction of ca.20 million tons of ore.

COST OF ICE BY MONITORS

Ice manufactured by theses very big units cost less than 1 kr/m^3. Smaller monitors naturally will increase the cost. The increase is however moderate as a major cost item is cost of pumping wich is directly proportional to the volumes of water that are used. In industrial installations at mines, ice costs probably never will excced

2.- kr/m^3 ice. The corresponding cost-addition is 2 - 4 NOK/tonn of ore.

Ores produced by the method described above plus the volumes of ice for auto-backfilling the emptied mine rooms will cost less than 60.- NOK/m^3, i.e. 20.- NOK/tonn of ore - varying with the specific gravity.

Open pit mining costs ca.10 NOK/tonn of ore and wasterock. Backfilling with ice becomes an economic attractive alternative when the ore/waste ratio is lower than 1/1.

When practically all ores in an orebody
can be extracted, the recoverable ore volume is enhanced by ca. 20%. Reduced mining
costs account for a similar increase in the
net values. The gained revenues are thus
ca. one third higher through the use of the
ice-backfill-method, compared to other mining methods.

If these gaines are spent in mining low
grade ores, faster return on investments,
prospecting, social benefit or wages that
is outside the scope of this paper.

The ice-backfill-method is an optional tool
to be used in exploiting:

1. Steeply dipping orebodies
2. Where winters are regular, but not
   necessarily very cold, and
3. where water is available for manufacturing of ice.

As examples of ore bodies that are suitable
for this mining method, the following places are given:

Kiruna, Malmberget and Viscaria of
LKAB, Näsliden of Boliden
Pyhäsalmi of Outokumpu
Bjørnevann and Bidjovagge of Sydvar-
anger.
Granduc in B.C. Canada
Kidd Creek in Ontario, Canada

Exploatation of the Isua deposit at West
Greenland or of ore bodies in the Ant-
arctic is hardly possible unless we join
forces with the glacial ice and use ice
as a partner instead of considering ice
an enemy.

By ice-backfilling, mining operations are
limited to the ore body itself. Encroachments upon nature are therefore very much
reduced. Disfiguring scars and waste dumps
are avoided.
Conservation of nature can be regarded,
while the needs of modern civilisation
for minerals and metals still can be met
in an efficient and resource-economic way.
The ice-backfill-method extends the available resources through full utilisation of
the deposits of ores and minerals.

AKNOWLEDGEMENTS

Stationary ice for support of lateral walls
in mines has been proposed by Hoberstorfer
and Norén through the THB-method in Sweden,
by Langerfeldt, Griffith and Barlett in the
U.S.A. and by Andreas Eriksen in Norway.
Their proposals may become feasable when
the economy of ice, manufactured by means
of natural winter conditions are employed.

Andreas Eriksen has given valuable advice
and encouragement during the preparation
of this new method that utilizes the dynamic and physical properties of ice in its
viscouos state, for the complete and economic exploatation of ores in one and one
only, mining process.

LITERATURE AND REFERENCES

Paterson, W.S.B. 1969: The Physics of
Glaciers, first edition, The Pergamon
Press, London   250 p.

Shumskii, P.A. 1964: Principles of
Structural Glaciology. Dover Publication
Inc. N.Y. 497 p.

Lewis, W.V:Norwegian Cirque Glaciers,
Royal Geogr. Soc. Research Series No 4
John Murray (Publishers) Ltd. London
104 p.

Glen, L.W. 1952: Experiments on the Deformation of Ice. Journal of Glaciology
vol. 2 . Oxford, 111 - 114 p.

Carslaw, H.S. & Jaeger 1959: Conduction
of Heat in Solids

Lindblom, Ulf ISBN-91-7284-072-2:
Jordvärmeboken, Ingeniörsförlaget,
Stockholm.

Fredrikson, H. 1978: Metallernas Gjut-
ning. Kompendium, KTH, Stockholm.

Various techical publications from
the U.S.A.C.R.R.E.L. Hannover, New Hamp.

Fangel, H.: Patents no's: N-126 092,
U.S.-3790-215, S.-7 116 199-6, Can.-961.
060, Arg. -201.841, Chi.-28.896,
SF-53016 and DK-1263/72.